Electric Transportation Systems in Smart Power Grids

The leading countries around the globe, including Australia, have taken serious steps to decarbonize their energy and transportation sectors as part of their obligations for a suitable future with fewer emissions and a better environment. The decarbonization plans in different countries have resulted in changes such as increases in the penetration level of renewable energy sources and the introduction of electric vehicles as a target for future transportation systems. This is the point where mobility meets electricity and brings new challenges and opportunities, especially in the integration with modern power systems. The main impacts would be on the demand-side and the distribution network. These impacts would also be reflected in the operation, control, security, and stability of transmission systems. This creates a new grid architecture characterized by growing variability and uncertainties. Moreover, the growth in the share of renewable energy in the total energy market is one of the major causes of the increasing fluctuations in the balance between generation and consumption in the whole system. Therefore, the key challenge lies in developing new concepts to ensure the effective integration of distributed energy resources and electric transportation systems, including EVs, into existing and future market structures.

Electric Transportation Systems in Smart Power Grids address how these issues—EVs, E-buses, and other smart appliances on the demand side—can be aggregated to form virtual power plants, which are considered an efficient solution to provide operational flexibility to the grid. The book also discusses how EV-based virtual power plants can also provide myriad services for distribution system operators, transmission system operators, and even local prosumers within the energy community.

FEATURES

- Describes the services required to power systems from EVs and the electric transportation sector
- Covers frequency control in modern power systems using aggregated EVs
- Discusses integration and interaction between EVs and smart grids
- Introduces electric vehicle aggregation methods for supporting power systems
- Highlights provided from electric transportation systems to the smart energy sector
- Discusses the penetration level of renewable energy sources and EVs

Electric Transportation Systems in Smart Power Grids

Integration, Aggregation, Ancillary Services, and Best Practices

Edited by
Hassan Haes Alhelou,
Ali Moradi Amani,
Samaneh Sadat Sajjadi,
and Mahdi Jalili

CRC Press
Taylor & Francis Group
Boca Raton London New York

CRC Press is an imprint of the
Taylor & Francis Group, an **informa** business

First edition published 2023
by CRC Press
6000 Broken Sound Parkway NW, Suite 300, Boca Raton, FL 33487-2742

and by CRC Press
4 Park Square, Milton Park, Abingdon, Oxon, OX14 4RN

CRC Press is an imprint of Taylor & Francis Group, LLC

ISBN: 978-1-032-27744-8 (hbk)
ISBN: 978-1-032-27767-7 (pbk)
ISBN: 978-1-003-29398-9 (ebk)

DOI: 10.1201/9781003293989

Typeset in Times New Roman
by Apex CoVantage, LLC

Contents

Contents

Preface

The energy security and environmental concerns are forcing governments to change both energy and transportation systems in the context of smart grids and smart cities. In this regard, a high interest in upgrading transportation systems using electric vehicles (EVs) and E-buses has been noticed. This is the point where mobility meets electricity and brings new challenges and opportunities, especially in the integration with modern power systems. The main impacts would be on the demand-side and the distribution network. These impacts would be also reflected in the operation, control, security, and stability of transmission systems. In the same context, the rapid and large-scale deployment of various renewable energy systems have considerably impacted the electric grid structure by moving from conventional generation to variable distributed energy production. This creates a new grid architecture characterized by growing variability and uncertainties. Moreover, the growth in the share of renewable energy in the total energy market is one of the major causes of the increasing fluctuations in the balance between generation and consumption in the whole system. Therefore, the key challenge lies in developing new concepts to ensure the effective integration of distributed energy resources (DERs) and electric transportation systems, including EVs, in existing and future market structures. To address these issues, EVs, E-buses, and other smart appliances on the demand-side can be aggregated to form virtual power plants, which are considered an efficient solution to provide operational flexibility to the grid. Based on aggregation and optimal control of DERs, an electric vehicle aggregator increases profits by guaranteeing various ancillary services for both low- and high-voltage distribution systems. The EV-based virtual power plants can also provide several services for distribution system operators (DSOs), transmission system operators (TSOs), and even local prosumers within the energy community.

MATLAB® is a registered trademark of The Math Works, Inc.
For product information, please contact:

The Math Works, Inc.
3 Apple Hill Drive
Natick, MA 01760-2098
Tel: 508-647-7000
Fax: 508-647-7001
E-mail: info@mathworks.com
Web: http://www.mathworks.com

Editors

Dr. Hassan Haes Alhelou is a senior member of IEEE. He is with the Department of Electrical and Computer Systems Engineering, Monash University, Clayton, VIC 3800, Australia. At the same time, He is a professor and faculty member at Tishreen University in Syria, and a consultant with Sultan Qaboos University (SQU) in Oman. Previously, he was with the School of Electrical and Electronic Engineering, University College Dublin (UCD), Dublin 4, Ireland between 2020 and 2021, and with Isfahan University of Technology (IUT), Iran. He completed his B.Sc. from Tishreen University in 2011 and his M.Sc. and Ph.D. from Isfahan University of Technology, Iran, all with honors. He was included in the 2018 & 2019 Publons and Web of Science (WoS) list of the top 1% best reviewers and researchers in the field of engineering and cross-fields around the world. He was the recipient of the Outstanding Reviewer Award from many journals, e.g., *Energy Conversion and Management* (ECM), *ISA Transactions*, and *Applied Energy*. He was the recipient of the best young researcher in the Arab Student Forum Creative among 61 researchers from 16 countries at Alexandria University, Egypt, 2011. He also received the Excellent Paper Award 2021/2022 from IEEE's *CSEE Journal of Power and Energy Systems* (SCI IF: 3.938; Q1). He has published more than 200 research papers in high-quality peer-reviewed journals and international conferences. His research papers received 2,550 citations with h-index of 31 and i-index of 65. He authored/edited 15 books published in reputed publishers such as Springer, IET, Wiley, Elsevier, and Taylor & Francis Group. He serves as an editor in a number of prestigious journals, such as *IEEE Systems Journal, Computers and Electrical Engineering* (CAEE-Elsevier), *IET Journal of Engineering*, and *Smart Cities*. He has also performed more than 800 reviews for high prestigious journals, including *IEEE Transactions on Power Systems*, *IEEE Transactions on Smart Grid*, *IEEE Transactions on Industrial Informatics*, *IEEE Transactions on Industrial Electronics*, *Energy Conversion and Management*, *Applied Energy*, and *International Journal of Electrical Power & Energy Systems*. He has participated in more than 15 international industrial projects around the globe. His major research interests are renewable energy systems, power systems, power system security, power system dynamics, power system cybersecurity, power system operation, control, dynamic state estimation, frequency control, smart grids, micro-grids, demand response, and load shedding.

Dr. Ali Moradi Amani was awarded his PhD from RMIT University in 2020. His graduate and postgraduate studies have been all in control engineering. He is currently a research fellow at RMIT. He had more than 12 years of industry experience before starting his PhD studies at RMIT. His main expertise is in control systems, control and scheduling of DER in distribution power grids, and complex dynamical networks. His publications have received more than 225 citations with an h-index of 9.

Dr. Samaneh Sadat Sajjadi received her B.Sc. and M.Sc. in Electrical and Mechanical Engineering from Sharif University of Technology, Tehran, Iran, in

2004 and 2007, respectively. She was awarded her PhD from Ferdowsi University of Mashhad, Iran, in 2015. Her graduate and postgraduate studies have all been in control engineering. She was an assistant professor with the Department of Electrical Engineering, Hakim Sabzevari University, Iran, from 2015 to 2020. She is currently a researcher at the School of Engineering, RMIT University, Melbourne, Australia. Her research interests include the modeling and optimization/optimal control of the nonlinear systems (power, renewable energy, and biology systems), data-driven control, computational methods of optimal control for nonlinear and fractional-order systems, as well as fractional dynamics and their application in science and engineering. Her publications have received more than 650 citations with an h-index of 11.

Dr. Mahdi Jalili is an associate professor of Electrical Engineering and AI at RMIT University, Melbourne, Australia. He holds a PhD in Computer, Communications, and Information Sciences from Swidd Federal Institute of Technology (EPFL), Lausanne, Switzerland. He was an Australia Research Council DECRA Fellow and an RMIT Vice-Chancellor Research Fellow. His research expertise is on complex network systems, machine learning, and energy analytics.

Contributors

Ali Abdali
University of Zanjan
Zanjan, Zanjan Province, Iran

Hassan Haes Alhelou
Monash University
Melbourne, Victoria, Australia

Yaseen Alwesabi
Binghamton University
Binghamton, New York

Ali Moradi Amani
RMIT University
Melbourne, Victoria, Australia

Arash Asrari
Southern Illinois University
Carbondale, Illinois

Amin Babazadeh
Sahand University of Technology
Sahand, East Azerbaijan Province, Iran

Mohammad Ali Baherifard
Sahand University of Technology
Sahand, East Azerbaijan Province, Iran

Najmeh Bazmohamadi
Aalborg University
Aalborg, Denmark

Jalil Boudjadar
Aarhus University
Aarhus, Denmark

Morris Brenna
Politecnico di Milano
Milan, Italy

Y. Bulatov
Bratsk State University
Bratsk, Irkutsk Oblast, Russia

Alessandro Casavola
University of Calabria
Arcavacata, Italy

Francesco Castelli Dezza
Politecnico di Milano
Milan, Italy

Reza Eslami
Sahand University of Technology
Sahand, East Azerbaijan Province, Iran

Federica Foiadelli
Politecnico di Milano
Milan, Italy

Giuseppe Franzè
University of Calabria
Arcavacata, Italy

Viswanathan Ganesh
Chalmers University of Technology
Gothenburg, Sweden

Meysam Gheisarnejad
Aarhus University
Aarhus, Denmark

S. Hemavathi
CSIR-Madras Complex
Chennai, Tamil Nadu, India

Seyed Hadi Hosseini
University of Zanjan
Zanjan, Zanjan Province, Iran

Sébastien Jacques
University of Tours
Tours, France

Mahdi Jalili
RMIT University
Melbourne, Victoria, Australia

Fatemeh Jozi
University of Zanjan
Zanjan, Zanjan Province, Iran

Hamed Jafari Kaleybar
Politecnico di Milano
Milan, Italy

Rasool Kazemzadeh
Sahand University of Technology
Sahand, East Azerbaijan Province, Iran

Mohammad Hassan Khooban
Aarhus University
Aarhus, Denmark

Ibrahim Kovan
Anhalt University of Applied Science
Bernburg, Germany

V. M. Ajay Krishna
Chalmers University of Technology
Gothenburg, Sweden

A. Kryukov
Irkutsk National Research Technical
    University
Irkutsk, Irkustsk Oblast, Russia

Chen Liu
RMIT University
Melbourne, Victoria, Australia

Kazem Mazlumi
University of Zanjan
Zanjan, Zanjan Province, Iran

Mohammad Bagher Menhaj
Amirkabir University of Technology
Tehran, Tehran Province, Iran

H. Mohammadian-Alirezachaei
Sahand University of Technology
Sahand, East Azerbaijan Province, Iran

Nabil Mohammed
Monash University
Melbourne, Victoria, Australia

Mehran Moradi
Tarbiat Modares University
Tehran, Tehran Province, Iran

Mahdi Mosayebi
Aarhus University
Aarhus, Denmark

Ehsan Naderi
Southern Illinois University
Carbondale, Illinois

Mehdi Rafiei
Aarhus University
Aarhus, Denmark

Samaneh Sadat Sajjadi
RMIT University
Melbourne, Victoria, Australia

Alberto Sendin
Comillas Pontifical University
Madrid, Spain

S. Senthilmurugan
SRM Institute of Science and
    Technology
Chennai, Tamil Nadu, India

C. Sharmeela
Anna University
Chennai, Tamil Nadu, India

Hui Song
RMIT University
Melbourne, Victoria, Australia

Amir Abolfazl Suratgar
Amirkabir University of Technology
Tehran, Tehran Province, Iran

K. Suslov
Irkutsk National Research Technical
   University
Irkutsk, Irkustsk Oblast, Russia

Francesco Tedesco
University of Calabria
Arcavacata, Italy

Stefan Twieg
Anhalt University of Applied Science
Bernbrug, Germany

R. Uthirasamy
KPR Institute of Engineering and
   Technology
Coimbatore, Tamil Nadu, India

Xinghuo Yu
RMIT University
Melbourne, Victoria, Australia

Reza Zamani
Tarbiat Modares University
Tehran, Tehran Province, Iran

Introduction

This book discusses the electric transportation sector, including electric vehicles and their roles and applications in modern and future smart energy systems, with focus given to power systems and their smart grid concept. Each chapter begins with the fundamental structure of the problem required for a rudimentary understanding of the methods described.

Chapter 1: Using fossil fuels as a source of energy often results in greenhouse gas emissions. The result is current global warming and melting icebergs due to climate change. To tackle this result, various preventive measures and climate policies are needed to reduce its effects. There are a variety of initiatives taking place to reduce transportation sector emissions. The electrification of transportation leads to many benefits for the distribution systems. With the presence of electric vehicles (EVs) energy security could be increased, economic growth would be promoted, and the environment would be protected due to the diversification of energy sources, the creation of new advanced industries, and reduction in tailpipe emissions, respectively. This chapter presents a comprehensive review of EVs and their implementation in the grid. In addition, their consequences and impacts in the smart distribution system have been addressed.

Chapter 2: With growing concerns about global warming, lack of fossil energy resources, and the need for clean energy generation, plug-in electric vehicles (PEV) in smart-grids have become an undeniable solution to existing challenges. One of the advantages of electric vehicles is the increase in the flexibility of the future smart grid with grid support. Alternatively, the aggregated EVs can balance deviations in demand predictions, consumption, and even distributed generation (DG) of the power system. As an economic agent in the electricity market, the EV aggregator must contend with different actors to sell/buy electricity in the daily markets by offering the best bidding strategies. To reach this goal, the aggregator must comprehend electric vehicle characteristics, such as driving patterns, charge status, or total capacity in real time to respond to network management during the charging term to respond to the network managing system. The main goal of the aggregator is the efficient aggregation of DERs in a unified entity that may be used as a storage device or generation system by providing grid-required capacity and power commodities or by operating as a profit-maximizing controlled load.

Chapter 3: This chapter includes generalized models that forecast the energy consumption of electrical vehicles (EV) using real-world datasets and machine learning (ML). The objective of the study is to predict the number of EV registrations and the amount of power consumption over the next three years. Combining real-world data and a variety of ML approaches to identify potential strategies of problem solving is the focus of this study. First, it forecasts the number of EV sales in the upcoming years using the ML model, which is developed considering the number of EV sales in past years. Second, the forecasting of future energy consumption is done by taking historical energy consumption into account. Two approaches are blended with various ML techniques, such as LSTM, SARIMAX, and NeuralProphet to make the model more flexible and fit real-world problems. Additionally, the study evaluates the performance, which considers the different assumptions about how much energy is demanded by EVs.

Chapter 4: Parking lots not only provide electric vehicle (EV) owners with the opportunity to charge their battery, they also contribute power to the grid via vehicle-to-grid (V2G) technology. This significantly improves the reliability of distribution systems. In addition, parking lots equipped with V2G capability are able to participate in the electricity market as a producer. Accordingly, in this chapter, nine optimal strategies for scheduling the process of EVs charging and discharging in a parking lot equipped with V2G capability are proposed, with the aim of maximizing parking profits. The proposed strategies include constraints on the amount of power exchange between the parking lot and the distribution system, as well as the random and stochastic nature of EV behavior. The results demonstrate that using the proposed optimal charging and discharging strategies scheduling increases the parking profit by 31%.

Chapter 5: Electric vehicle (EV) parking is a suitable solution for operation improvement of EVs. Parking is the proper option for charging and discharging services. Parking provides EV owners the opportunity to charge their batteries and power injection to the grid through vehicle-to-grid (V2G) technology, if owners are satisfied. This case improves the reliability of distribution systems. Additionally, V2G-equipped parking is able to participate in the electricity market as a producer. In this chapter, to validate the impact of V2G-equipped parking on reliability, different strategies for planning the charge and discharge process of EVs have been investigated with regards to the stochastic and uncertainty nature of various parameters. For each strategy, reliability indexes for the distribution system connected to parking can be evaluated and calculated. The results show the V2G charging and discharging strategies ameliorate the reliability indexes of the distributed system.

Chapter 6: This work proposes a framework to optimize the EVCS placement problem with the consideration of EV user charging discomfort. The proposed method is modeled on the number of EV charging demands as they have fluctuated in different time slots. Also, EV charging discomfort is considered. After that, in order to achieve the optimization objectives after EVCS installation, an online charging planning framework has been proposed. A series of algorithms are developed for the allocating, updating, and cancellation of the online charging request. Both proposed frameworks have been simulated with a large-scale real-world dataset. The result has shown under the optimal placement of EVCS that EV charging planning helps EV users allocate an EVC with a lower charging discomfort to achieve win-win outcomes.

Chapter 7: At the conclusion of the 2021 Conference of the Parties (COP26) in Glasgow (Scotland, UK), the green agenda is more important than ever and important decisions regarding transport have been taken. Although technological developments on electric vehicles have accelerated considerably, many questions remain unanswered regarding battery chargers. For the on-board type, many experts believe it is essential to develop cost-effective chargers while optimizing their footprint in the electric vehicle. Mid-range (e.g., 3.7 kW) on-board battery chargers typically consist of a large bulk electrolytic capacitor, which generates high inrush currents on the AC grid due to its charging process, resulting in both numerous reliability issues with the charger itself and excessive current stresses on the AC grid. This chapter discusses the power component control strategies currently implemented to limit inrush currents during battery charger startup. The use of thyristors and Triacs in these strategies are particularly tested and analyzed.

Chapter 8: The concept of intelligent power supply systems (IPSS) of railways is presented. A description of the active elements used to control the IPSS modes is given. The analysis of technical solutions for mode control devices is presented. The structures of the regime control systems are described. The results of computer studies aimed at solving the problems of controlling the modes of IPSS based on distributed generation units are presented. The results of the development of the concept of multi-agent control of the IPSS modes and fuzzy control algorithms of its individual elements are presented. The research results show that effective control of the modes of railroad power supply systems can be implemented using smart grid technologies.

Chapter 9: During the last several years, nations have increased the development of EVs and E-buses as an important strategy to reduce carbon emissions and the consumption of fossil fuels. However, supplying power to multitudinous charging stations may overload the utility grid and cause indirect emissions. The integration of electric railway systems (ERSs) and EV charging infrastructres at strategic points, with targeted exploitation of trains regenerative braking energy as an ancillary supply and taking advantages of renewable energy resources, can improve energy efficiency and reduce costs. However, according to the diversity of ERSs, different integration architectures can be defined utilizing the DC or AC energy hub concept. In this chapter, different DC and AC hub-based integration architectures are defined and analyzed. Meanwhile, the concept of train-to-vehicle (T2V) and vehicle-to-train (V2T) technologies together with challenges as the architecture of future power supply systems in sustainable transportation will be explained.

Chapter 10: The growth and dominance of electric vehicles has been enormous in the past five years. One of the primary reasons for the high demand is the ease of use and socioeconomic benefits gained. Developed nations have realized the need for electric vehicles and the industrial revolution in the upcoming years. On the other hand, developing nations are transforming the transportation sector from conventional internal combustion engines to electric vehicles in compliance with the United Nations' Sustainable Development Goals. This chapter introduces electric vehicle charging infrastructures and provides an example of DC fast charging for quick refueling of electric vehicle batteries and provides reliability in the conversion of conventional IC engine vehicles to greener electric vehicles.

Chapter 11: Electric vehicles (EVs) play a vital role all over the world in declining carbon emissions in the automotive industry. Charge infrastructure in particular caused issues with DC fast EVs, causing the issue of quick charging time, which leads to an overloading of the grid and an inaccurate forecast. In this sense, the use of bidirectional power flow chargers between EVs and grid systems (V2G and G2V) enhances grid stability through load leveling and load shaving. This chapter presents a detailed review of the promising DC-DC converter topologies for V2G/G2V technology, including the Snubber Assist Zero-Voltage Zero-Current converter and the Re-Lift Converter, along with models and application scenarios. These converters are intended for use in fast-charging stations for electric vehicles. They can utilize bidirectional DC-DC converter technology, and the architecture helps to design and to reduce the impact on power grids.

Chapter 12: This chapter reviews the state-of-the-art research in systems optimization of electric bus charging planning using dynamic wireless charging and the power loss problem in transportation electrification. Section 12.1 illustrates the trends of future public transportation electrification in smart cities. Section 12.2 illustrates the planning stages of future public transportation electrification. Electric bus charging stations are discussed in Section 12.3. Recent studies on optimization models of electric bus charging planning are reviewed in Section 12.4. Section 12.5 reviews electric bus scheduling methods. Section 12.6 illustrates the data collection process, while Section 12.7 demonstrates the mathematical model and an illustrative example of electric bus scheduling model considering the battery status constraints. Section 12.8 demonstrates the optimizations of reactive power in transportation systems. Finally, conclusions are drawn in Section 12.9.

Chapter 13: Electricity, as the service of power systems, is instrumental in modern societies. The evolution of power systems sits within the smart grid concept, and electric transportation systems both benefit from and are affected by it. The smart grid is a consequence of the interest of utilities to improve the operation of their grids, and a need of the widespread integration of distributed energy resources, including plug-in electric vehicles (PEVs). The challenge of enhancing the remote monitoring capabilities, the observability and controllability of the grid, is greater when the injection or demand of energy cannot be easily predicted, as it is in the case of PEVs. Telecommunication networks and services operated by utilities are hybrid (a mix of private telecommunication networks and public commercial services, wireline or wireless-based technologies, etc.). The challenge of pervasive high bandwidth, low latency, and high reliability telecommunication services at the customer edge is still to be realized.

Chapter 14: In this chapter, the problem of managing EV charging demand in power distribution grids is addressed. EV charging can significantly increase residential net power consumption, and simultaneous charging in a distribution grid may cause technical issues, such as transformer overload, if it is not well managed. There are two main approaches to implementing real-time smart charging using smart meter data. The first is to directly control the EV rate of charge based on the current consumption of the household and the desired consumption profile of the total residential grid. Another solution is to control the discharge rate of battery storage systems in the grid to compensate for the EV charging demand. In this chapter, the concept of real-time control of EV charging is reviewed and the performance of a model-based predictive controller in this context is assessed considering real-world data from smart meters of residential consumers in Victoria, Australia.

Chapter 15: In this chapter, we summarize our past research activity on the study of tertiary LFC supervisory solutions based on constrained control ideas aimed at developing wide-area LFC control schemes for supervising remotely located generation units, connected by distribution lines to remotely distributed loads, under coordination requirements that enforce pointwise-in-time constraints on the evolutions of relevant system variables. In particular, the proposed LFC scheme is able to directly enforce constraints on the maximum allowable frequency and power deviations from nominal values and has some reconfiguration capability under system faults.

Chapter 16: In this chapter, the class of constrained coordination and supervision problems described in Chapter 15 is generalized to master/slave distributed architectures that can better face the communication latency and packet drops. Here, the network latency is taken into account and is abstractly modeled as a time-varying time delay, which is allowed to become unbounded in the case of data loss. Amongst many, this approach has the merit of permitting the analysis and the design of networked control systems without being hampered by network and protocol details. In fact, it has a low impact on the existing control methodologies, which can be used without modifications apart from the presence of the time delay, expressed now as a multiple of the sampling time.

Chapter 17: In smart distribution systems, electric vehicles (EVs) play a significant role due to their ability to serve as flexible loads (leading to demand response) or mobile distributed storage assets (facilitating the flexible production). However, rapid improvement of smart grid structure along with ongoing development of electric transportation systems via EVs result in several challenges, including but not limited to a higher vulnerability of network to malicious cyberattacks. This chapter scrutinizes an organized attack framework, where an adversary targets EV aggregators by launching false data injection (FDI) cyberattack, resulting in system congestions and potentially cascading power outages. More importantly, a remedial action scheme (RAS) based on coordination of distribution-level market operator (DMO) and distribution generation aggregators is proposed in order to alleviate the impacts of the indicated attack using a secondary distribution-level electricity market. The effectiveness of the proposed framework is validated on the IEEE 33-bus distribution system, modified to contain distributed generation retailers and EV aggregators.

Chapter 18: Nowadays, the coastal communities around the world are faced to challenges related to the increasing energy consumption, raising energy costs, climate change, and so on. Therefore, many countries intend to implement different policies to develop clean energy production. There has been a new paradigm in policy from utilization of greenhouse gases (GHGs), particularly CO_2, toward sustainable energy resources to access a high level of security and reliability. This chapter discusses the new trends of hybrid marine power systems and analyzes various sustainable resources, such as PV, tidal turbine, and wind turbine. The restrictions and opportunities of combining the various technologies in the ship power systems have been investigated from both economic and environmental perspectives.

Chapter 19: These days the use of electric vehicles (EVs) increases every day. The main reason for that is systematic optimization of cost, pollution, and welfare. The EVs for charging use an electricity network. Connecting EVs to grids causes some critical problems. Usually for energy dispatch in grids an optimal power flow (OPF) problem is solved, and the solution is then used for optimal dispatch. In the near future, the usage of EVs will dramatically increase; therefore, charging and connecting them all significantly requires the solution of OPF in the grid, especially in micro grids. This problem can be considered from different points of view. First, it can be considered a distributed optimization problem with unknown nonlinear non-affine dynamics. It may be considered a hybrid switching system. If there are few EVs, the problem of plug and play can be considered a robust optimal control one. The EVs connection is considered as an uncertainty. The recent point of view is not suitable because plenty of EVs will be used in the near future. In this chapter, we will present some distributed optimization methods for unknown systems with input constraints as well as for the optimal consensus problem.

1 Electric Vehicles and Their Impact in Smart Distribution Systems

*Mehran Moradi, Reza Zamani,
and Hassan Haes Alhelou*

CONTENTS

DOI: 10.1201/9781003293989-1

Using fossil fuels as a source of energy often results in greenhouse gas emissions. The result is global warming and melting icebergs due to climate change. In order to tackle this result, various preventive measures and climate policies are needed to reduce its effects. There are a variety of initiatives taking place to reduce transportation sector emissions. The electrification of transportation leads to many benefits for distribution systems. With the presence of electric vehicles (EVs) energy security could be increased, economic growth would be promoted, and the environment would be protected due to the diversification of energy sources, the creation of new advanced industries, and the reduction of tailpipe emissions, respectively. This chapter presents a comprehensive review of EVs and their implementation in the grid. In addition, their consequences and impacts in the smart distribution system have been addressed.

1.1 INTRODUCTION

The International Energy Agency (IEA) has presented a scenario for a future energy system that seeks to limit world average temperature increase to two degrees Celsius by 2050 [62]. The transportation sector is expected to double its greenhouse gas (GHG) emissions by 2050, while contributing 25% of global GHG emissions in 2009 [62].

There are a variety of initiatives taking place to reduce transportation sector emissions. The objective is to minimize GHG emissions and enhance vehicle performance via the development of new fuels and the use of clean technology features. The electrification of transportation has the potential to provide significant advantages [66]. With EVs, energy security could be increased through the diversification of energy sources, economic growth would be promoted through the creation of new advanced industries, and the environment would be protected due to a reduction in tailpipe emissions. The superior performance of EVs over internal combustion engine vehicles (ICEVs) is attributed to their efficient motors and powertrains [8,68,69]. Around the world, several initiatives, policies, and programs are being implemented to promote the use of EVs. Various actions are being taken to promote EVs, including lowering the purchase cost of EVs, enhancing the charging infrastructure, and raising public consciousness of the vehicles' potential. As a result of these efforts, EVs are now becoming more popular with the public. According to the Worldwide EV Outlook concluded by the Electric Vehicle Initiative (EVI) and IEA, global EV production reached approximately 180,000 units by the end of 2012 [31]. As a consequence, EVs will account for 0.02% of the world's vehicle stock, while also enabling continued research and development to take place.

To boost EVs' capabilities and maintain their competitiveness, EV technologies must constantly be upgraded. As an example, a number of technologies have been developed to support powertrains, batteries, and charging infrastructures. The diversity of powertrain configurations is being designed throughout the EV development procedure in order to address a variety of needs, including series, parallel, and series-parallel [51]. They may enhance fuel efficiency and driving range as a result of the very efficient electric motors employed in these power train systems [8, 63, 67]. Battery technology has also been improved from lead-acid to nickel-based to ZE-BRA batteries to lithium-based types, which have been found to be lightweight, affordable, safe, and reliable, so they can be used for storage purposes [7]. Metal-air batteries have been examined, as they possess a remarkable energy density of up to 1,700 Wh/kg, which is equivalent to conventional ICEVs [39]. To alleviate range anxiety among drivers of EVs, high-power charging infrastructures, such as DC fast-charging stations, are gradually eliminating low-power chargers [62, 64]. The cost-effectiveness of EVs is heavily influenced by the generation mix used to charge them [45]. Since EVs are powered by power grid electricity, the cost of power generation has a significant impact on their price. The economic consequences of EV penetration may be examined based on their impact on the electricity grid and EV owners. Power grids will need greater capacity to satisfy EV demand, but EV owners must pay their high initial purchase costs, while EVs remain unaffordable at the moment. EV integration can be cost-effective for the power grid and EV owners by having coordinated charging, energy exchanging, and diverse electricity pricing policies. EV integration has a significant impact on the environment [33]. EVs have no greenhouse gas pollution, making them environmentally friendly and cleaner. The process of supplying electricity to EVs does result in GHG emissions, as EVs are powered by electricity from the power grid. EV use will have a diverse environmental impact according to the type of electricity source employed. EVs have been gaining popularity due to the widespread use of renewable energy that is more environmentally friendly than conventional CEVs [33, 44]. The connectivity of EVs to the power grid to obtain charges raises concerns about the negative implications of EV charging on power systems. EV charging is expected to cause harmonics, system losses, voltage drops, phase unbalances, overloads, and stability issues [9, 65].

1.2 EV HISTORY

The evolution of EVs has been dramatic since the nineteenth century. Robert Anderson designed the first electric-powered carriage between 1832 and 1839 that employed non-rechargeable batteries [21]. He then created a number of electric-powered carriage prototypes, but none were developed into dependable products because an effective electric motor and battery were not available.

Direct current electric motors (DC) and rechargeable batteries underwent a series of development between 1856 and 1881. Werner Siemens, Antonio Pacinotti, and Znobe Gramme all received credit for the development of the high-efficiency DC electric motor [21]. Gaston Plant designed the initial rechargeable lead-acid battery in 1859, which was subsequently introduced into the market by Camille Alphonse Faure in 1881 [62]. The development of DC electric motors and power-recharging batteries has revolutionized EV technology. In 1897, an electric taxi became the first commercially available EV in New York City. EVs accounted for 28.8% of road vehicles in three years and were the preferred mode of transportation [21]. A hand crank was no longer required to start gasoline-powered vehicles after Charles Kettering invented the electric starter in 1912. Gasoline-powered vehicles also have lower fuel costs compared to EVs because of the availability of cheap petrol. These factors led to a great deal of acceptance of gasoline-powered cars, while the popularity of EVs diminished. Consequently, EVs could only travel over short distances, and charging stations were few in number [62]. EVs ceased to be available around 1935.

A few decades later, gasoline-powered vehicle emissions and high oil prices sparked renewed interest in EVs. Since then the government has taken steps to minimize air pollution and encourage the development of electric and hybrid vehicles. It is becoming more and more common to see hybrid vehicles on the road. One of the first EVs was manufactured and leased by General Motors in 1996 [21]. Toyota introduced the world's first hybrid electric vehicle (HEV), the Prius, in Japan the following year, selling 18,000 units in its first year of production [62]. As oil prices increased, more automakers shifted their focus to EVs. As of 2010, battery electric vehicles (BEVs) and plug-in hybrid electric vehicles (PHEVs), such as the Nissan LEAF, the Mitsubishi I-MiEV, the Chevrolet Volt, and the Tesla Model S, have made their debut in the automotive industry [62].

1.3 IMPACTS OF EV DEPLOYMENT

Deployment of EVs has been extensively investigated. There are three key impacts to consider: economic, environmental, and power grid. In the following sections, the key findings of these three categories are described in detail.

1.3.1 ECONOMIC IMPACT

There are two perspectives to consider when evaluating the economic impact of EV deployment: the power grid's perspective and the EV owners' perspective [41]. From the power grid's perspective, EVs need a link to the grid in order to get charging. System costs will rise as a result of the increased fuel consumption necessary to generate electricity as a result of the massive growth in EV loads [28]. Increasing the EVs' demands causes increases in power losses during the transferring power in the transmission lines and distribution lines. Managing the charging of EVs is a straightforward solution to address this issue [13]. Controlling the charging of EVs plays an integral role in reducing system operation costs by saving up to 60% [57]. The cost reduction is even more remarkable when renewable energy sources are integrated into the power grid. EV owners perceive a low operational cost of EVs based on the efficient electric motor and low electricity prices [58]. Nevertheless, employing the battery in an EV is associated with a more costly initial investment compared to ICEVs. To make the initial investment of an EV affordable, numerous techniques are applied, including the production of electric vehicles in large quantities [14], using the notion of transactive energy (TE) [27], and establishing effective charging strategies [22].

In the short term, employing EVs is not profitable for power grids and EV owners. EVs require more power, which involves power grids to expand their generation capacity in order to meet their demands. Furthermore, the initial purchase costs of EVs are relatively high, and they are not affordable for most people. In reality, using EVs will be profitable for both distinct groups (EV owners and power grid owners) when the aforementioned techniques are implemented (producing EVs in large quantities, using the notion of TE, and developing effective charging strategies) [62].

1.3.2 Environmental Impact

Zero greenhouse gas emissions make EVs green and environmentally friendly. EVs require electricity to charge, but the generation of that electricity contributes to air pollution and GHG emissions. For comparison, the term "wells-to-wheel emissions" is applied to the greenhouse gas emissions of EVs and ICEVs. Including the emissions from the fuel and materials used to power a vehicle and the direct GHG emissions, well-to-wheel emissions calculate GHG emissions over a vehicle's lifetime, and EVs demonstrate the lowest emissions when calculated well-to-wheel [26].

EVs may produce more emissions from wells-to-wheels than ICEVs if they are charged with electricity generated from the combustion of fossil fuels. To illustrate, an analysis of the Texas power grid, which has a mix of coal- and natural gas–fired power plants, shows that EVs produce more carbon dioxide emissions than ICEVs [46]. EVs use of the Ohio power grid with coal-fired generation releases a more incredible amount of SO_2 and NO_2 emissions, whereas CO_2 emissions can be reduced by 24 percent from EVs versus ICEVs [56]. If EVs are charged by a power generated by burning fossil fuels, these results suggest that the vehicles are perhaps not eco-friendly. However, EVs will be more environmentally friendly if they are powered by different types of renewable energy. As a result, GHG emissions from wells-to-wheels of EVs will be minimized.

1.3.3 Impact on Power Grid

Concerns regarding the influence of EV charging on the electricity grid have arisen due to their implementation. The connectivity of immense EV fleets to the grid for charging purposes may have a detrimental consequence on the power grid, resulting in system losses, increased power consumption, voltage drop, equipment overloading, harmonics, phase imbalance, and stability challenges [17].

In addition, dynamic behavior and the variable charging rates of EVs further exacerbate the possible effects. EV charging has severe impacts on power grids, outlined further in this section.

1.3.3.1 Impact on Load Profile

A significant change in load profile is a result of incorporating EVs. Numerous nations have spent considerable time and money determining whether their present or future power supply capacity is sufficient to fulfill the growing demand for electric car charging [18]. Reference [56] analyzes how EVs impact the hourly load profile in the USA. When electric vehicle owners are given the freedom to charge their vehicles anywhere and at any time, peak hours and late afternoons will experience a surge in demand, which occurs as EVs arrive at work and return home from work. A delay-based charging control scheme is detailed in [56] to minimize peak-hour demand.

According to [36], the deployment of EVs will have a substantial influence on the load profile of the Korean grid. A few scenarios that consider the various characteristics of EVs, patterns of use, charging rates, and charging locations are evaluated based on the demand impact of EV charging. According to the research, adding EVs to the grid could raise the load profiles and make the system less reliable. A time-of-use (TOU) scheme can mitigate the issue by shifting EV loads from peak hours to off-peak times.

1.3.3.2 Impact on Power Grid Components

EVs must be charged from the grid with a large amount of electric power, which must be transported from the power plants and then over transmission and distribution lines. The existing system components might be overloaded because these components may not be designed to handle the additional EV load. Transformers and cables, which are regularly overloaded with electrical power, may pose a major barrier to the widespread adoption of EVs [20]. According to [61], increasing EV penetration in the power system will have a detrimental effect on the transformer lifespan. The study in [1] investigates the impact of EV charging on the power grid cables. According to [1], current wires can support just 15% of EV penetration rate for rapid charging and 25% for standard charging. According to this study, the present distribution network is incapable of sustaining high levels of EV penetration. Generally, EVs are generally additional loads that will affect the performance of system components. Consequently, to prepare for their adoption, load management and network planning should be applied.

1.3.3.3 Impact on Power Losses

To charge anEV, electricity must be delivered from power plants to charging devices. In the process, a greater percentage of system losses occurs across the grid, which makes utilities concerned [37]. To reduce power system losses, charging EV loads with local distributed generation may be an option [48].

1.3.3.4 Impact on Voltage Profile

A voltage drop and voltage deviation will result from charging an EV from the power grid. Therefore, charging many EVs at once may be incompatible with the voltage constraints for the network. According to [93], a penetration rate of 50% or more of EVs leads to system voltages exceeding the 7% voltage fluctuation tolerance.

1.3.3.5 Impact on Phase Imbalance

Single-phase AC charging contributes to phase unbalance, which is another consequence of EV charging [54]. If residential EV charging is not equally distributed throughout all three phases, it may lead to significant phase imbalance problems. The study in [42] connects all EVs to phase "a" in order to determine the effect of AC single-phase EV charging on phase imbalance, and the result indicates that phase imbalance issues exist. Effective load management may help avoid phase imbalances caused by charging EVs by appropriately dispersing EV loads throughout all three phases.

1.3.3.6 Harmonic Distortion

The electric vehicles are charged by power electronic components, which may have an impact on the quality of power; notably, harmonics are the main issues for the power grids [2]. The effect of harmonic distortion caused by charging an EV on the network is examined in [47]. During the charging of EVs, it was revealed in [47] that the system's harmonic distortion varied between 20% and 45%, indicating the need for a harmonic filter to avoid distortions.

1.3.3.7 Impact on Grid Stability

The capacity of a power system to revert to steady-state functioning after disturbances or transients is referred to as stability. Stabilizing the power grid is crucial to ensuring that power is supplied reliably. It is still unknown what impact EVs will have on the stability of distribution networks since EVs are a new load on the grid. Reference [34] claims that with the adoption of EVs, electricity grids have become more vulnerable to disturbances in the form of the period of time it takes to restore steady-state conditions. Two properties of the charger that contribute to this are the injection of harmonics and the absorption of reactive power. The effect of EV charging on grid voltage stability was studied in [10], where it was found that EV charging had a significant effect on steady-state voltage stability.

1.4 EVs IN THE SMART GRIDS

It was not necessary for a power grid to connect to the transportation sector in previous decades because the transportation sector was powered by gasoline and diesel. The development of EVs and environmental concerns, however, has prompted attention to be drawn to electric transportation, and the necessity of a connection between the electricity grid and EVs has become more evident. A previous section discussed the challenges of the power grid in relation to EVs. EVs have, however, created unique opportunities, such as vehicle-to-grid (V2G) technology and EV/RES integration.

1.4.1 SMART GRIDS

With smart grids, energy and data are exchanged bidirectionally; therefore, the components of the grid will be more interconnected and more cooperative [4]. Smart meters in smart grids will collect information, and utilities will automatically and intelligently monitor all activities of network-connected components on the basis of information collected [40]. The smart grid facilitates active consumer participation in the operation of the electricity network. Electricity consumption, tariffs, and incentive programs are available to consumers in real time, enabling them to adjust their consumption patterns accordingly [62]. The use of distributed generation (DG) in smart grids also increases network reliability and reduces the risk of attacks and natural disasters [40]. Smart grids are capable of improving the quality and reliability of electricity. Numerous smart grid projects are currently underway around the world, such as Smart Grid Smart City in Australia, Ontario Smart Grid Plan in Canada, Low Carbon London in the UK, ECAR Project in Ireland, Yokohama Smart City Project In Japan, the Jeju smart grid system in South Korea, and the Houston smart grid in the United States [62]. One major difference between smart grids and traditional grids is the ability for components to communicate bidirectionally. This feature facilitates the use of numerous applications, such as demand response, V2G, integrating EV and RES, and smart meters.

1.4.2 THE APPLICATION OF EVs IN SMART GRIDS

Given the capabilities and benefits of smart grids, power utilities throughout the world have made significant investments in modernizing their conventional power grids to smart grids. The evolution of smart grid technology has reached a point of maturity, which creates new opportunities and promotes the adoption of fresh applications. With the recent growth in EV adoption, improved smart grid technologies have assisted in the promotion of V2X technology, where X can be a grid, an EV, or a home. Furthermore, the V2X technology has the potential to assist the smart grid by making the integration of distributed generation into the system more convenient.

1.4.2.1 V2X Technology

Interactive cooperation between electric vehicles and the power grid, other EVs, and smart homes is a comprehensive view of V2X technology, which is illustrated in Figure 1.1. V2X technology has been developed in phases [43]. The first phase is unidirectional V2X (UV2X), in which EVs serve as consumers. Bidirectional V2X (BV2X) technology is another phase in which users will be able to play a role not just as consumers but also as producers. Both phases will be discussed in further detail in the following sections.

1.4.2.1.1 OV2X Technology

The implementation of UV2X is affordable, and it is limited to controlling the charge of EVs [48]. The charging process of an EV is controlled by UV2X, which regulates the charging rate in response to an energy management or incentive scheme, and setting the policy and regulation is the first step to employing UV2X [35]. When EV owners charge their EVs outside of peak hours, they will earn incentives. Controlling EV charging shifts their demand to off-peak hours, resulting in cost savings and power-loss reduction [48]. Figure 1.1a illustrates the UV2X framework. Consequently, UV2X technology can offer benefits to both EV owners and the power grid upon its deployment.

1.4.2.1.2 BV2X Technology

Recent attention has been drawn to BV2X due to its potential to provide bidirectional communications and energy trading capabilities. BV2X technology can be divided into three categories: vehicle-to-vehicle (V2V), vehicle-to-home (V2H), and vehicle-to-grid (V2G) [35].

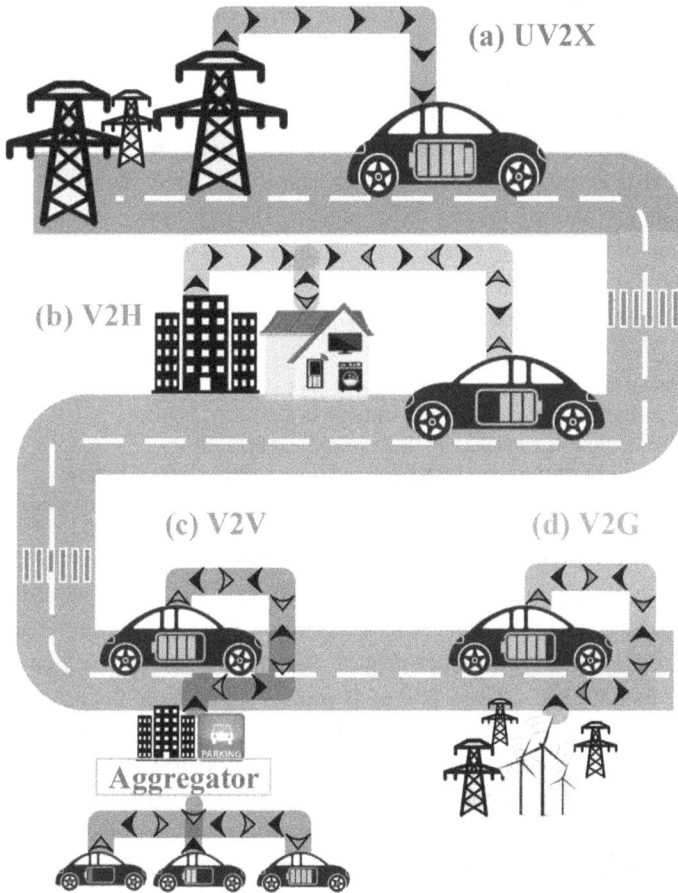

FIGURE 1.1 A comprehensive view of V2X technology.

1.4.2.1.2.1 V2H Technology The V2H technology used in smart homes allows EVs to serve as energy storage devices, storing excess energy produced by the house's distributed generation units and then discharging it to power other devices when the renewable energy production of the smart home is insufficient. V2H technology, when combined with renewable generation units, enables the creation of an energy-efficient, safe, and environmentally responsible living environment [3]. The V2H system shown in Figure 1.1b includes an EV and small-scale renewable generation in a smart house. Utilizing V2H technology, the following specific features can be achieved [24]:

Generally, V2H operates inside a single smart house equipped with a single EV.
Accelerating the deployment of smart grid.
Increasing the efficiency of residential renewable production via the use of EV batteries.
Enhancing the attractiveness of smart homes via the use of V2H technology. Simple installation.
Smoothing the daily load profile.

1.4.2.1.2.2 V2V Technology   V2V technology enables EVs that have excess energy to exchange it with other EVs that are in need of charging [25]. By increasing the penetration of EVs in communities, the community becomes more suited for V2V technology deployment on the power grid [59]. Power loss may be significantly decreased via the use of V2V technology. Furthermore, this technology is easily implemented from inside the current grid. The V2V system shown in Figure 1.1c comprises an EV parking lot and an aggregator. The aggregator, who is in charge of charging and/or discharging a group of EVs, may facilitate communication amongst all the EVs in a parking lot [24]. The V2V system facilitates the community grid to operate with the following specific capabilities [24]:

> In general, V2V entails the use of numerous EVs.
> V2V exchanges energy through parking lots and smart homes.
> V2V's structure is considerably less simple and flexible.
> V2V has low transmission losses and requires a simple infrastructure.
> V2V technology has the potential to accelerate the pace of the smart grid.

1.4.2.1.2.3 V2G Technology   Finally, V2G technology facilitates the discharging of EV storage devices into the grid in addition to the transfer of power from the grid to EVs [16]. Since an individual EV's battery storage capacity is limited, a single EV has a negligible impact on the power grid. As a result, the V2G design is centered on a collection of EVs and how energy is distributed between the grid and the EVs [13] [16]. A variety of benefits can be derived from V2G technology in the sense of peak shaving and voltage regulation with an efficient control system and monitoring system. Considering the immense number of grid-connected EVs, however, several uncertainties must be addressed, including the different state of the batteries and the dynamic probability of EVs connecting to the grid. In addition, unit commitment is an efficient way to manage the power flow between EVs and the grid.

The proposed V2G structure is shown in Figure 1.1d, which comprises power grid, renewable energy sources, and EVs. Regarding electric grid operation, the V2G structure has the following distinguishing characteristics [24]:

> V2G entails a significant number of EVs.
> V2G is capable of transferring energy via smart homes, parking lots, and quick-charging stations.
> For power allocation, EV aggregators are highly recommended.
> EVs play a significant part in V2G's active power distribution operation. Employing the internal capacitors, EV chargers may offer reactive power to the power system.
> The EV battery may provide active power assistance autonomously. V2G system optimization solutions are viable and flexible.

1.4.2.2 EV Services to Power Grid

Although the preceding sections discussed the challenges encountered by the power system when using EVs, a broad variety of capabilities and services may be supplied to the power system when utilizing EVs. By incorporating EVs into the electricity grid, the following benefits may be concluded:

1.4.2.2.1 Ancillary/Spinning Reserve

For making the grid stable and reliable, ancillary services are required [32]. The energy stored in EVs can be used to prepare ancillary services for the main grid, enabling the generation output to balance the demand when it is not able to meet it [49]. Ancillary service refers to the supporting service supplied to the power grid for the purpose of maintaining the reliability and sustainability of the power grid [32]. The energy stored in EV batteries is utilized as an additional generator to compensate for a power outage, enabling them to provide ancillary services to the power system [49]. EVs are able to provide failure recovery and reduce the amount of back-up generation capacity by utilizing the spinning reserve service [15].

1.4.2.2.2 Active Power Support and Peak Load Shaving

The peak power consumption of residential and commercial consumers occurs only during short intervals. Maintaining a flat load profile assists the electrical system by reducing component degradation, minimizing grid overloading, and increasing energy efficiency and economic advantages [60]. To accomplish these objectives, EVs can be deployed by using the excess energy stored in their batteries during peak hours and charging them during off-peak hours [55].

1.4.2.2.3 Voltage Regulation and Reactive Power Support

The grid will face numerous challenges due to the integration of more and more EVs into the system, which will negatively impact power efficiency and voltage stability [23]. Static reactive power compensators have historically been used to regulate voltage and correct power factors. However, through the use of EVs, the bidirectional charger corrects the power factor and regulates the voltage [53].

1.4.2.2.4 Harmonic Filtering

Smart grids contain several non-linear loads, which cause harmonics in the power grids. EV chargers generate harmonics in smart grids since they use converter switches [12]. Therefore, using the EV charger will have a negative impact on the power quality, which requires corrective actions. Since EV chargers are equipped with filters, they can filter out the harmonics generated by non-linear loads and EV chargers [6,12]. In the charger, a converter can function as a variable impedance, and the impedance will change depending on the harmonic frequency, and the harmonic problem is solved by filtering appropriately [6].

1.4.2.2.5 Deploying Renewable Energy Resources

Renewable energy resources face numerous problems, such as intermittency. EVs can contribute significantly toward solving this problem [11]. The EVs can be charged from RES when surplus power is available, but the EV is discharged to the grid when the RES does not provide enough energy. Moreover, EVs can regulate the voltage of the main grid, equipped with RESs. Therefore, EV are the ideal solution to cover the uncertainties of renewable energy sources, and their uncertainties can then be handled, allowing RESs to be integrated more and more [30].

Comparison among the UV2X and BV2X technology is demonstrated in Table 1.1.

1.5 RESEARCH BARRIERS

The power sector, as well as the environment, can derive benefits from the use of EVs. The absence of greenhouse gas emissions contributes to a cleaner environment, especially in the case of the environment. With respect to the power section, by adopting V2X technology and integrating more and more renewable energy sources into smart grids, EVs provide an opportunity for converting conventional grids into smart grids. In order to realize all the potential benefits of using EVs, we have to overcome specific barriers if we hope to achieve successful deployment of EVs [62]. Despite all the advancements in EVs, they are still not fully developed. In particular, the life cycle of EVs and the high cost of initial purchase still pose a challenge for their deployment. A few technologies for batteries have been developed with better performance than others; however, they are still under experimentation. Moreover, as a result of the high number of times that batteries are charged and discharged using the V2X technology, battery deterioration will be hastened. All in all, it is necessary to conduct more research in order to develop the technical and economical aspects of EV batteries [62]. The technology of the batteries are not the only issue; the technology related to the charger is also a problem, and the smart grids are not prepared to implement BV2X technology. There are mainly unidirectional EV chargers in the market, no matter how fast or how slowly they charge. In order to implement BV2X, however, bidirectional chargers are necessary, which are still in the experimental stages. It is also vital for EV employment to have a comprehensive EV infrastructure with enough EV chargers, but installing such a network requires considerable planning and investment. Consequently, it is imperative that comprehensive research is conducted on installing complex infrastructure and bidirectional charging technology [62]. The integration of EVs into the power grid allows V2X technology to be implemented, and more benefits, including integrating RESs, can be derived from the implementation of V2X technology. To implement V2X technology, EV owners need to participate actively, but many factors, such as battery degradation, hinder their involvement. In the absence of EV owners' tendency to participate in V2X technology, the technology cannot succeed. As a result, researchers need to

TABLE 1.1
Comparison of UV2X and BV2X Technology

V2X	Services	Benefits	Barriers
UV2X	• Ancillary services	• Minimizing the power loss • Minimizing the costs • Minimizing the emission	• Limited services
BV2X	• Ancillary services • Peak load shaving • Voltage regulation • Filtering the harmonics • Integrate more and more RES into the main grid	• Minimizing the power loss • Minimizing the costs • Minimizing the emission • Preventing grid overload • Regulating the voltage • Improving the load profile	• Social challenges • High infrastructure costs • Battery degradation

explore alternative solutions, such as energy management policy and incentive-based programs, to tackle this enormous social challenge [62].

1.6 EV TRIALS

1.6.1 MY ELECTRIC AVENUE PROJECT (MEA)

The MEA deployed almost 200 Nissan LEAFs with a 24 kWh battery throughout the United Kingdom, making it one of the biggest (if not the biggest) EV trials to date in terms of examining the obstacles and advantages associated with the usage of this technology at home (slow charging mode, at approximately 3.6 kW). The project's primary objective was to evaluate a solution (dubbed Esprit) for mitigating the effect that EVs may have on European low-voltage grids. To do this, the study analyzed EV data, developed models, investigated the impacts, and conducted management studies. MEA was the first initiative to concentrate on the optimal management practices for the local power network, while a significant number of EVs are charging simultaneously on the same street [38].

1.6.2 SWITCH EV TRIALS

The northeast of England is a pioneer in the demonstration of electric cars on a large scale in the United Kingdom. The Switch EV experiment is one of only eight in the United Kingdom to receive financing from the Technology Strategy Board's Ultra Low Carbon Vehicle (ULCV) Demonstrator Program. Forty-four fully electric cars have been installed with data loggers as part of the three-year study to give information into how early adopters are using their vehicles and, more crucially, their charging behavior with a high number of charging ports nearby. The trials link car-generated driving and performance data with charging behavior gleaned from vehicle loggers and EV charging infrastructure, as well as soft data gleaned from surveys, focus groups, and structured interviews with participants [5].

1.6.3 VICTORIAN EV TRIAL

The Victorian EV experiment brought together all the burgeoning EV market's components and provided low-risk, low-cost operating conditions. In order to streamline market development, the trial serves as a test bed for the implementation of novel technologies and business models. Approximately 70 corporate participants and 120 households have participated in the trial, which will be the base for the Victorian (and Australian) EV market. A preliminary report with limited and sparse statistics has been released in [19].

1.6.4 CSIRO TRIAL

The CSIRO issued reports on EVs and their influence on the electricity grid in 2011 in partnership with the University of Technology Sydney [50]. They estimated measured kilometers per hour for ICE cars throughout the week in Victoria,

Australia, using data from the "Victorian Integrated Survey of Travel and Activity 2007" [52]. The CSIRO utilized this data to build the average energy consumption curve for EVs.

1.6.5 WA EV TRIAL

A large-scale test of EVs in Australia was conducted by the WA EV Trial, with the goal of monitoring their performance, environmental benefits, policy, infrastructure, and practical implications for fleets adopting EVs [29]. The purpose of the WA EV Trial is as follows:

Demonstrate the operational features and environmental advantages of electric cars to the general population.

Create opportunities for local industrial growth in various sectors, as well as technology transfer to universities for a new significant industry.

Allow trial participants to determine if electric cars are suitable for their particular operating needs.

Identify concerns regarding the deployment of electric cars in Perth, including technical, regulatory, policy, and planning concerns.

1.7 CONCLUSION

The current status, challenges, impacts, and opportunities of EV deployment are demonstrated in this chapter. There were 180,000 EVs available in 2012, proving that EVs look pretty intriguing to a wide range of people. The implementation of incentive-based programs, promoting environmental awareness, and developing the infrastructure are factors that can lead to EV market penetration at higher levels. Additionally, the impacts of EV deployment are described in three distinct sections: environment, economics, and power grids. It is stated that without EV charging management, EV deployment will negatively impact the power grid. EVs, on the other hand, are a chance for the main grids to leverage technologies such as V2X and integrate more and more renewable energy sources.

REFERENCES

[1] E Akhavan-Rezai, MF Shaaban, EF El-Saadany, and Aboelsood Zidan. Uncoordinated charging impacts of electric vehicles on electric distribution grids: Normal and fast charging comparison. In *2012 IEEE Power and Energy Society General Meeting*, pages 1–7. IEEE, 2012.

[2] Richard Bass, Ronald Harley, Frank Lambert, Vinod Rajasekaran, and Jason Pierce. Residential harmonic loads and ev charging. In *2001 IEEE Power Engineering Society Winter Meeting, Conference Proceedings (Cat. No. 01CH37194)*, volume 2, pages 803–808. IEEE, 2001.

[3] Florence Berthold, Benjamin Blunier, David Bouquain, Sheldon Williamson, and Ab-dellatif Miraoui. Phev control strategy including vehicle to home (v2h) and home to vehicle (h2v) functionalities. In *2011 IEEE Vehicle Power and Propulsion Conference*, pages 1–6. IEEE, 2011.

[4] Jignesh Bhatt, Vipul Shah, and Omkar Jani. An instrumentation engineers review on smart grid: Critical applications and parameters. *Renewable and Sustainable Energy Reviews*, 40:1217–1239, 2014.

[5] PT Blythe, GA Hill, Y Hübner, V Suresh, J Austin, L Gray, and J Wardle. The north east england electric vehicle and infrastructure trials'. *evs*, 26, 2012.

[6] AR Boynuegri, Mehmet Uzunoglu, Ozan Erdinc, and E Gokalp. A new perspective in grid connection of electric vehicles: Different operating modes for elimination of energy quality problems. *Applied Energy*, 132:435–451, 2014.

[7] Michela Catenacci, Giulia Fiorese, Elena Verdolini, and Valentina Bosetti. Going electric: Expert survey on the future of battery technologies for electric vehicles. In *Innovation under Uncertainty*. Edward Elgar Publishing, 2015.

[8] Zahra Darabi, and Mehdi Ferdowsi. Impact of plug-in hybrid electric vehicles on electricity demand profile. In *Smart Power Grids 2011*, pages 319–349. Springer, 2012.

[9] HS Das, MM Rahman, S Li, and CW Tan. Electric vehicles standards, charging infrastructure, and impact on grid integration: A technological review. *Renewable and Sustainable Energy Reviews*, 120:109618, 2020.

[10] CH Dharmakeerthi, N Mithulananthan, and TK Saha. Impact of electric vehicle fast charging on power system voltage stability. *International Journal of Electrical Power & Energy Systems*, 57:241–249, 2014.

[11] Farivar Fazelpour, Majid Vafaeipour, Omid Rahbari, and Marc A Rosen. Intelligent optimization to integrate a plug-in hybrid electric vehicle smart parking lot with renewable energy resources and enhance grid characteristics. *Energy Conversion and Management*, 77:250–261, 2014.

[12] Javier Gallardo-Lozano, M Isabel Milanés-Montero, Miguel A Guerrero-Martínez, and Enrique Romero-Cadaval. Electric vehicle battery charger for smart grids. *Electric Power Systems Research*, 90:18–29, 2012.

[13] Shuang Gao, KT Chau, CC Chan, Chunhua Liu, and Diyun Wu. Optimal control framework and scheme for integrating plug-in hybrid electric vehicles into grid. *Journal of Asian Electric Vehicles*, 9(1):1473–1481, 2011.

[14] Viktoria Gass, Johannes Schmidt, and Erwin Schmid. Analysis of alternative policy instruments to promote electric vehicles in austria. *Renewable Energy*, 61:96–101, 2014.

[15] Taraneh Ghanbarzadeh, Sassan Goleijani, and Mohsen Parsa Moghaddam. Reliability constrained unit commitment with electric vehicle to grid using hybrid particle swarm optimization and ant colony optimization. In *2011 IEEE Power and Energy Society General Meeting*, pages 1–7. IEEE, 2011.

[16] Dipayan P Ghosh, Robert J Thomas, and Stephen B Wicker. A privacy-aware design for the vehicle-to-grid framework. In *2013 46th Hawaii International Conference on System Sciences*, pages 2283–2291. IEEE, 2013.

[17] Robert C Green II, Lingfeng Wang, and Mansoor Alam. The impact of plug-in hybrid electric vehicles on distribution networks: A review and outlook. *Renewable and Sustainable Energy Reviews*, 15(1):544–553, 2011.

[18] Stanton W Hadley, and Alexandra A Tsvetkova. Potential impacts of plug-in hybrid electric vehicles on regional power generation. *The Electricity Journal*, 22(10):56–68, 2009.

[19] K Handberg. Creating a market: Victorian electric vehicle trial mid-term report. *Victorian Government, June*, 2013.

[20] Alexander D Hilshey, Pooya Rezaei, Paul DH Hines, and Jeff Frolik. Electric vehicle charging: Transformer impacts and smart, decentralized solutions. In *2012 IEEE Power and Energy Society General Meeting*, pages 1–8. IEEE, 2012.

[21] Frank E Jamerson. History and prospects for electric vehicles and electric bikes: Pathway to sustainable carbon free energy and transportation. *SAE Technical Papers*, 2020:1–10, 2020.

[22] Orkun Karabasoglu, and Jeremy Michalek. Influence of driving patterns on life cycle cost and emissions of hybrid and plug-in electric vehicle powertrains. *Energy Policy*, 60:445–461, 2013.

[23] Mithat C Kisacikoglu, Burak Ozpineci, and Leon M Tolbert. Examination of a phev bidirectional charger system for v2g reactive power compensation. In *2010 Twenty-Fifth Annual IEEE Applied Power Electronics Conference and Exposition (APEC)*, pages 458–465. IEEE, 2010.

[24] Chunhua Liu, KT Chau, Diyun Wu, and Shuang Gao. Opportunities and challenges of vehicle-to-home, vehicle-to-vehicle, and vehicle-to-grid technologies. *Proceedings of the IEEE*, 101(11):2409–2427, 2013.

[25] Chunhua Liu, KT Chau, Diyun Wu, and Shuang Gao. Opportunities and challenges of vehicle-to-home, vehicle-to-vehicle, and vehicle-to-grid technologies. *Proceedings of the IEEE*, 101(11):2409–2427, 2013.

[26] Clemens Lorf, Ricardo F Martínez-Botas, David A Howey, Luca Lytton, and Ben Cussons. Comparative analysis of the energy consumption and co_2 emissions of 40 electric, plug-in hybrid electric, hybrid electric and internal combustion engine vehicles. *Trans-Portation Research Part D: Transport and Environment*, 23:12–19, 2013.

[27] Benedikt Lunz, Zexiong Yan, Jochen Bernhard Gerschler, and Dirk Uwe Sauer. Influence of plug-in hybrid electric vehicle charging strategies on charging and battery degradation costs. *Energy Policy*, 46:511–519, 2012.

[28] Thomas P Lyon, Mark Michelin, Arie Jongejan, and Thomas Leahy. Is smart charging policy for electric vehicles worthwhile? *Energy Policy*, 41:259–268, 2012.

[29] T Mader, and Thomas Braunl. Western australian electric vehicle trial 2010–2012 final report, 2013.

[30] Francesco Marra, Guangya Y Yang, Chresten Træholt, Esben Larsen, Jacob Østergaard, Boštjan Blažič, and Wim Deprez. Ev charging facilities and their application in lv feeders with photovoltaics. *IEEE Transactions on Smart Grid*, 4(3):1533–1540, 2013.

[31] Andrea Monacchi, and Wilfried Elmenreich. Assisted energy management in smart microgrids. *Journal of Ambient Intelligence and Humanized Computing*, 7(6):901–913, 2016.

[32] Jonathan Mullan, David Harries, Thomas Bräunl, and Stephen Whitely. The technical, economic and commercial viability of the vehicle-to-grid concept. *Energy Policy*, 48:394–406, 2012.

[33] Evanthia A Nanaki, and Christopher J Koroneos. Comparative economic and environmental analysis of conventional, hybrid and electric vehicles: The case study of greece. *Journal of Cleaner Production*, 53:261–266, 2013.

[34] Omer C Onar, and Alireza Khaligh. Grid interactions and stability analysis of distribution power network with high penetration of plug-in hybrid electric vehicles. In *2010 Twenty-Fifth Annual IEEE Applied Power Electronics Conference and Exposition (APEC)*, pages 1755–1762. IEEE, 2010.

[35] Niamh O Connell, Qiuwei Wu, Jacob Østergaard, Arne Hejde Nielsen, Seung Tae Cha, and Yi Ding. Day-ahead tariffs for the alleviation of distribution grid congestion from electric vehicles. *Electric Power Systems Research*, 92:106–114, 2012.

[36] Woo-Jae Park, Kyung-Bin Song, and Jung-Wook Park. Impact of electric vehicle penetration-based charging demand on load profile. *Journal of Electrical Engineering and Technology*, 8(2):244–251, 2013.

[37] Jayakrishnan R Pillai, and Birgitte Bak-Jensen. Impacts of electric vehicle loads on power distribution systems. In *2010 IEEE Vehicle Power and Propulsion Conference*, pages 1–6. IEEE, 2010.

[38] Jairo Quiros-Tortos, Luis Ochoa, and Timothy Butler. How electric vehicles and the grid work together: Lessons learned from one of the largest electric vehicle trials in the world. *IEEE Power and Energy Magazine*, 16(6):64–76, 2018.

[39] Md Rahman, Xiaojian Wang, Cuie Wen, et al. A review of high energy density lithium-air battery technology. *Journal of Applied Electrochemistry*, 44(1):5–22, 2014.

[40] KS Reddy, Madhusudan Kumar, TK Mallick, H Sharon, and S Lokeswaran. A review of integration, control, communication and metering (iccm) of renewable energy based smart grid. *Renewable and Sustainable Energy Reviews*, 38:180–192, 2014.

[41] David B Richardson. Electric vehicles and the electric grid: A review of modeling approaches, impacts, and renewable energy integration. *Renewable and Sustainable Energy Reviews*, 19:247–254, 2013.

[42] Peter Richardson, Damian Flynn, and Andrew Keane. Impact assessment of varying penetrations of electric vehicles on low voltage distribution systems. In *IEEE PES General Meeting*, pages 1–6. IEEE, 2010.

[43] Arup Sinha, S Neogi, RN Lahiri, S Chowdhury, SP Chowdhury, and N Chakraborty. Smart grid initiative for power distribution utility in india. In *2011 IEEE Power and Energy Society General Meeting*, pages 1–8. IEEE, 2011.

[44] Fereidoon P Sioshansi, Reza Zamani, and Mohsen Parsa Moghaddam. 1—energy transformation and decentralization in future power systems. In Mohsen Parsa Moghaddam, Reza Zamani, Hassan Haes Alhelou, and Pierluigi Siano, editors, *Decentralized Frameworks for Future Power Systems*, pages 1–18. Academic Press, 2022.

[45] Ramteen Sioshansi, Riccardo Fagiani, and Vincenzo Marano. Cost and emissions impacts of plug-in hybrid vehicles on the ohio power system. *Energy Policy*, 38(11):6703–6712, 2010.

[46] Ramteen Sioshansi, and Jacob Miller. Plug-in hybrid electric vehicles can be clean and economical in dirty power systems. *Energy Policy*, 39(10):6151–6161, 2011.

[47] FJ Soares, JA Pec̦as Lopes, and PM Rocha Almeida. A monte carlo method to evaluate electric vehicles impacts in distribution networks. In *2010 IEEE Conference on Innovative Technologies for an Efficient and Reliable Electricity Supply*, pages 365–372. IEEE, 2010.

[48] Eric Sortomme, and Mohamed A El-Sharkawi. Optimal charging strategies for unidirectional vehicle-to-grid. *IEEE Transactions on Smart Grid*, 2(1):131–138, 2010.

[49] Eric Sortomme, and Mohamed A El-Sharkawi. Optimal scheduling of vehicle-to-grid energy and ancillary services. *IEEE Transactions on Smart Grid*, 3(1):351–359, 2011.

[50] Stuart Speidel, and Thomas Bra̎unl. Driving and charging patterns of electric vehicles for energy usage. *Renewable and Sustainable Energy Reviews*, 40:97–110, 2014.

[51] Siang Fui Tie, and Chee Wei Tan. A review of energy sources and energy management system in electric vehicles. *Renewable and Sustainable Energy Reviews*, 20:82–102, 2013.

[52] Victorian Integrated Survey of Travel and Activity, Victoria Department of Transport, ISBN: 9780731187799, 2009.

[53] Iason Vittorias, Michael Metzger, Dennis Kunz, Matthias Gerlich, and Georg Bachmaier. A bidirectional battery charger for electric vehicles with v2g and v2h capability and active and reactive power control. In *2014 IEEE Transportation Electrification Conference and Expo (ITEC)*, pages 1–6. IEEE, 2014.

[54] Hui Wang, Qi Song, Liyan Zhang, Fushuan Wen, and Jiansheng Huang. Load characteristics of electric vehicles in charging and discharging states and impacts on distribution

systems, In *International Conference on Sustainable Power Generation and Supply (SUPERGEN 2012)*, 2012.

[55] Zhenpo Wang, and Shuo Wang. Grid power peak shaving and valley filling using vehicle-to-grid systems. *IEEE Transactions on Power Delivery*, 28(3):1822–1829, 2013.

[56] Claire Weiller. Plug-in hybrid electric vehicle impacts on hourly electricity demand in the united states. *Energy Policy*, 39(6):3766–3778, 2011.

[57] Allison Weis, Paulina Jaramillo, and Jeremy Michalek. Estimating the potential of controlled plug-in hybrid electric vehicle charging to reduce operational and capacity expansion costs for electric power systems with high wind penetration. *Applied Energy*, 115:190–204, 2014.

[58] Aaron Windecker, and Adam Ruder. Fuel economy, cost, and greenhouse gas results for alternative fuel vehicles in 2011. *Transportation Research Part D: Transport and Environment*, 23:34–40, 2013.

[59] Diyun Wu, KT Chau, and Shuang Gao. Multilayer framework for vehicle-to-grid operation. In *2010 IEEE Vehicle Power and Propulsion Conference*, pages 1–6. IEEE, 2010.

[60] Diyun Wu, KT Chau, and Shuang Gao. Multilayer framework for vehicle-to-grid operation. In *2010 IEEE Vehicle Power and Propulsion Conference*, pages 1–6. IEEE, 2010.

[61] Qin Yan, and Mladen Kezunovic. Impact analysis of electric vehicle charging on distribution system. In *2012 North American Power Symposium (NAPS)*, pages 1–6. IEEE, 2012.

[62] Jia Ying Yong, Vigna K Ramachandaramurthy, Kang Miao Tan, and Nadarajah Mithulananthan. A review on the state-of-the-art technologies of electric vehicle, its impacts and prospects. *Renewable and Sustainable Energy Reviews*, 49:365–385, 2015.

[63] Reza Zamani, Mohammad Esmail Hamedani Golshan, Hassan Haes Alhelou, and Nikos Hatziargyriou. A novel synchronous dgs islanding detection method based on online dynamic features extraction. *Electric Power Systems Research*, 195:107180, 2021.

[64] Reza Zamani, Mohsen Parsa Moghaddam, and Mahmoud-Reza Haghifam. Dynamic characteristics preserving data compressing algorithm for transactive energy management frameworks. In *IEEE Transactions on Industrial Informatics*, pages 1–1, 2022.

[65] Reza Zamani, Mohsen Parsa Moghaddam, Maryam Imani, Hassan Haes Alhelou, Mohammad Esmail Hamedani Golshan, and Pierluigi Siano. A novel improved hilbert-huang transform technique for implementation of power system local oscillation monitoring. In *2019 IEEE Milan PowerTech*, pages 1–6, 2019.

[66] Reza Zamani, Mohsen Parsa Moghaddam, Habib Panahi, and Majid Sanaye-Pasand. Fast islanding detection of nested grids including multiple resources based on phase criteria. *IEEE Transactions on Smart Grid*, 12(6):4962–4970, 2021.

[67] Reza Zamani, Mohsen Parsa Moghaddam, and Mahmoud-Reza Haghifam. Evaluating the impact of connectivity on transactive energy in smart grid. *IEEE Transactions on Smart Grid*, 13(3):2491–2494, 2022.

[68] Habib Panahi, Majid Sanaye-Pasand, Reza Zamani, and Hasan Mehrjerdi. A novel DC transmission system fault location technique for offshore renewable energy harvesting. *IEEE Transactions on Power Delivery*, 35(6):2885–2895, 2020.

[69] H Panahi, M Sanaye-Pasand, SHA Niaki, and R Zamani. fast low frequency fault location and section identification scheme for VSC-based multi-terminal HVDC systems. *IEEE Transactions on Power Delivery*, 37(3):2220–2229, June 2022, doi: 10.1109/TPWRD.2021.3107513.

2 Electric Vehicle Aggregation Methods

H. Mohammadian-Alirezachaei,
Amin Babazadeh, Mohammad Ali Baherifard,
Rasool Kazemzadeh, and Reza Eslami

CONTENTS

2.1 INTRODUCTION

Energy and power systems are fundamentally changing, transforming from a conventional carbon-based system to a more intelligent, dynamic, cleaner, and bidirectional model, the so-called smart grid.

One of the main components of smart grids is electric vehicles (EVs). Due to the heightened amount of energy consumed in the transportation sector, by electrification of this sector, dependence on fossil energies is reduced, and clean energy resources can be used to provide energy for the transport sector.

Electric vehicles can also operate as a new source and RESs to solve the problem of renewable energy uncertainties and some issues related to network reliability and stability, etc. [1,2].

Uncoordinated charging of EVs as significant loads can affect the distribution network. Therefore, charging management of EVs and actors in this process should be done by considering the capability of existing networks and future smart grids to prevent damages such as voltage drop and network losses, etc.

DOI: 10.1201/9781003293989-2

19

2.2 SMART GRIDS AND ELECTRIC VEHICLES

In a broad sense, EVs such as electric cars, boats, trains, airplanes, etc. are primarily powered by electricity and have at least one electric motor.

The first EV model was built in the 1830s. Nowadays, there are many types of EVs. Currently, EVs are usually divided into three classifications in the consumer market: 1. electric vehicles with batteries (BEV), 2. hybrid electric vehicles (HEV), and 3. plug-in hybrid electric vehicles (PHEV).

On the other hand, using V2X capabilities such as vehicle-to-grid (V2G) and grid-to-vehicle (G2V), EVs could be utilized as a distributed energy source and storage systems in case proper management can support power grid performance. By storing excess energy (G2V) and re-injecting it into the grid (V2G) when it is infrequent, they can help regulate supply and demand. Due to their high flexibility [2], EVs can participate in demand response programs and supply ancillary services that confirm sustainable network performance [3].

In general, EVs play a fundamental role in managing smart grids. The study of the application of electric vehicles in concepts other than transportation will guide future studies and research.

2.2.1 EVs' Advantage in Environmental Pollution

There are various opinions on the use of electric vehicles and their relevant advantages and disadvantages, some of which we will address.

One of the main incentives for utilizing electric vehicles is solving environmental problems. CO_2 emitted by gasoline vehicles causes global warming, and the release of CO and SO_2 from gasoline vehicles leads to severe air pollution and is generally not environmentally friendly. Particulate matter (PM) can cause lung problems and other diseases. Due to the lower price of electricity compared to gasoline, driving electric vehicles has a more affordable fuel cost. It can also be expected to have economic benefits with existing plans.

Reducing noise pollution is another benefit of using electric vehicles, encouraging the community to use electric vehicles more and more. Electric vehicles produce less noise pollution than gasoline vehicles due to having smaller internal combustion engines or electric engines.

In reference [67], a wireless method for communication between aggregators and electric vehicles in the V2G network is presented to reduce costs and energy consumption using long-range wide area network technology.

2.2.2 EVs' Disadvantages and Limitations in Environmental Pollution

The authors in [4] discuss some of the existing disadvantages. M. Yilmaz et al. mention the gap between EVs' charging points, the lack of charging stations in many areas, and the long time required to charge EVs as disadvantages of this plan. Thus, it may be difficult to charge the car on all paths in case of long travel. In addition, charging them with "Type 1" charging power (6 to 8 hours) takes a long time, while refueling a gasoline vehicle only takes a few minutes.

TABLE 2.1
The Multifaceted Characteristics of V2G Development and the Participants Involved

Sociotechnical dimension	Aggregators	EV industry	Electricity grid operator	EV owners, consumers	Government	Electricity producers
Technological	Develop algorithms to maximize privacy and minimize technical influences	Grow study into battery degradation, charger efficiency	Develop V2G security and privacy standards	Battery degradation approval	R&D investment in battery technology	No role or limited
Financial/ Business/ Economic	Develop a sound business model; persuade customers of economic benefits	Decrease the marginal cost of EVs and V2G in general	Avoid dual tax rules, define the role in business models	Rising willingness to pay for EVs in V2G, more increased attention of V2G earnings	Decrease regulatory borders to the business model, define actors' roles in the business model, incentivize EVs and V2G	Potentially contribute to the business model as associates in electricity markets
Administrative/ Political	Require clarification earlier; can participate in specific markets, barrier removal like double tax, the definition of obligations	Lobby government to take regulatory action and design markets for V2G, improve policies favoring EVs and V2G.	Develop clear and suitable regulations for the V2G participation in different markets	Beneficiaries of guidelines aimed to raise V2G participation and EV adoption	Assign and instruct TSOs and DSOs to develop sensible spaces for V2G to participate in their markets	May play a role as third parties in electricity market participation, such as responsible balance parties (RBPs)
Societal/ Consumer	Take a more active role in including the study of social borders during premature pilot projects, encourage user tinkering and attention, target specific user types	Permit and motivate users to test by standardizing V2X in EVs	Can recreate a role in knowledge campaigns with consideration to electricity markets	Active user engagement, tinkering and reinvention, rising consumer knowledge and approval	Engage with and target user types, provide guidelines to target consumers, provide assistance for social research	Can play a role in knowledge campaigns with considerations to electricity markets

Source: Noel et al. [58–59].

This time in, Type 3 chargers have reached less than 15 minutes, which requires relatively more time than refilling a gasoline vehicle. The infrastructure cost of this type of charger is also high [73].

EVs are more expensive than regular cars due to battery use and battery life cycle (depending on the type and use). However, prices are expected to be cheaper in the future due to the development of battery manufacturing techniques [5], [6], [7]. On the other hand, manufacturing batteries increases environmental pollution.

As shown in Figure 2.1, the costs associated with owning an electric vehicle are different from those of vehicles that still use fossil fuels.

2.2.3 VEHICLE-TO-GRID TECHNOLOGY (V2G)

V2G technology can use the capacity of EV batteries to react quickly to counteract the swing force generated by renewable energy sources. Batteries can act as a load or power supply depending on the system's performance. Power supplied from V2G reaches consumers through a grid connection, and if there is excess energy in the grid, it is stored in EVs. W. Kempton et al. in [9] state that the transmission system operator (TSO) decides to call an individual vehicle or fleet of EVs to transmit power via a control signal based on an internet connection, cellular network, or a power line carrier.

Figure 2.2 shows the V2G design with its main components.

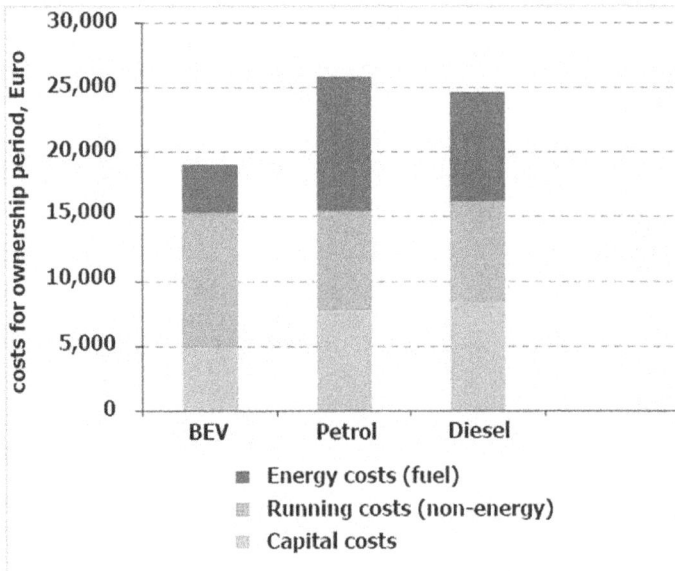

FIGURE 2.1   Comparison of relative ownership costs of battery-powered electric vehicles (BEV) compared to gasoline and diesel equivalents [8].

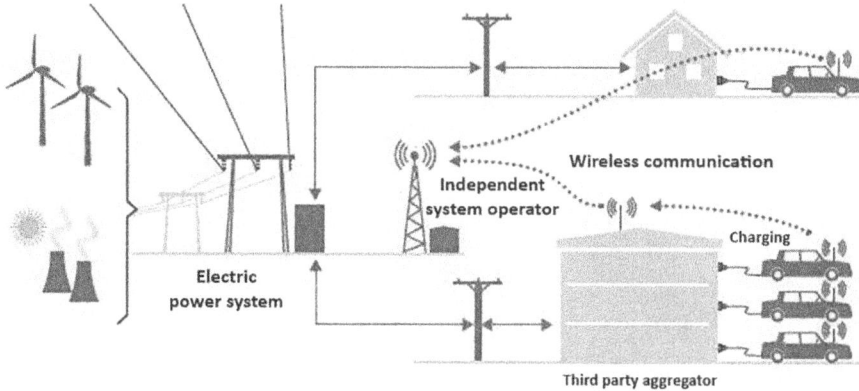

FIGURE 2.2 Components of a V2G system [10].

2.2.4 GRID-TO-VEHICLE TECHNOLOGY (G2V)

EVs as controllable loads make it possible to manage energy consumption and create concepts such as flattening consumption and load profiles. This issue reduces the necessity to increase power production and power plant construction.

2.2.5 DRIVING PATTERN AND AVAILABLE CHARGING TIMES

In the analysis [11,12] and as a case study, based on the data related to the driving pattern of Denmark, three databases are introduced: 1. AKTA data (GPS-based data that track vehicles), 2. MDCars (odometer reading database), and 3. TU data (Danish National Transport Survey). Such studies allow engineers to know EV users' behavior and optimally and intelligently manage EVs' parking lots. Advantages of intelligent charge/discharge management of EV parking lots include improving electrical network technical issues, such as power grid imbalances and reliability.

2.2.6 EV CHARGING PROCESS, CHALLENGES AND FLEXIBILITY

In addition to their benefits, electric vehicles can generate a variety of network problems if not properly planned. Therefore, in this section, we analyze the studies conducted in the field of electric vehicle charging and their existing effects and challenges.

EVs place a huge strain on the power system network, which, if viewed as a planned load, can help to enhance the system's reliability and efficiency [82]. To decrease CO2 emissions, electric vehicles must draw all or part of their electricity from renewable energy sources (RES). It has also been demonstrated in [91] that this is only achievable if electric vehicle charging is correctly controlled. To improve electrical energy usage, [92] proposes a distributed charging system program based on EV charger modeling. It goes into great detail about the influence of batteries on electric vehicle performance [93].

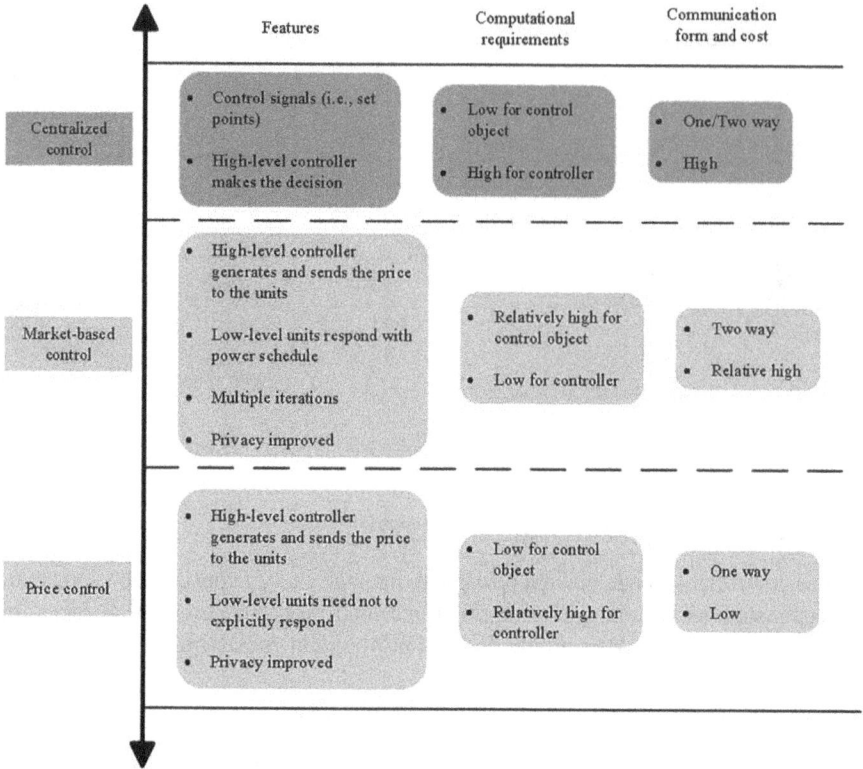

FIGURE 2.3 Overview of charging management strategies [18–34].

According to [94], there are four different types of electric car charging stations:

- Residential/Home charging stations
- Charging stations in the parking lot
- Open charging stations
- Battery replacement stations

According to [95], charging stations are classified into four groups in terms of speed:

- Slow charge (AC)
- Normal charge (AC)
- Fast charge (AC/DC)
- Super-fast charge (DC)

Another essential consideration is the function of EV charging stations in easing EV customers' everyday travel demands. Gnann et al. [96] used real-world fast charge data from Sweden and Norway to develop a model that can analyze the demands of future fast charge sites. Sun et al. [97] present an energy-transportation strategy that can show how the route choice of battery electric vehicles (BEVs), charging systems, and power pricing interact. Their modeling methodology includes charging and electrical networks, with the goal of lowering overall social costs.

The diverse consequences of electric cars are summarized in Figure 2.4.

Now we want to talk about the network consequences of vehicle charging, such as peak load, balance, harmonics, losses, frequency, stability, and so on.

In [98], the concept (V2G) was introduced, and it was said that in peak load conditions, EV could be used as a rotational storage to shave the peak load. The researchers evaluated the influence of plug-in EV charging on static voltage stability and offered reducing strategies such as correct PEV programming and voltage management in [99]. A phase imbalance might be caused by a high volume of electric vehicles charged in the same phase [100]. Also, an imbalance in current can lead to an imbalance in voltage. Gray and colleagues [101] suggested a Monte Carlo approach for assessing the impacts of electric vehicle charging demand on feeder voltage quality, including voltage imbalances and under/over voltages. In [102], the impact of a variety of charging techniques, including uncoordinated, tariff-driven, and coordinated charging, on the transformer's life is assessed, and it is found that uncoordinated charging of EVs is deleterious to the transformer's life. In [103], a harmonic analysis of the IEEE 34-node test system in the presence of an EV charge is presented. The authors suggest a coordinated EV charging technique in [104], which can decrease power losses while also increasing the load factor of the main network. According to Wu et al. [105], the V2G performance of electric vehicles can improve the power grid's transient stability. Improved frequency regulation is accomplished by rectifying the variation in the network frequency with the V2G technology [106]. Because of the foregoing interpretations and the necessity of charging electric vehicles, it is vital to regulate vehicle charging. As a result, we'll present a quick overview of the situation.

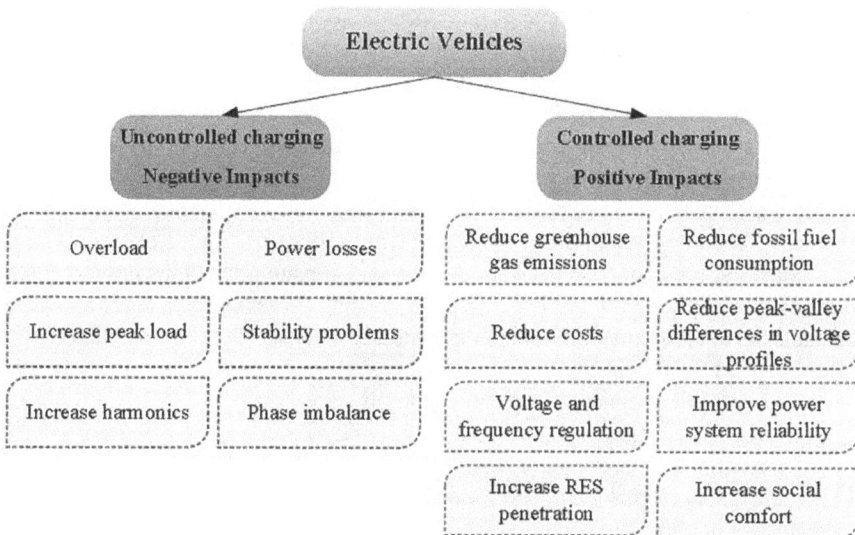

FIGURE 2.4   Different electric vehicles.

In general, it can be said that the charge control of EVs is divided into two categories with the focus on timing (direct method) and pricing (indirect method), which in the direct method is divided into three categories: centralized, decentralized, and hierarchical [107]. Later in this chapter, we describe the work done in this literature. Various centralized scheduling techniques handled optimization difficulties from both the network and the standpoint of EV users. The objectives of [108] and [109] were to maximize customer satisfaction while lowering system running expenses. The goal of [110] was to maximize the overall profit of charging stations while also increasing the quantity of EVs planned. The findings revealed that a high number of electric vehicles were planned, as well as a rise in total earnings. The authors of [111] concentrated on filling the network's valley and lowering client charging expenses. In [112], a charging protocol was proposed to reduce the computational load of EV controllers, increase the pricing mechanism's security, and protect customer privacy. An idea of charging choice was introduced in [113] in order to optimize EV user convenience without violating network restrictions. This idea dramatically minimized network impacts while safeguarding client privacy. In [114], a pricing plan for service providers to promote economic profitability, improve customer satisfaction, and reduce network effects was proposed. Rasheed et al. [90] presented a technique to decide the price of charging in different places, minimize expenses, stable the power grid, and enhance EV customer satisfaction levels.

2.3 EV AGGREGATION

PEVs have the potential to be a critical component of power administration strategies in residential homes because they can help production storage and coordination, smooth the load profile, or manage power consumption depending on the price at different hours. For several intentions other than this regional application, there is a general settlement on the significance of managing the aggregation of many PEVs to affect the network in other significant ways, as discussed in references [13] and [14]. This article analyzes the structure and optimal control of EV aggregators equipped with intelligent meters using the Benders analysis method and queuing theory [68].

The rapid increase in EV penetration will lead to the growth of EV aggregators, which will become a vital part of the ecosystem participating in the electricity market and lead to a balance between production and consumption. This issue requires studying the interaction of the EV fleet with existing power systems and developing mechanisms to ensure that this integration occurs in the best possible way. Therefore, it examines the factors in line with this goal.

The effect of PEV aggregations can be classified as [15]:

- **Centralized energy purchasing:** The considerable amount of power required to charge many PEV batteries reduces costs strategically in the market or gains financial benefits from power providers.
- **Demand-side management (DSM):** To charge PEV batteries, energy demand can be a challenge for the entire energy system. Nevertheless,

this type of load's "transferable" nature is valuable for implementing DSM administrations, primarily when many PEVs are handled.

- **State of Charge (SOC):** Since SOC of batteries is a crucial problem for using stored power for V2G goals, the high accumulation of batteries allows for managing the charge/discharge instructions for each PEV depending on the SOC of the battery. The minimum state of charge required is used in charge/discharge management by accessing the distance required by the EV user to reach the destination successfully.

- **Achievable participation in ancillary systems:** The megawatt power level is required to make these services desirable. The typical range of battery energy is shallow, so collecting them for this purpose is required.

Over the past years, the accumulation of PEVs in power systems has become increasingly important for the reasons mentioned. As a result, shortly, profiteers, the so-called PEV aggregators, are likely to expand as a business intermediary between the electrical market, DSOs, TSOs, and owners of PEVs. PEV aggregators will also be the major developer and manager of V2G auxiliary activities, such as frequency regulation.

The aggregator can make more precise decisions and better view the market than individual EVs by gathering information from the fleet of EVs and managing batteries. Since the introduction of aggregators, EV aggregators have been extensively studied in various configurations. In [16], R. J. Bessa et al. provide a comprehensive review of the EV aggregator structure and its relationship to existing infrastructure and electricity markets. In addition, in studies [16,17] and [18], detailed studies of the economic and technical aspects of EV aggregation can be found. The study [19] also reviews the artificial intelligence methods used to manage the fleet of EVs, including routing problems and congestion management.

M. D. Galus et al. in [20] identifies actors affected by the introduction of EVs in the planning and operation of power systems and suggests potential changes to mitigate these effects. They describe an EV integration framework that can be used for activities such as load management and service delivery. In reference [69], in addition to proposing EV aggregators to assist the network, an optimal planning model is designed to combine real-time and day-ahead plans to reduce operating costs and increase efficiency. In order to minimize costs, emissions, EV charging prices, etc., this study introduces an optimal energy management method [70]. With the aim of peak shaving in the power grid, a novel bidding model that addresses an optimization problem is proposed for EV involvement in the electricity market by considering the interconnection between the energy operator, the distribution system operator, and aggregators [61].

In paper [71], V2G is included in the framework by considering a virtual power plant that negotiates V2G contracts with drivers and tenders in the market the next day. V2G is also envisaged in [72], which includes battery damage compensation in their participation model in tomorrow and sub-markets.

Another aspect of EV aggregation, namely, the ability to respond to fleet demand, is considered in [3]; their work includes a multi-market scenario with a day-ahead and several examples of daily markets. In addition, the integration of EV aggregation with wind generation in [21] is considered.

All of this paved the way for the significant integration of EVs into future markets, but several constraints exist. First, the price effect is often modeled linearly, leading to a loss of accuracy. Second, some models address the market clearing process and introduce additional computational complexity.

The analysis in [22] states that a perceptual model for V2G performance is presented. An aggregator is introduced as a renewed factor that aims to gather large batteries of PEVs by absorbing and maintaining owners to manage enough MW capacity to influence the network and get the benefits. PEV aggregation may act as a source or as a controllable load. First and foremost, a peak-shaving goal may be achieved to reduce charging payments for PEVs and reduce adjustment requirements for generators. In the next case, the aggregation of PEVs can act up/down adjustment duties on the basis of each battery's SOC.

In references [22,23,24], aggregators are explained as instruments that create a cluster of EVs that can utilize their one-way and two-way power transmission capabilities as a source of energy distribution and positively impact the network. In addition, aggregators connect to various network entities through a communication interface, for example, an energy service provider (ESP), transmission system operator (TSO), or DSO, which allows it to evaluate its ability to supply electricity and power services.

EV aggregators can show great flexibility in balancing load and controlling the charging power of EVs by using V2G and G2V capabilities. To achieve this goal and manage the network load, the aggregator must be informed about the various characteristics of EVs, such as driving patterns, SOC, etc.

However, such an implementation would require a scheme that would give aggregators suitable motivations to attract and maintain enough EV owners whose energy capacity could interest an ESP or DSO. Thus, as noted in [21], in addition to meeting the demand for EVs, the services provided by EVs can be a supplemental source of revenue for aggregators, and part of this revenue can be used to improve preferential rates for EV owners.

Article [74] suggests aggregation and scheduling procedures for distributed EVs and the effect of their DR service load return on the futuristic scheduling pattern.

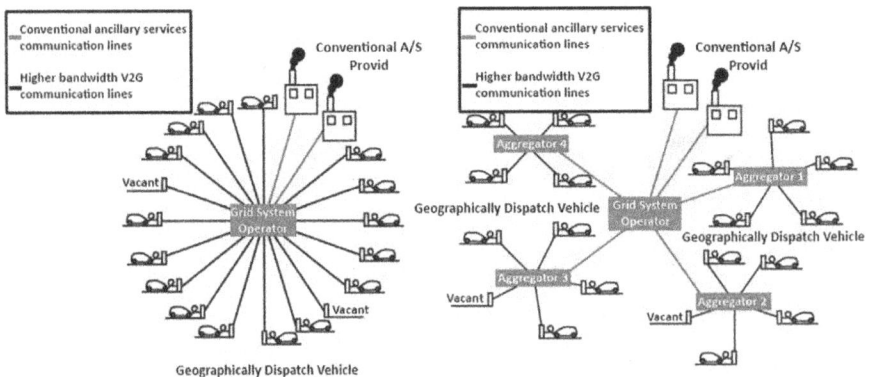

FIGURE 2.5 Conventional V2G centralized network framework (left panel) versus integrated and decentralized V2G framework (right panel) [25].

2.3.1 MARKET-BASED TECHNIQUES FOR EV AGGREGATORS

The function and objective of the aggregators may vary based on market scenario. Therefore, to perceive the involvement of EV aggregators in these market-based strategies, some related research is provided in Table 2.2. Depending on the design and formulation, these market-based strategies might be either centralized or decentralized.

Paper [2] suggests a mechanism for determining EVs' two-way energy storage capacity in the supplementary services market as a long-term contract to supply up/down regulations to the grid. Paper [11] develops a two-layer distributed optimization platform using the alternating direction method of multipliers (ADMM) as an exchange problem to solve the EV charge management problem (EVCMP). The proposed model creates a layer of coordination between EV aggregators, which

TABLE 2.2
Comparison of Different Market Methods for EV Aggregators [60]

TE	Price-based management	LFM	Objective	Method	Network constraints	Ref.
✓		✓	Cost and system peak load of DSO minimizing by buying flexibility services from the Evs aggregator	Centralized optimization	✓	[61]
		✓	Evs aggregator profit-maximizing while the DSO gets the descending reserve services cost	Multi-agent management, game theory	✓	[62]
	✓		Peak demand and Evs aggregator operational cost-minimizing	Two-stage robust optimization		[63]
	✓		Aggregators total power cost, system peak load, and meet charging requirements of Evs minimizing	Centralized optimization	✓	[64]
		✓	Framework for a DSO to manage local load constraints by achieving flexibility from aggregators	Multi-agent management	✓	[65]
✓			i EVs aggregator charging cost and DSO's technical losses minimizing	Centralized optimization	✓	[66]

increases the overall optimization efficiency while maintaining the independence of the aggregators. Paper [12] introduces a market-based planning model for the two options of potential storage investment and the purchase of proprietary energy storage systems, which consider both planning and operations to maximize EVA profits.

In several case examples in article [75], the development of a management system by an EV power aggregator in the electricity market context under a randomized mixed linear programming approach to collect the expected profit was proposed. In study [76], to reduce costs and effects on the network, the coordination of EV aggregators in the day-ahead electricity market was analyzed by employing the alternating direction method of multipliers. Article [77] proposes the dispatchable region of large-scale EVs scheduling instead of individual EVs. So microgrid bidding in both day-ahead and balancing markets allows MG operators to manage EV aggregation against profit maximization.

2.3.1.1 Price-Based Designs

DSO anticipates potential congestions in its networks over specific periods in price-based schemes and sends congestion pricing locally to aggregators for resource scheduling. However, according to [26], the use of a price-based method might cause a lot of uncertainty in the load due to unapproved aggregator plans in the DSO. Aggregators first calculate their optimal energy usage pattern, limited by their power and energy constraints. The DSO then provides electricity price motivations for each aggregator using a robust two-step optimization model. Z. Xu et al. in [64] are suggested a framework of cooperation between DSOs and EV aggregators to minimize the operating cost of the aggregator and the peak load of the system. Ea aggregator buys and sells power at TOU tariffs at retail prices to electric car owners for a profit.

In order to increase the economic aspect of using charging stations and aggregators, this paper uses a price-based and linear optimization model as a tool to coordinate the aforementioned infrastructures by using the vehicle charging policy [79]. Article [78] proposes an efficient planning method to consider electricity price uncertainty for EV aggregators to achieve the maximum expected profit using the stochastic p-robust optimization technique.

2.3.2 Providing Distribution Network Services by Aggregators

Depending on current and future grid requirements, the aggregator's services can be provided to the DSO through EVs and other distributed energy resources (DERs). Therefore, as considered in [27], addressing the association between DERs and EVs' managing schemes is necessary.

Studies show that uncontrolled large-scale charging of EVs will convey challenges such as peak power, voltage drops, network congestion, power losses, etc. to the distribution network. The cases of smart charging, uncoordinated charging, and smart charging/discharging (two-way V2G) are the discussed strategies that impact the distribution network performance in the studies [28,29].

Authors [30] have suggested a legal model for charging EVs, introducing two electricity market operators, an EV fleet collection operator and an EV charging manager. The charge manager is responsible for developing the charging infrastructure. The manager is in charge of delivering charging operations to the electric vehicle fleet as well as controlling the EV fleet to serve additional services. Considering the possibility of exploiting the concept of the fleet operator (FO), Bessa and Matos [13] provided an overview of studies related to the financial and technological management of an EV aggregation factor. The reviewed article [13] is divided into three thematic sections: 1. technical issues, 2. electricity market, and 3. economic EV.

In order to avoid unessential investments in the EV charging infrastructure, this study introduced a new framework with multi–time scale flexibility for accumulated EV load by considering various EV charging procedures under several electricity markets [80].

In study [34], similar objectives that FOs consider in their studies are:

- Ensure that EV owners' driving demands are met through efficient EV charging management.
- Using V2G technology, deliver the maximum power to the electrical network.
- Optimal allocation of EV resources to provide ancillary services to power system operators.

However, regarding the structure for power system operators, the services in question are 1. up and down adjustment, 2. rotating storage, and 3. non-rotating storage. That is why the tuning market has the highest price among all frequency service markets.

The proposed taxonomy in this study is one of the multiple probabilities obtained in the review of research [37,81,82]. As a result of the versatility of such sources and the desire to satisfy the network's technical goals, aggregator services can be separated into load support and voltage control based on the right division of network areas with the relevant aggregators, according to load standards.

Study [31] discusses the criteria for considering a suitable ancillary service performed by PEVs. Accordingly, the following four norms should be studied for this purpose:

- Supply time: This is when the service in question is called. The suitable supply time for PEVs is from a few minutes to a few hours.
- Redirection: Some services are highly unstable and change rapidly in both directions (several times per minute). PEVs can go up or down quickly at a lower cost than idle power plants, which is an advantage.
- Response rate: The time it takes for the ancillary service provider to start the service. Response times for various ancillary operations range from a few seconds to an hour, due to the nature of the occurrence.
- Service duty: Refers to ancillary services' periodic or continuous nature.

Reduction of voltage imbalance through charge/discharge optimization of EVs has been evaluated in [32]. Referring to the ancillary services supplied by the EVs fleet, a smart planning approach based on a nonlinear programming model (NLP) provides voltage and frequency regulation services through the integration of V2G-enabled EVs [33]. In this study, aggregators maximize the EV owners' profit, minimize charging costs and counteract reduced battery life when participating in both network services.

In the article [83], in order to expand a reduced EV aggregator model and to realize frequency regulation, the state-space method is proposed.

This goal is accomplished by a power adjustment computation model that optimizes the gain of the aggregator and the frequency tracking signal control implementation of the DSO.

The overhead studies have shown that EVs can contain a high share of up and down adjustment and rotary storage services. These services need a rapid response from sources (such as generators, DR programs, and EV). Also, a good telecommunications infrastructure is needed to help resource-responsive ancillary services.

Studies in [35] indicate that, if not enough attention is paid to the distribution of charge load during off-peak periods, even lower degrees of infiltration of EVs can produce fresh peak loads above the typical peak.

In [84], a strategy for dealing with active/reactive management of a distribution network with EVs is provided in order to optimize the voltage profile while obtaining the lowest operating cost.

In order to reduce the uncertainty of aggregators caused by EVs during frequency regulation, this study introduces a novel model that determines the exact stability space [84].

As mentioned in the analyses of [36], [37], and [38], one of the major challenges for setting up a power system with renewable energy sources such as solar and wind is their alternating treatment, which is affected by the nature of their energy sources randomness. With this in mind, EVs have been suggested as one of the most suitable solutions to decrease the effects of this alternating treatment, compared to other solutions such as using a backup production or centralized storage system, which involves heightened capital costs for power system operators. Electric vehicles can store excess energy from renewable sources in their batteries and use it for everyday driving. Some electric vehicles with V2G capabilities can return the energy to the grid.

In order to provide ancillary services to the electricity market, this article proposes a scheme for assessing two investment choices in storage and dedicated energy storage systems [85,86].

The results show that incorporating EVs and V2G capability to power systems allows the integration of much higher levels of renewable energy and dramatically decreases CO_2 emissions. The impact of EV charging on the German power system with a 50% share of renewable energy in 2030 (wind and solar) in [39] has been studied; in this paper, the charging strategy is achieved by responding to consumer price.

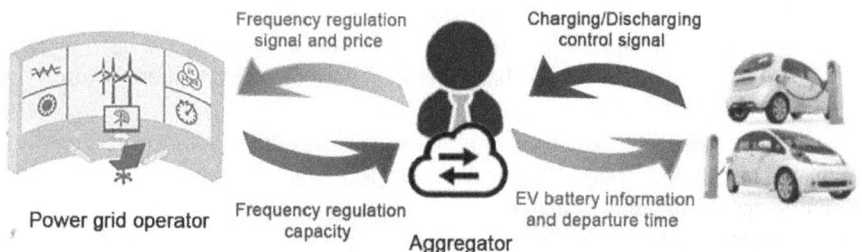

FIGURE 2.6 A simple scheme of centralized dispatch system.

In addition, this article concludes that EVs can play an essential part in reducing the alternating treatment of renewable energies.

In [33], a particle congestion optimization algorithm is proposed to consider integrating renewable energy sources and EVs in the power grid and investigate its impact. The authors conclude that using EVs with V2G capabilities in smart grids will help the grid by minimizing costs and emissions and reducing imbalances in the use of intermittent renewable energy.

Birnie [40] proposes utilizing photovoltaic (PV) panels to provide intraday charge for EVs in a parking lot, presuming New Jersey solar radiation to integrate EVs with solar energy. This study derived that PV panels could only meet the driving needs of EV owners during the summer. Authors [41] investigated the incorporation of PV panels in cooperation with heat pumps and EVs in the Kansai region in Japan.

In general, and according to studies, it can be seen that connecting more EVs in the network and controlling their charge rate can increase the share of renewable energy in the power system and reduce their erratic behavior. In addition, EVs with V2G capabilities can return excess renewable energy previously stored in batteries to the grid when power is required, such as during periods of high demand and low renewable energy production.

In [87], the EV FO is presumed to manage the electricity market participation of an EV fleet and provide a framework for the optimal charging of EVs. The consequence showed that the EVs' electricity bills are decreased. Also, the electricity price in the regulatory market, the daily market, and the driving patterns of the electric fleet are commonly recognized by the FO, which is assumed to be the transition price in electricity.

However, Christofferson et al. in [42] also examined EV management possibilities. The FO has a significant market share and can influence the price of electricity by changing loads through charge and discharge.

In addition to the aforementioned studies, reference [43] shows how the EV controller can use dynamic programming to create an optimal charging program considering the market price of electricity. In [44,45], the authors presented a strategy for an EV collector. V2G service can reduce the cost of charging for the benefit of owners of EVs.

2.4 ACCUMULATION-BASED PEV MODELING STRUCTURE

Aggregators can depend on the number of vehicles, accessibility, kind of battery, and storage capacity, as well as other extra factors, such as energy utilization rate (ECR; kWh/km), arrival/departure time, initial SOC, battery failure, traveled distance, and charge/discharge rate request for a specific power level from/to the network. However, the problem with stacked EV modeling is that most of these characteristics are related to EV owners' behavior uncertainty.

The most basic and most uncomplicated approach [46], [47], and [48] is to use a DTEB model for each PEV of a work sample and, therefore, demonstrate them at the aggregate level. For simplicity, some researchers [46,49,50] ignore the losses and consider that all EVs have the same capacity and battery performance and keep the charge or discharge rate constant. However, the authors in [39–51] offer more accurate models in which the parameters vary within a specific range. In [47], optimal bidding was used as a two-stage random linear programming problem and multiple scenarios to represent uncertainty (driving pattern). The issue with such methods is scalability, since optimization difficulties become insurmountable as the PEVF grows larger. In addition, this approach may be constrained by the lack of information, for example, scenarios for individual driving patterns based on historical data. Pillai [52] presents an EEC virtual battery model for analyzing the impact of V2G on load frequency control in western Denmark, and the virtual battery parameters in this paper are obtained based on simple assumptions. Also, it is concluded that 50% of the PEVs in the fleet are always available for V2G services. Starting with a first-order model to describe the evolution of the amount of energy of a PEV-like [88], then a charge requirement indicator was presented as the proportion of the charge time required at full capacity to the remaining time (as determined by the user) to charge. The flexibility of PEV batteries, as demonstrated by their limitations, can be combined to form cumulative flexibility [53]. In the approach [48–54], the authors consider a wide range of available data, including the number of consumers and EV demand, the production of photovoltaics, or local battery systems installed to facilitate calculations. In an analysis [55], a PEVF is used to provide both principal and ancillary frequency control. The dynamic response of an individual PEV is demonstrated as a first-order transfer function, and a participation factor is used for each PEV to indicate its availability for initial control. An average contribution coefficient is calculated at the total level, and an hourly indicator is used to indicate the total hourly accessibility.

In [89], the accumulation of large-scale EVs in real distribution networks was analyzed as a paradigm for several levels of planning of influence and simultaneous factors to analyze the relationship between investment and maintenance expenditures with overall energy consumption. Losses in HV, MV, and LV networks. As a result, the researchers stressed that if simultaneous charging factors are reduced (smart charging), up to 60–70% of the incremental investment needed to upgrade the network can be avoided. Therefore, network planning should be developed with simpler models that allow the TSO or DSO to evaluate the aggregation effect of large amounts of EVs.

Various studies have been performed in [42], [56], and [57] to appraise the effect of DERs, reduce power losses and voltage deviations, meet EV charging demand cost, as well as to evaluate EV battery depreciation. However, the drawbacks of this approach are the modeling details and the considerable amount of data needed, which makes its usage computationally more time consuming.

2.5 CONCLUSIONS

From a production point of view, there will be an increase in the penetration of renewable energy sources into the power grid, mainly through wind power use. On the other hand, electricity consumers will demand new services, including being able to charge/discharge their EVs regularly at home.

PEVs can serve an important part in energy management systems in local areas and distribution networks, as per the studies. This chapter examined the EV aggregation concept and the different methods presented in various papers for modeling accumulated EV demand. From this outline, forthcoming work should consider developing new approaches based on the totality of existing actors that enable effective modeling of EVs and additional DERs, thus providing tools for DSOs for operational and planning purposes. Along with several other studies, PEVs are appropriate for ancillary services such as voltage control as well as frequency regulation. However, many limitations need to be overcome to consider PEVs as an effective energy storage system for distribution network management.

Moreover, there is still a requirement to set up monitoring techniques to implement new factors like aggregators in network management.

Specifically, given the current growth of DERs and EVs in low voltage networks, it is necessary to continue with traditional distribution system planning to overcome the influx of more accurate and simpler models that take into account the increasing number of distributed and variable resources in distribution networks and allow DSO to estimate them.

REFERENCES

[1] Unter, W., J. Auer, and G. Erdmann, Integration of electric vehicles in the Austrian electricity system. 2013. 99(57254).

[2] Pasaoglu, G., et al., Projections for electric vehicle load profiles in Europe based on travel survey data. In Joint Research Centre of the European Commission: Petten, The Netherlands, 2013.

[3] Heydarian-Forushani, E., M.E. Golshan, and M. Shafie-khah, Flexible interaction of plug-in electric vehicle parking lots for efficient wind integration. Applied Energy, 2016. 179: p. 338–349.

[4] Yilmaz, M., and P.T. Krein, Review of battery charger topologies, charging power levels, and infrastructure for plug-in electric and hybrid vehicles. IEEE Transactions on Power Electronics, 2012. 28(5): p. 2151–2169.

[5] Burke, A.F., Batteries and ultracapacitors for electric, hybrid, and fuel cell vehicles. Proceedings of the IEEE, 2007. 95(4): p. 806–820.

[6] Dalla Chiara, B., et al., Competence area of commercial electric vehicles and relevance of an ITS support for transport and charging issues. In 21st World Congress on Intelligent Transport Systems, ITSWC 2014: Reinventing Transportation in Our Connected World, 2014.

[7] Energy Efficiency and Conservation Authority (EECA). New Zealand Energy Efficiency and Conservation Authority, EV battery life. Energy Efficiency and Conservation Authority (EECA). 2017. (April): p. 1–30. https://www.eeca.govt.nz/assets/EECA-Resources/Research-papers-guides/ev-lca-final-report-nov-2015.pdf

[8] Lund, H., et al., System and market integration of wind power in Denmark. Energy Strategy Reviews, 2013. 1(3): p. 143–156.

[9] Kempton, W. and J. Tomić, Vehicle-to-grid power fundamentals: Calculating capacity and net revenue. Journal of Power Sources, 2005. 144(1): p. 268–279.

[10] Meta, K., *Optimal aggregation of electric cars for their charging within smart distribution grids*. Politecnico di Torino, Italy, 2017.

[11] Wu, Q., et al. Driving pattern analysis for electric vehicle (EV) grid integration study. In 2010 IEEE Pes Innovative Smart Grid Technologies Conference Europe (Isgt Europe). 2010. IEEE.

[12] Technical University of Denmark. The Danish National Travel Survey. [Online]. 2013, Apr. www.dtu.dk/centre/Modelcenter/English/TU/Hvad%20er%20TU,-q-.aspx

[13] Bessa, R.J., and M.A. Matos, Economic and technical management of an aggregation agent for electric vehicles: A literature survey. European Transactions on Electrical Power, 2012. 22(3): p. 334–350.

[14] Xi, X. and R. Sioshansi, Using price-based signals to control plug-in electric vehicle fleet charging. IEEE Transactions on Smart Grid, 2014. 5(3): p. 1451–1464.

[15] González-Romera, E., et al. Overview of plug-in electric vehicles as providers of ancillary services. In 2015 9th International Conference on Compatibility and Power Electronics (CPE). 2015. IEEE.

[16] Bessa, R.J. and M.A. Matos, "The role of an aggregator agent for EV in the electricity market," *7th Mediterranean Conference and Exhibition on Power Generation, Transmission, Distribution and Energy Conversion (MedPower 2010), 2010*, pp. 1–9, doi: 10.1049/cp.2010.0866.

[17] Bessa, R.J., and M. Matos, Global against divided optimization for the participation of an EV aggregator in the day-ahead electricity market. Part I: Theory. Electric Power Systems Research, 2013. 95: p. 309–318.

[18] You, S., J. Hu, and C. Ziras, An overview of modeling approaches applied to aggregation-based fleet management and integration of plug-in electric vehicles. Energies, 2016. 9(11): p. 968.

[19] Rigas, E.S., S.D. Ramchurn, and N. Bassiliades, Managing electric vehicles in the smart grid using artificial intelligence: A survey. IEEE Transactions on Intelligent Transportation Systems, 2014. 16(4): p. 1619–1635.

[20] Galus, M.D., M. Zima, and G. Andersson, On integration of plug-in hybrid electric vehicles into existing power system structures. Energy Policy, 2010. 38(11): p. 6736–6745.

[21] Honarmand, M., A. Zakariazadeh, and S. Jadid, Integrated scheduling of renewable generation and electric vehicles parking lot in a smart microgrid. Energy Conversion and Management, 2014. 86: p. 745–755.

[22] Guille, C. and G. Gross, A conceptual framework for the vehicle-to-grid (V2G) implementation. Energy Policy, 2009. 37(11): p. 4379–4390.

[23] Battistelli, C., L. Baringo, and A. Conejo, Optimal energy management of small electric energy systems including V2G facilities and renewable energy sources. Electric Power Systems Research, 2012. 92: p. 50–59.

[24] Mwasilu, F., et al., Electric vehicles and smart grid interaction: A review on vehicle to grid and renewable energy sources integration. Renewable and Sustainable Energy Reviews, 2014. 34: p. 501–516.

[25] Quinn, C., D. Zimmerle, and T.H. Bradley, The effect of communication architecture on the availability, reliability, and economics of plug-in hybrid electric vehicle-to-grid ancillary services. Journal of Power Sources, 2010. 195(5): p. 1500–1509.

[26] Parhizi, S., A. Khodaei, and M. Shahidehpour, Market-based versus price-based microgrid optimal scheduling. IEEE Transactions on Smart Grid, 2016. 9(2): p. 615–623.

[27] Quiros-Tortos, J., L. Ochoa, and T. Butler, How electric vehicles and the grid work together: Lessons learned from one of the largest electric vehicle trials in the world. IEEE Power and Energy Magazine, 2018. 16(6): p. 64–76.

[28] García-Villalobos, J., et al., Plug-in electric vehicles in electric distribution networks: A review of smart charging approaches. Renewable and Sustainable Energy Reviews, 2014. 38: p. 717–731.

[29] Green II, R.C., L. Wang, and M. Alam, The impact of plug-in hybrid electric vehicles on distribution networks: A review and outlook. Renewable and Sustainable Energy Reviews, 2011. 15(1): p. 544–553.

[30] San Román, T.G., et al., Regulatory framework and business models for charging plug-in electric vehicles: Infrastructure, agents, and commercial relationships. Energy Policy, 2011. 39(10): p. 6360–6375.

[31] Moss, B., et al., Ancillary service revenue opportunities from electric vehicles via demand response. 2011.

[32] Farahani, H.F., Improving voltage unbalance of low-voltage distribution networks using plug-in electric vehicles. Journal of Cleaner Production, 2017. 148: p. 336–346.

[33] Amamra, S.-A., and J. Marco, Vehicle-to-grid aggregator to support power grid and reduce electric vehicle charging cost. IEEE Access, 2019. 7: p. 178528–178538.

[34] Bibak, B., and H. Tekiner-Moğulkoç, A comprehensive analysis of Vehicle to Grid (V2G) systems and scholarly literature on the application of such systems. Renewable Energy Focus, 2021. 36: p. 1–20.

[35] Rahman, S., and G. Shrestha, An investigation into the impact of electric vehicle load on the electric utility distribution system. IEEE Transactions on Power Delivery, 1993. 8(2): p. 591–597.

[36] Richardson, D.B., Electric vehicles and the electric grid: A review of modeling approaches, impacts, and renewable energy integration. Renewable and Sustainable Energy Reviews, 2013. 19: p. 247–254.

[37] Yong, J.Y., et al., A review on the state-of-the-art technologies of electric vehicle, its impacts and prospects. Renewable and Sustainable Energy Reviews, 2015. 49: p. 365–385.

[38] Rueda-Medina, A.C., A. Padilha-Feltrin, and J. Mantovani. Active power reserve for frequency control provided by distributed generators in distribution networks. In 2014 IEEE PES General Meeting| Conference & Exposition. 2014. IEEE.

[39] Dallinger, D., and M. Wietschel, Grid integration of intermittent renewable energy sources using price-responsive plug-in electric vehicles. Renewable and Sustainable Energy Reviews, 2012. 16(5): p. 3370–3382.

[40] Birnie III, D.P., Solar-to-vehicle (S2V) systems for powering commuters of the future. Journal of Power Sources, 2009. 186(2): p. 539–542.

[41] Zhang, Q., et al., Integration of PV power into future low-carbon smart electricity systems with EV and HP in Kansai Area, Japan. Renewable Energy, 2012. 44: p. 99–108.

[42] Kristoffersen, T.K., K. Capion, and P. Meibom, Optimal charging of electric drive vehicles in a market environment. Applied Energy, 2011. 88(5): p. 1940–1948.

[43] Rotering, N., and M. Ilic, Optimal charge control of plug-in hybrid electric vehicles in deregulated electricity markets. IEEE Transactions on Power Systems, 2010. 26(3): p. 1021–1029.

[44] Sortomme, E., and M.A. El-Sharkawi, Optimal scheduling of vehicle-to-grid energy and ancillary services. IEEE Transactions on Smart Grid, 2011. 3(1): p. 351–359.

[45] Al-Awami, A.T. and E. Sortomme, Coordinating vehicle-to-grid services with energy trading. IEEE Transactions on Smart Grid, 2011. 3(1): p. 453–462.

[46] He, Y., B. Venkatesh, and L. Guan, Optimal scheduling for charging and discharging of electric vehicles. IEEE Transactions on Smart Grid, 2012. 3(3): p. 1095–1105.

[47] Vagropoulos, S.I. and A.G. Bakirtzis, Optimal bidding strategy for electric vehicle aggregators in electricity markets. IEEE Transactions on Power Systems, 2013. 28(4): p. 4031–4041.

[48] Di Giorgio, A., F. Liberati, and S. Canale, Electric vehicles charging control in a smart grid: A model predictive control approach. Control Engineering Practice, 2014. 22: p. 147–162.

[49] Parks, K., P. Denholm, and T. Markel, Costs and emissions associated with plug-in hybrid electric vehicle charging in the Xcel Energy Colorado service territory. 2007, National Renewable Energy Lab. (NREL), Golden, CO (United States).

[50] Li, P., et al., Participation in the frequency regulation control of a resilient microgrid for a distribution network. International Journal of Integrated Energy Systems, 2009. 1(1): p. 61–67.

[51] Liu, L., et al., A review on electric vehicles interacting with renewable energy in smart grid. Renewable and Sustainable Energy Reviews, 2015. 51: p. 648–661.

[52] Pillai, J.R. and B. Bak-Jensen, Integration of vehicle-to-grid in the western Danish power system. IEEE Transactions on Sustainable Energy, 2010. 2(1): p. 12–19.

[53] Trangbaek, K. and J. Bendtsen, Exact constraint aggregation with applications to smart grids and resource distribution. In 2012 IEEE 51st IEEE Conference on Decision and Control (CDC). 2012. IEEE.

[54] Saber, A.Y., and G.K. Venayagamoorthy, resource scheduling under uncertainty in a smart grid with renewables and plug-in vehicles. IEEE Systems Journal, 2011. 6(1): p. 103–109.

[55] Izadkhast, S., P. Garcia-Gonzalez, and P. Frías, An aggregate model of plug-in electric vehicles for primary frequency control. IEEE Transactions on Power Systems, 2014. 30(3): p. 1475–1482.

[56] Lund, H. and W. Kempton, Integration of renewable energy into the transport and electricity sectors through V2G. Energy Policy, 2008. 36(9): p. 3578–3587.

[57] Hansen, H., et al., Coordination of system needs and provision of services. 2013.

[58] Noel, L., J. Kester, G. Zarazua de Rubens, B.K. Sovacool. *Vehicle-to-grid: A sociotechnical transition beyond electric mobility.* Basingstoke: Palgrave, 2019.

[59] Sovacool, B.K., et al., Actors, business models, and innovation activity systems for vehicle-to-grid (V2G) technology: A comprehensive review. Renewable and Sustainable Energy Reviews, 2020. 131: p. 109963.

[60] Borray, A.F.C., et al., A review of the population-based and individual-based approaches for electric vehicles in network energy studies. Electric Power Systems Research, 2020. 189: p. 106785.

[61] Masood, A., et al., Transactive energy for aggregated electric vehicles to reduce system peak load considering network constraints. IEEE Access, 2020. 8: p. 31519–31529.

[62] Khanekehdani, H.K., M.M. Tafreshi, and M. Khosravi, Modeling operation of electric vehicles aggregator in reserve services market by using game theory method. Journal of Renewable and Sustainable Energy, 2013. 5(6): p. 063127.

[63] Xu, Z., et al., Data-driven pricing strategy for demand-side resource aggregators. IEEE Transactions on Smart Grid, 2016. 9(1): p. 57–66.

[64] Xu, Z., et al., Coordination of PEVs charging across multiple aggregators. Applied Energy, 2014. 136: p. 582–589.

[65] Morstyn, T., A. Teytelboym, and M.D. McCulloch, Designing decentralized markets for distribution system flexibility. IEEE Transactions on Power Systems, 2018. 34(3): p. 2128–2139.

[66] Hu, J., et al., Application of network-constrained transactive control to electric vehicle charging for secure grid operation. IEEE Transactions on Sustainable Energy, 2016. 8(2): p. 505–515.

[67] Klaina, H., et al., Aggregator to electric vehicle LoRaWAN based communication analysis in vehicle-to-grid systems in smart cities. IEEE Access, 2020. 8: p. 124688–124701.

[68] Gao, S., and H. Jia, Integrated configuration and optimization of electric vehicle aggregators for charging facilities in power networks with renewables. IEEE Access, 2019. 7: p. 84690–84700.

[69] Jin, Y., et al., Optimal aggregation design for massive V2G participation in energy market. IEEE Access, 2020. 8: p. 211794–211808.

[70] Khan, M.W. and J. Wang, Multi-agents based optimal energy scheduling technique for electric vehicles aggregator in microgrids. International Journal of Electrical Power & Energy Systems, 2022. 134: p. 107346.

[71] Sousa, T., et al., Day-ahead resource scheduling in smart grids considering vehicle-to-grid and network constraints. Applied Energy, 2012. 96: p. 183–193.

[72] Sarker, M.R., Y. Dvorkin, and M.A. Ortega-Vazquez, Optimal participation of an electric vehicle aggregator in day-ahead energy and reserve markets. IEEE Transactions on Power Systems, 2015. 31(5): p. 3506–3515.

[73] Baherifard, M.A., et al., Intelligent charging planning for electric vehicle commercial parking lots and its impact on distribution network's imbalance indices. Sustainable Energy, Grids and Networks, 2022: p. 100620.

[74] Wei, C., et al., Aggregation and scheduling models for electric vehicles in distribution networks considering power fluctuations and load rebound. IEEE Transactions on Sustainable Energy, 2020. 11(4): p. 2755–2764.

[75] Gomes, I.L., R. Melicio, and V.M. Mendes. Electric vehicles aggregation in market environment: A stochastic grid-to-vehicle and vehicle-to-grid management. In Doctoral Conference on Computing, Electrical and Industrial Systems. 2019. Springer.

[76] Perez-Diaz, A., E. Gerding, and F. McGroarty, Decentralised coordination of electric vehicle aggregators. 2018.

[77] Zhou, M., et al., Forming dispatchable region of electric vehicle aggregation in microgrid bidding. IEEE Transactions on Industrial Informatics, 2020. 17(7): p. 4755–4765.

[78] Sriyakul, T., and K. Jermsittiparsert, Optimal economic management of an electric vehicles aggregator by using a stochastic p-robust optimization technique. Journal of Energy Storage, 2020. 32: p. 102006.

[79] Ding, Z., et al., Optimal coordinated operation scheduling for electric vehicle aggregator and charging stations in an integrated electricity-transportation system. International Journal of Electrical Power & Energy Systems, 2020. 121: p. 106040.

[80] Wang, B., et al., Aggregated electric vehicle load modeling in large-scale electric power systems. IEEE Transactions on Industry Applications, 2020. 56(5): p. 5796–5810.

[81] Sarabi, S., et al., Potential of vehicle-to-grid ancillary services considering the uncertainties in plug-in electric vehicle availability and service/localization limitations in distribution grids. Applied Energy, 2016. 171: p. 523–540.

[82] Habib, S., M. Kamran, and U. Rashid, Impact analysis of vehicle-to-grid technology and charging strategies of electric vehicles on distribution networks: A review. Journal of Power Sources, 2015. 277: p. 205–214.

[83] Wang, M., et al., Electric vehicle aggregator modeling and control for frequency regulation considering progressive state recovery. IEEE Transactions on Smart Grid, 2020. 11(5): p. 4176–4189.

[84] Sousa, T., et al., A multi-objective optimization of the active and reactive resource scheduling at a distribution level in a smart grid context. Energy, 2015. 85: p. 236–250.

[85] Dong, C., et al., Distorted stability space and instability triggering mechanism of EV aggregation delays in the secondary frequency regulation of electrical grid-electric vehicle system. IEEE Transactions on Smart Grid, 2020. 11(6): p. 5084–5098.

[86] Aldik, A., et al., A planning model for electric vehicle aggregators providing ancillary services. IEEE Access, 2018. 6: p. 70685–70697.

[87] Hu, J., et al., Coordinated charging of electric vehicles for congestion prevention in the distribution grid. IEEE Transactions on Smart Grid, 2013. 5(2): p. 703–711.

[88] Ma, Z., D.S. Callaway, and I. A. Hiskens, Decentralized charging control of large populations of plug-in electric vehicles. IEEE Transactions on Control Systems Technology, 2011. 21(1): p. 67–78.

[89] Ellabban, O., H. Abu-Rub, and F. Blaabjerg, Renewable energy resources: Current status, future prospects and their enabling technology. Renewable and Sustainable Energy Reviews, 2014. 39: p. 748–764.

[90] Rasheed, M.B., et al., An optimal scheduling and distributed pricing mechanism for multi-region electric vehicle charging in smart grid. IEEE Access, 2020. 8: p. 40298–40312.

[91] Xie, L., et al., Wind integration in power systems: Operational challenges and possible solutions. Proceedings of the IEEE, 2010. 99(1): p. 214–232.

[92] Karden, E., et al., Energy storage devices for future hybrid electric vehicles. Journal of Power Sources, 2007. 168(1): p. 2–11.

[93] Bendien, J.C., G. Fregien, and J.D. Van Wyk. High-efficiency on-board battery charger with transformer isolation, sinusoidal input current and maximum power factor. In IEE Proceedings B (Electric Power Applications). 1986. IET.

[94] Wu, T.H., et al. An optimization model for a battery swapping station in Hong Kong. In 2015 IEEE Transportation Electrification Conference and Expo (ITEC). 2015. IEEE.

[95] Falchetta, G., and M. Noussan, Electric vehicle charging network in Europe: An accessibility and deployment trends analysis. Transportation Research Part D: Transport and Environment, 2021. 94: p. 102813.

[96] Gnann, T., et al., Fast charging infrastructure for electric vehicles: Today's situation and future needs. Transportation Research Part D: Transport and Environment, 2018. 62: p. 314–329.

[97] Sun, X., Z. Chen, and Y. Yin, Integrated planning of static and dynamic charging infrastructure for electric vehicles. Transportation Research Part D: Transport and Environment, 2020. 83: p. 102331.

[98] Montazeri-Gh, M., A. Poursamad, and B. Ghalichi, Application of genetic algorithm for optimization of control strategy in parallel hybrid electric vehicles. Journal of the Franklin Institute, 2006. 343(4–5): p. 420–435.

[99] Dharmakeerthi, C., and N. Mithulananthan, PEV load and its impact on static voltage stability. In Plug in electric vehicles in smart grids. Springer, 2015, p. 221–248.

[100] Yong, J.Y., et al., A review on the state-of-the-art technologies of electric vehicle, its impacts and prospects. Renewable and Sustainable Energy Reviews, 2015. 49: p. 365–385.

[101] Gray, M.K. and W.G. Morsi, Power quality assessment in distribution systems embedded with plug-in hybrid and battery electric vehicles. IEEE Transactions on Power Systems, 2014. 30(2): p. 663–671.

[102] Grahn, P., et al. A method for evaluating the impact of electric vehicle charging on transformer hotspot temperature. In 2011 2nd IEEE PES International Conference and Exhibition on Innovative Smart Grid Technologies. 2011. IEEE.

[103] Xu, Y., et al. Harmonic analysis of electric vehicle loadings on distribution system. In 2014 IEEE International Conference on Control Science and Systems Engineering. 2014. IEEE.

[104] Clement-Nyns, K., E. Haesen, and J. Driesen, The impact of charging plug-in hybrid electric vehicles on a residential distribution grid. IEEE Transactions on Power Systems, 2009. 25(1): p. 371–380.

[105] Wu, D., et al., Transient stability analysis of SMES for smart grid with vehicle-to-grid operation. IEEE Transactions on Applied Superconductivity, 2011. 22(3): p. 5701105–5701105.

[106] Iqbal, S., et al. Aggregated electric vehicle-to-grid for primary frequency control in a microgrid: A review. In 2018 IEEE 2nd International Electrical and Energy Conference (CIEEC). 2018. IEEE.

[107] Song, B., U. Madawala, and C. Baguley. A review of grid impacts, demand side issues and planning related to electric vehicle charging. In 2021 IEEE Southern Power Electronics Conference (SPEC). 2021. IEEE.

[108] Shaaban, M.F., et al., Real-time PEV charging/discharging coordination in smart distribution systems. IEEE Transactions on Smart Grid, 2014. 5(4): p. 1797–1807.

[109] Hua, L., J. Wang, and C. Zhou, Adaptive electric vehicle charging coordination on distribution network. IEEE Transactions on Smart Grid, 2014. 5(6): p. 2666–2675.

[110] Gupta, V., et al., Multi aggregator collaborative electric vehicle charge scheduling under variable energy purchase and EV cancelation events. IEEE Transactions on Industrial Informatics, 2017. 14(7): p. 2894–2902.

[111] Crow, M., Cost-constrained dynamic optimal electric vehicle charging. IEEE Transactions on Sustainable Energy, 2016. 8(2): p. 716–724.

[112] Zhang, L., V. Kekatos, and G.B. Giannakis, Scalable electric vehicle charging protocols. IEEE Transactions on Power Systems, 2016. 32(2): p. 1451–1462.

[113] Wen, C.-K., et al., Decentralized plug-in electric vehicle charging selection algorithm in power systems. IEEE Transactions on Smart Grid, 2012. 3(4): p. 1779–1789.

[114] Luo, C., Y.-F. Huang, and V. Gupta, Stochastic dynamic pricing for EV charging stations with renewable integration and energy storage. IEEE Transactions on Smart Grid, 2017. 9(2): p. 1494–1505.

3 Forecasting the Energy Consumption Impact of Electric Vehicles by Means of Machine Learning Approaches

Ibrahim Kovan and Stefan Twieg

CONTENTS

DOI: 10.1201/9781003293989-3

3.1 INTRODUCTION

Machine learning (ML) is a method of problem-solving, a sub-branch of computer science and artificial intelligence (AI) that is statistically and mathematically based, dating back to the 1950s, and has increased in popularity in the past decade. Statistical discoveries that form the basis of ML date back to the 1700s, such as Bayes' theorem. ML was defined as a "field of study that gives the ability to learn without being explicitly programmed" by Arthur Samuel in 1959 (Samuel 1959). In terms of computer science, an input and a rule are presented to the computer, and an output is obtained per the designed rule in classical programming; a dataset (input and output) whose cause-and-effect relationship is unknown is presented to the computer as input, and by the help of various statistical and mathematical methods of ML the rules are adjusted by their weights for this particular dataset. Problem solving is performed in ML, which is established by the rule, obtained through the training of the computer: A data series whose input and output are known is trained by a computer to explore a rule between the input and the output by updating weights while training. When a random input is presented to the computer, it predicts an output based on the learning process. However, the answer is ultimately a statistical result engaged to the data that the rule has obtained.

ML approaches have taken their place in the literature, with the studies carried out as a solution to the problems encountered in electric vehicle (EV) technology. One of them is vehicle-to-grid (V2G) technology, where the results are obtained by establishing deep reinforcement learning methods, which is a sub-unit of ML and has been revealed to perform much better than other V2G systems (Kiaee 2020). In this chapter, it has been demonstrated by the authors that the model obtained by combining Q learning and deep neural network and the charge/discharge status of the EV can be systematized in a way that maximizes profit. Successful energy management for smart power grid systems, not only based on EV integration but also in general, will reveal a win-win situation in many aspects. In the simulation work, it has been observed that there is a decrease in energy costs in smart energy buildings with the Q learning–based energy management model compared to other models (Kim and Lim 2018). Another study, which deals with ML methods as an optimization method, is presented to determine when the EV would be charged to reduce the charging cost by using ML and deep neural network techniques by considering the current status of an EV (Lopez, Gagne, and Gardner 2019). In a study conducted in the field of EV charger behavior, it was stated that ML predictions were related to data entropy and data sparsity, and considering these two indicators, it was shown that the EV charging schedule was optimized with ML results (Chung et al. 2019). It is obvious that ML solutions partly provide more efficiency than traditional methods and can solve problems in many areas, including EV power grid integration in the future by taking the role of the optimizer in many areas from cost effectiveness to increase in performance.

3.2 THE CURRENT STATE OF ELECTRIC VEHICLES ON A GLOBAL BASIS

The idea of using electricity as a source for transport dates back to the end of the 19th century. Various approaches have been put forward to a proposal to develop the technology for EVs, which gained popularity in a short time after the idea had been introduced. One of them, Thomas Edison argued, by building better batteries EVs would be the superior mode of transportation. In 1901, Ferdinand Porsche, the founder of the sports car, created the world's first hybrid electric vehicle, namely, *Lohner Porsche Mixte*. This vehicle was powered by a battery and a gas engine. EVs, which remained popular until the middle of the 20th century, especially with the rise of gasoline prices, played an important role in the development of EVs. They became known around the world with the *Apollo Lunar Rover*, the electric vehicle NASA sent to the moon in 1971. Interests in EVs couldn't keep pace with the technology and the increase in performance and driving range of gas-powered motor vehicles. Due to that, EVs did not have a big impact in the last quarter of the 20th century. By the turn of the 21st century, though, EVs, along with studies of both EVs and batteries, were again becoming popular, signaling an active involvement in daily life. In 2006, Tesla Motors, a Silicon Valley start-up, announced that they would produce an electric sports car with a range of more than 200 as an environmentally friendly alternative to gas-powered cars. As can be seen, the EV and battery studies that continue all over the world have begun to reflect on daily life. Figure 3.1 compares the share of sales of EVs for the 15 highest percentages of prevalence among total vehicles.

The graph shows that, from a general perspective, EVs are more popular in Europe, especially in Norway, Iceland, Sweden, and the Netherlands. The share of EV sales in Norway increased the value by 230% in 5 years from 22.66% in 2015, retaining its peak position in 2020 at 74.75%. Based on the amount of increase, the highest change in 5 years belongs to Finland with a rate of 2,778%. Second on the list is Portugal, which was not even among the top 15 in 2015, with an increase of 2,355%. Globally, the rate was 0.78% in 2015 and increased to 4.61%, a 491% increase. It is observed that the demand for EVs is increasing with the increasing acceleration all over the world, especially in Europe.

The popularity of EVs has brought some imperatives. The amount of charging stations required for these vehicles is increasing along with this popularity. The change in the number of charging stations established for EVs by years is shown in Figure 3.2. The graph on the left compares the regions of the world with the highest number of charging stations, while the graph on the right includes the countries with the other highest number of charging stations.

The left-hand graph shows an upward acceleration in the number of electric car–charging stations in China, particularly after 2018. The number of charging stations in 2019 increased by 87.6% compared to 2018, and the rate is 39.3% for Europe. Worldwide, the total number of charging stations in 2019 grew by 62.6% compared to 2018. In 2020, the rate of increase slowed down compared to 2019 and was 46%. Table 3.1 shows the percentage of increase in regions compared to the previous year. The only region with a decrease in the number of charging stations between 2013 and 2020 compared to the previous year was Japan, which had 30,394 charging stations in 2019 and closed 1.8% of them in 2020, with 29,855 charging stations remaining. The increase continued for all the other regions every year.

Considering the increase in the electric vehicle sales share and the increase in the number of charging stations, it is predicted that these sharp changes would continue in the coming years until it reaches saturation, with the reflection of technological

EV sales share in percentage for cars by top 15 regions in 2020

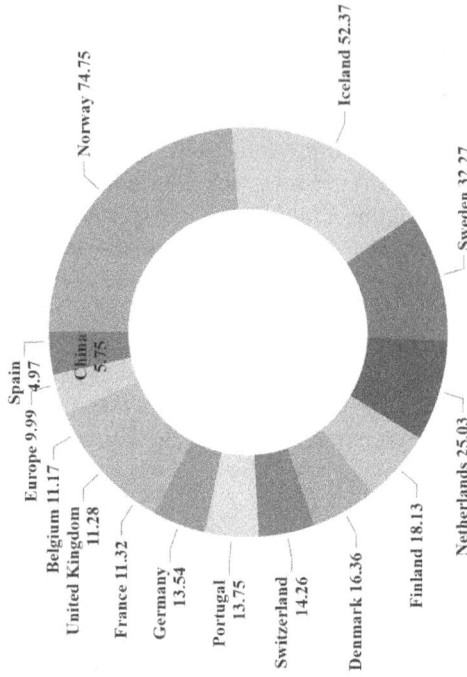

Norway 74.75

Iceland 52.37

Sweden 32.27

Netherlands 25.03

Finland 18.13

Denmark 16.36

Switzerland 14.26

Portugal 13.75

Germany 13.54

France 11.32

United Kingdom 11.28

Belgium 11.17

Europe 9.99

Spain 4.97

China 5.75

EV sales share in percentage for cars by top 15 regions in 2015

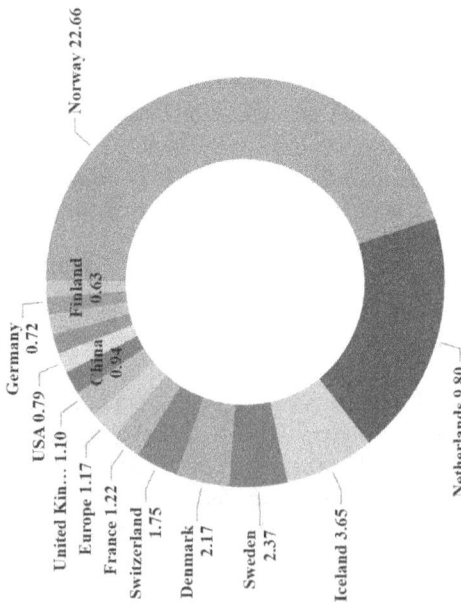

Norway 22.66

Germany 0.72

Finland 0.63

USA 0.79

China 0.94

United Kin... 1.10

Europe 1.17

France 1.22

Switzerland 1.75

Denmark 2.17

Sweden 2.37

Iceland 3.65

Netherlands 9.80

EVs refers to all electric vehicles (Battery electric vehicles+ plug-in hybrid electric vehicles)

FIGURE 3.1 Comparison of the share of electric vehicle sales in total vehicles in percentage for the top 15 regions in 2015 and 2020 (*Data from "Global EV Data Explorer—Analysis—IEA", 2021; Figure from Authors*).

The Number of EV Chargers

region ● China ◆ Europe ▲ World

The Number of EV Chargers

region ● France ● Germany ● Japan ● Korea ● Netherlands ● Norway ● United Kingdom

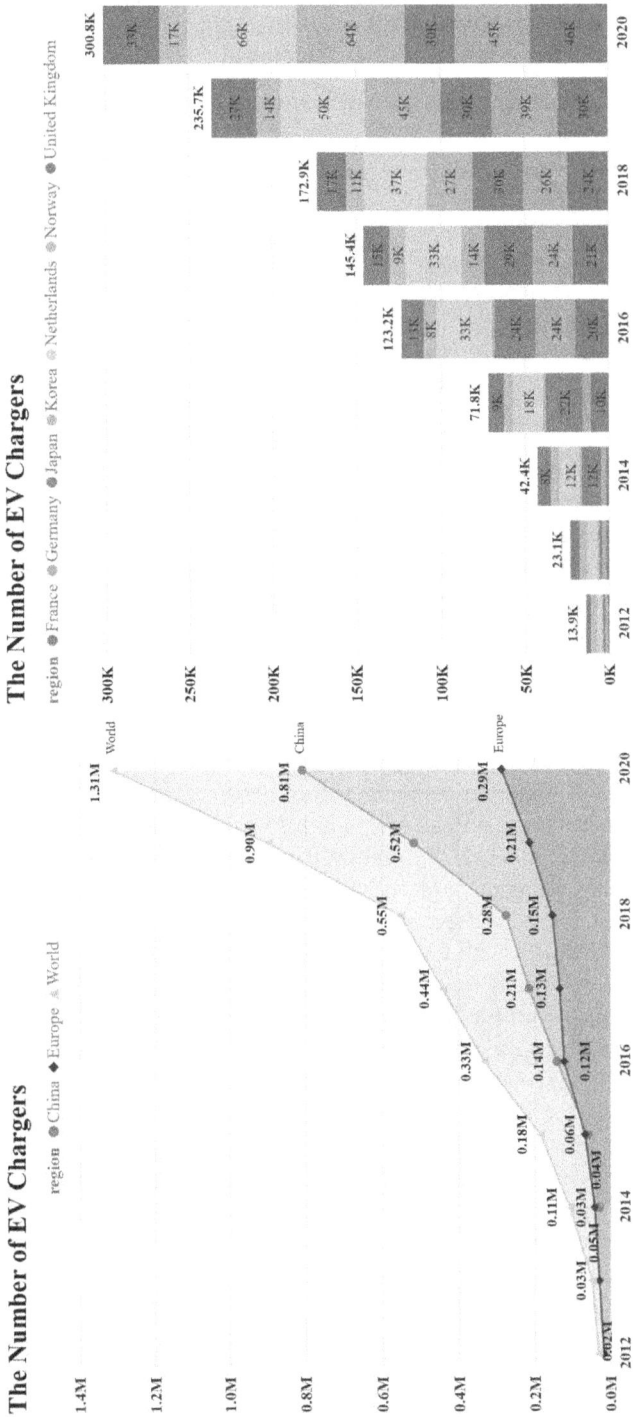

FIGURE 3.2   The number of electric vehicle chargers for the top 10 regions in 2012–2020 (*Data from "Global EV Data Explorer—Analysis—IEA", 2021; figure from authors*).

TABLE 3.1
The Percentage of Increase for EV Chargers in Regions Compared to the Previous Year

Year/ Region	France	Germany	Japan	Korea	Netherlands	Norway	UK	China	Europe	World
2013	122.7%	61.2%	29.9%	33.2%	106.6%	23.0%	100.4%	No Data	69.7%	51.8%
2014	0.7%	11.2%	541.6%	39.3%	106.9%	14.9%	35.4%	No Data	39.9%	119.7%
2015	475.8%	85.8%	91.9%	2.1%	50.3%	4.2%	19.9%	95.9%	61.2%	72.4%
2016	87.8%	372.5%	10.1%	40.3%	80.6%	36.8%	43.5%	140.4%	82.3%	80.0%
2017	8.0%	0.5%	18.3%	579.0%	2.3%	22.1%	14.9%	51.4%	9.4%	33.0%
2018	14.6%	10.2%	4.2%	100.0%	10.7%	14.7%	14.3%	28.6%	14.1%	24.8%
2019	22.3%	48.5%	1.4%	63.8%	36.2%	35.5%	55.5%	87.6%	39.3%	62.6%
2020	55.0%	13.7%	−1.8%	43.3%	30.9%	18.7%	23.5%	56.4%	35.0%	46.0%

developments in electric vehicles. The electricity demand brought by these increases should be taken into account by the countries. Due to the intensity that will arise from the electricity demand of electric vehicles, it is necessary to prevent the overload that may occur in power distribution units. In Figure 3.3, the change in electric demands according to the type of electric vehicles by years is shown.

It is seen that the electricity demand for electric buses, trucks, and vans in the Chinese region is much higher than compared to the other regions. On the other hand, one could say that in 2020, the most electric demand is observed in Europe, depending on the intensity of driving electric cars. It is also seen that the electric demand for cars is the highest in America before 2020.

It is an indisputable fact that fossil fuels cause damage to the environment through global warming and air pollution. It is known that this damage is indirectly common in internal combustion engine vehicles whose main energy source is fossil fuels. There would inevitably be a decrease in the use of fossil fuels with the replacement of fossil fuel vehicles with EVs. This reduction will help preserve the world's ecological balance, especially for the environment and human health. The change in this savings from the use of EVs from year to year is shown in Figure 3.4.

According to the data prepared based on the mobility model of the IEA (International Energy Agency), in 2015, 2.2 billion liters of gasoline were saved by the use of EVs, while this rate was calculated as 15.4 billion liters of gasoline in 2020. Given the electric demand shown in Figure 3.3, the demand for fossil fuels for heavy vehicles, such as buses and trucks, would be as high as the demand for electricity. Just as the electricity demand for electric buses, trucks, and vans in China is the highest, the amount of fossil fuels that would be saved by the use of these EVs is also the highest, as shown in Figure 3.4.

In the next sections, a detailed exemplary study is carried out on the Norway region, where the share of electric vehicles on a regional basis is the highest. The objective of the study is to predict the number of EV registrations in Norway and the amount of power consumption in Norway in the next three years. Combining completely real-world datasets and a variety of ML approaches to identify potential

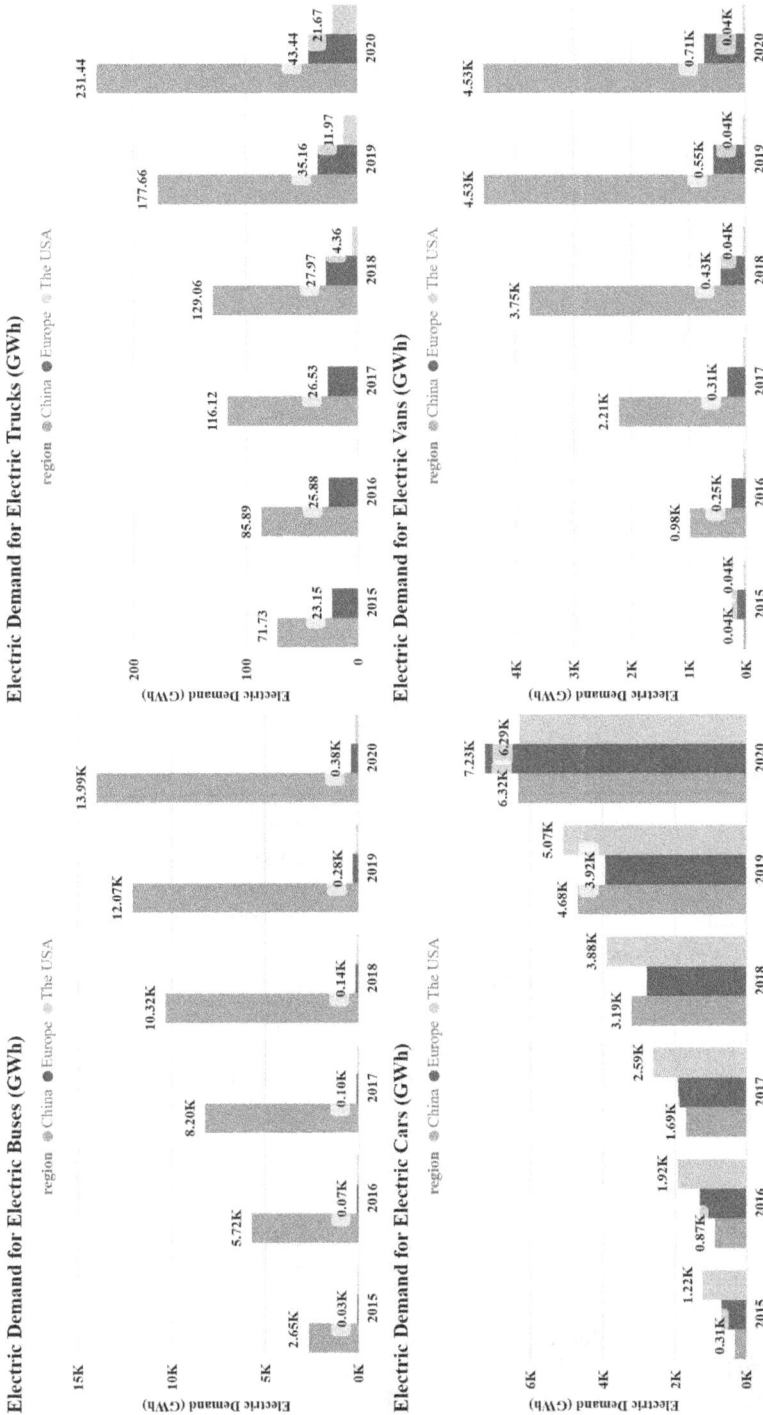

FIGURE 3.3 Electric demand for the electric vehicles by type of vehicle for three regions in 2015–2020 (*Data from "Global EV Data Explorer— Analysis—IEA", 2021.; Figure from Authors*).

The Amount of Saved Oil (Million Liters of Gasoline Equivalent)

region ● China ● Europe ● The USA ● The World

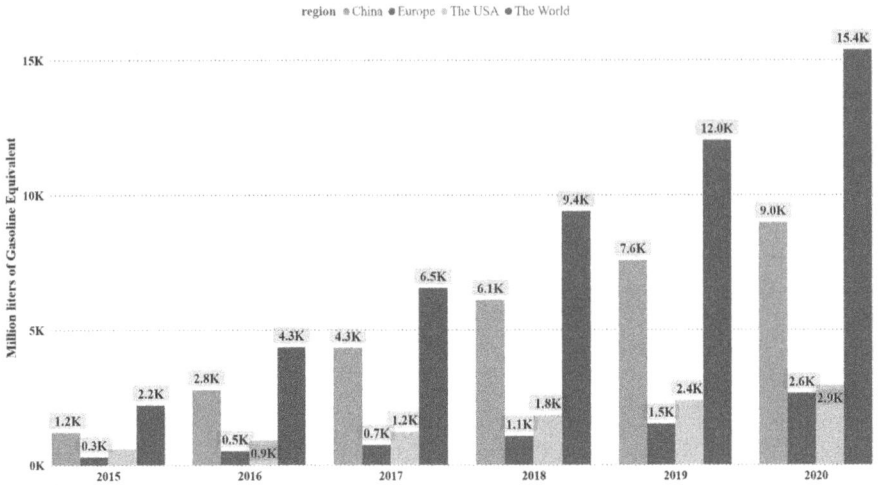

FIGURE 3.4 The amount of saved oil by the usage of electric vehicles in million liters of gasoline-equivalent for three regions and the world in 2015–2020 (*Data from "Global EV Data Explorer—Analysis—IEA", 2021; Figure from Authors*).

strategies for problem solving is the focus of this study. Moreover, the potential impact of EVs on the power consumption of the country is also revealed year by year.

3.3 PROBLEM DEFINITION

The use of EVs is becoming widespread throughout the world with increasing acceleration. Figure 3.5 illustrates that the electric fuel type has the highest share among all fuel types in Norway based on electric vehicle registration data ("New Car Sales in Norway | Kaggle" n.d.; "Norway EV Sales Data" n.d.). While the rate of fully electric vehicles (not hybrid) increased since 2017, the rate of hybrid vehicles decreased compared to the previous years. This indicates Norway's tendency toward fully electric vehicles.

As the first part of the problem definition, real-world datasets are used to predict the future number of EVs registered in Norway and the country's net power consumption.

3.3.1 ELECTRIC VEHICLE REGISTRATIONS IN NORWAY

The dataset includes monthly EV registration records in Norway between January 2011 and January 2022 ("New Car Sales in Norway | Kaggle" n.d.; "Norway EV Sales Data" n.d.).

As a result of the augmented Dickey–Fuller test, it is determined that the p-value is 0.998, that is, the dataset is non-stationary. The graph of the dataset, trend, seasonal, and residual values are visualized in Figure 3.6 for the statistical understanding of the dataset.

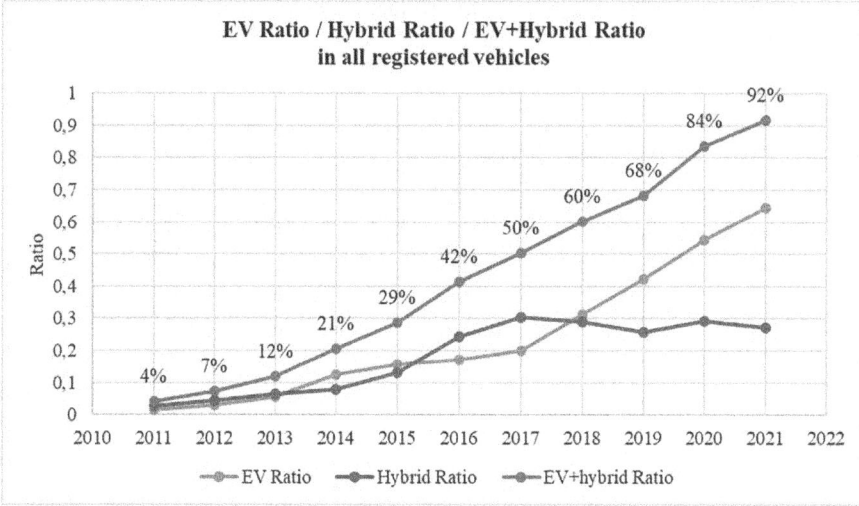

FIGURE 3.5 The proportion of electric vehicles in all fuel types in Norway.

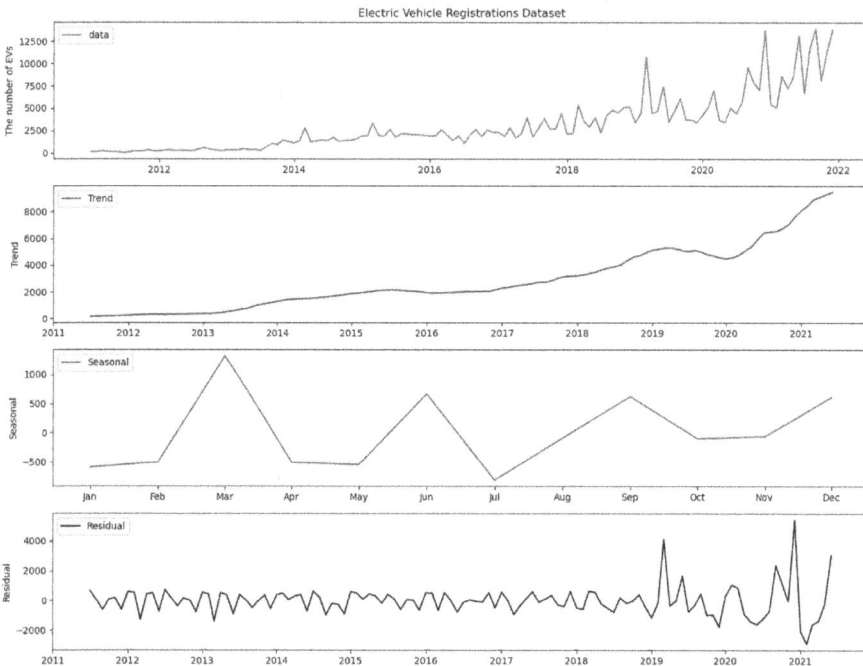

FIGURE 3.6 Statistical properties of electric vehicle registrations dataset in Norway.

The plain version of the EV registration dataset is visualized in red. The blue one indicates the trend value. As can be seen in the blue trend line, the number of EV registrations has gained upward acceleration after 2020 compared to the previous years. As the dataset is monthly based, the monthly recording trend is shown in the green graph. The last graph, the residual value, shows that the numbers of electric vehicle registrations show attitude random behavior after 2020 compared to the previous years.

3.3.2 POWER CONSUMPTION IN NORWAY

The dataset, merged by years, includes monthly net power consumption in Norway between January 2010 and January 2022 ("Market Data | Nord Pool" n.d.)

As a result of the augmented Dickey–Fuller test, it is understood that the p-value is 0.788, that is, the dataset is non-stationary. The reason why the p-value is lower than the EV registrations dataset is that the power consumption dataset draws a more regular graph, as seen in Figure 3.7.

The plain version of the power consumption dataset is visualized in red and appears to be more regular compared to the EV registration dataset. The blue one indicates that it is an uptrend. Looking at the green graph, it is seen that there is a seasonal decrease in summer months compared to winter months since household power consumption, such as heating, is less. The last graph shows that the power-consumption dataset has a certain pattern, albeit fluctuating, compared to the EV registration dataset. These comments also show why the p-value of the power consumption dataset is lower than the EV registration dataset.

In the second part of the problem definition, further real-world datasets are analyzed to predict average road traffic volumes. Furthermore, a methodology is

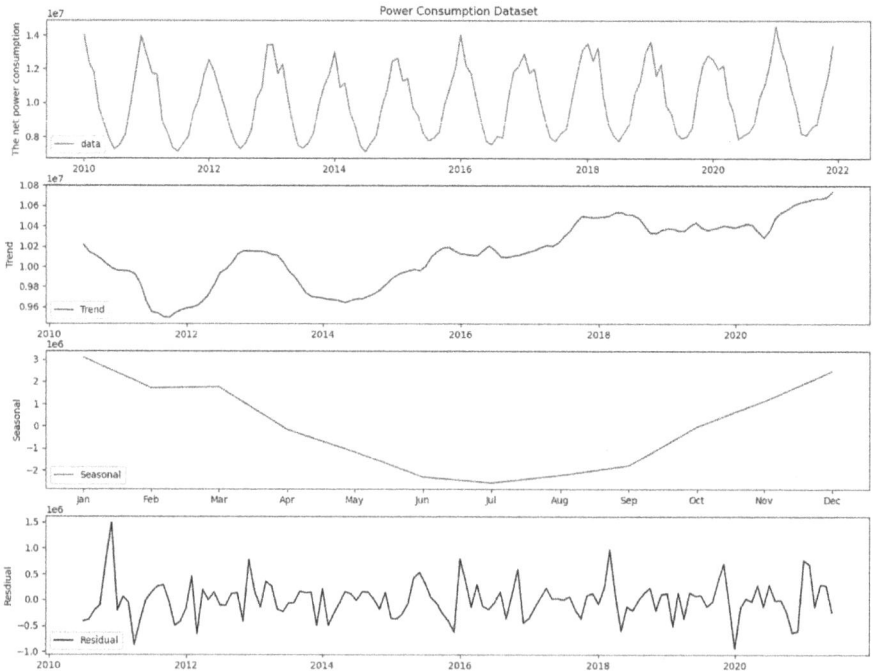

FIGURE 3.7 Statistical properties of the net power consumption dataset in Norway.

developed to handle and as well as to judge the ML algorithms with the help of some criteria.

3.3.3 AVERAGE ROAD TRAFFIC VOLUMES PER ELECTRIC VEHICLE

The dataset provides the average annual road traffic volumes of EVs between 2005 and 2021 ("12576: Road Traffic Volumes, by Home County of Vehicle Owner, Main Type of Vehicle and Type of Fuel (C) 2005–2020. Statbank Norway" n.d.).

Figure 3.8 shows the average road traffic volumes in kilometers per year an EV travels in Norway between 2005 and 2020. While a fluctuating draw is observed up to 8,000 km until 2013, it peaks after 2013 and blows past 10,000 km. The dataset indicates that EVs have become widespread in the country year by year, especially after 2013.

3.3.4 METHODOLOGY

In ML applications, especially for predicting the future values using time-series datasets, the large number of samples is a way to increase the acceptability of the model (Cui et al. 2019). The datasets of "EV registrations in Norway" and "Power consumption of Norway", which are essential for the study, are monthly based, thus they have an advantage in this regard. Figure 3.9 illustrates the number of EV registrations and the amount of net power consumption in Norway annually.

Forecasting using deep learning methods for time-series datasets relies roughly based on training the model by detecting the relationship between previous data points to predict the next data value (Chollet 2018). However, since the prediction of the near future would be made with the data points in the real dataset, it gives more realistic results. For example, assuming the model considers six historic values, the value of the

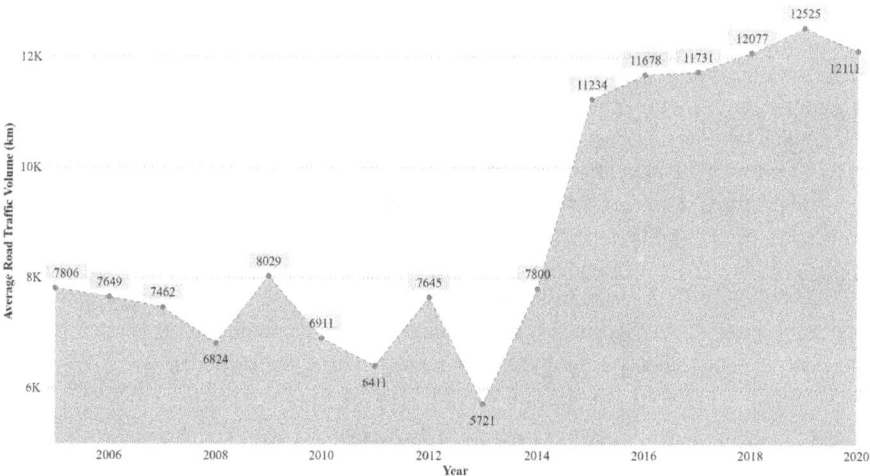

FIGURE 3.8 Average road traffic volumes per electric vehicle in Norway [2005–2020].

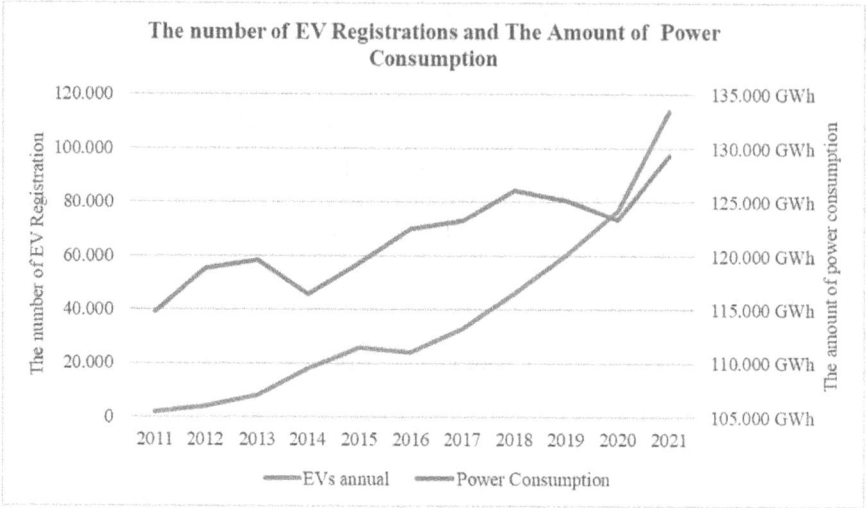

FIGURE 3.9 The number of electric vehicle registrations and the amount of power consumption in Norway [2011–2022].

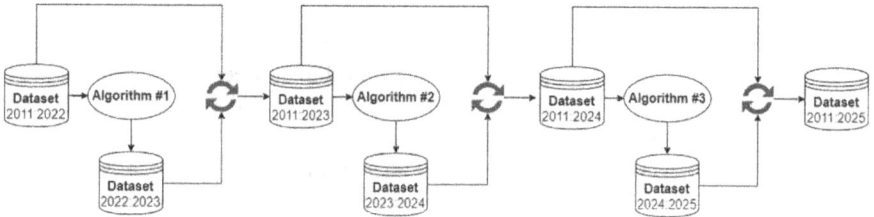

FIGURE 3.10 Methodology of the study.

January 2022 prediction would be based on the relationship between the data points of June 2021 and December 2021. Similarly, the values of the dates from June 2022 to December 2022 predicted by the model would be used to forecast the value of January 2023, and it would produce results that can reduce the reliability of the model. Therefore, the prediction of the next three years is made as in Figure 3.10, by combining three different algorithms instead of a single algorithm and updating the dataset according to the algorithm results. Again, to increase the reliability of the model, various algorithms are applied to the dataset in various sequences, averaging the obtained data on an annual basis to reach the final result. With this methodology, that is, by combining algorithms (regenerating updated ML weights each time), it is possible to obtain longer-term future predictions with higher accuracy than predictions with a single model.

3.3.5 Judgment Criteria and Development Environment

During the training process of the deep learning applications, updating the weights is carried out according to the optimizer and loss function that is set during the model compilation. Since the type of the study is also regression, mean squared error is preferred as a loss function. This adjustment is intended to minimize the difference between the estimated value and the actual value at each epoch.

The R^2 score, which is used to determine the model accuracy for regression algorithms, which indicates the variance relationship between the variables and is calculated as the square of the correlation coefficient (r) and root mean squared error (RMSE), which is related to the loss function, is chosen as the judging criteria.

The study was done using NVIDIA GeForce RTX 2060 GPU with CUDA implementation and the Intel® Core™ i7–9750H CPU @2.60GHz using Pandas, NumPy, Matplotlib, Seaborn, statsmodel, pmdarima, Tensorflow, Keras, Sklearn, NeuralProphet, and fbprophet libraries in Python.

3.4 FUNDAMENTALS OF FORECASTING IN TIME SERIES

The process of predicting individual outputs based on inputs is called regression in ML. In this chapter, the number of EV registrations, the amount of power consumption, and average road traffic volumes per EV value by 2025 are predicted individually year by year. But the reason why time series differ from normal regression is that the time axis does not make sense for the model. The x-axis is presented in the time domain that consists of sequential values, and the relationships between the y-axis values corresponding to these values are used to predict future values. Predictions are obtained with various complex methods, such as RNN-LSTM, ARIMA, etc. The main facts for recurrent neural network (RNN), long short-term memory (LSTM), Facebook Prophet, and autoregressive integrated moving average (ARIMA) are described in this section.

3.4.1 Recurrent Neural Network

Each data point at a time—t—takes place in a function based on the previous data points in the time-series dataset. For example, when the natural language process (NLP) is considered, the continuation of a sentence is shaped according to the words and/or integrity of the meaning of a whole text. Humans read an average of 3–4 words per second and the book can be read fluently. Despite thousands of words that have been read at the end of two hours of reading, it is not possible to repeat them by heart, unless the visual memory of a reader is unique. However, after two hours, the reading part can be understood and summarized easily by the reader. RNNs take charge to extract the essential pattern from such time-series (sequential) datasets. Let's continue with various instances; movies are the product of sequential frames, which means these are composed of frames and there are meaningful relationships between these frames according to the content of the movie. It is quite possible to predict the next scene or the following frames in the movie after having a grasp of the movie. RNNs could be used to forecast future values by recognizing the flowing pattern in the time-series (sequential) dataset. Judging by the examples currently in use, graphs that flow in time, such as stock market prediction and estimation of daily coronavirus cases can be predicted using this method. So, how can we express mathematically that forecasting process?

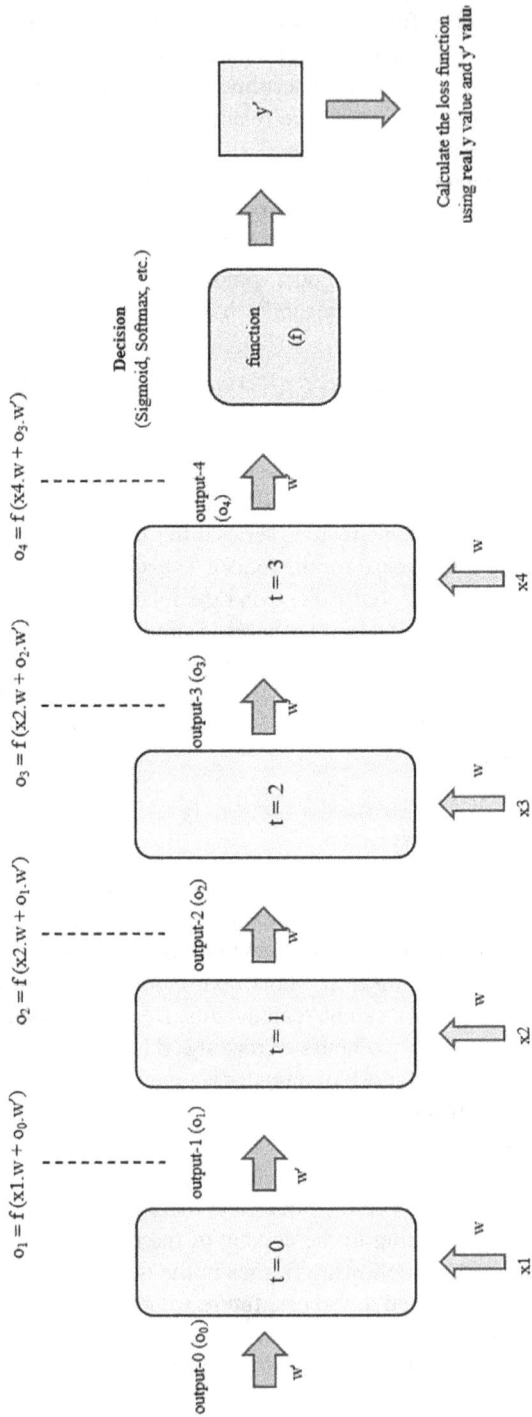

FIGURE 3.11 Schematic of recurrent neural network.

3.4.2 UNDER THE HOOD

Figure 3.11 illustrates an example of how the RNN works in a time-series (sequential) dataset with four data points in four seconds. Its fundamental working principle is similar to feedforward neural network. The system can be briefly summarized as follows:

The weights are assigned to the input values, and both are considered as input into the layers. Afterward, output values are subtracted from real outputs to determine the loss value, that is, the deviation value. The weight values are updated based on the loss value and this process is repeated until minimizing the loss value to detect optimum weight values.

RNN, on the other hand, is built sequentially, that is, backward, to discover the relationship between the data at time t and the data at a time (t−1). In fact, because of its backward-connected nature, its value at time t may be derived using the previous 10 values in addition to t and (t−1).

Let's suppose we have four data points (x1, x2, x3, and x4), as in Figure 3.11. Layers are presented with yellow structures. The equations for each output are shown above the yellow structures individually. It is seen that the output of each layer (yellow structure) is the input of the next layer sequentially. The final output of the whole system is created according to the activation function. After that, the loss value (real value—predicted final output) is calculated, and the weights are updated based on the loss value with the backpropagation process. The main structure of RNN works with this pattern; however, there is a significant disadvantage of this structure that prevents predicting correctly: the vanishing gradient problem (Konasani and Kadre 2015).

3.4.3 VANISHING GRADIENT DESCENT

With help of the gradient descent method, the weights are getting continuously updated. If the weights are adjusted to the global minimum, the model converges. Convergence stops for pretty small gradients, or in other words the weights will not get significantly updated—this is called vanishing gradient descent. RNN faces this problem when the weights are updated using the sigmoid function and which includes too many layers. The output of the sigmoid function gets values between 0 and 1. For this reason, the impact of the first inputs (first words for natural language process, first samples for any kind of time series) on the learning process is decreased since updated weights would become low level when the process reaches the first layer. In other words, the vanishing gradient descent causes problems considering the short-term previous values by almost ignoring the long-term previous values. The model that would be built without solving this problem would prevent realistic results in many studies.

If rectified linear activation unit (ReLu) is used instead of sigmoid function as an activation function, a situation opposite to vanishing gradient descent by updating the weights with high rates when they reach the first inputs because of the structure of ReLu, so that the main goal—reach the global minimum while updating loss function—cannot be achieved (Howard and Gugger 2020).

3.4.4 Long Short-Term Memory (LSTM)

Long short-term memory (LSTM) is designed to prevent the ignoring of the impact of the long-term previous inputs in the time series during backward propagation (Géron 2017).

Before starting the explanation, it must be known that:

- The output of the system takes values between 0 and 1 if the sigmoid (logistic) function is used as an activation function.
- The output of the system takes values between −1 and 1 if the tanh function is used as an activation function.

(Moroney 2019)

The current input and the output of the previous input are considered as input and the output is calculated by using the sigmoid function. It is mathematically inferred which of the past values is important and to what extent from the long-term states. The junction that calculates the final mathematical operation is called the *forget gate*.

The output of the previous data and the current input are considered as input again, and two different outputs are calculated using the sigmoid and tanh activation functions separately. These outputs are multiplied; therefore the importance (weight) of these outputs relative to each other is discovered. The place where this mathematical multiplication takes place is called the *input gate*.

In the last step, obtained pieces of information from the forget gate and input gate are compiled. The first inference comes from adding the output of the forget gate to the output of the input gate. The result obtained from the previous addition is regularized by using the tangent hyperbolic activation function and the value is multiplied by the new value, which comes from the result of two inputs: The output of previous data and the current input using the sigmoid function. Consequently, the short-term result is returned and at the *output gate*, the impact of the long-term previous values is sent to the system.

3.4.5 Seasonal Autoregressive Integrated Moving Average with Exogenous Factors (SARIMAX)

Among the autoregressive model types, which is a statistics-based method, SARIMAX, which examines seasonality, is chosen because datasets are monthly based. Looking at the timeline of the method, ARIMA, which is a modern version of autoregressive models is met in the base. ARIMA models combining AR and MA are defined as ARIMA (p, d, q), where p and q are the orders of AR and MA models, respectively, and d represents the degree of series integration (Banaś and Utnik-Banaś 2021).

- Autoregression (AR) (p): Regression equation for time series using previous values.
- Integration (I) (d): make the time series stationary.
- Moving average (MA) (q): A moving average represents a process in which an observation is a function of the previous time step.

(Jensen 1990)

The version for the seasonal effect of the ARIMA model is the seasonal autoregressive integrated moving average (SARIMA). The SARIMA model is set with the hyperparameters (p, d, q) x (P, D, Q). (p, d, q) refers to the non-seasonal component, while (P, D, Q) represents the seasonal component of the model (Vagropoulos et al. 2016). SARIMAX, on the other hand, is a different variant of the SARIMA model that increases forecasting performance by including explanatory variables in the evaluation.

3.4.6 FACEBOOK PROPHET

Facebook Prophet is a Bayesian-based forecasting method developed by Facebook in 2017 by analyzing the statistical features of the dataset (Kumar Jha and Pande 2021). It produces successful results, especially for datasets that change seasonally. FB Prophet creates its forecasts by taking into account the holidays; accordingly, there are studies in which very effective results are obtained for a time series. The algorithm is preferred because it allows prediction without requiring too much data for the appropriate datasets. Predicting the future values of the dataset with FB Prophet is done in the following way:

A combination of a trend function that indicates non-periodic changes in the model, a function that indicates periodic changes (daily, weekly, yearly, seasonally, etc.), a function that outputs the effect of holidays, and a function that indicates the error value used (Taylor and Letham 2018). In its output, FB Prophet provides the future forecast value as well as different values such as minimum and maximum trend values for each data point. The forecasting of the number of newly registered EVs in Norway between 2022 and 2025 is made with FB Prophet, and the results are shown in Figure 3.12.

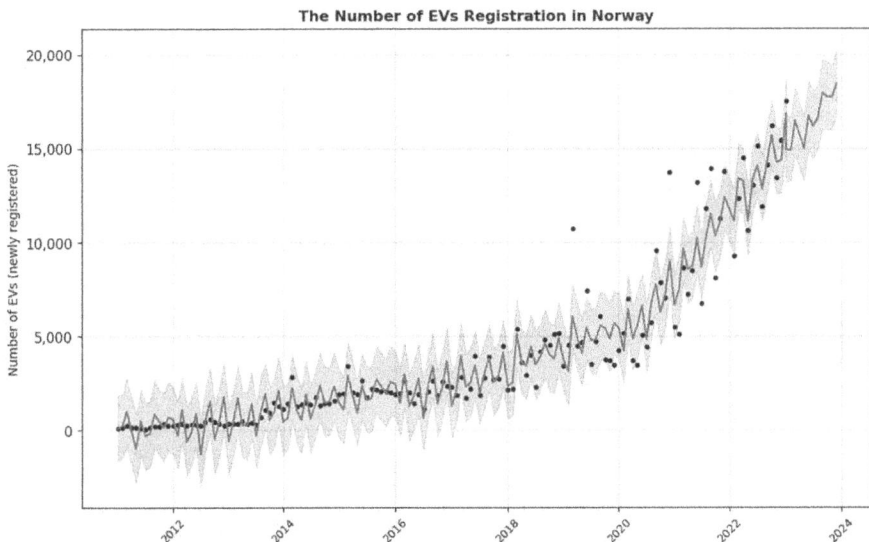

FIGURE 3.12 Forecasting result of Facebook Prophet.

The black dots indicate the real data points and show how much deviation there is when regularizing the dataset. The part with a blue shadow indicates the range of minimum and maximum values of predicting samples. The blue line shows the predicted number of individual electric vehicles.

3.5 FORECASTING OF ELECTRIC VEHICLE REGISTRATIONS IN NORWAY

Four different datasets until 2025 are obtained using four different model combinations, as seen in Figure 3.13, for the next three years by using the EV registration dataset until 2022 in Norway. Mainly, LSTM and its derivatives, which can cope with the vanishing gradient problem of RNN from forecasting algorithms, are preferred. In addition, the ARIMA model and FB Prophet are added to the study to increase the reliability of the results. According to the first forecasting results, the dataset between 2022 and 2023 is predicted using Stacked LSTM. Then, the dataset is updated and the number of EV registrations between 2023 and 2024 is predicted using FB Prophet. The dataset is updated again, and finally, the model is trained with NeuralProphet to predict the number of EV registrations between 2024 and 2025.

Afterward, the models are combined by averaging the results obtained annually. In other words, the number of EV registrations between 2022 and 2023 is the average of the number of EV registrations obtained as a result of four different algorithms: Stacked LSTM, NeuralProphet, GRU, and SARIMAX. The same procedures are applied for the following years.

3.5.1 COMPARISON OF FORECASTING RESULTS

Forecasting values of the number of EVs to be registered between 2022 and 2025 of the models are shown in Figure 3.14 by year. Considering the year 2022, SARIMAX predicts 194,678 EV registrations, while the LSTM derivative GRU predicts approximately 16% fewer, with 162,784 EV registrations. Similarly, looking at the year 2024, 217,075 EV registrations are predicted with the GRU, while 241,535 are predicted with the NeuralProphet.

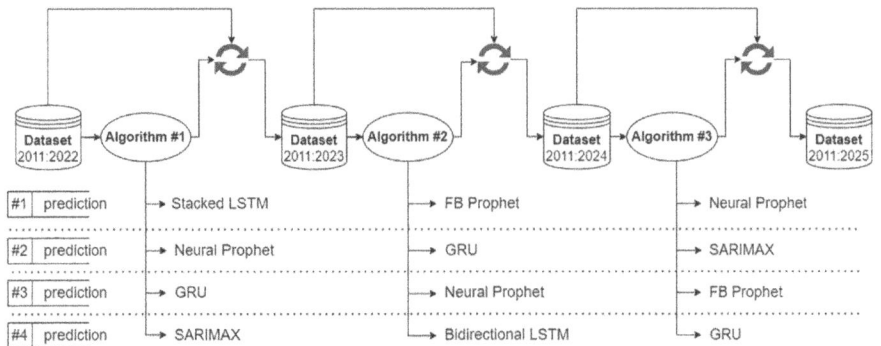

FIGURE 3.13 Model combinations of electric vehicle registration predictions for Norway.

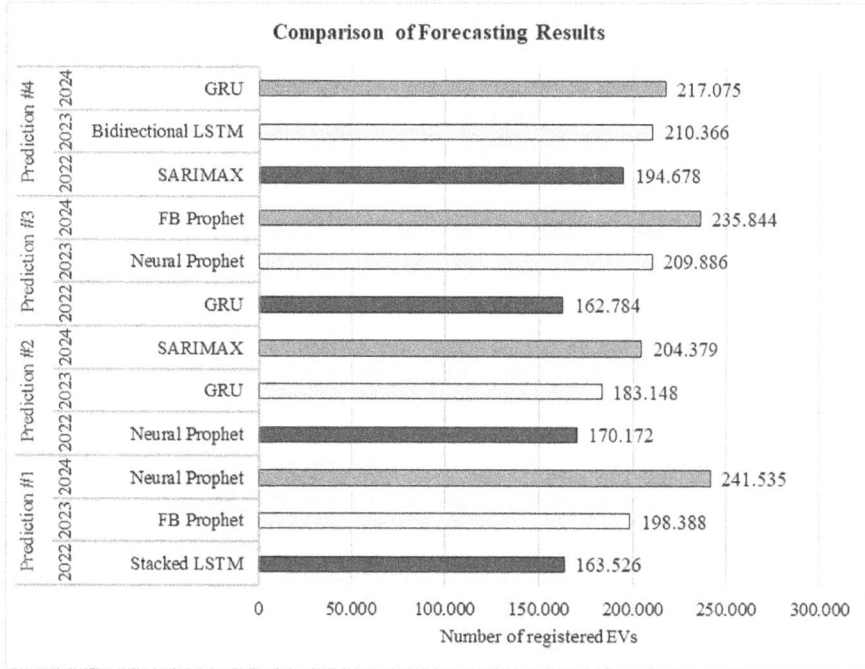

FIGURE 3.14 Comparison of the results of algorithms for electric vehicle registrations in Norway.

Since the prediction performances of the models are directly related to the previous prediction value, the orders of the SARIMAX, LSTM, and NeuralProphet algorithms are taken into consideration while creating the prediction models. Therefore, to obtain balanced and more reliable results, the combination process is carried out by averaging the value of the models over the years.

3.5.2 Prediction Results Trend of Registered EVs

The final result is obtained by averaging the different prediction results given by each model in line with its estimation ability, according to years. Using a single method would limit the predictive flexibility of the model. The average value would give a reasonable trend for the development of numbers. Table 3.2 shows the standard deviation and average values of the models.

The annual average number of forecasted EV registrations of the models and previous years are shown in Figure 3.15. Accordingly, the number of EVs is estimated to be 172,790 in 2022, 200,447 in 2023, and 224,708 in 2024. As expected, all algorithms are showing a continuous increase in demand in the coming years.

3.6 FORECASTING OF POWER CONSUMPTION IN NORWAY

The procedures for combining the models described in the previous sections are also applied to the power consumption dataset. The power consumption dataset has

TABLE 3.2
Comparison of EV Registration Prediction Results

Predictions/Years	2022	2023	2024
Prediction #1	163,526	198,388	241,535
Prediction #2	170,172	183,148	204,379
Prediction #3	162,784	209,886	235,884
Prediction #4	194,678	210,366	217,075
Standard deviation	12,960	11,079	14,829
Average	172,790	200,447	224,708

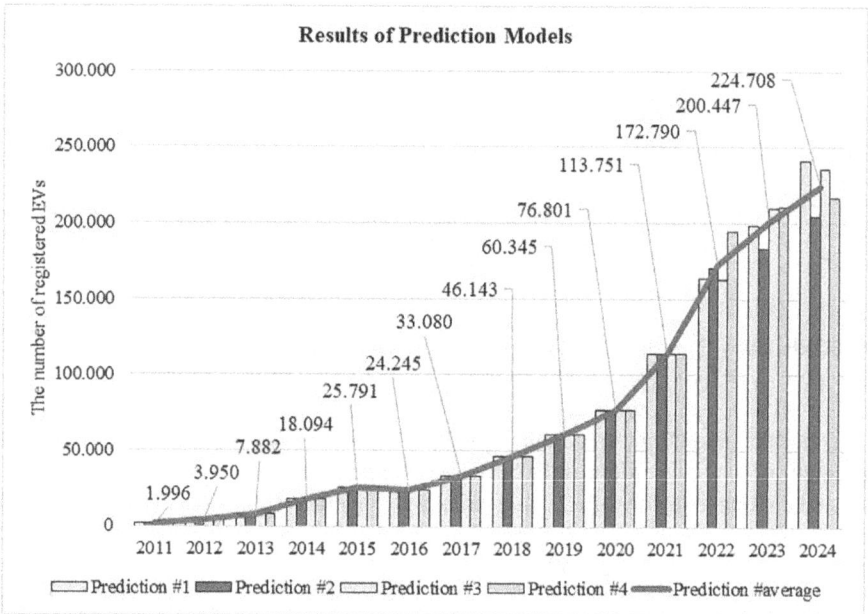

FIGURE 3.15 The predicted number of electric vehicle registrations annually for Norway.

a more understandable seasonality and rising trend because the country's annual electricity consumption is dependent on many things, and these variables move more slowly compared to the number of EVs registered in the dataset. The final result is achieved by combining three models as well as the previous one. The combinations are obtained by using different forecasting methods in different sequences and averaging the values annually, just as was done earlier. The algorithms and their orders are shown in Figure 3.16.

3.6.1 COMPARISON OF FORECASTING RESULTS

The forecasting results of the power consumption of the algorithms are shown in Figure 3.17. The fact that the country's net power consumption has a more regular graph is also reflected in the forecast values of the models. For 2022, SARIMAX predicts 132,776 GWh, while Stacked LSTM predicts 128,498 GWh. The difference is approximately 3%. Similarly, in 2024, the GRU predicts 127,911 GWh, while FB Prophet predicts 132,439 GWh at a rate of 3.5%. However, a more reliable generalized model is created by averaging the algorithms annually, just as was done earlier.

3.6.2 Forecasting Results of Net Power Consumption of Norway

By using the averaging method, as in the prediction of the future values of EVs, the average values and standard deviation values of different models are shown in Table 3.3 for the power consumption dataset. The prediction results of the models are much closer

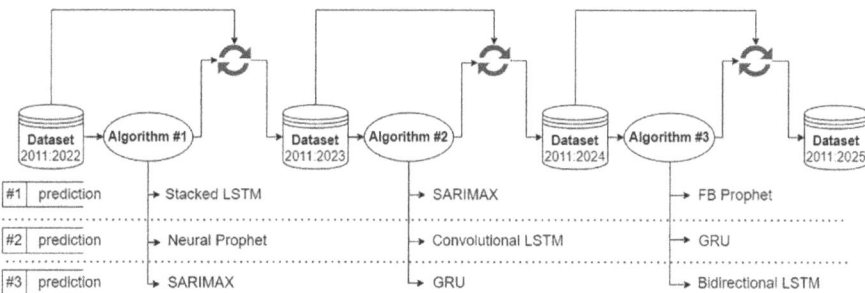

FIGURE 3.16   Model combinations of power consumption predictions for Norway.

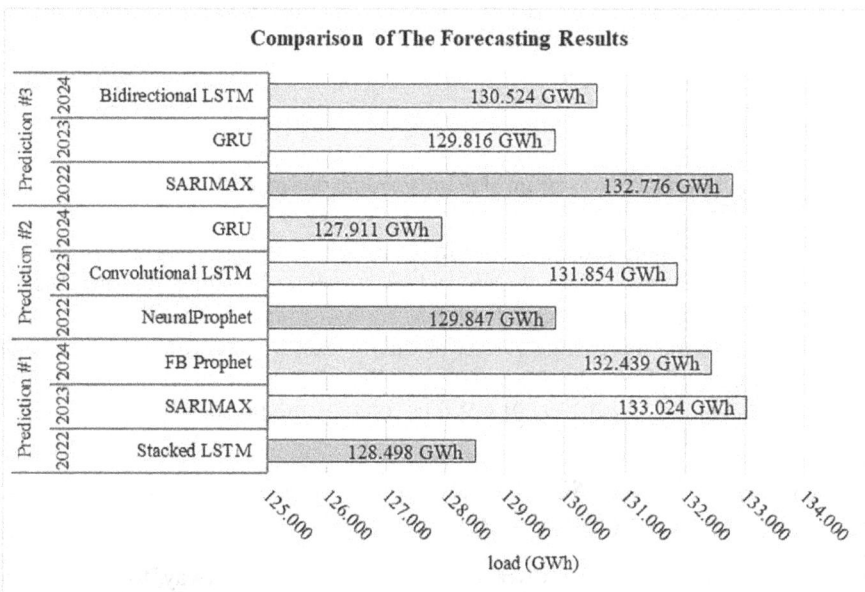

FIGURE 3.17   Comparison of the results of algorithms for net power consumption in Norway.

TABLE 3.3
Comparison of Power Consumption Prediction Results

Predictions/Years	2022	2023	2024
Prediction #1	128,498 GWh	133,024 GWh	132,439 GWh
Prediction #2	129,847 GWh	131,854 GWh	127,911 GWh
Prediction #3	132,776 GWh	129,816 GWh	130,524 GWh
Standard deviation	1,786	1,326	1,856
Average	130,374 GWh	131,565 GWh	130,291 GWh

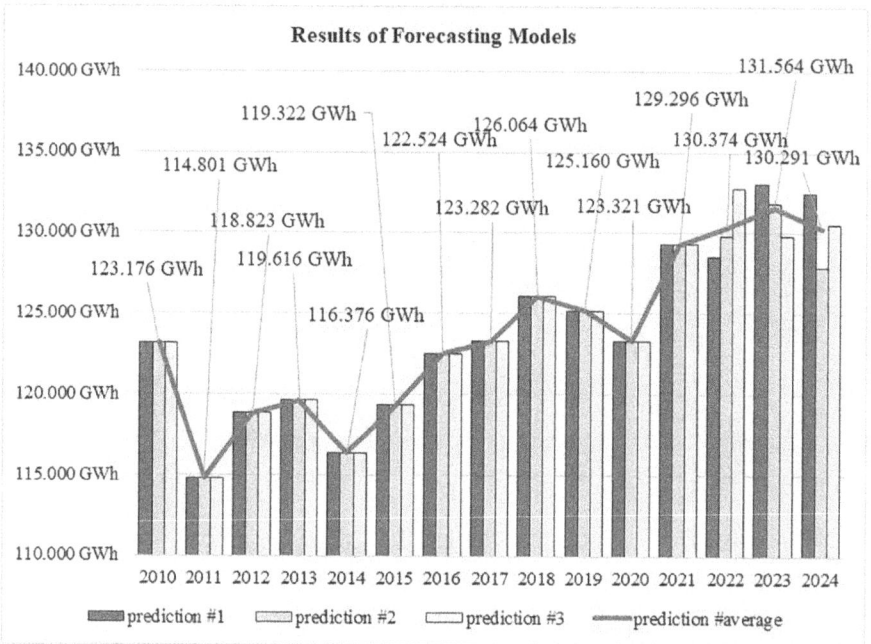

FIGURE 3.18 The predicted amount of power consumption annually for Norway.

to each other compared to the EV registration dataset. The reason for this is that the power consumption dataset is under seasonal effect, as seen in Figure 3.7, which means it shows a more regular graph compared to the EV registrations dataset.

According to the forecasting results of the models, the net power consumption values between 2022 and 2025 are shown in Figure 3.18. In 2022, 130,374 GWh; 131,564 GWh in 2023; and 130,291 GWh in 2024 are predicted.

3.7 THE POWER CONSUMPTION OF ELECTRIC VEHICLES IN NORWAY

Using the annual average traffic volumes dataset of EVs in Norway, forecasting for the next three years is made using NeuralProphet and FB Prophet. The deep learning method, LSTM, is not used for this dataset because it is presented on an annual basis.

Between 2005 and 2021, applying LSTM and its derivatives with a dataset containing only 16 samples would not yield realistic results due to the lack of samples. For this reason, predictions are made using NeuralProphet, a deep learning–based algorithm that also provides support for Bayesian-based FB Prophet and autoregression. In addition, the AR model could also be used here. Figure 3.19 shows the graph of the dataset and the prediction results.

Average road traffic volumes per EV are determined as 12,884 km for 2021, 13,295 km for 2022, 13,695 km for 2023, and 12,849 km for 2024, according to prediction results. The purpose of using this dataset is to precisely determine the annual total power demand of EVs in Norway.

3.7.1 ASSUMPTIONS

It is seen that the power consumption of EVs is roughly between 10 kWh/100 km and 30 kWh/100 km ("Energieverbrauch von Elektroautos Cheatsheet—EV Database" n.d.). Looking at concrete examples from the market, VW ID3 Pro S consumes around 13.7 kWh/100km ("Der ID.3 | Elektrofahrzeug | Modelle | Volkswagen Deutschland" n.d.), and Tesla Model 3 consumes around 17kWh/100km ("CO2-Emissionswerte Und Stromverbrauch" n.d.). However, some tests by ADAC (General German Automobile Club) have shown greater power consumption than expected ("Bordcomputer: Wie Genau Ist Die Verbrauchsanzeige? | ADAC" n.d.). Therefore, calculations are made according to consumption values of 15kWh/100km and 22kWh/100km to make this study inclusive and general.

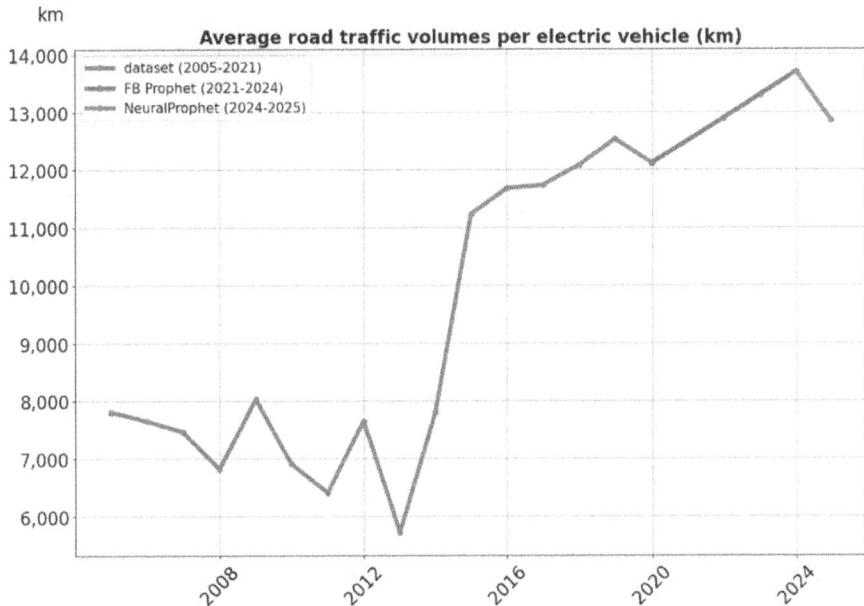

FIGURE 3.19   The average road traffic volumes per electric vehicle in Norway.

Assuming that newly registered EVs in Norway have a power demand of 15kWh/100km, the annual power consumed by these newly registered vehicles is calculated with the following formula:

$$Power\ Demand\ of\ EV = Number\ of\ EVs * \frac{15kWh}{100km}$$

$$* The\ Average\ Road\ Traffic\ Volumes\ (km) \qquad (3.1)$$

In addition, the annual EV registration dataset has been studied in previous sections, and the formula provided earlier only calculates the annual power demand of newly registered EVs. Assuming that the vehicles remain in traffic for about eight years, the total power consumption of newly registered EVs in the country in the last eight years would provide more realistic results. In Figure 3.20, both the annual demand for newly registered vehicles and the eight-year cumulative total of EVs are shown.

By blending the prediction outcomes of previous sections, the net power consumption in Norway, the newly registered EVs' power consumption, and the power demand of EVs assumed to have been in traffic for the last eight years are shown in Table 3.4. When the year 2021 is set as a reference point, the power demand for EVs would increase by 43% in 2022, 94% in 2023, and 132% in 2024. In other words, it

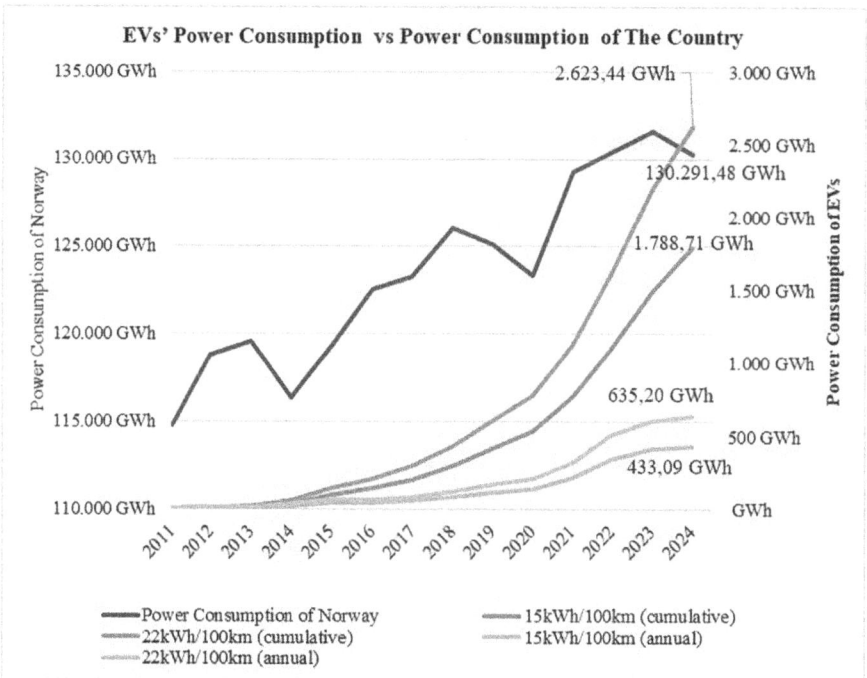

FIGURE 3.20 Comparison of the power consumption of electric vehicles in Norway and Norway's net power consumption by years.

is predicted that the demand for EVs is going to double between 2021 and the first quarter of 2024.

Figure 3.21 shows the rate of increase in power demands of EVs with new EV registrations compared to the previous year. It can be interpreted that the rate of increase in power demand has declined since 2021, which is our model's prediction, to reach saturation levels. The Norwegian Parliament had set a national goal that by 2025 all newly sold vehicles should be zero-emission (electric or hydrogen) ("Norwegian EV Policy—Norsk Elbilforening" n.d.). The predicted results of our model show that the downtrend, which began in 2021, is sharply 5.18% in 2024, increasing the reliability of our model in line with Norway's national goal.

TABLE 3.4
Forecasting Results of EVs' Power Demand and Norway's Power Demand

Years	Power demand of Norway	Power demand of EVs (cumulative value of last 8 years)		Power demand of EVs (annual)	
	Power consumption	15kWh/100km	22kWh/100km	15kWh/100km	22kWh/100km
2021	129,295 GWh	769 GWh	1,128 GWh	220 GWh	322 GWh
2022	130,373 GWh	1,102 GWh	1,617 GWh	345 GWh	505 GWh
2023	131,564 GWh	1,494 GWh	2,192 GWh	412 GWh	604 GWh
2024	130,291 GWh	1,788 GWh	2,623 GWh	433 GWh	635 GWh

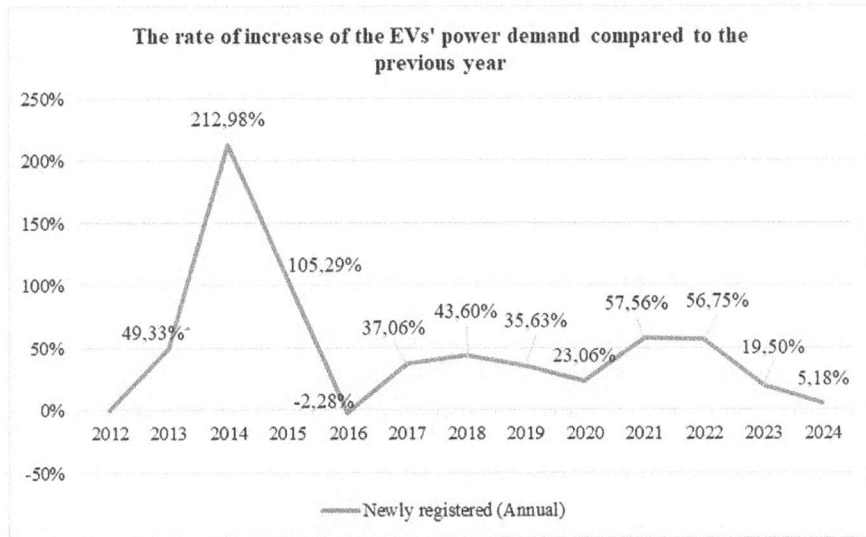

FIGURE 3.21 The rate of change in the power demand of electric vehicles compared to the previous year.

3.8 CONCLUSION

The number of EVs in Norway has been predicted by combining various algorithms until the year 2025. Norway's net power consumption has also been estimated until the year 2025 and similar results have been obtained with the increase in EVs. The annual power demand of the vehicles is calculated using two different approaches, 15kWh/100km, and 22kWh/100km, using the dataset that is the average road traffic load per EV. When the forecasting datasets are compiled, the share of the electricity consumption of the EVs in the country is calculated as follows and visualized in Figure 3.22:

$$ratio = \frac{EVs'power\ consumption}{Net\ power\ consumption\ of\ Norway} \qquad (3.2)$$

Both the number of EVs and net power consumption of Norway resulted in an upward trend for the coming years. The ratio of power demand for EVs to net power demand has been calculated for each year, for the years predicted in Table 3.5. It also shows that the share of EVs' power demand in the net power consumption of Norway would increase as a result of much higher power demand due to the increased rate of EVs despite increasing net power consumption.

Considering the demand for EVs, the problems that they will bring with them also need to be solved. Let's assume the extreme situation, where all charging stations are locally active at the same time, let alone increasing their share of the country's net power consumption. In this case, the electrical distribution system can be overloaded and cause problems. Of course, the assumption made for the charging station seems very likely in the near future. Utilizing ML-based predictions, as shown in this study,

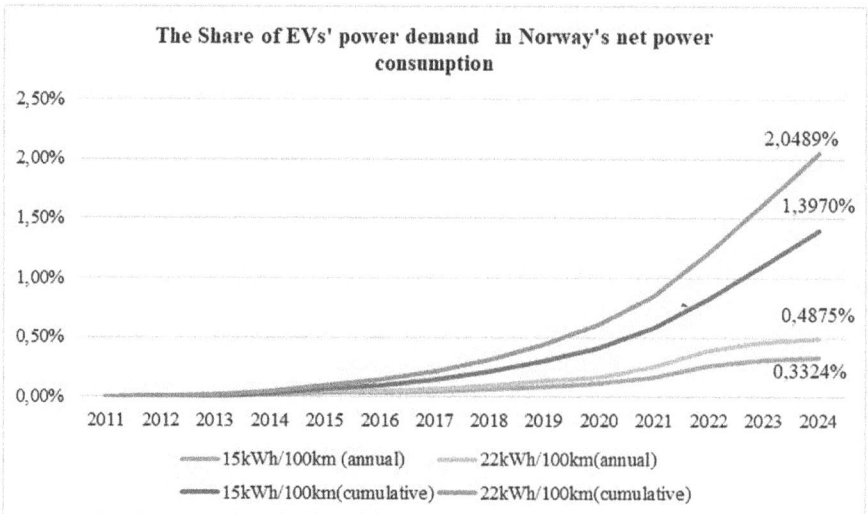

FIGURE 3.22 The share of power demand for electric vehicles in Norway's net power consumption

TABLE 3.5
The Impact of the EVs' Power Demand on Norway's Power Demand (Annually and Cumulative for the Last Eight Years)

Years	The share of EVs' power demand on power consumption (annual)		The share of EVs' power demand on power consumption (cumulative value of last 8 years)	
	15kWh/100km (annual)	22kWh/100km (annual)	15kWh/100km (cumulative)	22kWh/100km (cumulative)
2021	0.1700%	0.2494%	0.5765%	0.8456%
2022	0.2643%	0.3876%	0.8227%	1.2066%
2023	0.3130%	0.4590%	1.0992%	1.6122%
2024	0.3324%	0.4875%	1.3970%	2.0489%

forecasts can be made and used to control and manage the demand—therefore, prediction can be applied in a wide time range, from seconds to years.

Another challenge is that the ratio of the power consumption of EVs over the total country would vary according to the source of that country's total energy demand. With the help of ML, decisions and strategies can be developed to handle these challenges. In this study, only the model developed for Norway is presented. However, it shows that the investigated methodology can be used for other regions or datasets accordingly.

REFERENCES

"12576: Road Traffic Volumes, by Home County of Vehicle Owner, Main Type of Vehicle and Type of Fuel (C) 2005–2020: Statbank Norway." n.d. Accessed February 9, 2022. www.ssb.no/en/statbank/table/12576/.

Banaś, Jan, and Katarzyna Utnik-Banaś. 2021. "Evaluating a Seasonal Autoregressive Moving Average Model with an Exogenous Variable for Short-Term Timber Price Forecasting." *Forest Policy and Economics* 131 (October): 102564. https://doi.org/10.1016/J.FORPOL.2021.102564.

"Bordcomputer: Wie Genau Ist Die Verbrauchsanzeige? | ADAC." n.d. Accessed February 10, 2022. www.adac.de/rund-ums-fahrzeug/tests/autotest/bordcomputer-verbrauchswerte/.

Chollet, François. 2018. *Deep Learning with Phyton: In Manning, Book.* http://faculty.neu.edu.cn/yury/AAI/Textbook/Deep Learning with Python.pdf.

Chung, Yu Wei, Behnam Khaki, Tianyi Li, Chicheng Chu, and Rajit Gadh. 2019. "Ensemble Machine Learning-Based Algorithm for Electric Vehicle User Behavior Prediction." *Applied Energy* 254 (November): 113732. https://doi.org/10.1016/J.APENERGY.2019.113732.

"CO2-Emissionswerte Und Stromverbrauch." n.d. Accessed February 10, 2022. www.tesla.com/de_DE/support/european-union-energy-label.

Cui, Yin, Menglin Jia, Tsung Yi Lin, Yang Song, and Serge Belongie. 2019. "Class-Balanced Loss Based on Effective Number of Samples." *Proceedings of the IEEE Computer Society Conference on Computer Vision and Pattern Recognition* 2019 June: 9260–9269. https://doi.org/10.1109/CVPR.2019.00949.

"Der ID.3 | Elektrofahrzeug | Modelle | Volkswagen Deutschland." n.d. Accessed February 10, 2022. www.volkswagen.de/de/modelle/id3.html.

"Energieverbrauch von Elektroautos Cheatsheet—EV Database." n.d. Accessed February 10, 2022. https://ev-database.de/cheatsheet/energy-consumption-electric-car.

Géron, Aurélien. 2017. *Hands-on Machine Learning with Scikit-Learn, Keras, and TensorFlow (2019, O'reilly).* O'Reilly Media.

Howard, Jeremy, and Sylvain Gugger. 2020. "Deep Learning for Coders with Fastai and PyTorch." *O'Reilly Media* 66 (3): 622.

International Energy Agency. 2021. "Global EV Data Explorer." Articles. www.iea.org/articles/global-ev-data-explorer.

Jensen, Louise. 1990. "Guidelines for the Application of Arima Models in Time Series." *Research in Nursing & Health* 13 (6): 429–435. https://doi.org/10.1002/NUR.4770130611.

Kiaee, Farkhondeh. 2020. "Integration of Electric Vehicles in Smart Grid Using Deep Reinforcement Learning." *2020 11th International Conference on Information and Knowledge Technology, IKT 2020,* December, 40–44. https://doi.org/10.1109/IKT51791.2020.9345625.

Kim, Sunyong, and Hyuk Lim. 2018. "Reinforcement Learning Based Energy Management Algorithm for Smart Energy Buildings." *Energies 2018* 11(8): 2010. https://doi.org/10.3390/EN11082010.

Konasani, Venkat Reddy, and Shailendra Kadre. 2015. *Time-Series Analysis and Forecasting.* Practical Business Analytics Using SAS. https://doi.org/10.1007/978-1-4842-0043-8_12.

Kumar Jha, Bineet, and Shilpa Pande. 2021. "Time Series Forecasting Model for Supermarket Sales Using FB-Prophet." *Proceedings: 5th International Conference on Computing Methodologies and Communication, ICCMC 2021,* April: 547–554. https://doi.org/10.1109/ICCMC51019.2021.9418033.

Lopez, Karol Lina, Christian Gagne, and Marc Andre Gardner. 2019. "Demand-Side Management Using Deep Learning for Smart Charging of Electric Vehicles." *IEEE Transactions on Smart Grid* 10 (3): 2683–2691. https://doi.org/10.1109/TSG.2018.2808247.

"Market Data | Nord Pool." n.d. Accessed February 14, 2022. www.nordpoolgroup.com/Market-data1/#/nordic/table.

Moroney, Laurence. 2019. "AI and Machine Learning for Coders: A Programmer's Guide to Artificial Intelligence." *Journal of Chemical Information and Modeling* 53 (9): 100–120.

"New Car Sales in Norway | Kaggle." n.d. Accessed February 9, 2022. www.kaggle.com/dmi3kno/newcarsalesnorway.

"Norway EV Sales Data." n.d. Accessed February 9, 2022. https://robbieandrew.github.io/EV/.

"Norwegian EV Policy: Norsk Elbilforening." n.d. Accessed February 9, 2022. https://elbil.no/english/norwegian-ev-policy/.

Samuel, A. L. 1959. "Some Studies in Machine Learning Using the Game of Checkers." *IBM Journal of Research and Development* 44 (1–2): 207–219. https://doi.org/10.1147/rd.441.0206.

Taylor, Sean J., and Benjamin Letham. 2018. "Forecasting at Scale." *American Statistician* 72 (1): 37–45. https://doi.org/10.1080/00031305.2017.1380080.

Vagropoulos, Stylianos I., G. I. Chouliaras, E. G. Kardakos, C. K. Simoglou, and A. G. Bakirtzis. 2016. "Comparison of SARIMAX, SARIMA, Modified SARIMA and ANN-Based Models for Short-Term PV Generation Forecasting." *2016 IEEE International Energy Conference, ENERGYCON 2016,* July. https://doi.org/10.1109/ENERGYCON.2016.7514029.

4 Comprehensive Bidirectional EV Charging Scheduling Considering the Economic Aspects

Fatemeh Jozi, Ali Abdali, Kazem Mazlumi, and Seyed Hadi Hosseini

CONTENTS

DOI: 10.1201/9781003293989-4

4.1 INTRODUCTION

The essential reduction of fossil fuels and environmental pollutants have led to the consideration of electric vehicles (EVs) as a viable alternative to combustion vehicles [1, 2]. In order to increase the application of EVs, various types of EVs have been included, including hybrid electric vehicles (HEVs), plug-in hybrid electric vehicles (PHEVs) and battery electric vehicles (BEVs) [3, 4]. The following Figure 4.1 shows different standards followed by EVs [5].

Vehicle-to-grid (V2G) technology, which can only be implemented via EV, is the vehicle's ability to allow power flow to the distribution network. EVs can store energy and supply it to the grid and, if needed, can participate in power injection into the grid and act as a distributed storage set and as emergency generators on the distribution side. Figure 4.2 shows a simple schematic of V2G.

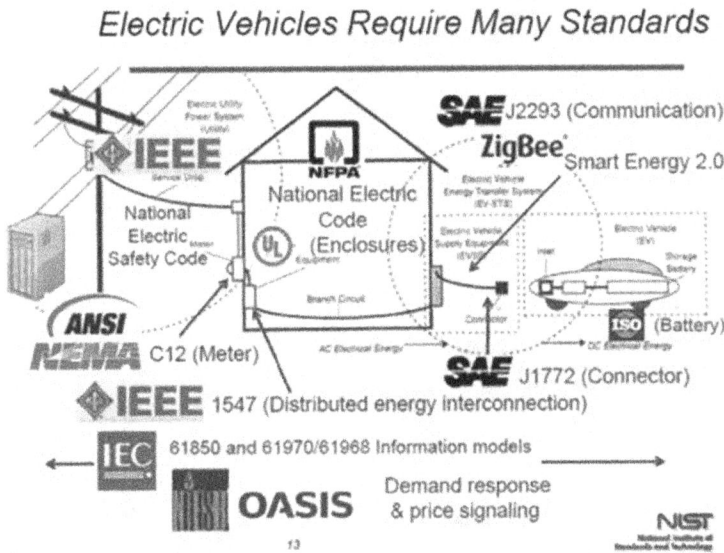

FIGURE 4.1 Different standards followed by EVs.

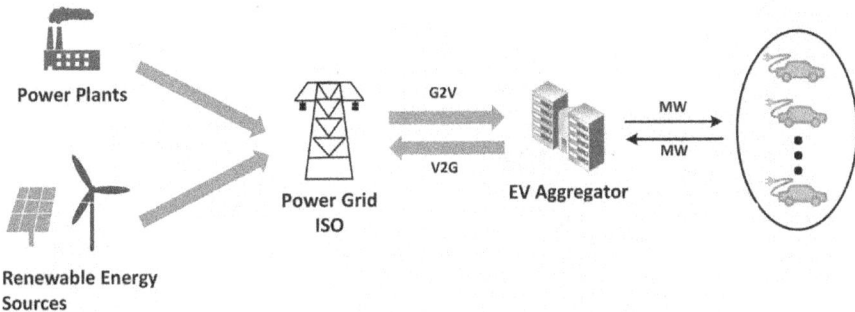

FIGURE 4.2 The schematic of V2G.

V2G-equipped EVs offer a variety of benefits, such as active power regulation, reactive power support, current harmonic filtering, and load peak correction. Also, EVs with V2G technology provide a support source for renewable resources. These features provide support services, including spinning reserve and voltage and frequency control. These factors improve the efficiency and reliability of the system. Furthermore, vehicles can participate in electricity markets and provide many economic benefits to their owners and the grid [5].

Figure 4.3 shows the general implementation of a V2G system, its requirements and the flow of power by a V2G system. The whole V2G system consists of charging locations and its infrastructure; aggregator and independent system operator (ISO); renewable energy resources and an electric utility; intelligent communication and control in terms of smart metering; whole battery pack and EV itself; two-way power flow, i.e. unidirectional and bidirectional; and communication between EV and ISO.

4.1.1 Charging and Discharging Strategy

It should be noted that the benefits of V2G depend on how the charging and discharging of EVs are performed [5–6]. If the charging process is not controlled, charging times will interfere with the load peak of the system, and the distribution network will encounter problems such as overloading, excessive loss of power and violation of voltage limitations [7–9].

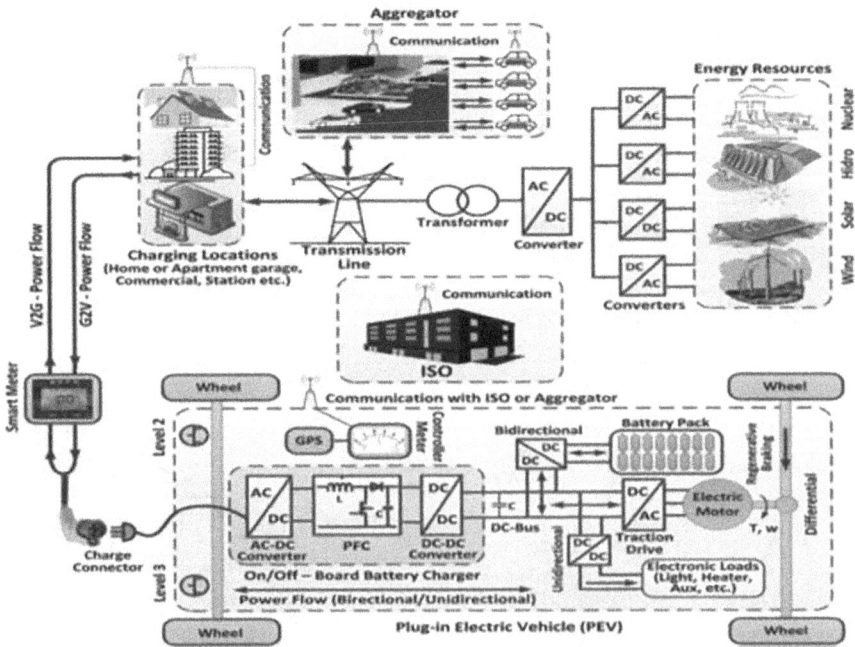

FIGURE 4.3 General implementation of a V2G system, its components and power flow of V2G system [5].

4.1.1.1 Unscheduled Charging and Discharging Strategy

In this strategy, EVs start charging as soon as they are parked, and there is no smart scheduling in this strategy. This allows car owners to charge EVs regardless of time. In this situation, with the increasing number of EVs, the power system loading reaches its highest level and the indices of power quality and network reliability are threatened.

4.1.1.2 Scheduled Charging and Discharging Strategy

With the expansion of the application of EVs, the parking lots of EVs have been considered in the distribution system [10–12]. Figure 4.4 shows an EV parking lot.

In addition to providing the energy required by EVs, these parking lots can store energy and send it to the grid. Accordingly, when charging EVs, the parking lot is considered as load, and when discharging EVs, the parking lot is considered as a distributed generation (DG). As mentioned, V2G technology improves the reliability of the distribution system, but it should always be borne in mind that parking lots can improve system reliability if charging and discharging strategies are properly designed. On the other hand, the parking lot contributes to the transmission of power to the grid where the received profit is appropriate; therefore, in the first stage, the charging and discharging strategy should be designed to convince the parking lot to contribute to improving the reliability of the distribution system with suitable parking lot profit.

With the advent of V2G technology, EVs can participate in electricity markets. Accordingly, many studies have been conducted on the scheduling of the power supply required by EVs as well as how they participate in the electricity markets. The purpose of these studies is to reduce the cost of charging and increase the received revenue for the sale of energy to the grid.

The main factor that affects the distribution systems during the charging process of EVs is the charging profile; therefore, it is necessary to study the approach and methods of controlling the charging process of EVs. For this purpose, various

FIGURE 4.4 EV parking lot.

methods for managing the charging time and frequency of EVs have been proposed [13].

In [8], a second-order mixed-integer linear programming (MILP) method for the charging process scheduling of EVs in unbalanced electrical distribution systems has been proposed. In this method, DG and loads imbalance are considered. This developed method defines an optimal charging scheme by taking into account the time of arrival and departure, the battery SoC at the time of arrival, as well as the energy sharing of EVs equipped with V2G capability. The purpose of this scheduling is to minimize the cost of energy provided by substations and DGs, as well as to reduce EV energy shortages if EVs cannot be fully charged. The results show that by using this method, EV will not suffer from an energy shortage and also the operation constraints, including voltage limitation, active and reactive power generation limitation and maximum current limitation, are met. In addition to these advantages, this method is well adapted to the arrival of new loads. In this method, the EV exit time is determined by its user. If this does not happen, it is assumed that the EV will remain connected to the grid for a certain period.

In [14], a control strategy was presented in order to optimize the charging and discharging process of EVs in the presence of renewable sources. The goal of the control strategy was to schedule the charging and discharging process so that the power imbalance in the network was minimized. This optimization was calculated frequently during the control period to correct the grid-to-vehicle (G2V) and V2G rates in order to deal with the predicted data errors, including load demand and renewable energy generation. In addition, EV user requirements and time schedules were included in the optimization process to ensure customer satisfaction and improve control performance. The results showed that the control strategy has reduced the power imbalance in the network. However, power losses and economic aspects have not been considered in this strategy.

In [15], a method based on a new metaheuristic optimization called water cycle algorithm (WCA) was presented to schedule the charging process of EVs. This method of scheduling the charging process minimizes the total cost of energy distributed to the EVs and the peak-to-average ratio (PAR) along with the reactive power compensation in the distribution network. The results demonstrate that using G2V and V2G technology, for both slow- and fast-charging methods, more EVs can be placed in charging stations without overloading the distribution transformer. In addition, the integration of EVs as a reactive power compensation device improves the grid voltage profile without reducing battery life. In this method of charging process scheduling, the unpredictable nature of the duration of each EV in the parking lot is not addressed, and it is assumed that as soon as the EVs arrive, the owners first send their data, including battery capacity, initial SOC, as well as the duration of their park, to the integrator. Accordingly, when data is reported by EV owners, the EFDAM integrator optimizes the EV charging scheduling process.

In [16], the impact of EV charging process scheduling with V2G technology was investigated in order to maximize profits for EV owners and minimize system costs. In this scheduling method, in addition to the formulation of battery destruction cost, according to the depth of discharge (DOD), in order to analyze in real time at any time interval, the firefly algorithm was used to optimize the system cost. In this paper, three scenarios with economic scheduling approach, including unscheduled charging process, scheduled charging process and scheduled charging process with renewable energy sources, were presented. The results showed that the scheduled charging process improved the voltage profile of the system under study. In addition, the total cost and losses of the system were reduced by using the scheduled charging scenario in the presence of renewable energy sources. However, in these scenarios, the amount of energy exchange between the EVs and the network was minimal, which made the EV owners' profits small and the system costs very high. For this reason, charging and discharging strategies should be designed in such a way that in addition to providing economic benefits to owners, they can also improve the rate of energy exchange to the distribution system.

In [17], a scheduled management structure for EVs in a low-voltage residential network was proposed, to support the power network against overloading and voltage sag. Charging and discharging EVs through an integrator was optimally scheduled to minimize the energy costs of EV owners while respecting local network constraints. The proposed structure transferred decision-making authority to the EV owner, using multi-agent systems (MAS). This scheduling method used second-order mixed-integer programming (MIP) optimization to minimize the energy costs of EV owners. The results showed that by using this strategy, the energy cost of EV owners was also reduced while reducing network overloading and maintaining voltage. However, it should be noted that because the process of charging and discharging EVs is done through the integrator and parking, charging and discharging strategies must be able to meet the economic benefits of parking owners in order to convince parking lots to participate in energy exchange with the distribution network.

4.1.2 INTERACTION OF EV OWNER AND PARKING LOT

In general, three types of interaction can be considered for the EV owner and the parking lot:

4.1.2.1 Interaction 1
The EV owner does not allow the parking lot to charge and discharge the EV and will only pay for the EV parking in the parking lot. Therefore, there is no need to design a charging and discharging strategy in this interaction.

4.1.2.2 Interaction 2
The EV owner allows the parking lot to charge and discharge the EV, and if the EV is not charged to the minimum required capacity when leaving the parking lot, the parking lot will be subject to a penalty to the EV owner. In this interaction, no information is available about how long the EV has been in the parking lot. In such a situation, the parking lot must adopt a suitable strategy for the charging and discharging process scheduling of EVs in order to increase its profit.

4.1.2.3 Interaction 3
The EV owner allows the parking lot to charge and discharge the EV. It also determines the EV exit time. In this interaction, if the EV is not charged to the minimum required capacity at the time of leaving the parking lot, the parking lot is required to

pay a penalty to its owner. On the other hand, the EV owner wants to provide facilities by the parking lot in order to determine the time of EV leaving. In such cases, in addition to charging and discharging process scheduling, it is necessary to determine the value of penalties that the EV owner must pay to the parking lot in case of violation of the provided information.

In this chapter, different strategies are proposed for scheduling the charging and discharging process of EVs in a V2G-equipped parking lot, aiming at maximizing the profit of parking. Restrictions on the amount of power exchange between the parking lot and the distribution network are studied. Also, the random and unpredictable nature of the quantities, such as the time of arrival and departure of the EVs to the parking lot and their state of charge (SoC) while arriving at the parking lot in the proposed strategies, have been considered. Therefore, charging and discharging strategies are designed independent of the duration of EVs' presence in the parking lot and considering the limitations of power exchange with the distribution system.

In Section 4.2, the function of the parking lot profit and charging and discharging strategies are presented in order to increase parking lot profit according to network constraints. In Section 4.3, the parking lot profit is studied. Finally, Section 4.4 provides the conclusions.

4.2 PROPOSED CHARGING AND DISCHARGING PROCESS SCHEDULING OF EVS

4.2.1 PARKING PROFIT FUNCTION

In order to determine the parking profit in each of the strategies, it is necessary to introduce a function to calculate the parking profit. The parking profit with a maximum capacity of N_{ev} EV (N_{ev} represents the number of EVs), which has been operated in the period [0-T], is calculated according to Eq. (4.1). The time interval [0-T] is divided into t time intervals of one hour.

$$W_{total} = C_p - C_n \qquad (4.1)$$

In Eq. (4.1), C_p equals total parking revenue and C_n equals total parking costs.

$$C_p = C_{gdch} + C_{park} + C_{ech} + C_{pen} \qquad (4.2)$$

$$C_n = C_{gch} + C_{edch} + C_{lim} + C_{sh} \qquad (4.3)$$

In the previous equations:

C_{gdch}: Total revenue received by the parking from the grid for the sale of energy to the grid

C_{park}: Total parking revenue received from EV owners for parking their EV in the parking lot

C_{ech}: Total parking revenue received from EV owners for increasing the EV battery charge when leaving the parking lot (received for selling energy to EV)

C_{pen}: Total parking revenue received from EV owners for infringement (if any agreement)

C_{gch}: Total parking cost paid to the network for energy purchase from the network

C_{edch}: Total parking cost paid to EV owners for reduced EV battery charging when leaving the parking lot (cost paid for purchasing energy from EV)

C_{lim}: Total parking cost paid to EV owners for not providing the minimum required EV charging (if any agreement)

C_{sh}: Total parking cost paid to EV owners for EV owner's share of V2G profits (if any agreement)

In the following, the equations of the components of the parking profit function are presented.

Parking using V2G capability can send power from the EV batteries in the parking lot to the network. Equations (4.4) and (4.5) calculate the parking revenue received for energy sales to the grid.

$$C_{gdch} = \sum_{i=1}^{Nev} C_{gdch,i} \tag{4.4}$$

$$C_{gdch,i} = \sum_{t \in R_i} P_{dch}(t) \cdot price_{dch,t} \tag{4.5}$$

In equations (4.4) and (4.5), $C_{gdch,i}$ is the receipt of the parking lot for the sale of energy to the grid from the battery discharging of EV No. I, and R_i is the time step when the EV No. i is present in the parking lot.

$P_{dch}(t)$ is the actual power that the parking lot sends to the grid from the battery discharging point of the EV No. i at time t, and $price_{dch,t}$ is equal to the selling price of energy at time t.

EV owners have to pay a cost to the parking owner depending on how long the EV has been in the parking lot. Due to the difference in the time of each EV presence in the parking lot, this cost will be different for each EV. This cost can be calculated using Eq. (4.6).

$$C_{park} = \sum_{i=1}^{Nev} C_{park,i} \tag{4.6}$$

In Eq. (4.6), $C_{park,i}$ is equal to the cost that EV No. i pays for parking the EV. Eq. (4.7) shows how to calculate the parking cost per EV.

$$C_{park,i} = \begin{cases} at_{park,i} & t_{park,i} \le t_0 \\ a\left(t_0 + \sum_{t=1}^{t_{park,i}-t_0}(1+bt)\right) & t_{park,i} > t_0 \end{cases} \tag{4.7}$$

According to [27], the duration of the presence of EVs in the parking lot is a random and stochastic quantity with a nonuniform distribution. In Eq. (4.7), $t_{park,i}$ is equal to the duration of the presence of EV No. i in the parking lot. Eq. (4.7) states that if the presence duration of the EV_i in the parking lot be less than t_0, it must pay a fixed cost per hour, in accordance with a factor; however, if the presence duration of the EV_i in the parking lot exceeds t_0, it will have to pay a variable cost per hour of parking in addition to the previous cost. This variable cost is determined by a factor of b. The values of the constant coefficients a, b and t_0 are selected by the parking lot.

In addition to selling energy to the grid, parking lots can make money by charging the EV battery and selling energy to the EV owner. Therefore, if the amount of SoC of EV battery has increased compared to its condition when entering the parking lot, the EV owner must pay a cost according to the amount of SoC increment to the parking lot. Equations (4.8) and (4.9) show how this cost is calculated.

$$C_{ech} = \sum_{i=1}^{Nev} C_{ech,i} \tag{4.8}$$

$$C_{ech,i} = R_1 \cdot \left(SoC_{out,i} - SoC_{in,i} \right) \tag{4.9}$$

In Eq. (4.8) and (4.9), $C_{ech,i}$ is equal to the cost that EV No. i pay for parking its SoC increment. SoC_{in} and SoC_{out} indicate the battery-charging status of EV No. i when entering and exiting the parking lot, respectively. R_1 is also the energy sales tariff to the EV owner.

If the EV owner provides the parking lot with information about how long EV has been in the parking lot, the parking lot will be able to make more detailed scheduling; however, if the EV owner acts contrary to his information, the parking scheduling will be disrupted and the parking lot will suffer financial losses. In order to avoid the financial loss of the parking lot, if any agreement, the EV must pay a cost for the parking violation. Obviously, the amount of payment for each EV varies according to the amount of the violation, and the total parking cost will be calculated from Eq. (4.10).

$$C_{pen} = \sum_{i=1}^{Nev} C_{pen,i} \tag{4.10}$$

In Eq. (4.10), $C_{pen,i}$ is equal to the receipt of parking received for the EV No. i violation. The required equations to calculate the amount of $C_{pen,i}$ will vary depending on the parking agreement and the owner of the EV_i. Therefore, how to calculate $C_{pen,i}$ will be explained in Section 3.3.9.

G2V capability allows the parking lot to receive the battery-charging energy of existing EVs from the grid. Equations (4.11) and (4.12) show how to calculate the cost of parking paid to the grid for energy purchases.

$$C_{gch} = \sum_{i=1}^{Nev} C_{gch,i} \tag{4.11}$$

$$C_{gch,i} = \sum_{t \in R} P_{ch}(t) \cdot price_{ch,t} \qquad (4.12)$$

In Eq. (4.11) and (4.12), $C_{gch,i}$ is the parking cost for purchasing energy from the grid from EV No. i battery charging. $p_{ch}(t)$ is the actual power that the parking lot receives from the grid for charging EV No. i at time t, and $price_{ch,t}$ is equal to the purchase price of energy from the grid at time t.

Sometimes the SoC of the EV battery not only does not increase when leaving but also decreases relative to the amount when entering the parking lot. In other words, the battery energy is discharged. In this case, the parking lot is obliged to pay a cost to the EV owner according to the amount of SoC reduction.

$$C_{edch} = \sum_{i=1}^{Nev} C_{edch,i} \qquad (4.13)$$

$$C_{edch,i} = R_2 \cdot (SoC_{in,i} - SoC_{out,i}) \qquad (4.14)$$

In Eq. (4.13) and (4.14), $C_{edch,i}$ is equal to the cost that the parking lot pays to the owner for reducing the SoC of the EV No. i battery. R_2 is also the energy purchase tariff from the EV owner.

The parking lot is required to charge the EV batteries to the specified minimum capacity. If the SoC of the battery of EV No. i is less than the specified minimum capacity at the time of exiting, the parking lot must pay a cost to the owner of EV No. i, if any agreement.

$$C_{lim} = \sum_{i=1}^{Nev} C_{lim,i} \qquad (4.15)$$

$$C_{lim,i} = R_3 \cdot (SoC_{lim,i} - SoC_{out,i}) \qquad (4.16)$$

In Eq. (4.15) and (4.16), $C_{lim,i}$ is equal to the cost that the parking lot pays to the owner of EV No. i for not providing the specified minimum capacity. $SoC_{lim,i}$ also indicates the minimum capacity specified for SoC_i.

Continuous charging and discharging affect the battery life of EVs. To compensate for this damage, if the parking lot benefits from the participation of EV No. i in the V2G process, it can provide a part of the profit from EV No. i to the owner under a specific agreement.

$$C_{sh} = \sum_{i=1}^{Nev} C_{sh,i} \qquad (4.17)$$

$$C_{sh,i} = K \cdot (C_{gdch,i} - C_{gch,i}) \qquad (4.18)$$

In Eq. (4.17) and (4.18), $C_{sh,i}$ is equal to the cost that the parking lot pays to the owner for the benefit of EV No. i. The contribution of EVs to the profit is determined by a coefficient of K.

4.2.2 PARKING LOT REQUIREMENTS

4.2.2.1 Study Period [18]

Considering the behavioral patterns of EVs, the period when the EV is parked and can be charged often begins in the last hours of the day and lasts until the next day. Therefore, the full EV charging period will be separated by two calendar days. Figure 4.5 shows the extended study period.

4.2.2.2 EV Behavior

EV behavior has three main characteristics, including the time of entering the parking lot, the time of leaving it and the amount of initial energy when entering the parking lot.

The time the EV enters and leaves the parking lot is random and unpredictable. Therefore, in each time step t, using stochastic methods, the EVs entering the parking lot and the cars leaving the parking lot are determined in time step t.

Eq. (19) is used to calculate EV energy when entering the parking lot, SoC_{in}.

$$SoC_{in} = E_s - (d_d \cdot C_{veh})$$

(4.19)

In Eq. (4.19), E_s stands for EV battery capacity. Also, d_d indicates the mileage by EV and C_{veh}, which indicates the rate of energy consumption per kilometer. The mileage is also a stochastic and unpredictable quantity; therefore, d_d for each EV is set stochastically.

4.2.2.3 Allowable Charging and Discharging Time-Step Determination

Due to the change in grid consumption over a day, energy prices also change as a function of consumption at each hour of the day. Therefore $price_{ch,t}$ and $price_{dch,t}$ will

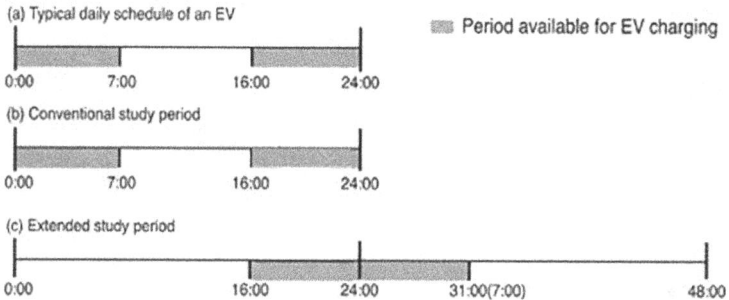

FIGURE 4.5 Extended study period.

take different values at each time step t. Obviously, as the $price_{ch,t}$ decreases and the $price_{dch,t}$ increases, the parking profit will increase. As a consequence, the first step in scheduling the charging and discharging process in a parking lot is to determine the allowable charging and discharging time step.

In the allowable charging and discharging time-step determination in a parking lot, it is necessary to pay attention to the following:

4.2.2.3.1 Received Parking Profit

Purchasing energy from the grid in the time steps when the energy price is lower and selling it to the grid in the time steps when the energy price is higher will increase the parking profit. Therefore, in the [0-T] interval, the allowable charging and discharging time steps must be determined in such a way that the $price_{dch}$ always has a higher value than the $price_{ch}$.

The median index is calculated for energy price data over a day because the median index divides the data into two groups with the same amount of data. If the median index for energy prices in a day is equal to M, in 12 time steps the energy price will be more than M and in the other 12 time steps the energy price will be less than M.

4.2.2.3.2 The Main Task of the Parking Lot

The main task of the parking lot is to charge the EVs and increase their SoC level when leaving the parking lot. Therefore, the number of allowable charging time steps must be more than the number of allowable discharging time steps. Therefore, the value of M is increased to a minimum amount, which leads to an increase in the number of allowable charging time steps relative to the number of allowable discharging time steps.

Thus, R_{ch} is allowable time steps for charging, and R_{dch} is allowable time steps for discharging. According to Eq. (4.20) and (4.21), R_{ch} consists of n discrete time steps for the charging process, and R_{dch} consists of m discrete time steps for the discharging process.

$$R_{ch} = \left[n_1, \ldots, n_j, \ldots, n_n \right] \tag{4.20}$$

$$R_{dch} = \left[m_1, \ldots, m_k, \ldots, m_m \right] \tag{4.21}$$

In Eq. (4.20), n_j is equal to the jth allowable charging discrete time step and in Eq. (4.21), m_k is equal to the kth allowable discharging discrete time step.

Each of the allowable charging and discharging discrete time steps consists of continuous time steps. According to Eq. (4.22) and (4.23), n_j consists of x allowable charging continuous time steps, and m_k consists of y allowable discharging continuous time steps.

$$n_j = \left[t_{n_j,1} \quad t_{n_j,x} \right] \tag{4.22}$$

$$m_k = \left[t_{m_k,1} \quad t_{m_k,y} \right] \tag{4.23}$$

In Eq. (4.22), $t_{nj,1}$ represents the beginning of the n_j time step, and $t_{nj,x}$ represents the end of the n_j time step. In Eq. (4.23), $t_{mk,1}$ represents the beginning of the m_k time step, and $t_{mk,y}$ represents the end of the m_k time step.

4.2.2.4 Parking Limitation on Power Exchange

The amount of power that is exchanged between the grid and the parking lot at each time step does not depend only on the number of EVs in the parking lot and their SoC. Parking also has limitations, and it is not possible to exchange more power than allowed for parking. Accordingly, the maximum number of EVs that the parking lot can charge or discharge at any time is determined according to Eq. (4.24) and (4.25).

$$M_{ch}\left(t\right) = \frac{P_{ch}^{max}\left(t\right)}{P_{ch}\left(t\right)} \tag{4.24}$$

$$M_{dch}\left(t\right) = \frac{P_{dch}^{max}\left(t\right)}{P_{dch}\left(t\right)} \tag{4.25}$$

In Eq. (4.24) and (4.25), $M_{ch}(t)$ is equal to the maximum number of EVs that the parking lot can charge in time step t, and $M_{dch}(t)$ is also equal to the maximum number of EVs that the parking lot can discharge in time step t. $P_{ch}^{max}(t)$ is equal to the maximum power that the parking lot can receive from the grid at time step t, and $P_{dch}^{max}(t)$ is also equal to the maximum power that the parking lot can send to the grid at time step t.

4.2.3 CHARGING AND DISCHARGING STRATEGY

The charging and discharging strategy refer to the control method that manages the charging and discharging process in the parking lot of EVs, in interactions 2 and 3. Here are nine different strategies for the charging and discharging process scheduling of EVs in the parking lot. Figure 4.6 shows the classification of proposed strategies.

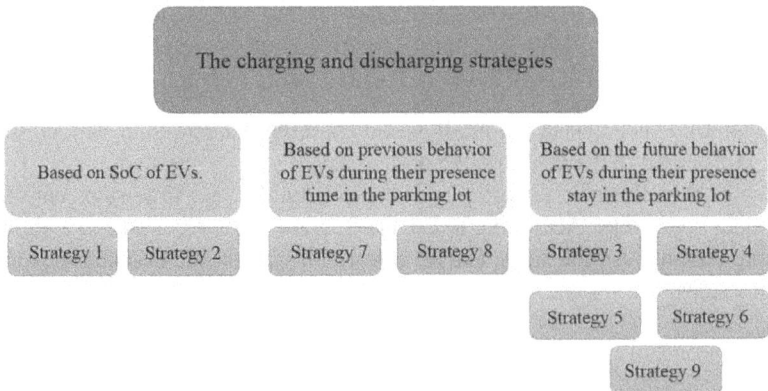

FIGURE 4.6   The classification of proposed strategies.

4.2.3.1 Strategy Based on SoC$_{lim}$ Supply in Time Step t (SoC$_{lim}$-supply-in-t)

4.2.3.1.1 Charging Strategy

To schedule the EV charging process at each allowable charging time step, it should be noted that due to the limitation on the amount of power sent from the grid, the parking lot will not be able to charge all EVs in the parking lot. Therefore, determining priority for EV charging has a very important role in the parking profit. The charging strategy at each time step determines the charging priority of the EVs.

In this strategy, the charging priority of EVs in each charging time step is determined only by their SoC level in the same time step, and the behavior of EVs in the previous charging and discharging time steps has no role in determining the charging priority. In the following, the required equations for calculating J_{ch} (t) are presented.

$$\text{MAX } J_{ch}(t) = \sum_{i=1}^{Nev(t)} n_i(t) \cdot F_i(t) \tag{4.26}$$

$$F_i(t) = E_s - SoC_i(t-1) \tag{4.27}$$

$$n_i(t) \in \{0,1\} \tag{4.28}$$

In Eq. (4.26), *Nev(t)* indicates the number of EVs present in the parking lot in the time step t, and in Eq. (4.27), *SoCi(t–1)* indicates the battery-charging status of the EV$_i$ in the time step $(t–1)$. Due to the limitation on the maximum number of EVs that the parking lot can charge at any time step, the coefficient $n_i(t)$ is considered in Eq. (4.26). $n_i(t)$ indicates the presence or absence of EVi in the charging process. Therefore, in Eq. (4.28) it is defined as a binary variable.

Charging-strategy constraints include Eq. (4.29) to (4.31).

$$0 \le SoC_i(t) \le E_s \tag{4.29}$$

$$SoC_i(t) = SoC_i(t-1) + n_i(t)p_{ch}(t) \tag{4.30}$$

$$0 \le \sum_{i=1}^{Nev(t)} n_i(t) \le M_{ch}(t) \tag{4.31}$$

Eq. (4.28) to (4.31) are true in all charging strategies. The charging process scheduling up to strategy performance-past-discharge will be done by the J_{ch} (t) function.

4.2.3.1.1.1 Control Filters
In the following, control filters g(x), s(x) and y(x) will be used to express the equations of each strategy.

$$g(x) = \begin{cases} 0 & x < 0 \\ x & x \ge 0 \end{cases} \tag{4.32}$$

$$s(x) = \begin{cases} 0 & x < 0 \\ 1 & x \geq 0 \end{cases} \tag{4.33}$$

$$y(x) = \begin{cases} 0 & x \leq 0 \\ 1 & x > 0 \end{cases} \tag{4.34}$$

4.2.3.1.2 Discharging Strategy

In scheduling the discharging process, the existence of a limitation on the amount of power that the parking lot can send to the grid at any time step has led to a limitation on the number of EVs that can be discharged at any time step. Therefore, in order to increase parking profits, the priority of discharging EVs must be determined at each time step.

In this strategy, the priority of discharging EVs in each time step t is determined by comparing the parking receipt from the grid, for sending power to the grid (via EV$_i$), in the time step t and the parking cost paid to the EVi owner, for not providing SoC$_{lim,i}$ in the time step t. In fact, the discharging function of strategy "SoC$_{lim}$-supply-in-t", $J_{dch}(t)$, according to the amount of profit from the discharging of EVs in the time step t and the limitation on the amount of power sent to the grid determines their discharging priority. In the following, the required equations for calculating $J_{dch}(t)$ are presented:

$$\text{MAX } J_{dch}(t) = \sum_{i=1}^{Nev(t)} n_i(t) \cdot F_i(t) \tag{4.35}$$

$$F_i(t) = C_{g,i}(t) - C_{lim,i}(t) \tag{4.36}$$

$$n_i(t) \in \{0,1\} \tag{4.37}$$

Due to the limitation on the maximum number of EVs that the parking lot can discharge at any time step t, the coefficient $n_i(t)$ is considered in Eq. (4.35). $n_i(t)$ indicates the presence or absence of EV$_i$ in the discharging process. In Eq. (4.36), $C_{g,i}(t)$ is the parking receipt from the grid for sending power via EVi in time step t. $C_{lim,i}(t)$ is equal to the parking cost paid to the EVi owner for not providing SoC$_{lim,i}$ at time step t.

The required equations to calculate $C_{g,i}(t)$ and $C_{lim,i}(t)$ in this strategy are introduced next (discharging rate is the same for all EVs).

$$C_{g,i}(t) = p_{dch}(t) \cdot price_{dch,t} \tag{4.38}$$

$$C_{lim,i}(t) = s\left(SoC_{lim,i} - \left(SoC_i(t-1) - p_{dch}(t)\right)\right) \cdot k_i(t) \tag{4.39}$$

According to Eq. (4.39), if the SoC level is still higher than $SoC_{lim,i}$, despite discharging EV$_i$ in time step t, parking does not pay a cost in time step t. Otherwise, the cost of parking in time step t is equal to $k_i(t)$.

As can be seen in Eq. (4.39), the amount of parking receipt for EVi discharging in the previous discharging time steps is not included in the calculation of $C_{g,i}(t)$. Therefore, the cost to the parking lot of not providing $SoC_{lim,i}$ at the previous discharging step time should not re-affect the EV discharging priority at time step t.

If EV_i has not been discharged by the parking lot in any of the previous discharging time steps, $k_i(t)$ is calculated by Eq. (4.40).

$$k_i(t) = \left(SoC_{lim,i} - \left(SoC_i(t-1) - p_{dch}(t) \right) \right) \cdot R_3 \qquad (4.40)$$

If EV_i has been discharged by the parking lot during the previous discharging time steps, to calculate $k_i(t)$, it is necessary to calculate how much the parking lot has paid in the previous discharging time steps. As mentioned before, the EV_i time step in the parking lot consists of charging and discharging discrete time steps. If the EV_i passes the charging discrete time step before reaching the time step t, it may be possible to compensate part of the cost incurred in the previous discharging time steps by charging the EV_i. However, due to restrictions on M_{ch}, part of this cost is not compensated, therefore $k_i(t)$ is calculated by Eq. (4.41).

$$k_i(t) = \left(SoC_{lim,i} - \left(SoC_i(t-1) - p_{dch}(t) \right) \right) \cdot R_3 - C_{n-s,i}(t) \qquad (4.41)$$

In Eq. (4.41), $C_{n-s,i}(t)$ represents the uncompensated payment up to the time step t. Eq. (4.42) calculates $C_{n-s,i}(t)$.

$$C_{n-s,i}(t) = g\left(\sum_{t_i=t_{m_1,1}}^{t-1} C_{l,i}(t_i) - \left(p_{ch}(t_i) \cdot R_3 \right) \right) + \dots g\left(\sum_{t_i=t_{m_k,1}}^{t-1} C_{l,i}(t_i) - \left(p_{ch}(t_i) \cdot R_3 \right) \right)$$
$$+ \dots g\left(\sum_{t_i=t_{m_m,1}}^{t-1} C_{l,i}(t_i) - \left(p_{ch}(t_i) \cdot R_3 \right) \right) \qquad (4.42)$$

In Eq. (4.42), $t_{mk,1}$ is equivalent to the first continuous time step in the kth of the discharging discrete time step.

Discharging strategy constraints include Eq. (4.43) to (4.45). These equations are true of all discharging strategies.

$$0 \le SoC_i(t) \le E_s \qquad (4.43)$$

$$SoC_i(t) = SoC_i(t-1) - n_i(t) p_{ch}(t) \qquad (4.44)$$

$$0 \le \sum_{i=1}^{Nev(t)} n_i(t) \le M_{dch}(t) \qquad (4.45)$$

4.2.3.2 Strategy Based on SoC$_{lim}$ Supply Up to Time Step t (SoC$_{lim}$-supply-up-to-t)

4.2.3.2.1 Discharging Strategy

In this strategy, the simplification and acceleration of the EV discharging process scheduling is considered. In this strategy, the priority of discharging EVs in each time step t, by comparing the total parking received profit from the grid, for power exchange with the grid (via EV$_i$), from the first allowable discharging time step of $t_{m1,1}$ to the time step t and the total parking cost paid to EV$_i$ owner for not providing SoC$_{lim,i}$ will be determined up to time step t. Therefore, the discharging function of strategy "SoC$_{lim}$-supply-upto-t", $J_{dch,}(t)$, examines the total received parking profit from the grid up to time step t and the total parking payment cost for not providing SoC$_{lim,i}$, if EVi is discharged at time step t.

$$MAX \quad J_{dch} = \sum_{i=1}^{Nev(t)} n_i(t) \cdot F_i(t) \tag{4.46}$$

$$F_i(t) = C_{g,i}(t) - C_{lim,i}(t) \tag{4.47}$$

In Eq. (4.47), $C_{g,i}(t)$ represents the total received profit by the parking lot from the grid for the exchange of power through EV$_i$, from the time step $t_{m1,1}$ to the time step t. $C_{lim,i}(t)$ represents the total cost of parking for not providing SoC$_{lim,i}$ up to time step t. According to the previous points, $C_{g,i}(t)$ is defined as Eq. (4.48).

$$C_{g,i}(t)\begin{cases} (t-t_{m_1,1}) \cdot P_{dch}(t) \cdot price_{dch,t} & t \in m_1 \\ \vdots & \vdots \\ (t-t_{m_k,1}) \cdot P_{dch}(t) \cdot price_{dch,t} & t \in m_k \\ \vdots & \vdots \\ (t-t_{m_m,1}) \cdot P_{dch}(t) \cdot price_{dch,t} & t \in m_m \end{cases} \tag{4.48}$$

In relation to (4.48), m_k is equivalent to $k_m{}^t$ of the discharging discrete time step, and $t_{mk,1}$ is its first continuous time step. Eq. (4.49) shows how $C_{lim,i}(t)$ is calculated.

$$C_{lim,i}(t) = g\left(SoC_{lim,i} - \left(SoC_i(t-1) - P_{dch}(t)\right)\right) \cdot R_3 \tag{4.49}$$

4.2.3.3 Strategy Based on Reducing the Total Cost Paid to EV Owners Assuming the Possibility of Charging on Time with the Lowest Energy Price (EV Owners-ch-l-p)

4.2.3.3.1 Discharging Strategy

In this strategy, at each time step, the parking receipt for EV discharging is compared with the lost receipt of parking for EV discharging, and if the parking receipt is more than the lost receipt, EV discharging will be possible. In other words, the discharging function of strategy "EV owners-ch-l-p", $J_{dch_i}(t)$, examines what the parking lot receives when the EV is discharged and also examines what the parking lot receives if the EV is not discharged. In the following, the required equations for calculating $J_{dch}(t)$ are presented.

$$MAX \ J_{dch}(t) = \sum_{i=1}^{Nev(t)} n_i(t) \cdot F_i(t) \tag{4.50}$$

In Eq. (4.50), $F_i(t)$ represents the difference between the parking receipt and its lost receipt (via EV_i) at time step t. In order to compare parking receipts and lost receipts, it must be determined at what time step the EV owner will pay or receive parking costs. Pay or receive parking costs by the EV owner is first determined by comparing the battery status to the amount of $SoC_{in,i}$. Accordingly, $F_i(t)$ is defined as Eq. (4.51).

$$F_i(t) = s\left(\left(SoC_i(t-1) - p_{dch}(t)\right) - SoC_{in,i}\right) \cdot k_{i,1}(t)$$
$$+ y\left(SoC_{in,i} - \left(SoC_i(t-1) - p_{dch}(t)\right)\right) \cdot k_{i,2}(t) \tag{4.51}$$

4.2.3.3.1.1 Calculation of $k_{i,1}(t)$ In order to calculate the parking cost for the EV owner at time step t, if, despite the EV discharging, the battery energy is still higher than the $SoC_{in,i}$, it is necessary to consider the costs of not providing the $SoC_{lim,i}$, thus $k_{i,1}$ is defined as Eq. (4.52).

$$k_{i,1}(t) = s\left(\left(SoC_i(t-1) - p_{dch}(t)\right) - SoC_{lim,i}\right) \cdot f_{i,1-1}(t)$$
$$+ y\left(SoC_{lim,i} - \left(SoC_i(t-1) - p_{dch}(t)\right)\right) \cdot f_{i,2-1}(t) \tag{4.52}$$

According to Eq. (4.52), the parking profit is equal to $f_{i,1-1}(t)$ in a situation where, despite the discharging of the battery, $SoC_i(t)$ is still higher than $SoC_{in,i}$ and $SoC_{lim,i}$. Equations (4.53) to (4.55) calculate $f_{i,1-1}(t)$.

$$f_{i,1-1}(t) = C_{g,i}(t) - C_{i,1}(t) \tag{4.53}$$

$$C_{g,i}(t) = p_{dch}(t) \cdot price_{dch,t} \tag{4.54}$$

$$C_{i,1}(t) = p_{dch}(t) \cdot R_1 \tag{4.55}$$

If with the EV discharging, battery energy is less than $SoC_{lim,i}$ at time step t, it should be noted that the amount of discharging at time step t to what extent decreases the value of $SoC_{lim,i}$. Therefore, the value of $SoC_i(t-1)$ should be considered to determine the extent of EV discharging effect on the not providing of $SoC_{lim,i}$ at time step t. Accordingly, $f_{i,2-1}(t)$ is defined as Eq. (4.56).

$$f_{i,2-1}(t) = s\left(SoC_i(t-1) - SoC_{lim,i}\right) \cdot h_{i,1-2-1}(t)$$
$$+ y\left(SoC_{lim,i} - SoC_i(t-1)\right) \cdot h_{i,2-2-1}(t) \tag{4.56}$$

According to Eq. (4.56), the parking profit is equal to $h_{i,1-2-1}(t)$ in a situation where, despite the battery discharging $SoC_i(t)$ is higher than the $SoC_{in,i}$, but the EV discharging at time step t causes the battery energy to decrease beyond the limitation set for the $SoC_{lim,i}$. Equations (4.57) and (4.58) calculate $h_{i,1-2-1}(t)$.

$$h_{i,1-2-1}(t) = C_{i,g}(t) - C_{i,2}(t) \tag{4.57}$$

$$C_{i,2}(t) = P_{dch}(t) \cdot R_1 + \left(SoC_{lim,i} - \left(soc_i(t-1) - P_{dch}(t)\right)\right) \cdot R_3 \tag{4.58}$$

According to Eq. (4.56), the parking profit is equal to $h_{i,2-2-1}(t)$, in a situation where despite the battery discharging, the $SoC_i(t)$ is higher than the amount of $SoC_{in,i}$, but the EV discharging in the time step t causes the battery energy to decreases more than previous time steps compared to $SoC_{lim,i}$. Equations (4.59) and (4.60) calculate $h_{i,2-2-1}(t)$.

$$h_{i,2-2-1}(t) = C_{i,g}(t) - C_{i,3}(t) \tag{4.59}$$

$$C_{i,3}(t) = P_{dch}(t) \cdot R_1 + P_{dch}(t) \cdot R_3 \tag{4.60}$$

4.2.3.3.1.2 Calculation of $k_{i,2}(t)$ In addition to the previous sections, it is necessary to investigate the cases in which EV discharging at time step t causes the battery energy to decrease relative to the amount of $SoC_{in,i}$. Parking lost receipts will vary due to EV discharging at time step t, depending on different conditions.

In the first step, it should be noted that if EV is not discharged in the time step t, is the value of E_s and $SoC_i(t-1)$ such that it is possible to charge EV in future time steps by purchasing energy at the lowest price from the grid and selling it to the EV owner to make a profit for the parking lot. Thus $k_{i,2}(t)$ is defined as Eq. (4.61).

$$k_{i,2}(t) = y\left(\left(SoC_i(t-1) + P_{ch}\right) - E_s\right) \cdot f_{i,1-2}(t)$$
$$+ s\left(E_s - \left(SoC_i(t-1) + P_{ch}\right)\right) \cdot f_{i,2-2}(t) \tag{4.61}$$

According to Eq. (4.61), the parking profit is equal to $f_{i,1-2}(t)$, where EV discharging at time step t reduces battery energy compared to $SoC_{in,i}$, but EV not discharging does not lead to purchase energy from the grid and sell it to the EV owner. Equations (4.62) and (4.63) calculate $h_{i,1-2-1}(t)$.

$$f_{i,1-2}(t) = C_{i,g}(t) - C_{i,4}(t) \tag{4.62}$$

$$C_{i,4}(t) = p_{dch}(t) \cdot R_2 \tag{4.63}$$

In order to calculate the parking costs to the EV owner at time step t, it is necessary to consider the costs of not providing $SoC_{lim,i}$, so $f_{i,2-2}(t)$ is defined as Eq. (4.64).

$$\begin{aligned} f_{i,2-2}(t) = s\Big(\big(SoC_i(t-1) - p_{dch}(t)\big) - SoC_{lim,i}\Big) \cdot w_{i,1}(t) \\ + y\Big(SoC_{lim,i} - \big(SoC_i(t-1) - p_{dch}(t)\big)\Big) \cdot w_{i,2}(t) \end{aligned} \tag{4.64}$$

According to Eq. (4.64), the parking received profit is equal to $w_{i,1}(t)$, when EV discharging at time step t reduces the battery energy compared to $SoC_{in,i}$ and on the other hand, if EV is not discharged, it is possible to purchase from the grid and sell it to the EV owner.

$$w_{i,1}(t) = C_{i,g}(t) - C_{i,5}(t) \tag{4.65}$$

In this case, the parking lot is obliged to pay for the energy purchase from the EV owner. In addition, the parking lot will lose the profit from the sale of energy to the EV owner. The amount of this profit is calculated according to $p_{dch}(t)(R_1 - R_{min})$. R_{min} is equal to the lowest energy purchase rate from the grid. Thus $C_{i,5}(t)$ is as Eq. (4.66).

$$C_{i,5}(t) = p_{dch}(t) \cdot R_2 + p_{dch}(t) \cdot (R_1 - R_{min}) \tag{4.66}$$

If with the EV discharging, battery energy is less than $SoC_{lim,i}$ at time step t, it should be noted that the amount of discharging at time step t to what extent decreases the value of $SoC_{lim,i}$. Therefore, the value of $SoC_i(t-1)$ should be considered to determine the extent of EV discharging effect on the not providing of $SoC_{lim,i}$ at time step t. Accordingly, $w_{i,2}(t)$ is defined as Eq. (4.67).

$$\begin{aligned} w_{i,2}(t) = s\big(SoC_i(t-1) - SoC_{lim,i}\big) \cdot w_{i,1-2}(t) \\ + y\big(SoC_{lim,i} - SoC_i(t-1)\big) \cdot w_{i,2-2}(t) \end{aligned} \tag{4.67}$$

According to Eq. (4.67), the parking profit is equal to $w_{i,1-2}(t)$ in a situation where EV discharging at time step t, in addition to reducing the battery energy relative to $SoC_{in,i}$, causes the battery energy to decrease beyond the limitation set for the $SoC_{lim,i}$. Equations (4.68) and (4.69) calculate $w_{i,1-2}(t)$.

$$w_{i,1-2}(t) = C_{i,g}(t) - C_{i,6}(t) \tag{4.68}$$

$$C_{i,6}\left(t\right) = p_{dch}\left(t\right)\cdot R_2 + p_{dch}\left(t\right)\cdot\left(R_1 - R_{min}\right)$$
$$+\left(SoC_{lim,i} - \left(SoC_i\left(t-1\right) - p_{dch}\left(t\right)\right)\right)\cdot R_3 \tag{4.69}$$

According to Eq. (4.67), the parking profit is equal to $w_{i,2-2}(t)$ in a situation EV discharging at time step t, in addition to reducing the battery energy relative to the $SoC_{in,i}$, causes the battery energy to decrease more than at previous time steps relative to the $SoC_{lim,i}$. Equations (4.70) and (4.71) calculate $w_{i,2-2}(t)$.

$$w_{i,2-2}\left(t\right) = C_{i,8}\left(t\right) - C_{i,7}\left(t\right) \tag{4.70}$$

$$C_{i,7}\left(t\right) = p_{dch}\left(t\right)\cdot R_2 + p_{dch}\left(t\right)\cdot\left(R_1 - R_{min}\right) + p_{dch}\left(t\right)\cdot R_3 \tag{4.71}$$

4.2.3.4 Strategy Based on Reducing the Total Cost Paid to EV Owners Assuming the Possibility of Charging on Time with the Highest Energy Price (EV Owners-ch-h-p)

4.2.3.4.1 Discharging Strategy
All the equations of this strategy are in accordance with strategy "EV owners-ch-l-p", and the only difference is the use of R_{max}, which is the highest purchase price of energy from the grid.

4.2.3.5 Strategy Based on Reducing the Total Costs Paid to EV Owners and Increasing the Amount of Participation in the Discharge Process, Assuming the Possibility of Charging on Time with the Lowest Energy Price (EV Owners-Discharge-ch-l-p)

4.2.3.5.1 Discharging Strategy
In this strategy, in order to increase the participation of EVs in the discharging process, a new constraint is added to the $C_{ech}(t)$ function. This constraint determines that the EV owner only needs to pay the cost of increasing SoC_{in} to the extent of $SoC_{lim,I}$ and will not be paid for higher amounts. Consequently, $C_{ech,i}(t)$ is defined as Eq. (4.72).

$$C_{ech,i} = \begin{cases} R_1\cdot\left(SoC_{out,i} - SoC_{in,i}\right) & SoC_{in,i} \leq SoC_{out,i} <= SoC_{lim,i} \\ R_1\cdot\left(SoC_{lim,i} - SoC_{in,i}\right) & SoC_{in,i} \leq SoC_{lim,i} <= SoC_{out,i} \end{cases} \tag{4.72}$$

In the following, the required equations for calculating the discharging function of strategy EV owners-discharge-ch-l-p, $J_{dch}(t)$, are presented.

$$MAX\ J_{dch}\left(t\right) = \sum_{i=1}^{Nev(t)} n_i\left(t\right)\cdot F_i\left(t\right) \tag{4.73}$$

In this strategy, it must also be determined in the first step, if EV is discharged in time step t, what will be the state of its battery energy compared to $SoC_{in,i}$. Thus, $F_i(t)$ is defined as Eq. (4.74).

$$F_i(t) = s\left(\left(SoC_i(t-1) - p_{dch}(t)\right) - SoC_{in,i}\right) \cdot k_{i,1}(t)$$
$$+ y\left(SoC_{in,i} - \left(SoC_i(t-1) - p_{dch}(t)\right)\right) \cdot k_{i,2}(t) \tag{4.74}$$

4.2.3.5.1.1 Calculation of $k_{i,1}(t)$ By changing the $C_{ech,i}(t)$ in strategy 5, if the EV is discharged it is necessary to determine what the battery energy will be compared to the $SoC_{lim,i}$. Thus, $k_{i,1}(t)$ is defined as Eq. (4.75).

$$k_{i,1}(t) = s\left(\left(SoC_i(t-1) - p_{dch}(t)\right) - SoC_{lim,i}\right) \cdot f_{i,1-1}(t)$$
$$+ y\left(SoC_{lim,i} - \left(SoC_i(t-1) - p_{dch}(t)\right)\right) \cdot f_{i,2-1}(t) \tag{4.75}$$

According to Eq. (4.75), the profit from EV discharging is equal to $f_{i,1-1}(t)$ in a situation that despite EV discharging, the amount of battery energy is more than $SoC_{in,i}$ and $SoC_{lim,i}$. Equations (4.76) to (4.78) calculate $f_{i,1-1}(t)$.

$$f_{i,1-1}(t) = C_{i,g}(t) - C_{i,1}(t) \tag{4.76}$$

$$C_{i,g}(t) = p_{dch}(t) \cdot price_{dch,t} \tag{4.77}$$

$$C_{i,1}(t) = 0 \tag{4.78}$$

Discharging EV_i at time step t may reduce battery energy compared to $SoC_{lim,i}$. In such cases, it must first be determined that the discharging at time step t has caused to what extent a reduction in the value of $SoC_{lim,i}$. Thus, $f_{i,2-1}(t)$ is defined as Eq. (4.79).

$$f_{i,2-1}(t) = s\left(SoC_i(t-1) - SoC_{lim,i}\right) \cdot h_{i,1-2-1}(t)$$
$$+ y\left(SoC_{lim,i} - SoC_i(t-1)\right) \cdot h_{i,2-2-1}(t) \tag{4.79}$$

According to Eq. (4.79), the parking profit is equal to $h_{i,1-2-i}(t)$ in the case that despite discharging the battery, $SoC_i(t)$ is higher than $SoC_{in,i}$, but discharging EV at time step t causes the battery energy to decrease beyond the limitation set for $SoC_{lim,i}$. Eq. (4.80) and (4.81) show how $h_{i,1-2-i}(t)$ are calculated.

$$h_{i,1-2-1}(t) = C_{i,g}(t) - C_{i,2}(t) \tag{4.80}$$

$$C_{i,2}(t) = \left(SoC_{lim,i} - \left(SoC_i(t-1) - p_{dch}(t)\right)\right) \cdot (R_3 + R_1) \tag{4.81}$$

According to Eq. (4.79), the parking profit is equal to $h_{i,2-2-l}(t)$ in situations where despite discharging of the battery, $SoC_i(t)$ is higher than $SoC_{in,i}$, but discharging of EV at time step t causes the battery energy to decrease more than the previous time steps compared to $SoC_{lim,i}$. Equations (4.82) and (4.83) show how $h_{i,2-2-l}(t)$ are calculated.

$$h_{i,2-2-1} = C_{i,g}(t) - C_{i,3}(t) \tag{4.82}$$

$$C_{i,3}(t) = p_{dch}(t) \cdot (R_1 + R_3) \tag{4.83}$$

4.2.3.5.1.2 Calculation of $k_{i,2}(t)$

According to the equations presented for the calculation of $C_{ech,i}(t)$, the parking received profit will vary in different cases if EV discharging leads to a decrease in battery energy from the amount of $SoC_{in,i}$. In the first step, if EV is not discharged in time step t, the value of Es and $SoC_i(t-1)$ is such that it is possible to have EV charging in the future time steps by purchasing energy at the lowest price from the grid and selling it to the EV owner to make a profit for the parking lot. Thus, $k_{i,2}(t)$ is defined as Eq. (4.84).

$$k_{i,2}(t) = y\big((SoC_i(t-1) + p_{ch}) - E_s\big) \cdot f_{i,l-2}(t)$$
$$+ s\big(E_s - (SoC_i(t-1) + p_{ch})\big) \cdot f_{i,2-2}(t) \tag{4.84}$$

According to Eq. (4.84), the parking profit is equal to $f_{i,l-2}(t)$ in a situation that EV discharging in time step t reduces the battery energy compared to $SoC_{in,i}$, but EV not discharging does not lead to purchase energy from the grid and sale of it to the EV owner. Equations (4.85) and (4.86) depict how to calculate $f_{i,l-2}(t)$.

$$f_{i,l-2}(t) = C_{i,g}(t) - C_{i,4}(t) \tag{4.85}$$

$$C_{i,4}(t) = p_{dch}(t) \cdot R_2 \tag{4.86}$$

EVi discharging at time step t may reduce battery energy compared to $SoC_{lim,i}$. In such cases, it must first be determined how much the discharging at time step t has resulted in a decrease relative to the value of $SoC_{lim,i}$. Thus, $f_{i,2-2}(t)$ is defined as Eq. (4.87).

$$f_{i,2-2}(t) = s\big((SoC_i(t-1) - p_{dch}(t)) - SoC_{lim,i}\big) \cdot w_{i,1}(t)$$
$$+ y\big(SoC_{lim,i} - (SoC_i(t-1) - p_{dch}(t))\big) \cdot w_{i,2}(t) \tag{4.87}$$

According to Eq. (4.87), the parking lot profit is equal to $w_{i,l}(t)$, where EV discharging at time step t, despite the reduction of battery energy from the $SoC_{in,i}$ amount, does not decrease the from the amount of $SoC_{lim,i}$.

$$w_{i,1}(t) = C_{i,g}(t) - C_{i,5}(t) \tag{4.88}$$

In this case, although the battery capacity allows the parking lot to make a profit by purchasing energy from the grid and selling it to the EV owner if the EV is not discharged, but because the amount of $SoC_i(t-1)$ is more than $SoC_{lim,i}$, with charging the EV the parking lot will not receive a cost from the EV owner. Thus, $C_{i,5}(t)$ is as Eq. (4.89).

$$C_{i,5}(t) = p_{dch}(t) \cdot R_2 \tag{4.89}$$

In cases where EV discharging at time step t causes the battery energy to decrease relative to the set value for $SoC_{lim,i}$, it is necessary to determine how much EV discharging at time step t decreases relative to $SoC_{lim,i}$. Thus, $w_{i,2}(t)$ is defined as Eq. (4.90).

$$w_{i,2}(t) = s\left(SoC_i(t-1) - SoC_{lim,i}\right) \cdot w_{i,1-2}(t)$$
$$+ y\left(SoC_{lim,i} - SoC_i(t-1)\right) \cdot w_{i,2-2}(t) \tag{4.90}$$

According to Eq. (4.90), the parking profit is equal to $w_{i,1-2}(t)$, while discharging EV at time step t reduces the battery energy from the amount of $SoC_{in,i}$ and $SoC_{lim,i}$.

$$w_{i,1-2}(t) = C_{i,g}(t) - C_{i,6}(t) \tag{4.91}$$

In this case, because the amount of $SoC_i(t-1)$ is more than $SoC_{lim,i}$, by charging the EV, the parking lot will not be able to receive a cost from the EV owner. Thus, $C_{i,6}(t)$ is as Eq. (4.92).

$$C_{i,8}(t) = p_{dch}(t) \cdot R_2 + p_{dch}(t) \cdot (R_1 - R_{min}) + p_{dch}(t) \cdot R_3 \tag{4.92}$$

In cases where the $SoC_i(t-1)$ is lower than the $SoC_{lim,i}$, the lack of EV discharging allows the parking lot to receive a cost from the EV owner by purchasing energy from the grid and increasing the battery energy. Parking receipts for energy sales to EV owners depend on $SoC_i(t-1)$. Therefore, $w_{i,2-2}(t)$ is defined as Eq. (4.93).

$$w_{i,2-2} = s\left(\left(SoC_i(t-1) + p_{ch}\right) - SoC_{lim,i}\right) \cdot v_{i,1}$$
$$+ y\left(SoC_{lim,i} - \left(SoC_i(t-1) + p_{ch}\right)\right) \cdot v_{i,2} \tag{4.93}$$

According to Eq. (4.93), the parking profit is equal to $v_{i,1}(t)$, while EV discharging at time step t reduces the battery energy from the amount of $SoC_{in,i}$, and the amount of deficiency of $SoC_i(t-1)$ from $SoC_{lim,i}$ is less than the amount of rechargeable energy at one time step.

$$v_{i,1} = C_{i,g}(t) - C_{i,7}(t) \tag{4.94}$$

In this case, as the battery energy charging increases by one step compared to $SoC_{lim,i}$, the profit from the sale of energy to the grid will be equal to $(SoC_{lim,i}\text{-}SoC_i(t-1))$. $(R_1\text{-}R_{min})$. Therefore, $C_{i,7}(t)$ is as Eq. (4.95).

$$C_{i,7}(t) = p_{dch}(t) \cdot R_2 + p_{dch}(t) \cdot R_3 + \left(SoC_{lim,i} - SoC_i(t-1)\right) \cdot \left(R_1 - R_{min}\right) \tag{4.95}$$

According to Eq. (4.93), the parking profit is equal to $v_{i,2}(t)$, while EV discharging at time step t reduces the battery energy from the amount of $SoC_{in,i}$, and the amount of deficiency of $SoC_i(t-1)$ from $SoC_{lim,i}$ is less than the amount of rechargeable energy at one time step.

$$v_{i,2} = C_{i,g}(t) - C_{i,8}(t) \tag{4.96}$$

In this case, because the battery energy charging remains less than the specified value of $SoC_{lim,i}$ per step, the profit from the sale of energy to the grid will be equal to $p_{dch}(t).(R_1\text{-}R_{min})$. Thus, $C_{i,8}(t)$ is as Eq. (4.97).

$$C_{i,8}(t) = p_{dch}(t) \cdot R_2 + p_{dch}(t) \cdot \left(R_1 - R_{min}\right) + p_{dch}(t) \cdot R_3 \tag{4.97}$$

4.2.3.6 Strategy Based on Reducing the Total Costs Paid to EV Owners and Increasing the Amount of Participation in the Discharge Process, Assuming the Possibility of Charging on Time with the Highest Energy Price (EV Owners-Discharge-ch-h-p)

4.2.3.6.1 Discharging Strategy

All the equations of this strategy are in accordance with strategy "EV owners-discharge-ch-l-p", and the only difference is the use of R_{max}, which is the highest purchase price of energy from the grid.

4.2.3.7 Strategy Based on Performance of EVs in the Past to Schedule the Discharging Process (Performance-Past-Discharge)

4.2.3.7.1 Discharging Strategy

In this strategy, the priority of discharging EVs at each time step is determined according to the behavior of EVs during the previous charging and discharging time steps. Accordingly, the number of times that EV has participated in the charging process in the previous time steps and also the number of times that EV has not participated in the discharging process in the previous time steps despite being in the parking lot plays an essential role in determining EV discharging priority. This is because EV_i's participation in the charging process, as well as its non-participation in the discharging process in the previous time steps, reduces the penalty paid to its owner.

EV_i performance during the charging and discharging time steps will affect the amount of cost paid for $C_{lim,i}$ (according to Eq. (4.15)). Therefore, the discharging function of strategy performance-past-discharge, J_{dch} (t), first examines: If EV_i is discharged in time step t, how it will operate during the time of entering the parking lot until the end of time step t, what will be the share of the amount of cost paid for $C_{lim,i}$? Then, by comparing this amount of contribution (paid contribution) and the maximum possible amount for it, the EV_i discharging priority is determined. The following equations are required to calculate J_{dch} (t):

$$\text{MAX } J_{dch}\left(t\right) = \sum_{i=1}^{Nev(t)} n_i\left(t\right) \cdot F_i\left(t\right) \qquad (4.98)$$

The parking payment contribution at time step t depends on the maximum possible amount for SoC_i from the entering time of EV_i to the parking lot to time step t. Thus, $F_i(t)$ is defined as Eq. (4.99).

$$F_i\left(t\right) = s\left(SoC_{max,i}\left(t\right) - SoC_{lim,i}\right) \cdot k_{i,1}\left(t\right) + y\left(SoC_{lim,i} - SoC_{max,i}\left(t\right)\right) \cdot k_{i,2}\left(t\right) \qquad (4.99)$$

In Eq. (4.99), $SoC_{max,i}(t)$ is the maximum possible value for SoC_i in time step t. Eq. (4.100) shows how to calculate $k_{i,1}(t)$.

$$\begin{aligned} k_{i,1}\left(t\right) = s\left(\left(SoC_i\left(t-1\right) - p_{dch}\left(t\right)\right) - SoC_{lim,i}\right) \cdot f_{i,1}\left(t\right) \\ + y\left(SoC_{lim,i} - \left(SoC_i\left(t-1\right) - p_{dch}\left(t\right)\right)\right) \cdot f_{i,2}\left(t\right) \end{aligned} \qquad (4.100)$$

In the following, the required equations for calculating $f_{i,1}(t)$, $f_{i,2}(t)$ and $k_{i,2}(t)$ are presented.

In the case where it is possible to obtain $SoC_{lim,i}$ for EV_i and despite the discharging of EV_i in time step t, the amount of SoC_i is higher than $SoC_{lim,i}$, the paid contribution of parking is as Eq. (4.101).

$$C_{i,1}\left(t\right) = 0 - \alpha\left(SoC_i\left(t-1\right)\right) \qquad (4.101)$$

In cases where it is possible to obtain $SoC_{lim,i}$ for EV_i and EV_i discharging at time t reduces the SoC_i from $SoC_{lim,i}$, the paid contribution of the parking lot is as Eq. (4.102).

$$C_{i,2}\left(t\right) = \left(SoC_{lim,i} - \left(SoC_i\left(t-1\right) - p_{dch}\left(t\right)\right)\right) \cdot R_3 - \alpha\left(SoC_i\left(t-1\right)\right) \qquad (4.102)$$

If it is not possible to obtain $SoC_{lim,i}$ for EV_i, the parking paid contribution is equal to the cost of not providing $SoC_{max,i}(t)$. Therefore, the paid contribution of the parking lot is in the form of Eq. (4.103).

$$C_{i,3}(t) = \left(SoC_{max,i}(t-1) - \left(SoC_i(t-1) - p_{dch}(t)\right)\right) \cdot R_3 - \alpha\left(SoC_i(t-1)\right) \quad (4.103)$$

According to the previous point, in case the parking paid cost is the same for different EVs, the EV with a higher $SoC_i(t-1)$ will have a higher priority. Therefore, $\alpha.(SoC_i(t-1))$ will be taken into account in the calculations. By performing trial and error calculations, α is equal to 0.0001.

On the other hand, the maximum values of $SoC_{max,i}(t)$, $SoC_{lim,I}$ and $SoC_i(t-1)$ are equal to E_s, and the minimum values are zero. Therefore, the range of changes $C_{i,2}(t)$ and $C_{i,3}(t)$ will be in the range $-E_s.R_3$ to $E_s.R_3$. Accordingly, $f_{i,1}(t), f_{i,2}(t)$ and $k_{i,2}(t)$ are defined as equations (4.104) to (4.106).

$$f_{i,1}(t) = E_s \cdot R_3 - C_{i,1}(t) \quad (4.104)$$

$$f_{i,2}(t) = E_s \cdot R_3 - C_{i,2}(t) \quad (4.105)$$

$$k_{i,2}(t) = E_s \cdot R_3 - C_{i,3}(t) \quad (4.106)$$

4.2.3.8 Strategy Based on Performance of EVs in the Past (Performance-Past)

In the performance-past strategy, the charging and discharging process of EVs is scheduled according to EVs' performances in the past time steps.

4.2.3.8.1 Charging Strategy

In this strategy, the priority of charging EVs at each time step is determined according to the behavior of EVs during the previous charging and discharging time steps. In other words, the number of times the EV did not participate in the charging process during the previous allowable charging time steps despite being in the parking lot, as well as the number of times the EV participated in the charging process during the previously allowable charging time steps, play a key role in determining EV charging priority. Therefore, the charging function of performance-past, $J_{ch}(t)$, first examines how the EV_i will function from the time it takes to enter the parking lot to the start of time step t, what will contribute to the amount of energy shortage compared to SoC_{lim}. Then, by comparing this amount of contribution and the maximum possible amount for it, the EV_i charging priority is determined. The following equations are required to calculate $J_{ch}(t)$.

$$MAX \ J_{ch}(t) = \sum_{i=1}^{Nev(t)} n_i(t) \cdot F_i(t) \quad (4.107)$$

The amount of not supplied contribution at time step t depends on the maximum possible value for SoC_i from the time EVi enters the parking lot until time step t. Thus, $F_i(t)$ is defined as Eq. (4.108).

$$F_i(t) = s\left(SoC_{max,i}(t) - SoC_{lim,i}\right) \cdot k_{i,1}(t) + y\left(SoC_{lim,i} - SoC_{max,i}(t)\right) \cdot k_{i,2}(t) \quad (4.108)$$

In the following, the required equations for calculating $k_{i,1}(t)$ and $k_{i,2}(t)$ are presented. For the calculation of $k_{i,1}(t)$ first $C_{i,1}(t)$ and for the calculation of $k_{i,2}(t)$ first $C_{i,2}(t)$ are calculated to determine the extent of EV_i energy shortage due to non-participation in the charging process, and also participation in the discharging process is encountered.

In case it is possible to obtain $SoC_{lim,i}$ for EV_i, its not supplied contribution is in the form of Eq. (4.109).

$$C_{i,1}(t) = \left(SoC_i(t-1) - SoC_{lim}\right) + \alpha\left(SoC_i(t-1)\right) \quad (4.109)$$

In case it is not possible to obtain $SoC_{lim,i}$ for EV_i, its not supplied contribution is in the form of Eq. (4.110).

$$C_{i,2}(t) = \left(SoC_i(t-1) - SoC_{max,i}(t)\right) + \alpha\left(SoC_i(t-1)\right) \quad (4.110)$$

According to the previous point, in case the amount of not supplied contribution is the same for different EVs, the EV with a higher $SoC_i(t-1)$ will have a higher priority. Therefore, $\alpha.(SoC_i(t-1))$ will be taken into account in the calculations. By performing trial and error calculations, α is equal to 0.0001.

On the other hand, the maximum values of $SoC_{max,i}(t)$, $SoC_{lim,I}$ and $SoC_i(t-1)$ are equal to E_s, and the minimum values are zero. Therefore, the range of changes $C_{i,1}(t)$ and $C_{i,2}(t)$ will be in the range $-E_s$ to E_s. Accordingly, $E_s\text{-}C_{i,1}(t)$ and $E_s\text{-}C_{i,2}(t)$ will always have a non-negative value, and as a result, it is possible for all EVs to participate in the charging process, even if their SoC value is higher than $SoC_{lim,i}$. Thus, $k_{i,1}(t)$ and $k_{i,2}(t)$ are defined as Eq. (4.111) and (4.112).

$$k_{i,1}(t) = E_s - C_{i,1}(t) \quad (4.111)$$

$$k_{i,2}(t) = E_s - C_{i,2}(t) \quad (4.112)$$

4.2.3.8.2 Discharging Strategy

As mentioned, strategy "performance-past-discharge" determines EV priorities based on their performance during their time in the parking lot. Therefore, the $J_{dch}(t)$ function of strategy "performance-past-discharge" is used to schedule the discharging process.

4.2.3.9 Strategy Based on the Performance of EVs in the Allotted Time for Their Presence in the Parking Lot (Performance-Allotted Time)

4.2.3.9.1 C_{park} Calculation in Interaction 3

In interaction 3, the EV owner determines the exit time, so the parking lot is required to schedule the charging and discharging according to the EV leaving time. As mentioned, the EV owner is asking for parking facilities in exchange for determining the duration of the EV's presence in the parking lot. Accordingly, the value of $C_{park,i}$ in interaction 3 is calculated according to Eq. (4.113). In this equation, $r<a$ must be specified.

$$C_{park,i} = \begin{cases} rt & t \le t_0 \\ rt_0 + \sum_1^{t-t_0} r(1+bt) & t \ge t_0 \end{cases} \tag{4.113}$$

4.2.3.9.2 C_{pen} Calculation in Interaction 3

Charging and discharging scheduling is done according to how long the EV will be in the parking lot, and providing incorrect information by the EV owner will reduce the parking profit. Therefore, the EV owner is required to pay a cost if he does not leave the parking lot at a specified time.

If the EV leaves the parking lot earlier than scheduled, the parking lot will not be able to attain the benefits of selling energy to the grid despite scheduling, so the EV owner is required to compensate the cost. This cost is determined by Eq. (4.114).

$$C_{pen,i} = \left(t_{d,i} - t_{out,i}\right) \cdot P_{dch} \cdot R_4 \tag{4.114}$$

In Eq. (4. 114), $t_{d,i}$ is the allotted time for EV_i to be present in the parking lot. Also, $t_{out,i}$ is the length of time EV_i has been in the parking lot.

If the EV leaves the parking lot later than scheduled, due to the parking lot not being aware of how long the EV will be in the parking lot, EV charging and discharging cannot be done based on the length of time in the parking lot and the parking profit will be reduced. Accordingly, the EV owner is required to pay the EV parking cost in the parking lot for a period longer than the allotted time, in accordance with the amount determined in interaction 2. This cost is determined by Eq. (4.115).

$$\left(\frac{a}{r} - 1\right) \cdot \left(C_{park,i}\left(t_{out,i}\right) - C_{park,i}\left(t_{d,i}\right)\right) \tag{4.115}$$

4.2.3.9.3 Charging Strategy

In this strategy, the charging priority of EVs at each time step is determined based on their future behavior. The charging function of strategy performance-allotted time, $J_{ch,9}(t)$ determines the EV_i charging priority according to the remaining time for its presence in the parking lot and its SoC level. The following equations are required to calculate $J_{ch}(t)$.

$$MAX\ J_{ch}\left(t\right) = \sum_{i=1}^{Nev(t)} n_i\left(t\right) \cdot F_i\left(t\right) \tag{4.116}$$

In order to satisfy the customer as well as the possibility of participation of EVs in the sale of energy to the grid, EVs whose energy level, even if EV is charged at all allowable charging time steps and not discharged at all allowable discharging future time steps, is less than $SoC_{lim,i}$, they have a higher priority than other EVs to participate in the charging process. Thus, $F_i(t)$ is defined as Eq. (4.117).

$$F_i\left(t\right) = s\left(\left(n_{ch,i} - 1\right)p_{ch}\left(t\right) + SoC_i\left(t-1\right) - SoC_{lim,i}\right) \cdot k_{i,1}\left(t\right)$$
$$+ y\left(SoC_{lim,i} - \left(\left(n_{ch,i} - 1\right)p_{ch}\left(t\right) + SoC_i\left(t-1\right)\right)\right) \cdot k_{i,2}\left(t\right) \tag{4.117}$$

The expected $SoC_{out,i}$ and $SoC_{exp,i}$, are calculated according to the conditions in time step t through Eq. (4.118).

$$SoC_{exp,i}\left(t\right) = n_{ch,i}\,p_{ch}\left(t\right) - n_{dch,i}\,p_{dch}\left(t\right) + SoC_i\left(t-1\right) \tag{4.118}$$

If there is no limitation on the battery capacity, the maximum possible value for SoC_{exp} is equal to E_{max}. Therefore, for calculating $J_{ch,9}(t)$, Eq. (4.119) and (4.120) are used.

$$k_{i,1}\left(t\right) = E_{max} - SoC_{exp,i}\left(t-1\right) \tag{4.119}$$

$$k_{i,2}\left(t\right) = E_{max} - \beta \cdot SoC_{exp,i}\left(t-1\right) \tag{4.120}$$

According to Eq. (4.119) and (4.120), the values of $k_{i,1}(t)$ and $k_{i,2}(t)$ will always be non-negative. The coefficient β is also determined 0.1 by performing calculations.

4.2.3.9.4 Discharging Strategy

In this strategy, the priority of discharging EVs in time step t is determined based on their behavior in the future, i.e., how long they will be in the parking lot. The strategy performance-allotted time discharging function, $J_{dch}\ (t)$, determines the priority of EV$_i$ discharging according to the remaining time for its presence in the parking lot and its SoC level. The following equations are required to calculate $J_{dch}\ (t)$.

$$MAX\ J_{dch}\left(t\right) = \sum_{i=1}^{Nev(t)} n_i\left(t\right) \cdot F_i\left(t\right) \tag{4.121}$$

In order to satisfy the EV owner, if the parking lot presence time and the amount of $SoC_i(t-1)$ are such that if the EV is discharged at time step t and charged at all allowable future time steps, the amount of battery energy will decrease compared to $SoC_{lim,i}$, the parking lot is not allowed to discharge it. Because of the limitation in $M_{ch}(t)$, EV_i may not be able to participate in all steps of charging, so if the EV is discharged, the difference between $SoC_{out,i}$ and $SoC_{lim,i}$ will increase. According to this, $F_i(t)$ is defined as Eq. (4.122).

$$F_i(t) = s\left(n_{ch,i} P_{ch}(t) - p_{dch}(t) + SoC_i(t-1) - SoC_{lim,i}\right) \cdot k_{i,1}(t) \qquad (4.122)$$

EVs that are present in the parking lot for such a time step that even if the EV is discharged in all allowable discharging time steps and it is not charged in all allowable charging time steps, its energy level still exceeds the value of $SoC_{lim,i}$, have higher priority and otherwise, they have less priority to participate in the discharging process. Thus, $k_{i,1}(t)$ is defined as follows:

$$
\begin{aligned}
k_{i,1}(t) &= s\left(SoC_i(t-1) - n_{dch,i} P_{dch}(t) - SoC_{lim,i}\right) \cdot h_{i,1}(t) \\
&+ y\left(SoC_{lim,i} - \left(SoC_i(t-1) - n_{ch,i} P_{ch}(t)\right)\right) \cdot h_{i,2}(t)
\end{aligned}
\qquad (4.123)
$$

$$h_{i,1}(t) = \mu \cdot SoC_{exp,i}(t-1) \qquad (4.124)$$

$$h_{i,2}(t) = SoC_{exp,i}(t-1) \qquad (4.125)$$

Where coefficient $\mu \gg 1$.

4.2.4 COMPARISON OF STRATEGIES

Table 4.1 has investigated the pros and cons of the proposed strategies.

4.2.5 CONCLUSION

In order to increase the profit of parking and improve the reliability of the distribution system, various charging and discharging strategies are presented. Each of these strategies schedules the charging and discharging process of the EVs in the parking lot based on the information that the EV owner provides to the parking lot. The existence of a parking limitation based on the maximum amount of power exchange with the distribution system has led different strategies to use different criteria in order to prioritize charging and discharging EVs in the parking lot in order to increase the parking profit.

4.3 REVIEW AND ANALYSIS OF RESULTS

In this chapter, according to the type of interaction between EVs and parking lot owners, there are different strategies for scheduling the charging and discharging process. In each of the strategies, the charging and discharging priority of EVs is determined by optimizing the $J_{ch,k}(t)$ and $J_{dch,k}(t)$ functions. Each strategy leads to a certain amount of profit and load behavior.

TABLE 4.1
Pros and Cons of the Proposed Strategies

Strategy	Pros	Cons
SoC$_{lim}$-supply-in-t	Unification of SoCs in the charging process Reducing in paid cost for $C_{lim,i}$	Need to calculate the uncompensated paid cost in the previous time steps
SoC$_{lim}$-supply-up-to-t	Acceleration and facility in discharging scheduling	Using approximate equations and simplification hypotheses
EV owners-ch-l-p	Reducing in paid costs for $C_{lim,i}$ and	Need to use complex relationships
EV owners-ch-h-p	$C_{edch,i}$ Increasing in profit for $C_{ech,i}$	Need to access R_{max} and R_{min} values
EV owners-discharge-ch-l-p EV owners-discharge-ch-h-p	Increasing the participation of EVs in the discharging process compared to strategies EV owners-ch-l-p and EV owners-ch-h-p	Decreasing in profit due to a change in the calculation way of $C_{ech,i}$
performance-past-discharge	Discharging process scheduling according to EV behavior during its presence in the parking lot Not being the priority of SoCs of EV as the only determinant of discharging	Need to investigate EV behavior during the time in the parking lot
performance-past	Charging process scheduling according to EV behavior during its presence in the parking lot Not being the priority of SoCs of EV as the only determinant of charging	Need to investigate EV behavior during the time in the parking lot
performance-allotted time	Accurate scheduling and reduction of costs due to knowledge of how long EV is in the parking lot	Reducing receipts due to C_{park} Possibility of violating the information provided by EV owners

Due to the probabilistic nature of EV behavior, the parking profit and load behavior of each strategy are random. The Monte Carlo method is used to obtain the profit and load behavior of each strategy. Figure 4.7 shows the flowchart of calculating the parking profit. In this study, the number of strategies is equal to 9 (S.N = 9). Also, the number of iterations for each strategy is equal to 100 (iterations= 100).

In this chapter, energy price values for a time step of 48 hours were obtained from [19] and according to Table 4.2.

4.3.1 CALCULATION OF PARKING PROFIT

According to the explanations mentioned in Section 4.2.2.3, first, the allowable charging time steps and the allowable discharging time steps must be determined according to the energy price. The parking situation at different time steps is also shown in Table 4.2.

The results of the scheduled charging and discharging strategy can lead to logical analysis when the driving patterns of the EVs are properly modeled. Accordingly, numerical results and equations in [20] are used to model the driving patterns of EVs.

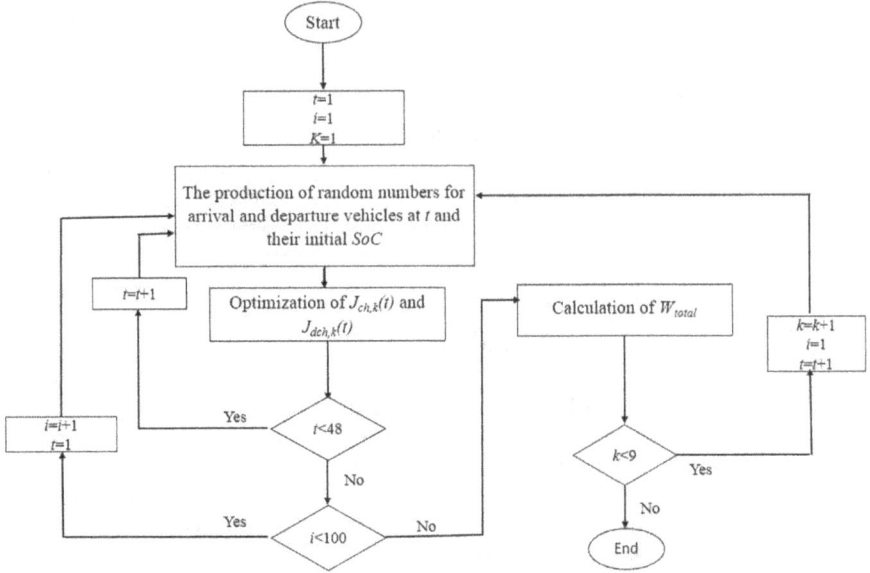

FIGURE 4.7 Flowchart of calculating the parking profit.

TABLE 4.2
Energy Price and Parking State

	Time step			
	1–9 33–25	10–14 34–38	15–19 39–43	20–24 44–48
Energy price (€/kWh)	0.032 0.062	0.073 0.076	0.062 0.066	0.080 0.094
Parking state	charging	discharging	charging	discharging

In this study, a parking lot with a capacity of 100 EVs has been used to investigate the proposed strategies. The characteristics of EVs and under-study parking lots are shown in Table 4.3.

In order to increase the participation of EVs in sending power to the grid and thus improve the reliability of the distribution system, energy purchase and sale tariffs from the EV owner as well as SoC_{lim} not supplied tariffs are set out in Table 4.4.

- Interaction 1: As mentioned in the previous section, in this situation the parking lot is not allowed to charge and discharge EVs. Therefore, the only source of profit for parking is the profit from parking EVs inside the parking lot.
- Interaction 2: In this interaction, the parking lot can charge and discharge the parked EVs in the parking lot. Therefore, in addition to the profit from parking EVs in the parking lot, the parking lot can make a profit by selling energy to the grid, i.e., discharging EVs, as well as selling energy to EV owners, i.e.,

TABLE 4.3
EV and Parking Lot Specifications

Value	Parameter
3.2 (kW)	$p_{ch}(t)=p_{dch}(t)$
32 (kWh)	E_s
24 (kWh)	SoC_{lim}
192 (kW)	$P^{max}_{ch}(t)=P^{max}_{dch}(t)$

TABLE 4.4
Economic Specifications of the Parking Lot

Value	Parameter
0.073 (€/kWh)	R_1
0.03 (€/kWh)	R_2
0.03 (€/kWh)	R_3
70%*0.094 (€/h)	a
20%	b
10%	k

charging EVs. In addition, parking in this situation, according to the functions described in the previous section, is required to pay costs to the grid as well as the owner of the EV. For this reason, various charging and discharging strategies are offered in order to increase parking profits. Strategies "SoC$_{lim}$-supply-in-t" to "performance-past schedule" the charging and discharging process of EVs in the parking lot according to interaction 2.

- Interaction 3: In this interaction, EV owners determine how long EVs will be in the parking lot. As a result, the parking lot will be able to schedule the charging and discharging process according to the EV's departure time. The "performance-allotted time" strategy schedules the process of charging and discharging EVs in the parking lot according to interaction 3.

In the following, the average values of cost functions are evaluated. The average values of cost functions are shown in Table 4.5.

Investigating Table 4.5, it can be seen that the received profit by the parking lot in interaction 1 has the lowest value compared to the others due to the lack of charging and discharging of the EVs in the parking lot. In strategies "EV owners-ch-l-p" and "EV owners-discharge-ch-h-p", due to the reduction of energy sales to the grid, the received profit from the parking lot has decreased. Although the amount of energy sales to the grid has increased in strategies "EV owners-discharge-ch-l-p" and "EV owners-discharge-ch-h-p" compared to strategies "EV owners-ch-l-p" and "EV owners-ch-h-p", changing and reducing the C_{ech} function will reduce the parking profit. In strategy "performance-past-discharge", due to the increase in the amount of energy purchased from the grid, the parking lot was able to make more profit than

TABLE 4.5
Parking Profit Functions

Strategy	Function (€)								
	C_{gdch}	C_{park}	C_{ech}	C_{gch}	C_{edch}	C_{lim}	C_{sh}	C_{pen}	W_{total}
Interaction1	0	209.73	0	0	0	0	0	0	209.73
SoC_{lim}-supply-in-t	131.77	209.73	57.69	111.43	5.50	8.46	4.42	0	269.39
SoC_{lim}-supply-up-to-t	136.28	209.73	56.58	112.82	5.80	9.03	4.54	0	270.40
EV owners-ch-l-p	41.35	209.73	82.13	85.36	0.85	0.15	0.73	0	246.13
EV owners-ch-h-p	93.46	209.73	76.05	103.59	3.83	3.08	3.03	0	265.70
EV owners-discharge-ch-l-p	64.15	209.73	61.47	92.03	2.24	1.41	1.68	0	238
EV owners-discharge-ch-h-p	104.89	209.73	60.51	105.91	4.89	4.40	3.58	0	256.33
performance-past-discharge	138.90	209.73	55.82	114.62	5.46	8.89	4.31	0	271.16
performance-past	138.90	209.73	53.01	113.30	5.24	9.71	4.21	0	269.17
performance-allotted time	119.25	194.75	68.65	115.73	2.59	0	3.13	13.76	275.01

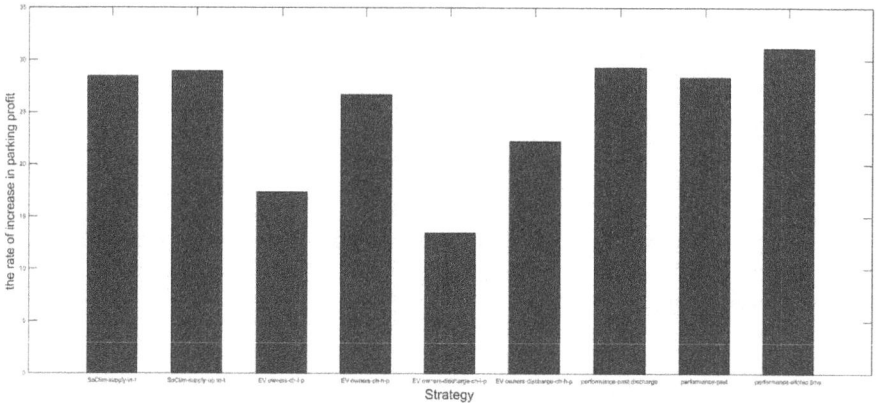

FIGURE 4.8 The rate of increase in parking profits in different strategies compared to interaction 1.

strategy "performance-past", by selling energy to EV owners. In interaction 3, the parking lot can make more detailed schedules according to the duration of the presence of EVs in the parking lot. This has reduced the cost of parking to zero for C_{lim}. If the parking strategy does not require a reduction in parking costs, the parking profit will increase much more than the W_{total} amount. Figure 4.8 shows the rate of increase in parking profits in different strategies compared to interaction 1. Figure 4.9 shows the total SoC increment in each of the strategies.

As can be seen in Figure 4.8, the parking lot was able to increase its profitability in all strategies compared to interaction 1 by changing the amount of SoC of the EVs. The maximum profit increment in interaction 2 is 29%, and in interaction 3 it is 31%. Having enough information about the duration of EVs in the parking lot has improved scheduling and increased profits in interaction 3.

Figure 4.9 shows that scheduling the charging and discharging process, in addition to increasing the parking profit, also increases the total SoC of EVs.

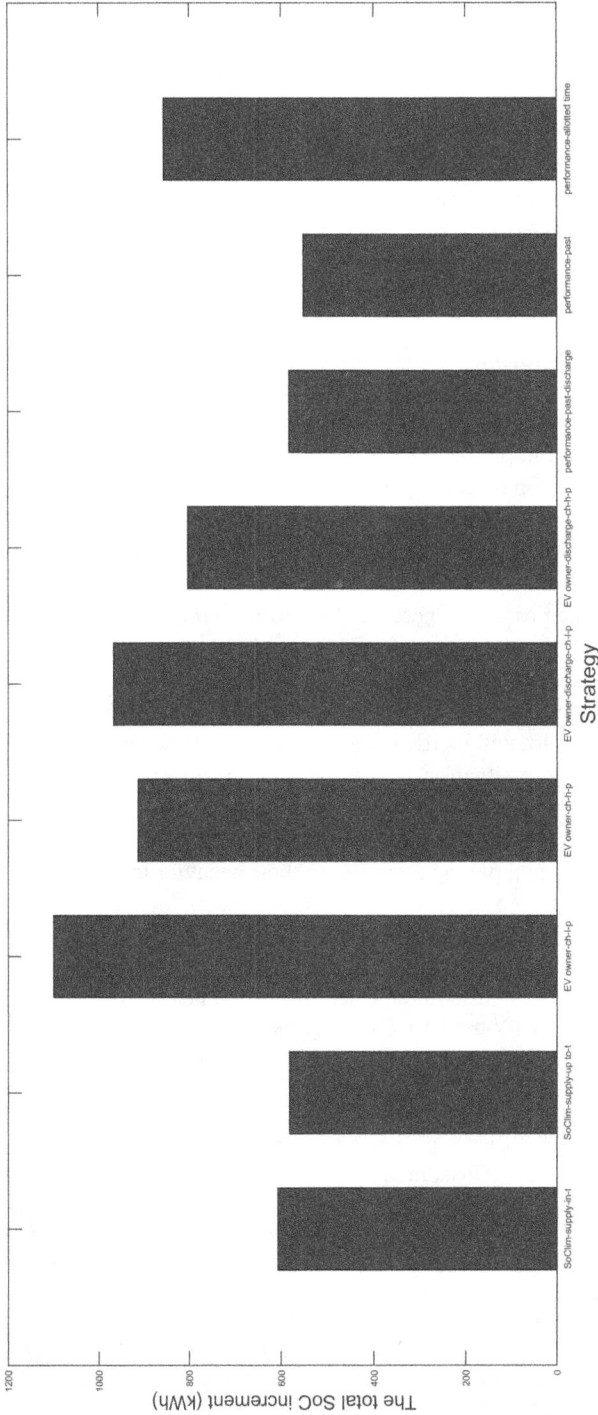

FIGURE 4.9 The total SoC increment in each of the strategies.

4.4 CONCLUSION

Increasing parking profit is a very important step in satisfying the parking lot to participate in the V2G process and improve the reliability of the distribution system. Due to the important point that energy pricing per hour is directly related to the amount of power consumption, scheduling the charging and discharging process based on energy prices also improves the reliability of the system. Accordingly, in this chapter, according to the receipt and payment of parking and its profit function, strategies are introduced to schedule the charging and discharging process. In the second step, according to the energy price, the allowable time steps for charging (time steps when the energy price is lower) and the allowable time steps for discharging are selected.

In designing the strategies, two serious limitations are considered in the scheduling of the charging and discharging process of EVs:

- Stochastic and unpredictable nature of EV entry and exit times and their energy level when entering the parking lot
- Limitation on the amount of power sent from the distribution system to the parking lot and vice versa

The proposed strategies try to increase the parking profit by increasing the parking receipts and reducing the amount of paid costs.

By evaluating the economic benefits of each strategy, it is considered that if there is appropriate information about the duration of EV in the parking lot, the received profit by the parking lot will increase due to more precise scheduling and no payment for C_{lim}. However, in a situation where the parking lot does not have information about the exit time of the EVs, strategy "performance-past-discharge" increases the profit of the parking lot to an acceptable level by unifying the conditions of all EVs in the charging process and paying attention to their behavior in past time steps.

REFERENCES

[1] A. Abdali, K. Mazlumi, and J. M. Guerrero, "Integrated control and protection architecture for islanded PV-battery DC microgrids: Design, analysis and experimental verification," *Appl. Sci. (Basel)*, vol. 10, no. 24, p. 8847, 2020.
[2] A. Abdali, K. Mazlumi, and R. Noroozian, "A precise fault location scheme for low-voltage dc microgrids systems using multi-layer perceptron neural network," *Sigma J. Eng. Nat. Sci*, vol. 36, no. 3, pp. 821–834, 2017.
[3] D. van der Meer, G. R. Chandra Mouli, G. Morales-Espana Mouli, L. R. Elizondo, and P. Bauer, "Energy management system with PV power forecast to optimally charge EVs at the workplace," *IEEE Trans. Industr. Inform.*, vol. 14, no. 1, pp. 311–320, 2018.
[4] F. Jozi, A. Abdali, K. Mazlumi, and S. H. Hosseini, "Reliability improvement of the smart distribution grid incorporating EVs and BESS via optimal charging and discharging process scheduling," *Front. Energy Res.*, vol. 10, 2022.
[5] S. Habib, M. Kamran, and U. Rashid, "Impact analysis of vehicle-to-grid technology and charging strategies of electric vehicles on distribution networks: A review," *J. Power Sources*, vol. 277, pp. 205–214, 2015.

[6] N. Z. Xu, and C. Y. Chung, "Reliability evaluation of distribution systems including vehicle-to-home and vehicle-to-grid," *IEEE Trans. Power Syst.*, vol. 31, no. 1, pp. 759–768, 2016.

[7] A. Alahyari, M. Fotuhi-Firuzabad, and M. Rastegar, "Incorporating customer reliability cost in PEV charge scheduling schemes considering vehicle to home capability," *IEEE Trans. Veh. Technol.*, pp. 1–1, 2014.

[8] C. Sabillon Antunez, J. F. Franco, M. J. Rider, and R. Romero, "A new methodology for the optimal charging coordination of electric vehicles considering vehicle-to-grid technology," *IEEE Trans. Sustain. Energy*, vol. 7, no. 2, pp. 596–607, 2016.

[9] L. Hua, J. Wang, and C. Zhou, "Adaptive electric vehicle charging coordination on distribution network," *IEEE Trans. Smart Grid*, vol. 5, no. 6, pp. 2666–2675, 2014.

[10] E. Akhavan-Rezai, M. F. Shaaban, E. F. El-Saadany, and F. Karray, "New EMS to incorporate smart parking lots into demand response," *IEEE Trans. Smart Grid*, vol. 9, no. 2, pp. 1376–1386, 2018.

[11] T. Sattarpour, and M. Farsadi, "Parking lot allocation with maximum economic benefit in a distribution network: Electric vehicle," *Int. trans. electr. energy syst.*, vol. 27, no. 1, p. e2234, 2017.

[12] E. Hadian, H. Akbari, M. Farzinfar, and S. Saeed, "Optimal allocation of electric vehicle charging stations with adopted smart charging/discharging schedule," *IEEE Access*, vol. 8, pp. 196908–196919, 2020.

[13] Y. He, B. Venkatesh, and L. Guan, "Optimal scheduling for charging and discharging of electric vehicles," *IEEE Trans. Smart Grid*, vol. 3, no. 3, pp. 1095–1105, 2012.

[14] H. N. T. Nguyen, C. Zhang, and M. A. Mahmud, "Optimal coordination of G2V and V2G to support power grids with high penetration of renewable energy," *IEEE Trans. Transp. Electrif.*, vol. 1, no. 2, pp. 188–195, 2015.

[15] M. Mazumder, and S. Debbarma, "EV charging stations with a provision of V2G and voltage support in a distribution network," *IEEE Syst. J.*, vol. 15, no. 1, pp. 662–671, 2021.

[16] M. Sufyan, N. A. Rahim, M. A. Muhammad, C. K. Tan, S. R. S. Raihan, and A. H. A. Bakar, "Charge coordination and battery lifecycle analysis of electric vehicles with V2G implementation," *Electric Power Syst. Res.*, vol. 184, no. 106307, p. 106307, 2020.

[17] M. S. H. Nizami, M. J. Hossain, and K. Mahmud, "A coordinated electric vehicle management system for grid-support services in residential networks," *IEEE Syst. J.*, vol. 15, no. 2, pp. 2066–2077, 2021.

[18] N. Z. Xu, and C. Y. Chung, "Well-being analysis of generating systems considering electric vehicle charging," *IEEE Trans. Power Syst.*, vol. 29, no. 5, pp. 2311–2320, 2014.

[19] L. Igualada, C. Corchero, M. Cruz-Zambrano, and F.-J. Heredia, "Optimal energy management for a residential microgrid including a vehicle-to-grid system," *IEEE Trans. Smart Grid*, vol. 5, no. 4, pp. 2163–2172, 2014.

[20] D. Dallinger, D. Krampe, and M. Wietschel, "Vehicle-to-grid regulation reserves based on a dynamic simulation of mobility behavior," *IEEE Trans. Smart Grid*, vol. 2, no. 2, pp. 302–313, 2011.

5 Reliability Analysis of Radially Connected Distribution Networks in the Presence of Bidirectional Charging

Fatemeh Jozi, Ali Abdali, Kazem Mazlumi, and Seyed Hadi Hosseini

CONTENTS

DOI: 10.1201/9781003293989-5

5.1 INTRODUCTION

The arrival of electric vehicles (EVs) opens a new pathway to improve the reliability of the distribution system by utilizing EVs as a source of power for distribution systems. The transportation sector is moving from conventional fuel/gas vehicles toward EVs to reduce emissions and make the environment clean. The number of EVs is increasing exponentially around the world, from 6,000 in 2010 to 750,000 in 2016, and the light-duty EVs are predicted to reach 150 million by 2030 [1].

EVs equipped with appropriate discharging facilities are the new bidirectional assets of the electricity grid. They exist as non-stationary loads during the charging period (grid-to-vehicles G2V) as well as distributed storage units during the discharging period (vehicle-to-grid [V2G]). Figure 5.1 shows the schematic of V2G.

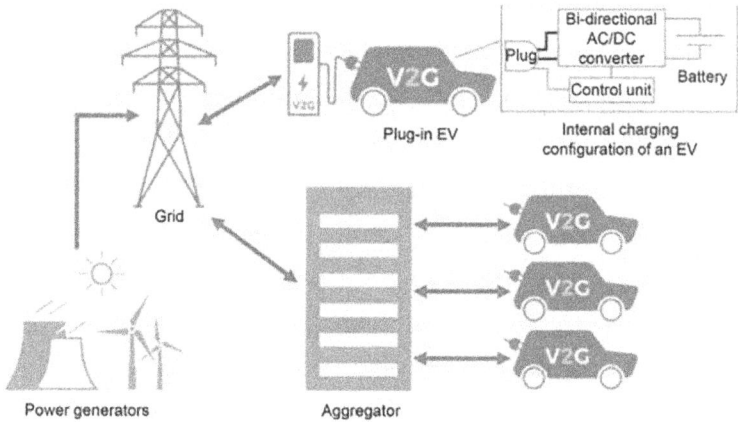

FIGURE 5.1 The schematic of V2G.

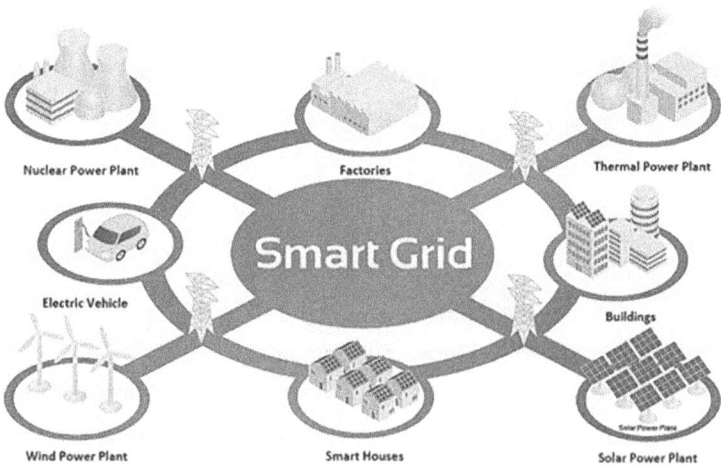

FIGURE 5.2 The energy producer and consumer resources in the smart grid.

It provides several benefits to the drivers, parking lot owners and utilities if they are adequately managed [2–3]. As such, management and control of EV operation modes has become one of the most challenging tasks of the evolving smart grids [4–5]. Figure 5.2 shows the energy producer and consumer resources in the smart grid. Figure 5.3 shows the main features and goal of smart grid [6].

The reliability of a system is the ability of that system to perform the task under certain environmental and operation conditions for a certain period of time.

Typically, reliability issues fall within two particular classes:

- Power quality disturbance and
- Short-term shutdown and long-term shutdown (blackouts).

Also, there are three main issues used to assess reliability in electrical systems [7]:

- The number of components (consumers) influenced
- The recurrence whenever blackouts or voltage disturbances occur and
- The duration of failures.

Correspondingly, the reliability of electrical components is gauged in terms of how often possible reliability issues occur (failures and duration) to determine how many customers or components are influenced. Figure 5.4 shows the area related to reliability performance in a power system [8].

In the case of the distribution system, reliability is related to the power outage of the customers and the disruption of the equipment. In this regard, reliability indicators have been defined to assess the reliability of distribution systems.

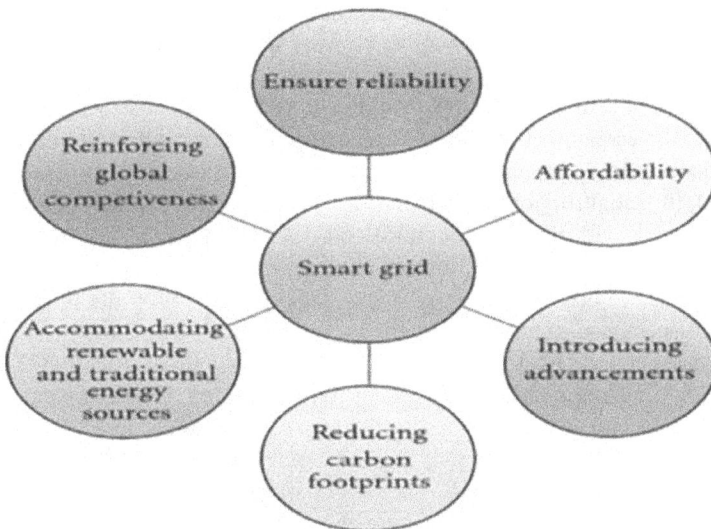

FIGURE 5.3   Smart grid features and goals.

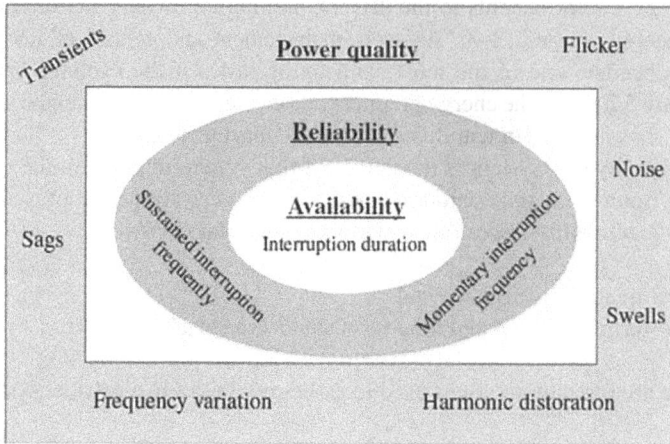

FIGURE 5.4   Reliability area performance in power system.

The EVs can be used as a source of energy for the grid in the form of V2G by appropriate planning and modeling strategies [9–12]. Therefore, to enhance the reliability of the grids by suitable planning and modeling of the EVs, appropriate optimization is required to integrate the EVs and resolve the operational problems of distribution systems [13].

As mentioned, V2G technology improves the reliability of the distribution system, but it should always be borne in mind that parking lots can improve system reliability if charging and discharging strategies are properly designed. On the other hand, parking lots contribute to the transmission of power to the grid where the received profit is appropriate; therefore, in Chapter 4, the charging and discharging strategies are designed to convince the parking lot to contribute to improving the reliability of the distribution system with suitable parking lot profit.

With the advent of V2G technology, EVs can participate in electricity markets. In addition to the economic benefits, V2G technology has enabled EVs to reduce interrupt loads by sending power to the network during an outage.

In [14], the reliability of the distribution system in the presence of EVs is evaluated. For each load point, two centralized and distributed charging topologies are considered. In the event of an islanding situation, residential loads can benefit from local vehicle to home (V2H) or V2G, depending on the type of charging topology. In grid-connected situations, inter-area V2G reduces the amount of energy not supplied (ENS). The results show that taking advantage of V2H and V2G capabilities improves system reliability. In addition, increasing the capacity of the batteries, as well as increasing the allowable power exchange limitations, further improves the reliability of the system. However, in the previous cases, the distribution system can only use V2G capability for one continuous interrupt period, and it is not possible to use this feature in several discrete interrupt periods.

In [15], a comprehensive structure for providing reliability-differentiated services in a residential distribution network has been presented; therefore, customers can benefit from different levels of reliability. A reliability-differentiated pricing mechanism is also proposed to motivate customers to consume power by creating economic incentives to improve the reliability of the power system. The results show that this method can improve the reliability of the global residential distribution system, increase the economic benefits of PHEV and improve the power quality in the residential distribution system. However, in this method, customers are classified based on the amount of load demand and the level of reliability; therefore, if customers have a significant amount of demand compared to others, more energy will be allocated to it through V2G. This will reduce the reliability of other consumers. In addition, this method ignores the unpredicted and random nature of entry and exit times and the amount of energy of plug-in hybrid electric vehicles (PHEVs).

Unforeseen errors can disrupt the performance of the power grid and consumers. In [13], the charging and discharging process scheduling of EVs and the optimal location for the parking lot to increase the reliability of the distribution system are presented. The purpose of this study is to ensure the continuity of the power supply of customers in the definite times and peak hours of consumption, by the energy stored in the batteries of EVs. Numerous factors that have a large impact on the energy capacity of the distribution system and EVs such as charging/discharging rate and time, daily distance and arrival and departure time have been considered for modeling. In this paper, second-order mixed-integer nonlinear programming (MINLP) to maximize the reliability of the distribution system (improving ENS, system average interruption duration index [SAIDI] and system average interruption frequency index [SAIFI]), minimizing losses and the distribution system voltage regulation using PSO optimization is presented. It should always be noted that charging and discharging strategies will be successful in increasing the reliability of the distribution system if they convince parking lot owners by increasing parking profit.

In [16], the reliability of the power system is evaluated by considering the forced outage rates (FORs) of generators, transmission lines and transformers. In this study, a probabilistic method for load modeling and EVs with different penetration levels is presented. PHEV and full electric vehicle (FEV) modeling includes important features such as battery capacity, charging reduction distance, charging rate, recharging time, vehicle entry and exit time, discharging rate and mileage. In addition, various charging strategies, i.e., unscheduled charging and scheduled charging with and without V2G technology, have been used to assess the reliability of the power system. The results show that controlled charging strategies improve reliability and also increase the loss of load probability (LOLP) with increasing load.

In [17], the effects of the EV parking lots on the reliability of the distribution system are studied. Accordingly, it is assumed that the parking lots will be used as a backup provider for the network in cases of non-supply or failure of feeder components. The simulation results show that the reliability of the feeder and load point is improved by 26% and 44%. In addition, the economic benefits of this support can reach up to $1,100 per month depending on pricing conditions. In this study, EVs are used on average 5–7 times a year as backup resources for the network, and it is not possible to use EVs daily and continuously as a power supply unit. In addition, the desire of EV owners to provide support services to the distribution system is not taken into account.

This chapter has investigated the reliability (SAIFI, SAIDI, ASAI (average service availability index) and ENS) of a radial distribution network in the presence of EVs. In addition, the improvement of the reliability of the distribution system and the economic benefits of V2G are examined simultaneously.

5.1.1 DISTRIBUTION SYSTEM PERFORMANCE INDICES [18]

Before introducing reliability indicators, it is necessary to mention the three basic parameters that are of special importance in the study of the reliability of distribution systems.

These three basic parameters are:

- λ: Failure rate [failure/year]
- r: Duration of interruption [hour]
- U: Annual unavailability or duration of interruption [hour/year]

In the following, the reliability indicators of the distribution system are examined.

5.1.1.1 SAIFI: System Average Interruption Frequency Index (Interruption/Customer/Year)

$$\text{SAIFI} = \frac{\text{Total number of customers interruptions}}{\text{Total number of customers}} = \frac{\sum_{z \in R} \lambda_z N_z}{\sum_{z \in R} N_z} \qquad (5.1)$$

The SAIFI is one of the most important indicators of customer-based system reliability assessment. This indicator shows that a customer has experienced an average of several outages during the reporting period. The larger the value of this index, the weaker the distribution system in terms of reliability. A high SAIFI rate directly increases customer dissatisfaction.

5.1.1.2 SAIDI: System Average Interruption Duration Index (Hour/Customer/Year)

$$\text{SAIDI} = \frac{\text{Total duration of customers interruptions}}{\text{Total number of customers}} = \frac{\sum_{z \in R} U_z N_z}{\sum_{z \in R} N_z} \qquad (5.2)$$

This index indicates the average length of time that each customer does not have access to electrical energy, which is not related to transient outages. With this index, it is possible to find out how long each customer spends off time, on average, in a certain period. The lower the value of this index, the higher the reliability of the distribution system in terms of interruption time per customer.

5.1.1.3 ASAI: Average System Availability Index

$$ASAI = \frac{\text{Total annual hours of customers access to electricity}}{\text{Total hours of the year for all customers}}$$

$$= \frac{\sum_{z \in R} 8760 N_z - \sum_{z \in R} U_z N_z}{\sum_{z \in R} 8760 N_z} \tag{5.3}$$

During the year, it is ideal for each customer to have 8,760 hours of electricity, but due to technical and non-technical reasons and the reliability of the networks under service, part of these hours are spent. The ASAI index shows reliability in terms of interruption duration. The closer this index is to 1, the higher the reliability of the distribution system.

5.1.1.4 ENS: Energy Not Supplied (kWh/Year)

$$ENS = \sum_{z \in R} P_z \cdot U_z \tag{5.4}$$

One of the most important indicators based on load and energy is ENS. By obtaining this index, the amount of energy that has not been delivered to the customers can be determined.

In the previous equations:

1. λ_z: Failure rate at busbar z
2. U_z: Annual unavailability or duration of interruption at busbar z
3. N_z: Number of customers at busbar z
4. P_z: The average power consumption at busbar z

In Section 5.2, the reliability indices of the distribution system in charging mode and discharging mode are presented. In Section 5.3, the reliability indices of the distribution system in the strategies of Chapter 4 are studied. Finally, Section 5.4 provides the conclusions.

5.2 RELIABILITY EVALUATION

As mentioned, the use of V2G technology improves the reliability of the distribution system. It is worth mentioning that improving the reliability of the distribution system depends on how the EVs are charged and discharged. Therefore, in each of the strategies, it is necessary to evaluate the reliability of the distribution system in which the parking lot is located so that in addition to increasing the economic benefits, the reliability of the distribution system is also improved. Accordingly, first, the charging and discharging strategies discussed in Chapter 4 are reviewed.

5.2.1 Proposed Charging and Discharging Process Scheduling of EVs

In Chapter 4, nine strategies are proposed for scheduling the charging and discharging process of EVs in a V2G-equipped parking lot, aiming at maximizing the profit of parking. In Chapter 4, in order to schedule the charging and discharging process in different strategies, the functions $J_{ch}(t)$ and $J_{dch}(t)$ are used, respectively. Furthermore, how to schedule and prioritize the charging and discharging of EVs in the parking lot in each of the strategies is described in detail in Section 4.2.3 of Chapter 4.

TABLE 5.1

Strategy Based on SoC$_{lim}$ Supply in Time Step t (SoC$_{lim}$-Supply-in-t)

Charging strategy

$$\text{MAX } J_{ch}(t) = \sum_{i=1}^{Nev(t)} n_i(t) \cdot F_i(t)$$

$$F_i(t) = E_s - SoC_i(t-1)$$

Discharging strategy

$$\text{MAX } J_{dch}(t) = \sum_{i=1}^{Nev(t)} n_i(t) \cdot G_i(t)$$

$$G_i(t) = C_{g,i}(t) - C_{lim,i}(t)$$

$$C_{g,i}(t) = p_{dch}(t) \cdot price_{dch,t}$$

$$C_{lim,i}(t) = s\left(SoC_{lim,i} - \left(SoC_i(t-1) - p_{dch}(t)\right)\right) \cdot k_i(t)$$

Calculation of k$_i$(t)

Conditions	Equations

$$SoC_i(t-1) \geq SoC_{in,i} \qquad k_i(t) = \left(SoC_{lim,i} - \left(SoC_i(t-1) - p_{dch}(t)\right)\right) \cdot R_3$$

$$k_i(t) = \left(SoC_{lim,i} - \left(SoC_i(t-1) - p_{dch}(t)\right)\right) \cdot R_3 - C_{n-s,i}(t)$$

$$SoC_i(t-1) \geq SoC_{lim,i}$$

$$SoC_i(t-1) < SoC_{in,i}$$

$$C_{n-s,i}(t) = g\left(\sum_{t_i=t_{m_j,1}}^{t-1} C_{1,i}(t_i) - \left(p_{ch}(t_i) \cdot R_3\right)\right)$$

$$+ \dots g\left(\sum_{t_i=t_{m_k,1}}^{t-1} C_{1,i}(t_i) - \left(p_{ch}(t_i) \cdot R_3\right)\right)$$

$$+ \dots g\left(\sum_{t_i=t_{m_k,1}}^{t-1} C_{1,i}(t_i) - \left(p_{ch}(t_i) \cdot R_3\right)\right)$$

TABLE 5.2

Strategy Based on SoC$_{lim}$ Supply up to Time Step t (SoC$_{lim}$-Supply-up to-t)

Charging strategy	Discharging strategy
MAX $J_{ch}(t) = \sum_{i=1}^{Nev(t)} n_i(t) \cdot F_i(t)$	MAX $J_{dch}(t) = \sum_{i=1}^{Nev(t)} n_i(t) \cdot G_i(t)$
$F_i(t) = E_s - SoC_i(t-1)$	$G_i(t) = C_{g,i}(t) - C_{lim,i}(t)$

$$C_{g,i}(t) \begin{cases} (t - t_{m_1,1}) \cdot P_{dch}(t) \cdot price_{dch,t} & t \in m_1 \\ M & M \\ (t - t_{m_k,1}) \cdot P_{dch}(t) \cdot price_{dch,t} & t \in m_k \\ M & M \\ (t - t_{m_m,1}) \cdot P_{dch}(t) \cdot price_{dch,t} & t \in m_m \end{cases}$$

$$C_{lim,i}(t) = g\left(SoC_{lim,i} - \left(SoC_i(t-1) - P_{dch}(t)\right)\right) \cdot R_3$$

Accordingly, Tables 5.1–5.8 summarize how to schedule charge and discharge in each of the strategies.

5.2.2 CALCULATION OF RELIABILITY INDICES

In this chapter, the reliability of the radial distribution system introduced in [18] is evaluated under different strategies, and it is assumed that the parking lot is located in busbar C. Figure 5.5 shows the studied radial distribution system.

Considering the change in the amount of load consumed as well as the power produced by the parking lot during each hour, reliability indices should be calculated per hour.

5.2.2.1 Charging Mode

If the parking lot is in charging mode, the reliability indices are calculated as follows:

$$SAIFI(t) = \sum_{N=0}^{M_{ch}} (P_p(t)) \cdot \frac{\lambda_A N_A + \lambda_B N_B + \lambda_C N_C + N(t)}{N_T + N(t)} \tag{5.5}$$

$$SAIDI(t) = \sum_{N=0}^{M_{ch}} (P_p(t)) \cdot \frac{U_A N_A + U_B N_B + U_C N_C + N(t)}{N_T + N(t)} \tag{5.6}$$

TABLE 5.3

Strategy Based on Reducing the Total Cost Paid to EV Owners Assuming the Possibility of Charging on Time with the Lowest/Highest Energy Price (EV Owners-ch-l/h-p)

Charging strategy	Discharging strategy
$\text{MAX } J_{ch}(t) = \sum_{i=1}^{Nev(t)} n_i(t) \cdot F_i(t)$	$\text{MAX } J_{dch}(t) = \sum_{i=1}^{Nev(t)} n_i(t) \cdot G_i(t)$
$F_i(t) = E_s - SoC_i(t-1)$	$G_i(t) = C_i(t)$

Calculation of $C_i(t)$

Conditions			Equations
$SoC_{d,i}(t)^l \geq SoC_{in,i}$	$SoC_{d,i}(t) \geq SoC_{lim,i}$		$C_{i,1}(t) = p_{dch}(t) \cdot R_1$
	$SoC_{d,i}(t) < SoC_{lim,i}$	$SoC_i(t-1) \geq SoC_{lim,i}$	$C_{i,2}(t) = p_{dch}(t) \cdot R_1 + (SoC_{lim,i} - SoC_{d,i}(t)) \cdot R_3$
		$SoC_i(t-1) < SoC_{lim,i}$	$C_{i,3}(t) = p_{dch}(t) \cdot (R_1 + R_3)$
$SoC_{d,i}(t) < SoC_{in,i}$	$SoC_{c,i}(t)^2 > E_s$		$C_{i,4}(t) = p_{dch}(t) \cdot R_2$
	$SoC_{c,i}(t) \leq E_s$	$SoC_{d,i}(t) \geq SoC_{lim,i}$	$C_{i,5}(t) = p_{dch}(t) \cdot R_2 + p_{dch}(t) \cdot (R_1 - R_{price})$
		$SoC_{d,i}(t) < SoC_{lim,i}$, $SoC_i(t-1) \geq SoC_{lim,i}$	$C_{i,6}(t) = p_{dch}(t) \cdot (R_2 + (R_1 - R_{min})) + (SoC_{lim,i} - SoC_{d,i}(t)) \cdot R_3$
		$SoC_i(t-1) < SoC_{lim,i}$	$C_{i,7}(t) = p_{dch}(t) \cdot (R_2 + (R_1 - R_{min}) + R_3)$

Calculation of R_{price}

Strategy "EV owners-ch-l-p"	$R_{price} = R_{min}$
Strategy "EV owners-ch-h-p"	$R_{price} = R_{max}$

$$ASAI(t) = \sum_{N=0}^{M_{ch}} (P_P(t)) \cdot \frac{T \cdot (N_T + N(t)) - (U_A N_A + U_B N_B + U_C (N_C + N(t)))}{T \cdot (N_T + N(t))} \quad (5.7)$$

$$ENS(t) = \sum_{P=0}^{P_{ch}^{max}} (P_P(t)) \cdot (P_A \cdot U_A + P_B \cdot U_B + (P_C + P(t)) \cdot U_C) \quad (5.8)$$

TABLE 5.4
Strategy Based on Reducing the Total Costs Paid to EV Owners and Increasing the Amount of Participation in the Discharge Process, Assuming the Possibility of Charging on Time with the Lowest/Highest Energy Price (EV Owners-Discharge-ch-l/h-p)

Charging strategy	Discharging strategy
$\text{MAX } J_{ch}(t) = \sum_{i=1}^{Nev(t)} n_i(t) \cdot F_i(t)$	$\text{MAX } J_{dch}(t) = \sum_{i=1}^{Nev(t)} n_i(t) \cdot G_i(t)$
$F_i(t) = E_s - SoC_i(t-1)$	$G_i(t) = C_i(t)$

Calculation of C_i(t)

Conditions				Equations
		$SoC_{d,i}(t) \geq SoC_{lim,i}$		$C_{i,1}(t) = 0$
$SoC_{d,i}(t)$ $\geq SoC_{in,i}$	$SoC_{d,i}(t) < SoC_{lim,i}$	$SoC_i(t-1)$ $\geq SoC_{lim,i}$		$C_{i,2}(t) = \left(SoC_{lim,i} - SoC_{d,i}(t) \right) \cdot (R_3 + R_1)$
		$SoC_i(t-1)$ $< SoC_{lim,i}$		$C_{i,3}(t) = p_{dch}(t) \cdot (R_1 + R_3)$
$SoC_{d,i}(t)$ $< SoC_{in,i}$	$SoC_{e,i}(t)$ $\leq E_s$	$SoC_{d,i}(t)$ $< SoC_{lim,i}$	$SoC_{e,i}(t)$ $\geq SoC_{lim,i}$	$C_{i,4}(t) = p_{dch}(t) \cdot R_2$
				$C_{i,5}(t) = p_{dch}(t) \cdot R_2$
			$SoC_i(t-1) \geq SoC_{lim,i}$	$C_{i,6}(t) = p_{dch}(t) \cdot R_2$ $+ \left(SoC_{lim,i} - SoC_{d,i}(t) \right) \cdot R_3$
			$SoC_i(t-1)$ $< SoC_{lim,i}$	$C_{i,7}(t) = p_{dch}(t) \cdot (R_2 + R_3)$ $+ \left(SoC_{lim,i} - SoC_i(t-1) \right) \cdot$ $\left(R_1 - R_{price} \right)$
			$SoC_{e,i}(t)$ $< SoC_{lim,i}$	$C_{i,8}(t) = p_{dch}(t) \cdot \left(R_2 + \left(R_1 - R_{min} \right) + R_3 \right)$

The conditions row: $SoC_{e,i}(t) > E_s$ gives $C_{i,4}(t) = p_{dch}(t) \cdot R_2$ and $SoC_{d,i}(t) \geq SoC_{lim,i}$ gives $C_{i,5}(t) = p_{dch}(t) \cdot R_2$.

Calculation of R_price

Strategy EV owners-discharge-ch-l-p	$R_{price} = R_{min}$
Strategy EV owners-discharge-ch-h-p	$R_{price} = R_{max}$

In equations (5.5) to (5.8), $N(t)$ is the number of EVs that are charged by the parking in time step t. Also, $P(t)$ is the parking power consumption and $Pp(t)$ is the probability of occurrence of any of the conditions. According to reference [18], T is equivalent to 8,760 hours.

TABLE 5.5

Strategy Based on Performance of EVs in the Past to Schedule the Discharging Process (Performance-Past-Discharge)

Charging strategy	Discharging strategy
$\text{MAX } J_{ch}(t) = \sum_{i=1}^{Nev(t)} n_i(t) \cdot F_i(t)$	$\text{MAX } J_{dch}(t) = \sum_{i=1}^{Nev(t)} n_i(t) \cdot G_i(t)$
$F_i(t) = E_s - SoC_i(t-1)$	

Calculation of $G_i(t)$

Conditions	Equations
$SoC_{max,i}(t) \geq SoC_{lim,i}$	$SoC_{d,i}(t) \geq SoC_{lim,i} \quad G_i(t) = E_s \cdot R_3 - \left(0 - \alpha\left(SoC_i(t-1)\right)\right)$
	$SoC_{d,i}(t) < SoC_{lim,i} \quad G_i(t) = E_s \cdot R_3 - \left(\left(SoC_{lim,i} - SoC_{d,i}(t)\right) \cdot R_3 - \alpha\left(SoC_i(t-1)\right)\right)$
$SoC_{max,i}(t) < SoC_{lim,i}$	$G_i(t) = E_s \cdot R_3 - \left(\left(SoC_{max,i}(t) - SoC_{d,i}(t)\right) \cdot R_3 \right.$ $\left. -\alpha\left(SoC_i(t-1)\right)\right)$

TABLE 5.6

Strategy Based on Performance of EVs in the Past (Performance-Past)

Charging strategy	Discharging strategy
$\text{MAX } J_{ch}(t) = \sum_{i=1}^{Nev(t)} n_i(t) \cdot F_i(t)$	Strategy "performance-past-discharge"

Calculation of $F_i(t)$

Conditions	Equations
$SoC_{max,i}(t) \geq SoC_{lim,i}$	$F_i(t) = E_s - \left(\left(SoC_i(t-1) - SoC_{lim}\right) + \alpha\left(SoC_i(t-1)\right)\right)$
$SoC_{max,i}(t) < SoC_{lim,i}$	$F_i(t) = E_s - \left(\left(SoC_i(t-1) - SoC_{max,i}(t)\right) + \alpha\left(SoC_i(t-1)\right)\right)$

TABLE 5.7

Charging Strategy Based on the Performance of EVs in the Allotted Time for Their Presence in the Parking Lot (Performance-Allotted Time)

$$\text{MAX } J_{ch}(t) = \sum_{i=1}^{Nev(t)} n_i(t) \cdot F_i(t)$$

Conditions	Equations
$(n_{ch,i} - 1)p_{ch}(t) + SoC_i(t-1) \geq SoC_{lim,i}$	$F_i(t) = E_{max} - SoC_{exp,i}(t-1)$
$(n_{ch,i} - 1)p_{ch}(t) + SoC_i(t-1) < SoC_{lim,i}$	$F_i(t) = E_{max} - \beta \cdot SoC_{exp,i}(t-1)$

TABLE 5.8
Discharging Strategy Based on the Performance of EVs in the Allotted Time for Their Presence in the Parking Lot (Performance-Allotted Time)

$$\text{MAX} \quad J_{dch}(t) = \sum_{i=1}^{Nev(t)} n_i(t) \cdot G_i(t)$$

Conditions	Equations
$n_{ch,i}P_{ch}(t) - P_{dch}(t) + SoC_i(t-1) \geq SoC_{lim,i}$	$SoC_i(t-1) - n_{dch,i}P_{dch}(t) \geq SoC_{lim} \quad G_i(t) = \mu \cdot SoC_{exp,i}(t-1)$ $SoC_i(t-1) - n_{dch,i}P_{dch}(t) < SoC_{lim} \quad G_i(t) = SoC_{exp,i}(t-1)$

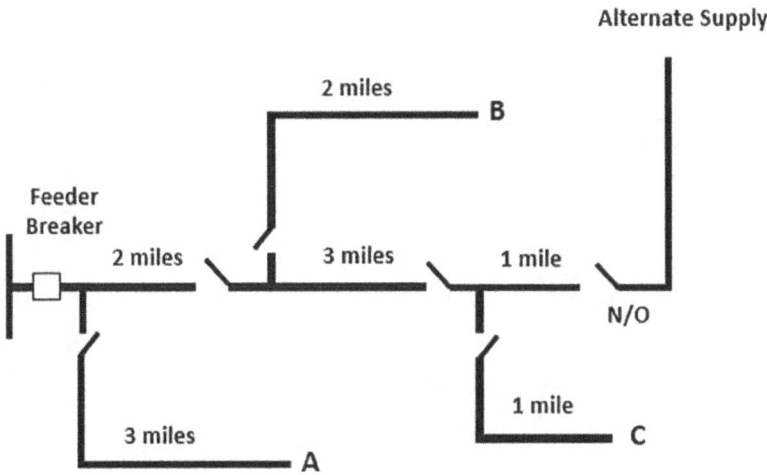

FIGURE 5.5   The studied radial distribution system [28].

5.2.2.2 Discharging Mode

When the parking lot acts as a generator, if the generating power exceeds the load on the C busbar, this excess can be transferred to the B busbar via the CB line. In this case, the reliability indices are calculated as follows:

$$\text{SAIFI}(t) = \sum_{P=0}^{P_{dch}^{max}} (P_P(t)) \cdot \text{SAIFI}(t, P(t)) \tag{5.9}$$

$$\text{SAIFI}(t, P(t)) = \begin{array}{ll} \text{SAIFI}_1(t, P(t)) & P(t) \quad P_C(t) \\ \text{SAIFI}_2(t, P(t)) & P(t) > P_C(t) \end{array} \tag{5.10}$$

$$\text{SAIFI}_1(t, P(t)) = \frac{\lambda_A N_A + \lambda_B N_B + \lambda_C (N_C - M_C(t))}{N_T} \tag{5.11}$$

$$\text{SAIFI}_2\left(t, P(t)\right) = P_{\text{CB-R}} \cdot \frac{\lambda_A N_A + \lambda_B \left(N_B - M_B(t)\right)}{N_T} + P_{\text{CB-U}} \cdot \frac{\lambda_A N_A + \lambda_B N_B}{N_T} \quad (5.12)$$

$$p_C(t) = \frac{P_C(t)}{N_C} \quad (5.13)$$

$$M_C(t) = \frac{P(t)}{p_C(t)} \quad (5.14)$$

$$p_B(t) = \frac{P_B(t)}{N_B} \quad (5.15)$$

$$M_B(t) = \frac{P(t) \quad P_C(t)}{p_B(t)} \quad (5.16)$$

In the equations (5.9) to (5.16), $p_C(t)$ is the average power consumption of each C busbar customer, and $p_B(t)$ is the average power consumption of each B busbar customer. $M_C(t)$ and $M_B(t)$ also represent the number of busbar C and B customers, respectively, fed by the parking lot.

$$\text{SAIDI}(t) = \sum_{P=0}^{P_{\max}} \left(P_P(t)\right) \cdot \text{SAIDI}\left(t, P(t)\right) \quad (5.17)$$

$$\text{SAIDI}\left(t, P(t)\right) = \begin{array}{ll} \text{SAIDI}_1\left(t, P(t)\right) & P(t) \quad P_C(t) \\ \text{SAIDI}_2\left(t, P(t)\right) & P(t) > P_C(t) \end{array} \quad (5.18)$$

$$\text{SAIDI}_1\left(t, P(t)\right) = \frac{U_A N_A + U_B N_B + U_C (N_C - M_C(t))}{N_T} \quad (5.19)$$

$$\text{SAIDI}_2\left(t, P(t)\right) = P_{\text{CB-R}} \cdot \frac{U \cdot N_A + U_B \left(N_B - M_B(t)\right)}{N_T}$$
$$+ P_{\text{CB-U}} \cdot \frac{U_A N_A + U_B N_B}{N_T} \quad (5.20)$$

$$\text{ASAI}(t) = \sum_{P=0}^{P_{ch}} \left(P_P(t)\right) \cdot \text{ASAI}\left(t, P(t)\right) \quad (5.21)$$

$$\text{ASAI}\left(t, P(t)\right) = \begin{array}{ll} \text{ASAI}_1\left(t, P(t)\right) & P(t) \quad P_C(t) \\ \text{ASAI}_2\left(t, P(t)\right) & P(t) > P_C(t) \end{array} \quad (5.22)$$

$$\text{ASAI}_1\left(t, P(t)\right) = \frac{T \cdot N_T - \left(U_A N_A + U_B N_B + U_C \left(N_C - M_C(t)\right)\right)}{T \cdot N_T} \quad (5.23)$$

$$ASAI_2\left(t, P(t)\right) = P_{CB-R} \cdot \frac{T \cdot N_T - \left(U_A \cdot N_A + U_B \left(N_B - M_B(t)\right)\right)}{T \cdot N_T}$$

$$+ P_{CB-U} \cdot \frac{T \cdot N_T - \left(U_A \cdot N_A + U_B N_B\right)}{T \cdot N_T} \tag{5.24}$$

$$ENS(t) = \sum_{P=0}^{P_{ch}} \left(P_P(t)\right) \cdot ENS\left(t, P(t)\right) \tag{5.25}$$

$$ENS\left(t, P(t)\right) = \begin{array}{ll} ENS_1\left(t, P(t)\right) & P(t) \quad P_C(t) \\ ENS_2\left(t, P(t)\right) & P(t) > P_C(t) \end{array} \tag{5.26}$$

$$ENS_1\left(t, P(t)\right) = U_A P_A + U_B P_B + U_C (P_C \quad P(t)) \tag{5.27}$$

$$ENS_2\left(t, P(t)\right) = P_{CB-R} \cdot \left(U \cdot P_A + U_B \left(P_B - P_{BC}(t)\right)\right)$$

$$+ P_{CB-U} \cdot \left(U_A \cdot P_A + U_B P_B\right) \tag{5.28}$$

5.3 REVIEW AND ANALYSIS OF RESULT

Charging time steps are included in lower energy prices and discharging time steps in higher energy prices so that the parking lot can purchase energy at low prices and sell it to the EV owner, as well as store energy and sell it to the grid during hours when the price of energy is higher to make a profit. On the other hand, the price of energy at different times of the day and night is directly related to the amount of load per hour on the distribution system. Therefore, the proposed scheduling will also improve the reliability of the distribution system because in the under-loading hours, the parking lot will act as a load in the distribution system. On the other hand, when the load on the distribution system increases and the grid is unable to supply the load, the parking lot will act as a power generator. As a result, not only is the load resulting from EV charging not imposed on the grid but also the power generation of parking reduces the load interruption of the distribution system.

The parking lot is not able to charge or discharge all the EVs in the parking lot every hour due to the limitation of power exchange with the distribution grid. Therefore, in order to schedule the charging and discharging process to increase the parking profit, it is necessary to determine which EVs should participate in the charging and discharging process at each time step. On the other hand, the parking lot is not able to predict when EVs will enter and leave the parking lot, as well as the amount of EV battery energy when entering the parking lot. As a result, in the previous chapter, various strategies are proposed to prioritize charging and discharging EVs.

5.3.1 CALCULATION OF DISTRIBUTION SYSTEM RELIABILITY

In the Chapter 4, the charging and discharging priority of EVs are determined by optimizing the $J_{ch,k}(t)$ and $J_{dch,k}(t)$ functions. Each strategy leads to a certain load

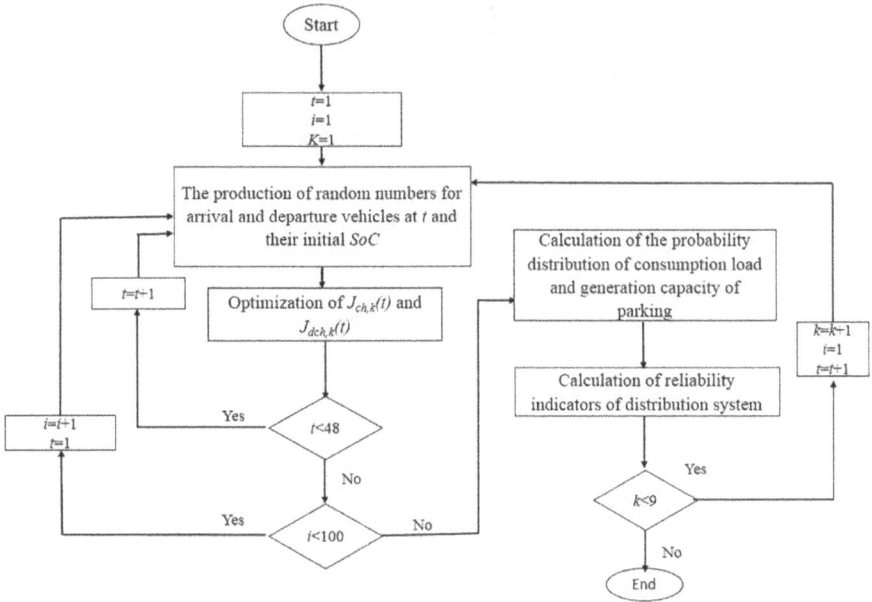

FIGURE 5.6 Flowchart of calculating the reliability of the distribution system.

TABLE 5.9
Parking State

	Time step			
	1–9 33–25	10–14 34–38	15–19 39–43	20–24 44–48
Parking state	charging	discharging	charging	discharging

behavior. Therefore, the load probabilistic distribution and parking power generation in different strategies will be different. These probabilistic distribution diagrams are used to evaluate the reliability of the distribution system.

Due to the probabilistic nature of EV behavior, the load behavior of each strategy is random. The Monte Carlo method is used to obtain the load behavior of each strategy. Figure 5.6 shows the flowchart of calculating the reliability of the distribution system. In this chapter, the number of strategies is equal to 9 (S.N = 9). Also, the number of iterations for each strategy is equal to 100 (iterations = 100).
The parking situation at different time steps is shown in Table 5.9.

In this chapter, the load duration curve (LDC) introduced in [19] is also used to determine the peak load per hour. Peak of power consumption of the studied distribution system per hour is shown in Table 5.10.

TABLE 5.10
Peak of Power Consumption of the Studied Distribution System per Hour

Time step	Peak power consumption (kWh)
1	630
2	620
3	600
4	580
5	590
6	650
7	720
8	850
9	950
10	990
11	1,000
12	990
13	930
14	920
15	900
16	880
17	900
18	920
19	960
20	980
21	960
22	900
23	800
24	700

5.3.1.1 Probability Distribution of Consumption Load and Power Generation of Parking

Charging and discharging process scheduling is different in different strategies. As a result, first, in order to investigate the reliability indices in each of the strategies, it should be noted that the load probabilistic distribution of parking in charging time intervals and the power generation probabilistic distribution of parking in discharging time steps will be different in each strategy. Figures 5.7–5.15 show the probability distribution of parking consumption load in the period 17:00–17:59 in different strategies.

Figures 5.16–5.24 show the probability distribution of parking generation power in the period 19:00–19:59 in different strategies.

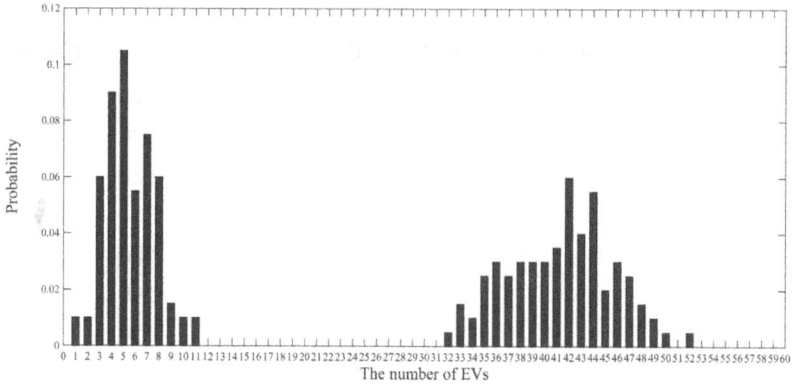

FIGURE 5.7 Probability distribution of parking consumption load (in 17:00–17:59) in strategy "SoC$_{lim}$-supply-in-t".

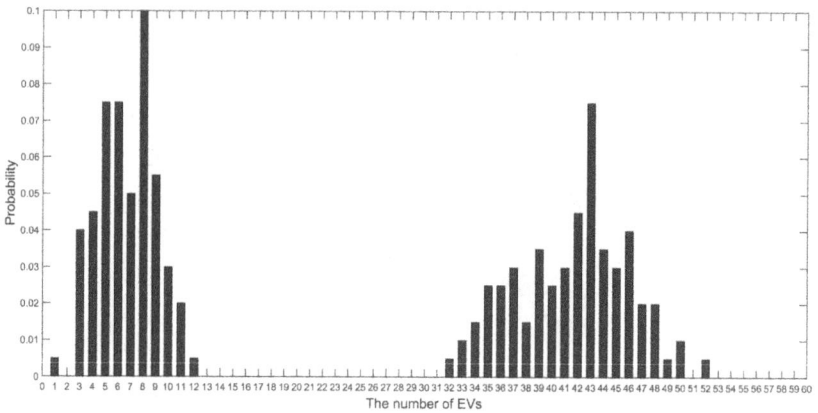

FIGURE 5.8 Probability distribution of parking consumption load (in 17:00–17:59) in strategy "SoC$_{lim}$-supply-up to-t".

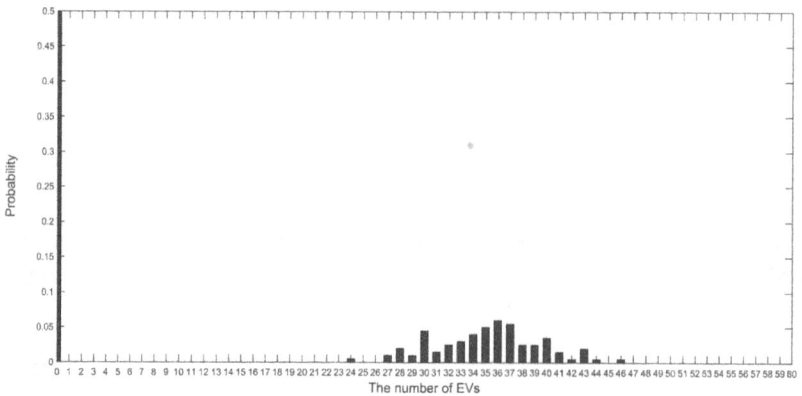

FIGURE 5.9 Probability distribution of parking consumption load (in 17:00–17:59) in strategy "EV owners-ch-l-p".

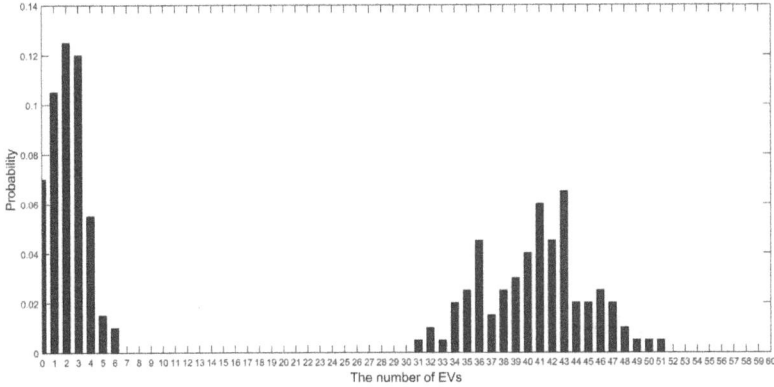

FIGURE 5.10 Probability distribution of parking consumption load (in 17:00–17:59) in strategy "EV owners-ch-h-p".

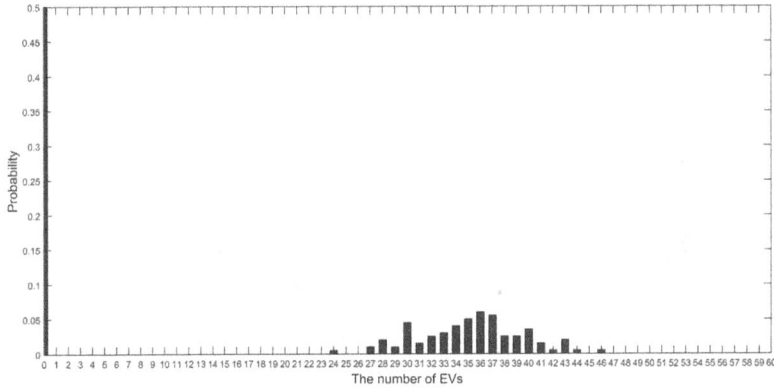

FIGURE 5.11 Probability distribution of parking consumption load (in 17:00–17:59) in strategy "EV owners-discharge-ch-l-p".

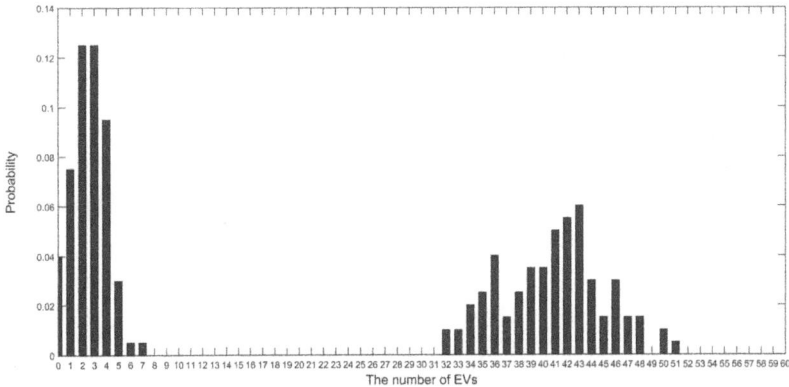

FIGURE 5.12 Probability distribution of parking consumption load (in 17:00–17:59) in strategy "EV owners-discharge-ch-h-p".

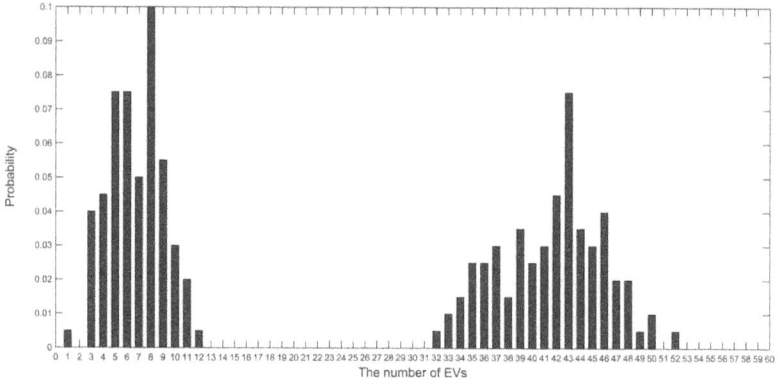

FIGURE 5.13 Probability distribution of parking consumption load (in 17:00–17:59) in strategy "performance-past-discharge".

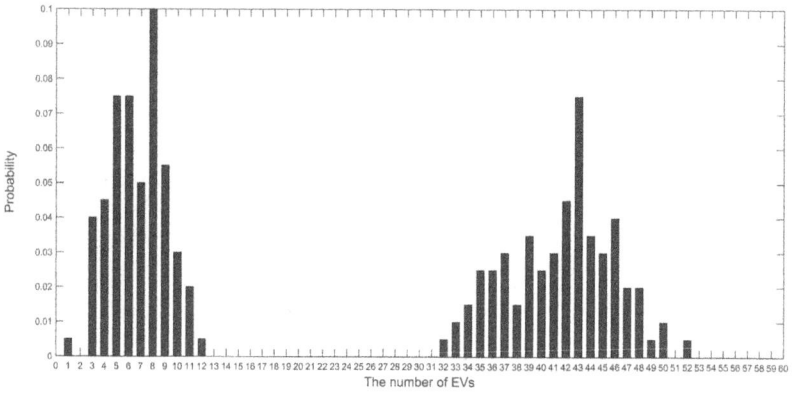

FIGURE 5.14 Probability distribution of parking consumption load (in 17:00–17:59) in strategy "performance-past".

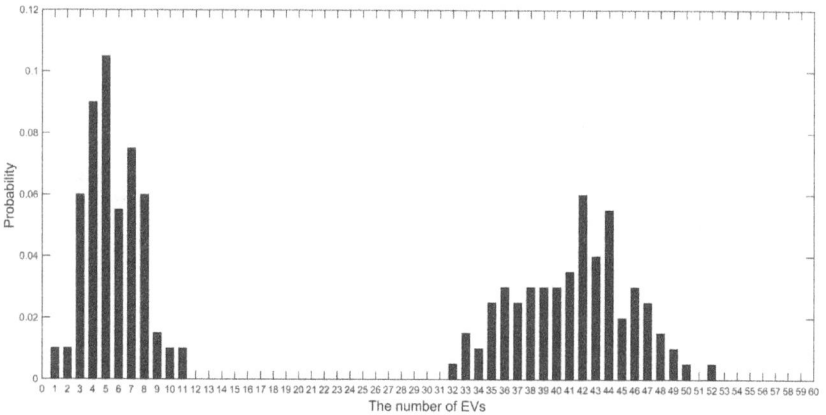

FIGURE 5.15 Probability distribution of parking consumption load (in 17:00–17:59) in strategy "performance-allotted time".

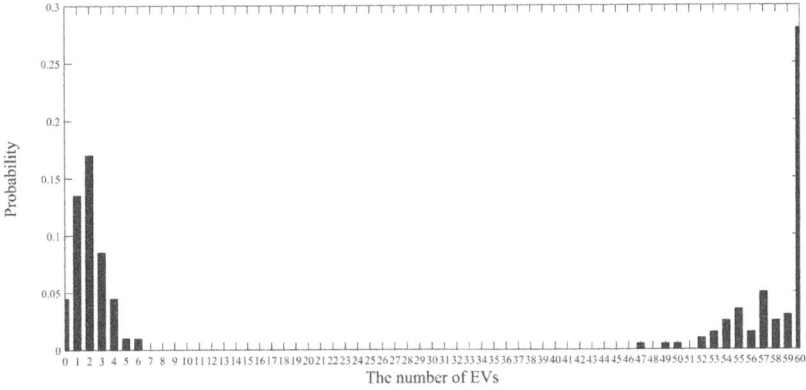

FIGURE 5.16 Probability distribution of parking generation power (in 19:00–19:59) in strategy "SoC_{lim}-supply-in-t".

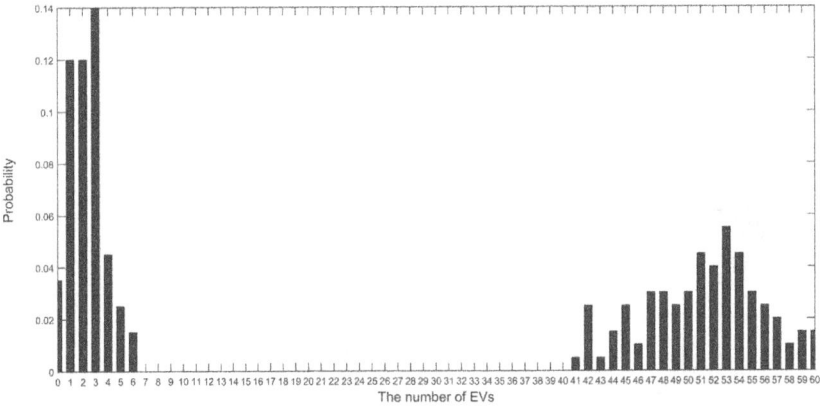

FIGURE 5.17 Probability distribution of parking generation power (in 19:00–19:59) in strategy "SoC_{lim}-supply-up to-t".

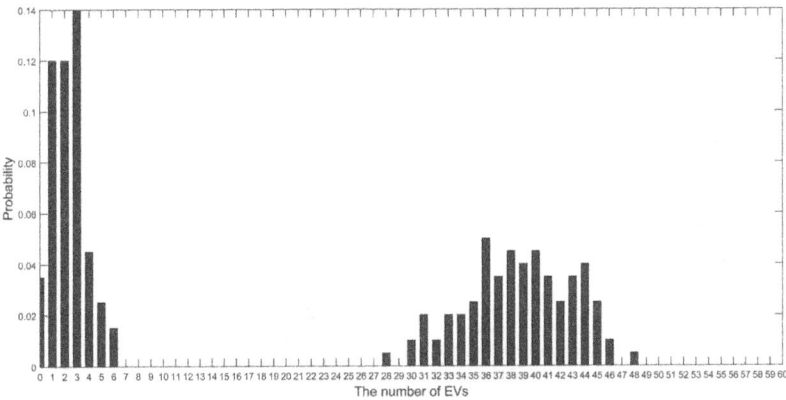

FIGURE 5.18 Probability distribution of parking generation power (in 19:00–19:59) in strategy "EV owners-ch-l-p".

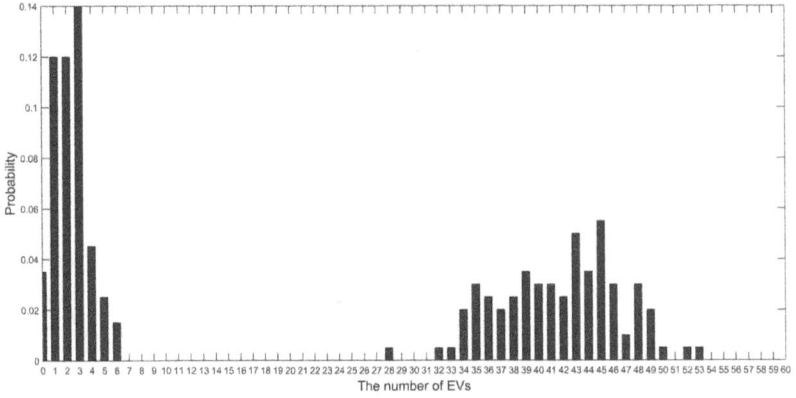

FIGURE 5.19 Probability distribution of parking generation power (in 19:00–19:59) in strategy "EV owners-ch-h-p".

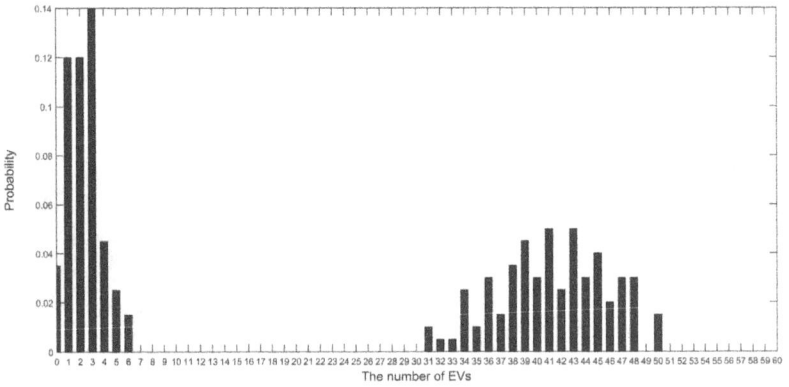

FIGURE 5.20 Probability distribution of parking generation power (in 19:00–19:59) in strategy "EV owners-discharge-ch-l-p".

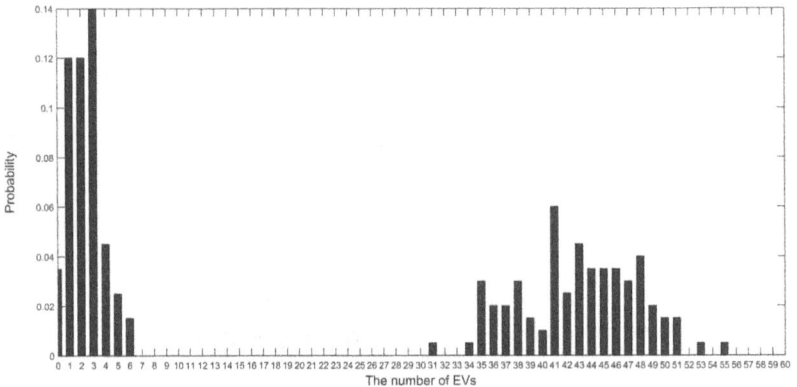

FIGURE 5.21 Probability distribution of parking generation power (in 19:00–19:59) in strategy "EV owners-discharge-ch-h-p".

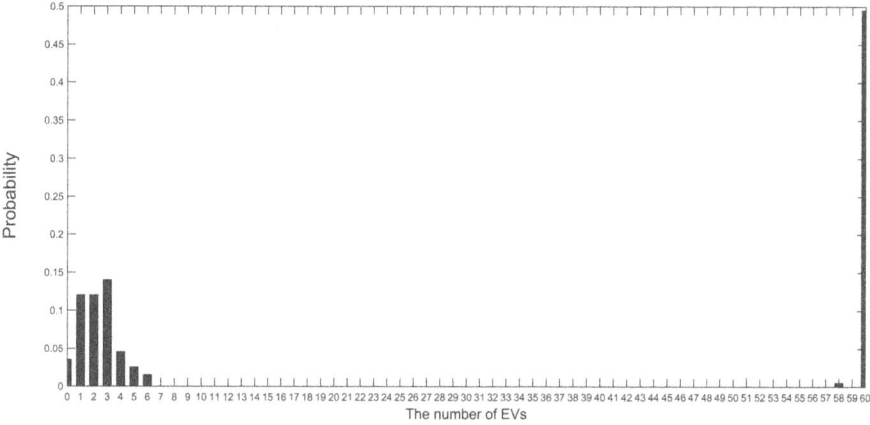

FIGURE 5.22 Probability distribution of parking generation power (in 19:00–19:59) in strategy "performance-past-discharge".

5.3.1.2 Calculation of Distribution System Reliability Indices

Calculation reliability indices—*ENS, ASAI, SAIFI* and *SAIDI*—are very necessary in order to investigate the impact of each strategy on the reliability of the system, in cases where parking is not equipped with V2G technology and in addition, the process of charging the EVs in the parking lot is performed in a random and unplanned way. Figure 5.25 shows the mean values of ENS in charging conditions without scheduling (the random) and scheduled charging and discharging.

As seen in Figure 5.25, the availability of V2G capabilities during high-load hours will reduce the amount of energy not supplied in the system.

Figure 5.26 shows the average ASAI values in charging conditions without scheduling and scheduled charging and discharging.

As observed in Figure 5.26, having a V2G capability and scheduling the charge-discharge process will increase the probability of system availability in all scenarios relative to random and unscheduled mode; because with the generation power of the parking lot, a number of customers are fed on the C and B buses every hour and as a result, they will not be interrupted by the system interruption.

Figure 5.27 compares the amount of received energy from the grid, the amount of energy sent to the grid and the amount of SoC increment in each of the strategies.

Figures 5.28–5.31 show the values of the SAIFI, SAIDI, ASAI and ENS indices in different interactions.

According to Figures 5.27 and 5.28, it should be noted that when the parking lot is in power-generation condition, as the amount of energy sent to the grid increases, the number of not supplied customers decreases and the SAIFI index improves. Accordingly, strategy "performance-past-discharge" has improved the SAIFI index more than other strategies.

According to Figures 5.27, 5.29 and 5.30, if the parking lot can compensate more for its consumption load, the SAIDI and ASAI indices will improve. Accordingly, in strategy EV owners-ch-l-p, because the amount of energy allocated to increasing

FIGURE 5.23   Probability distribution of parking generation power (in 19:00–19:59) in strategy "performance-past".

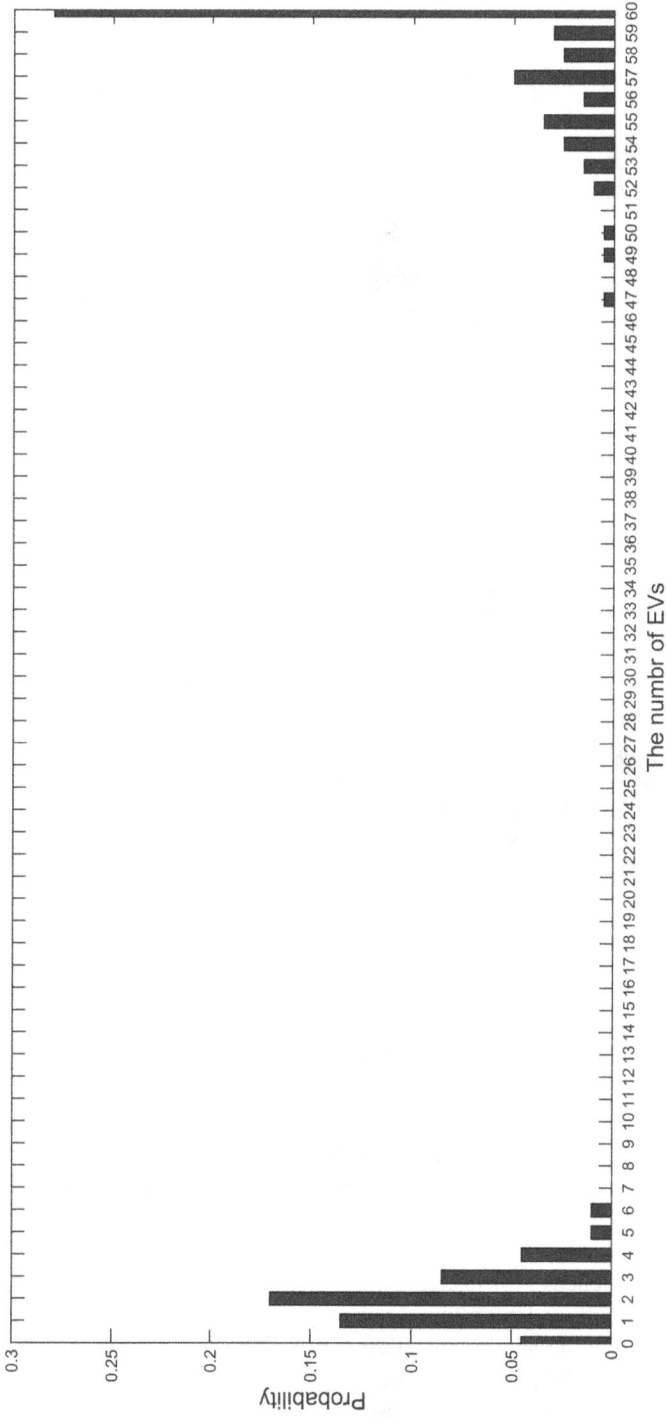

FIGURE 5.24 Probability distribution of parking generation power (in 19:00–19:59) in strategy "performance-allotted time".

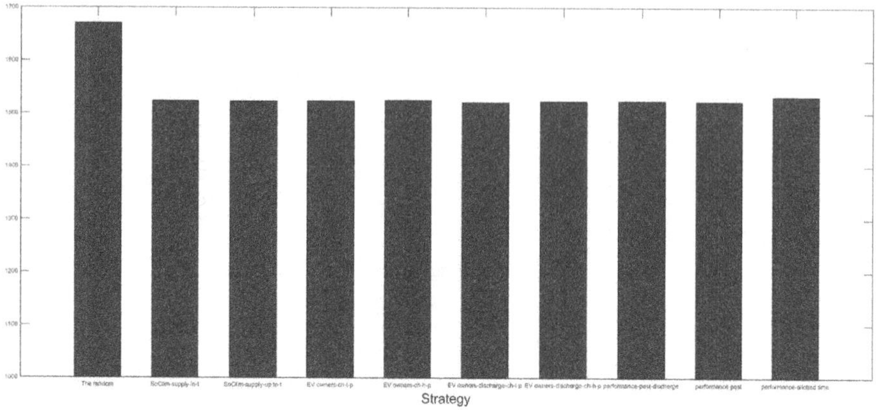

FIGURE 5.25 ENS index.

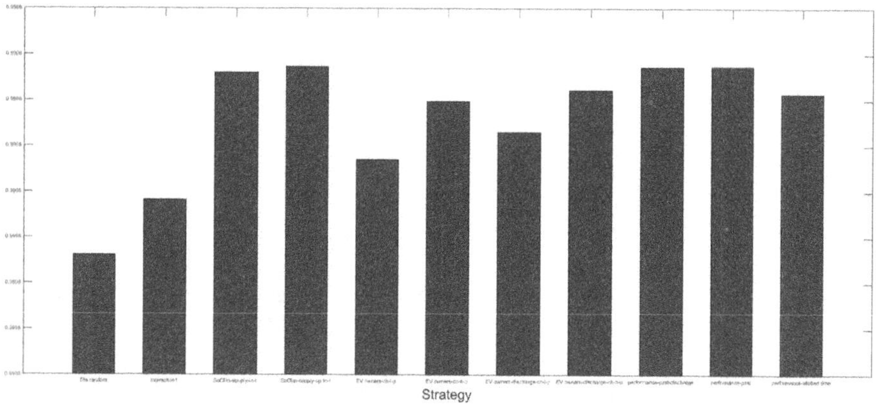

FIGURE 5.26 ASAI index.

the SoC is more than other strategies, the rate of improvement of SAIDI and ASAI in this strategy is less than other strategies.

According to Figures 5.27 and 5.31, the ENS index increases with increasing energy consumption and non-compensation. Therefore, because in the "performance-allotted time" strategy, the total energy received from the grid and the energy allocated to increasing the SoC (energy not supplied to the grid) is more than other strategies, ENS has been significantly increased compared to other strategies.

With the comprehensive investigation of all strategies, it is determined that in the situation that the parking lot is allowed to charge and discharge the EVs present in the parking lot because the number of interrupted customers decreases during the discharge time steps, SAIFI, SAIDI and ASAI indices compared to interaction 1 are also improved. But the presence of a new customer increases the ENS index compared to

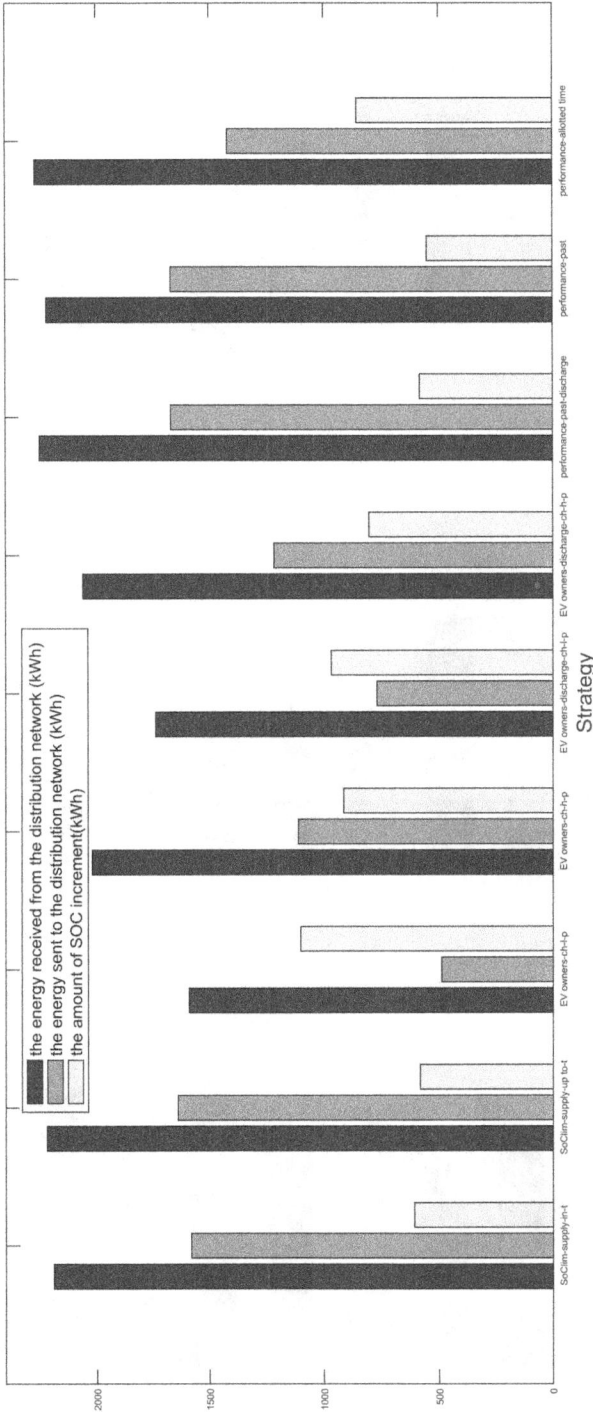

FIGURE 5.27 Comparison of the amount of received energy from the grid, the amount of energy sent to the grid and the amount of SoC increment in each of the strategies.

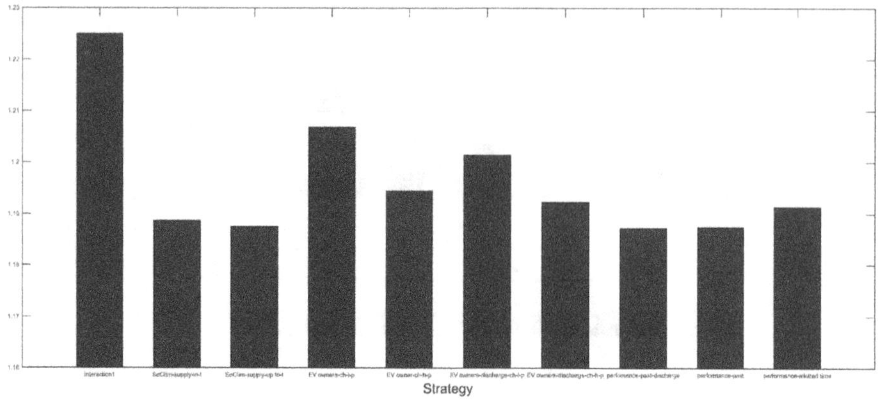

FIGURE 5.28   Values of the SAIFI index.

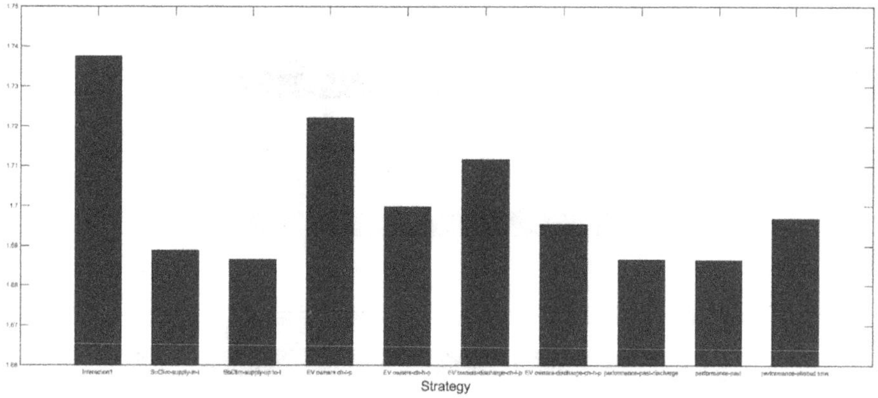

FIGURE 5.29   Values of the SAIDI index.

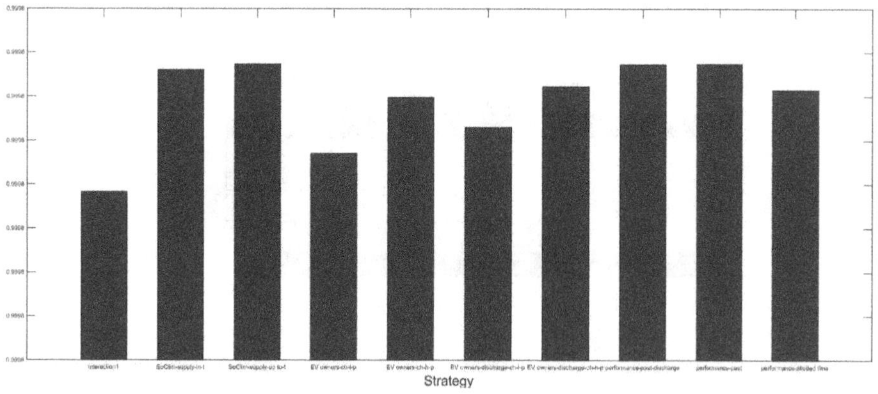

FIGURE 5.30   Values of the ASAI index.

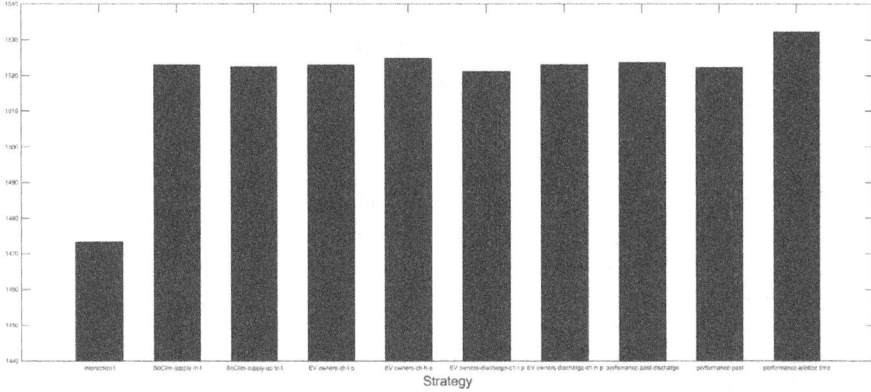

FIGURE 5.31 Values of the ENS index.

interaction 1. Due to the reduction of parking power generation in the strategies "EV owners-ch-l-p" and "EV owners-discharge-ch-l-p", the amount of improvement of reliability indices will also decrease. On the other hand, in the "performance-allotted time" strategy, increasing energy purchases from the grid and reducing the amount of exchange with the grid causes ENS to increase compared to other strategies.

5.4 CONCLUSION

Increasing the parking profit is a very important step in satisfying the parking lot to participate in the V2G process and improve the reliability of the distribution system. Due to the important point that energy pricing per hour is directly related to the amount of power consumption, scheduling the charging and discharging process based on energy prices also improves the reliability of the system.

The results show that the strategies "SoC$_{lim}$-supply-in-t", "SoC$_{lim}$-supply-up to-t", "performance-past" and "performance-past-discharge" improve the SAIFI index more than other strategies because they increase the energy exchange between the parking lot and the grid. Also, the strategies "SoC$_{lim}$-supply-in-t", "performance-past-discharge", "SoC$_{lim}$-supply-up to-t" and "performance-past" will improve SAIDI and ASAI because the parking lot sends more energy than the benefit perceived from the grid through V2G toward the grid. The improvement rate of the ENS index in strategy "performance-allotted time" is less because the total energy received from the grid and the energy delivered to the owners is more than other strategies. Given this, the "performance-past-discharge" strategy, because it enhances the contribution of EVs in the sale of energy to the grid, can improve reliability indicators more than other strategies in addition to the parking lot profit.

The results of each of the strategies show that taking advantage of V2G technology and also scheduling the process of charging and discharging EVs in the parking lot improves the reliability indices of the distribution network. In addition, the strategies that generate profits through the sale of energy to the grid can simultaneously improve parking revenue and the reliability of the distribution network.

It is essential to take advantage of the strategies proposed in order to improve the reliability of the distribution network because it increases the parking owner's profit and encourages the owner of the parking lot to participate in the electricity market.

REFERENCES

[1] A. Abdali, K. Mazlumi, and J. M. Guerrero, "Integrated control and protection architecture for islanded PV-battery DC microgrids: Design, analysis and experimental verification," *Appl. Sci. (Basel)*, vol. 10, no. 24, p. 8847, 2020.

[2] C A. Abdali, K. Mazlumi, and R. Noroozian, "A precise fault location scheme for low-voltage dc microgrids systems using multi-layer perceptron neural network," *Sigma J. Eng. Nat. Sci*, vol. 36, no. 3, pp. 821–834, 2017.

[3] D. Wang, J. Coignard, T. Zeng, C. Zhang, and S. Saxena, "Quantifying electric vehicle battery degradation from driving vs. vehicle-to-grid services," *J. Power Sources*, vol. 332, pp. 193–203, 2016.

[4] B. Knopf, P. Nahmmacher, and E. Schmid, "The European renewable energy target for 2030: An impact assessment of the electricity sector," *Energy Policy*, vol. 85, pp. 50–60, 2015.

[5] N. I. Nimalsiri, E. L. Ratnam, C. P. Mediwaththe, D. B. Smith, and S. K. Halgamuge, "Coordinated charging and discharging control of electric vehicles to manage supply voltages in distribution networks: Assessing the customer benefit," *Appl. Energy*, vol. 291, no. 116857, p. 116857, 2021.

[6] S. Habib, M. Kamran, and U. Rashid, "Impact analysis of vehicle-to-grid technology and charging strategies of electric vehicles on distribution networks: A review," *J. Power Sources*, vol. 277, pp. 205–214, 2015.

[7] R. Allan, "Power system reliability assessment: A conceptual and historical review." Reliability Engineering & System Safety, vol. 46, no. 1, pp. 3–13, 1994 Jan 1.

[8] F. Jozi, A. Abdali, K. Mazlumi, and S. H. Hosseini, "Reliability improvement of the smart distribution grid incorporating EVs and BESS via optimal charging and discharging process scheduling," *Front. Energy Res*, vol. 10, 2022.

[9] D. P. Tuttle, and R. Baldick, "The evolution of plug-in electric vehicle-grid interactions," *IEEE Trans. Smart Grid*, vol. 3, no. 1, pp. 500–505, 2012.

[10] W. Yao, J. Zhao, F. Wen, Y. Xue, and G. Ledwich, "A hierarchical decomposition approach for coordinated dispatch of plug-in electric vehicles," *IEEE Trans. Power Syst.*, vol. 28, no. 3, pp. 2768–2778, 2013.

[11] Z. Liu, F. Wen, and G. Ledwich, "Optimal planning of electric-vehicle charging stations in distribution systems," *IEEE Trans. Power Deliv.*, vol. 28, no. 1, pp. 102–110, 2013.

[12] H. Huang, C. Y. Chung, K. W. Chan, and H. Chen, "Quasi-Monte Carlo based probabilistic small signal stability analysis for power systems with plug-in electric vehicle and wind power integration," *IEEE Trans. Power Syst.*, vol. 28, no. 3, pp. 3335–3343, 2013.

[13] R. Khan, *et al.*, "An optimization-based reliability enhancement scheme for active distribution systems utilizing electric vehicles," *IEEE Access*, vol. 9, pp. 157247–157258, 2021.

[14] N. Z. Xu, C. Y. Chung, and S. Member, "Reliability evaluation of distribution systems including vehicle-to-home and vehicle-to-grid," *IEEE Trans. Power Syst.*, pp. 1–10, 2015.

[15] J. Tan, and L. Wang, "Enabling reliability-differentiated service in residential distribution networks with PHEVs: A hierarchical game approach," *IEEE Trans. Smart Grid*, vol. 7, no. 2, pp. 684–694, 2016.

[16] S. Shafiq, U. B. Irshad, M. Al-Muhaini, S. Z. Djokic, and U. Akram, "Reliability evaluation of composite power systems: Evaluating the impact of full and plug-in hybrid electric vehicles," *IEEE Access*, vol. 8, pp. 114305–114314, 2020.

[17] S. Guner, and A. Ozdemir, "Reliability improvement of distribution system considering EV parking lots," *Electric Power Syst. Res.*, vol. 185, no. 106353, p. 106353, 2020.

[18] Billinton, and W. Li, Eds., *Reliability assessment of electric power systems using Monte Carlo methods.* New York, NY: Springer, 2014.

[19] C. Barrows *et al.*, "The IEEE reliability test system: A proposed 2019 update," *IEEE Trans. Power Syst.*, vol. 35, no. 1, pp. 119–127, 2020.

NOMENCLATURE

Parameter	Descriptions	Parameter	Descriptions
$ASAI$	Average System Availability Index	$SoC_{lim,i}$	Minimum capacity specified for SoC_i
$C_{g,i}(t)$	The parking receipt from the grid for sending power via EV_i in time step t	$SoC_{max,i}(t)$	The maximum possible value for SoC_i in the time step t
$C_{lim,i}(t)$	The parking cost paid to the EV_i owner for not providing $SoC_{lim,i}$ at time step t.	$SoC_{out,i}$	SoC when EVi go out of parking
E_s	Battery capacity	t	Time step
$F_{q,i}(t)$	The main function to prioritize the EV_i charge in $J_{ch,q}(t)$	$t_{mk,1}$	Beginning of the m_k time step
$g(x)$, $s(x)$ and $y(x)$	Control filters	$t_{mk,y}$	End of the m_k time step
$G_{q,i}(t)$	The main function to prioritize the EV_i discharge in $J_{dch,q}(t)$	$t_{nj,1}$	Beginning of the n_j time step
$J_{ch,q}(t)$	The function to schedule the EVs charging process in strategy number q at time step t	$t_{nj,x}$	End of the n_j time step
$J_{dch,q}(t)$	The function to schedule the EVs discharging process in strategy number q at time step t	U_z	Annual availability or duration of interruption at busbar z
m	The number of discrete time steps for the discharging process	$n_i(t)$	The binary variable indicates the presence or absence of EVi in the charging/discharging process at time t
m_z	The number of customers at busbar z	λ_z	Failure rate at busbar z
n	The number of discrete time steps for the charging process	ch	Charge
$n_{ch,i}$	The number of authorized charging steps during the time the EVi will be in the parking lot	dch	Discharge
$n_{dch,i}$	The number of authorized discharging steps during the time the EVi will be in the parking lot	i	EV number

(*Continued*)

(*Continued*)

Parameter	Descriptions	Parameter	Descriptions
N_T	Total number of distribution system customers	j	Allowable charging discrete time step number
$P(t)$	The parking power at time step t	k	Allowable discharging discrete time step number
$p_{ch}(t)$	The actual power that the parking lot receives from the grid for charging EV_i at time t	mcs	Minimum capacity specified
$p_{dch}(t)$	The actual power that the parking lot sends to the grid from the battery discharging point of the EV_i at time t	m_k	k^{th} allowable discharging discrete time step
$P_p(t)$	The probability of occurrence of any of the conditions	mpv	Maximum possible value
$price_{ch,t}$	The purchase price of energy from the grid at time t.	n_j	j^{th} allowable charging discrete time step
$price_{dch,t}$	The selling price of energy to grid at time t	z	Busbar name
p_z	The average power consumption of each Z busbar customer	*EV owners-ch-h-p*	Based on reducing the total cost paid to EV owners assuming the possibility of charging on time with the highest energy price
P_z	The average power consumption at busbar Z	*EV owners-ch-l-p*	Based on reducing the total cost paid to EV owners assuming the possibility of charging on time with the lowest energy price
R_1	The energy sales tariff to the EV owner	*EV owners-discharge-ch-h-p*	Based on reducing the total costs paid to EV owners and increasing the amount of participation in the discharge process, assuming the possibility of charging on time with the highest energy price
R_2	The energy purchase tariff from the EV owner	*EV owners-discharge-ch-l-p*	Based on reducing the total costs paid to EV owners and increasing the amount of participation in the discharge process, assuming the possibility of charging on time with the lowest energy price
R_{ch}	Allowable time steps for charging	*performance-allotted time*	Based on the performance of EVs in the allotted time for their presence in the parking lot

Parameter	Descriptions	Parameter	Descriptions
R_{dch}	Allowable time steps for discharging	*performance-past*	Based on the performance of EVs in the past
SoC	State of charge	*performance-past-discharge*	Strategy based on the performance of EVs in the past to schedule the discharging process
$SoC_{exp,i}$	Expected SoC for EVi considering its duration in the parking lot	$SoC_{lim.}$ *supply-in—t*	Based on SoC_{lim} supply in time step t
SoC_i	SoC of EV_i	$SoC_{lim.}$ *supply-up to—t*	Based on SoC_{lim} supply up to time step t
$SoC_{in,i}$	Initial SoC_i at the entrance to the parking lot		

6 Optimal Electric Vehicle Charging Station Placement with Online Charging Navigation Strategy in Urban Areas

Chen Liu, Hui Song, and Xinghuo Yu

CONTENTS

6.1 INTRODUCTION

Electric vehicles (EVs) are a new important component in the ground transportation system and play a key role in people's daily lives. The rapid development of EVs relieves the oil shortage burden and air and noise pollution. At the same time the EV charging station (EVCS) brings several new features compared to gas stations, such as being operated by electricity rather than gasoline; however, EV charging takes much longer than gasoline car refueling. With the significant growth of the EV population, EVCS placement problem is important for a wide range of scientific and industrial processes. Also, with the consideration of various EV driving

DOI: 10.1201/9781003293989-6

decision entirely at their discretion. Without online information, they are not aware of the decisions of other drivers and thus the charging queues of each EV charger (EVC) currently and/or in the near future. They may incline to select nearer and cheaper EVCs. This may introduce long waiting queues in some EVCs while others are unused. In order to satisfy EV drivers' charging demands with a minimized queuing time, an EV charging navigation framework should be developed.

In recent years, many works studying the EVCS sitting and sizing optimization problem have been published with the consideration of investment cost [1], operation cost [2], voltage deviation [3], traffic flow [4], etc. In [5], the binary lighting search algorithm is applied for optimal EVCS placement with the minimized power transportation loss cost, EVCS construction cost, and substation energy loss cost. In [6], the optimal EVCS deployment problem is considered with the distribution of generations and shunt capacitors. A two-stage framework is proposed, and the distribution of generations and shunt capacitors are optimized at the first stage. Then, the grasshopper optimization algorithm is proposed to distribute EVCSs based on the first-stage decision. The annual time opportunity cost, traveling cost, construction cost, and operating cost are modeled in a multi-objective function for the optimization framework to solve the EVCS placement optimization problem in [7], where the genetic algorithm is applied to obtain the solution. The EVCS distribution problem is widely modeled as a mixed-integer nonlinear problem [8, 9] to minimize both EVCS and power operation cost. However, the recent works consider the cost on the EVCS operator side only, and the EV users' charging activities and cost should be modeled as well. EV user charging behavior can be optimized based on the current EVCS locations and charging capacities. Namely, the EV users make decisions following EVCS operators' decisions. Thus, the optimization problem for EVCS operators and EV users can be described with a bilevel model.

Another popular category of EV-related study is the EV charging navigation problem. A number of works in the literature have been proposed to address this topic with the consideration of route selection [10–12], the benefit of the transportation system [13], power grid operation [14], etc. [15–17]. In [18], an intelligent transport system is proposed to navigate EV to find a charger at peak load hours. The proposed system includes four components: a power system control center, an ITS center, charging stations, and EV terminals that consider both demands and privacies of EV drivers. In [19], a dynamic transit services system is modeled into a multi-objective optimization model to distribute idle charging to the EV, which has the charging demand. Time-of-use (TOU) price is a major factor in the EV charging navigation problem [13, 19, 20], such as in [13], where the hierarchical game theory is applied to model the problem as a non-cooperative game. As mentioned in the introduction section, EV charging activity takes a couple of hours, and the queuing time brings disappointed user experience to EV users. Hence, the queuing time should be modeled in the navigation system. Also, the increasing EV and EVCS populations lead to higher charging demand. The charging navigation system may not be the optimal decision for EV users. A series of the online algorithms for processing the EV charging request should be developed.

The remainder of this chapter is organized as follows. Section 6.2 introduces the bilevel programming for the EVCS deployment problem. The online EV charging navigation strategy is presented and discussed in Section 6.3. The empirical study reports are presented in Section 6.4. Finally, the paper is concluded in Section 6.5.

6.2 BILEVEL PROGRAMMING FOR EVCS PLACEMENT

In the optimal EVCS placement problem, on the EV charging facility investor side, they would achieve the maximal profit by providing a number of charging services. However, on the EV user side, they would select an EV charger based on the most convenient. Thus, the EVCS investor should consider the distribution of EV charging demands and the features of the charging activity. Generally, the EV users will select an EVC with minimized charging discomfort, which includes the distance to the EVC, the charging and queuing time, and the charging cost.

In order to optimize the profit of EVCS investors and the charging discomfort of EV users, a bilevel optimization model is applied in Eq. (6.1) through (6.4), where A is the set of all candidate locations for EVCS installation, which includes commercial building car parks and parking lots on roadsides. J is the set of EVs with charging demand at time slot t.

$$
\text{UP}: \max_{x,y} \phi(x,y) = \sum_{i \in A} \left\{ \sum_{t \in T} \left[\pi_i \left(x_i - \left| x_i - \sum_{j \in J} y_{itj} \right|_+ \right) \right. \right.
$$

$$
\left. \left. - \left(\rho \left| \sum_{j \in J} y_{itj} - x_i \right|_+ \right) \right] - \omega_i x_i \right\}
\tag{6.1}
$$

$$
\text{s.t. } 0 \le x_i \le x_i^{max}, \forall i \in A,
\tag{6.2}
$$

$$
x_i \in \mathbb{Z}, \forall i \in A,
\tag{6.3}
$$

$$
\text{LP}: \min_y \mathcal{F}(x,y) = \alpha \sum_{i \in A} \sum_{t \in T} \sum_{j \in J} l_{ij} y_{itj} + \beta \sum_{i \in A} \sum_{t \in T} \sum_{j \in J} \pi_i y_{itj}
$$

$$
+ \gamma \sum_{i \in A} \sum_{t \in T} \sum_{j \in J} p_{it} y_{itj} + \delta \sum_{i \in A} \sum_{t \in T} \left| \sigma x_i - \sum_{j \in J} y_{itj} \right|,
\tag{6.4a}
$$

$$
\text{s.t. } \sum_{i \in A} \sum_{j \in J} y_{itj} = D_t, \forall t \in T,
\tag{6.4b}
$$

$$
\sum_{i \in A} y_{itj} = 1, \forall t \in T, \forall j \in J
\tag{6.4c}
$$

$$
y_{itj} \in \{0,1\}, \forall i \in A, \forall t \in T, \forall j \in J.
\tag{6.4d}
$$

The upper level of the optimization problem indicates maximizing the EVCS operation profit by providing the EV charging service. x_i^{max} is the maximum capacity of the EVC that can be installed at location i, which is identified by the number of the parking lot at the car park and the maximal power transmission from the power grid to the EVCS at time slot t. x_i is the decision variable that denotes the number of

EVCs that should be installed at the candidate location i. π_i is the charging price at the EVCS i; ω_i is the operation cost of EVCSs at i. $y_{it\,j}$ is a binary decision variable, $y_{it\,j} = 1$ indicates the EV j is charged at EVCS i at time slot t, otherwise, $y_{it\,j} = 0$.

In the lower level, the objective function explains the EV users' charging discomfort. l_{ij} is the traveling distance of EV j to a charging facility i. In this work, we assume the EV users only select the surrounding EVCS. In other words, the traveling distance should not be more than a threshold value $d(kNN)$, which is determined by the average traveling distance to k petrol stations by one gasoline vehicle [21]. p_{it} is the difference between charging time and average parking time at EVCS i. q_{it} is the queuing time at EVCS i at time slot t. The value of q_{it} is identified by the ratio of current charging demand and the capacity of the EVC. Eq. (6.4b) requires that all charging demand D_t at t should be satisfied. Eq. (6.4c) explains that for each EV, j is going to one EVCS only.

In the lower-level objective function, there are four components that describe the EV user charging discomfort together: (i) $\sum_{i\in A}\sum_{t\in T}\sum_{j\in J} l_{ij}y_{it\,j}$ denotes the overall traveling distance to charge the EVs at an EVCS; (ii) $\sum_{i\in A}\sum_{t\in T}\sum_{j\in J}\pi_i y_{it\,j}$ represents the overall charging monetary cost; (iii) $\sum_{i\in A}\sum_{t\in T}\sum_{j\in J} p_{it}y_{it\,j}$ indicates the overall extra time cost on EV charging; and (iv) $\sum_{i\in A}\sum_{t\in T}(\sigma x_i - \sum_{j\in J} y_{it\,j})$ is the overall queuing time. We use α, β, γ, and δ to identify the weight of the four components for different EV users. For different users may have different preference and understanding for the charging discomfort. σ is the average ratio between the EVCS capacity and the number of charging demand at this EVCS at a time slot. As the best result, all EVCs in the EVCSs are accessed, and there are no EVs waiting for charge. Hence, $\sigma = len(t)/acTime$ where $len(t)$ is the length of time period t and $acTime$ is the average charging time. In particular, $len(t)$ is not less than $acTime$ to ensure an EV can complete charging in a time period. In this study, $acTime = 4$ hours.

ALGORITHM 1
Alternating Method for Bilevel Optimization Programming (6.1)–(6.4)

```
 1:  procedure
 2:      for each location xi for i ∈ 𝒜 do
 3:          Randomly initialize xi;
 4:      end for
 5:      t = 1;
 6:      repeat
 7:          Solving lower-level optimization problem:
```

$$y^{t+1} = \arg\min_y \mathcal{F}\left(x^t, y\right) \text{ s.t.} \qquad \text{Eq. (6.4)}.$$

```
 8:          Solving upper-level optimization problem:
```

$$x^{t+1} = \arg\min_x -\phi\left(x, y^{t+1}\right) \text{ s.t.} \qquad \text{Eq. (6.1)–(6.3)}.$$

```
 9:          t = t + 1;
10:      until a stopping criterion is satisfied
11:  end procedure
```

The proposed bilevel problem is a NP-hard problem; it is difficult to obtain the optimal solution. To solve the problem, we relax the upper-level and lower-level problem, respectively. First, we replace the x_i from integer variables with continuous variables on the upper level. Then, we round the optimal x_i value as the solution. The lower-level problem is binary programming. The constraint that $y_{it\,j}$ is a binary and solving the corresponding optimization programming is removed. So, the constraints (6.4c) and (6.4d) are replaced by

$$\sum_{t \in T}\sum_{i \in A} y_{it\,j} = 1, \forall j \in J, \tag{6.5a}$$

$$y_{it\,j} \geq 0, \forall i \in A, \forall t \in T, \forall j \in J. \tag{6.5b}$$

With the relaxed bilevel programming problem, we proposed an alternating framework in Algorithm 1 to obtain the local minima solution that is proofed in the the previous work [21]. The stopping criterion is set as $-0.001 \leq \dfrac{x_i^{t+1} - x_i^t}{x_i^{t+1}} \leq 0.001$ for all $i \in \mathscr{A}$.

6.3 ONLINE CHARGING PLANNING

After installation of the EVCs as planned, the online charging planning services can be provided to EV drivers to minimize driver discomfort. Recall the charging facilities have been allocated to EV charging demands in the granularity level of regions as a part of the EV charging facility deployment solution. Specifically, it indicates the charging facilities in which regions should be allocated to satisfy the charging demands in each particular region. However, once the charging facilities have been physically deployed, such allocation schemes are less optimal for two reasons. First, such allocation schemes are based on a particular distribution of charging demands. However, the online charging planning concerns the current distribution of charging demand, which keeps change from time to time and from day to day. Second, such allocation schemes are on a regional basis. However, the online charging planning concerns the allocation between individual EV drivers and individual charging facilities.

The online charging planning service aims to allocate each EV that currently demands charging service so that driver discomfort can be minimized. Different drivers may have different charging preferences and the understanding of discomfort; for example, some EV drivers would like to pay more for charging at a convenient place, while other drivers are more concerned about the charging cost. In the simple scenario, only one charging request is in the system. The optimal charging facility can minimize driver discomfort for this EV. If there are a number of charging requests concurrently in the system, the optimal allocation of charging facilities can minimize overall driver discomfort.

The framework of online charging planning is shown in Figure 6.1. Each EV decides at its discretion to submit a request to an online charging planning

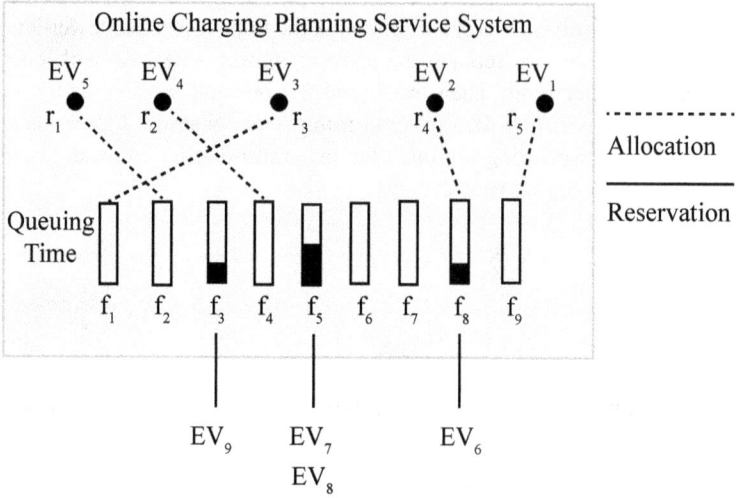

FIGURE 6.1 The framework of online charging planning.

service or not. The charging requests from EV drivers flow continuously into the service system. When a charging facility is allocated to an EV request, it is not actually allocated until reserved by the EV. Note that we also allow a charging facility to be reserved by other EVs in the online charging planning service. If multiple reservations take place concurrently, the allocated EV has higher priority than other EVs in the system, which in turn have higher priority than EVs not in the system. Once a charging facility is reserved, the queuing time increases by a unit, and once a request has reserved a charging facility, it is removed from the system. The online charging planning must efficiently maintain the optimal allocation of charging facilities all the time, i.e., update the allocation when the number of charging requests and the availability of charging facilities change over time.

6.3.1 PROBLEM FORMULATION

Given a set of charging facilities $\mathscr{F} = \{f_1, f_2, ..., f_n\}$ and a set of charging requests $\mathscr{R} = \{r_1, r_2, ..., r_m\}$, the objective of online charging planning can be formulated as:

$$\min f(y) = \sum_{i \in \mathscr{R}} \left\{ \beta_{1i} \sum_{j \in \mathscr{F}} s_{ij} y_{ij} + \beta_{2i} \sum_{j \in \mathscr{F}} c_j y_{ij} \beta_{3i} \sum_{j \in \mathscr{F}} p_j y_{ij} + \beta_{4i} \sum_{j \in \mathscr{F}} \pi_j y_{ij} \right\}, \quad (6.6)$$

$$\text{s.t.} \sum_{j \in \mathscr{F}} y_{ij} = 1, \forall i \in \mathscr{R}, \quad (6.7)$$

$$y_{ij} = \{0,1\} \cdot \forall i \in \mathscr{R}, \forall j \in \mathscr{F}. \quad (6.8)$$

$$p_j = p_j^a + \Delta t \sum_{k \in \mathcal{R}'} y_{kj}, \forall j \in \mathcal{F} \tag{6.9}$$

where $y_{ij} = 1$ means charging facility f_j is allocated to charging request r_i and $y_{i,j} = 0$ means not; Δt is the average charging time; $\beta_{1i}, \beta_{2i}, \beta_{3i},$ and β_{4i} define the driver discomfort of charging request r_i. In particular, $\beta_{1i} + \beta_{2i} + \beta_{3i} + \beta_{4i} = 1$. β_{1i} is the weight of $s_{i,j}$, which is the traveling time from the current location of charging request r_i to the location of charging facility f_j. It is possible to consider the current traffic situation. β_{2i} is the weight of c_i, which is the difference between parking time and charging time in the region around charging facility f_j; β_{3i} is the weight of p_j, which is the queuing time for charging facility f_j; the queuing time is calculated by the initialized queuing time p_j^a and reserved requests $\mathcal{R}'$ in facility f_j before request r_i; β_{4i} is the weight of π_j, which is the charging rate of facility f_j.

The objective is to minimize overall driver discomfort (Eq. (1.6)) at a particular time. The challenges lies in two aspects: (i) the objective is subject to that one EVC is allocated to only one charging facility (Eq. (1.7) and (1.8)); the problem is an integer linear programming, which is difficult to solve unless $\mathbf{NP = P}$; (ii) the online charging planning service system must be able to efficiently maintain the optimal allocation of charging facilities all the time, i.e., update the allocation when the number of charging requests and the availability of charging facilities change over time. To handle these two challenges, we propose an online planning algorithm that is a variant of greedy strategy.

6.3.2 Online Planning Algorithm

Given a set of charging facilities $\mathcal{F} = \{f_1, f_2, \ldots, f_n\}$, the current queuing time of f_i is denoted as p_i based on the number of drivers who have reserved f_i for charging. Given a set of charging requests $\mathcal{R} = \{r_1, r_2, \ldots, r_m\}$, if r_i is allocated to f_j, p_j does not increase until the driver related with r_i has reserved f_j.

At any time, the online planning algorithm allocates each $r_i \in \mathcal{R}$ to one $f_j \in \mathcal{F}$ where each f_j is associated with a value p_j^a, initialized to p_j. If charging request r_i is allocated to f_j in the charging plan, the driver discomfort is

$$g(r_i, f_j) = \beta_{1i} s_{ij} + \beta_{2i} c_j + \beta_{3i} p_j^a + \beta_{4i} \pi_j. \tag{6.10}$$

Then, p_j^a is updated by a unit for the following charging plan. If another charging request r_i' follows r_i in the queue of f_j in the charging plan, the queuing time of r_i' is the updated p_j^a.

The online planning algorithm takes one iteration for one charging request. In one iteration, the driver discomfort $g(r_i, f_j)$ is calculated for each possible allocation $< r_i, f_j >$. The one with the minimum driver discomfort is adopted and included in the charging plan. Suppose it is $< r_i, f_{j*} >$. Then, r_i is removed from $\mathcal{R}$ and f_{j*} gets p_{j*}^a updated. The online planning algorithm continues until all charging requests have been allocated. The pseudocode of the greedy algorithm is presented in Algorithm 2.

In the algorithm, $< g_{max}(\mathscr{R}) \cdot r, g_{max}(\mathscr{R}) \cdot f >$ denotes the current best allocation considering all unallocated charging requests and all charging facilities after previous allocations. The current best allocation is inserted into the solution. Then, $\mathscr{F}$ and $\mathscr{R}$ are updated. The solution A_l is measured by

$$f(A_l) = \sum_{i \in A_l} \frac{1}{g(r_i, f_j)},\tag{6.11}$$

where f_j is the charging facility allocated to charging request r_i, and $\dfrac{1}{g(r_i, f_j)}$ is the *driver comfort* if f_j is allocated to r_i. In the greedy algorithm, the solution is an empty set A_0 initially, and, in iteration i, the solution is augmented with the allocation, which maximizes the discrete derivative $\Delta(a|A_i)$ (i.e., $f(A_i \cup a) - f(A_i)$).

$$A_i = A_{i-1} \cup \arg\max_a \Delta(a | A_{i-1}).\tag{6.12}$$

The time complexity of greedy Algorithm 2 is $O(nm^2)$.

ALGORITHM 2
Online Charging Planning

1: **procedure** ALLOCATION $(\mathscr{F}, \mathscr{R})$
2:    $l \leftarrow 0$;
3:    $A_l \leftarrow 0$;
4:    $count \leftarrow 0$;
5:    **for** $f_i \in \mathscr{F}$ **do**
6:       $p_i^a \leftarrow p_i$;
7:    **end for**
8:    **for** $\mathscr{R} = \phi$ **do**
9:       **for each** $r_i \in \mathscr{R}$ **do**
10:         $g_{max}(r_i) \leftarrow max_{f_j \in \mathscr{F}}\left[g(r_i, f_j)\right]$;
11:      **end for**
12:      $g_{max}(\mathscr{R}) \leftarrow max_{r_i \in \mathscr{R}}\left[g_{max}(r_i)\right]$;
13:      $l \leftarrow l + 1$;
14:      $A_l \leftarrow A_{l-1} \cup < g_{max}(\mathscr{R}) \cdot r, g_{max}(\mathscr{R}) \cdot f$;
15:      increase queuing time p_j^a of $g_{max}(\mathscr{R}) \cdot f$ by Δt;
16:      removing $g_{max}(\mathscr{R}) \cdot r$ in $\mathscr{R}$;
17:   **end for**
18:   **return** A_l;
19: **end procedure**

6.3.3 APPROXIMATION FACTOR

The objective function $f(A)$ in Eq. (1.11) is submodular. Let X be a finite set. A function $H : 2^X \rightarrow \mathbb{R}$ is called submodular if for all subsets $S \subset T \subset X$ and all $x \in X \setminus T$

$$H(S \cup \{x\}) - H(S) \geq H(T \cup \{x\}) - H(T). \qquad (6.13)$$

In other words, if H measures benefit, then the marginal benefit of adding x to S is at least as high as the marginal benefit of adding it to T. Since $S \subset T$ and x are all arbitrary, it means that adding x to a bigger set cannot be better than adding it to a smaller set.

THEOREM 6.1

The function $f(A)$ in Eq. (1.11) is submodular, monotone, and non-negative.   ∎

Proof 1 *$f(A)$ is the summation of driver discomforts which are non-negative. So, $f(A)$ is monotone and non-negative. Let X be a finite set. For any subset $S \subset T \subset X$ and any $x \in X \setminus T$, let $\Delta(x|T)$ be $f(T \cup x) - f(T)$ and $\Delta(x|S)$ be $f(S \cup x) - f(S)$. Suppose x is the driver discomfort of request r_x allocating to facility f_x. In the situation of S, r_x searches all possible allocations for the one with the minimum driver discomfort $g(x_i, f_i)$. In the situation of T, the search space of possible allocations is no more than that in the situation of S. So, $\Delta(x|S) \geq \Delta(x|T)$, i.e., $f(A)$ is submodular.*

THEOREM 6.2

([22]) Fix a non-negative monotone submodular function $f : 2^X \to \mathbb{R}$ and let $\{A_l\}_{l>0}$ be the greedily selected sets defined in Eq. (1.12). For any positive integer k

$$f(A_l) \geq \left(1 - e^{-l/k}\right) \max_{S:|S| \leq k} f(S). \qquad (6.14)$$

In particular, for $l = k, f(A_l) \geq (1 - 1/e) \max_{S:|S| \leq k} f(S)$.   ∎

6.3.4 ONLINE UPDATE

The allocation solution A_l using the greedy algorithm is generated. As an online service, new charging requests keep flowing in the system and the charging facilities can be reserved at any time. As a response to the changes, allocation A_l needs to be updated. To avoid recomputing A_l from scratch, the efficient allocation solution update is necessary.

6.3.4.1 New Request

When a new request flows in, the update process is illustrated using an example shown in Figure 6.2. The current A_l is shown in Figure 6.2 (a). To update new charging request r_7, it takes the following steps.

1. The first allocation a_1 in A_l is checked as shown in Figure 6.2 (b). The best allocation of r_7 is f_2 and the driver discomfort is $g(r_7, f_2)$. Since $g(r_7, f_2) > g(r_2, f_4)$, it means the driver discomfort of $< r_2, f_4 >$ is less than that of $< r_7, f_2 >$; so $< r_2, f_4 >$ is not impacted by the new request and it is fixed.
2. The second allocation a_2 in A_l is checked as shown in Figure 6.2 (b). The best allocation of r_7 is f_2 and the driver discomfort is $g(r_7, f_2)$. Since $g(r_7, f_2)$

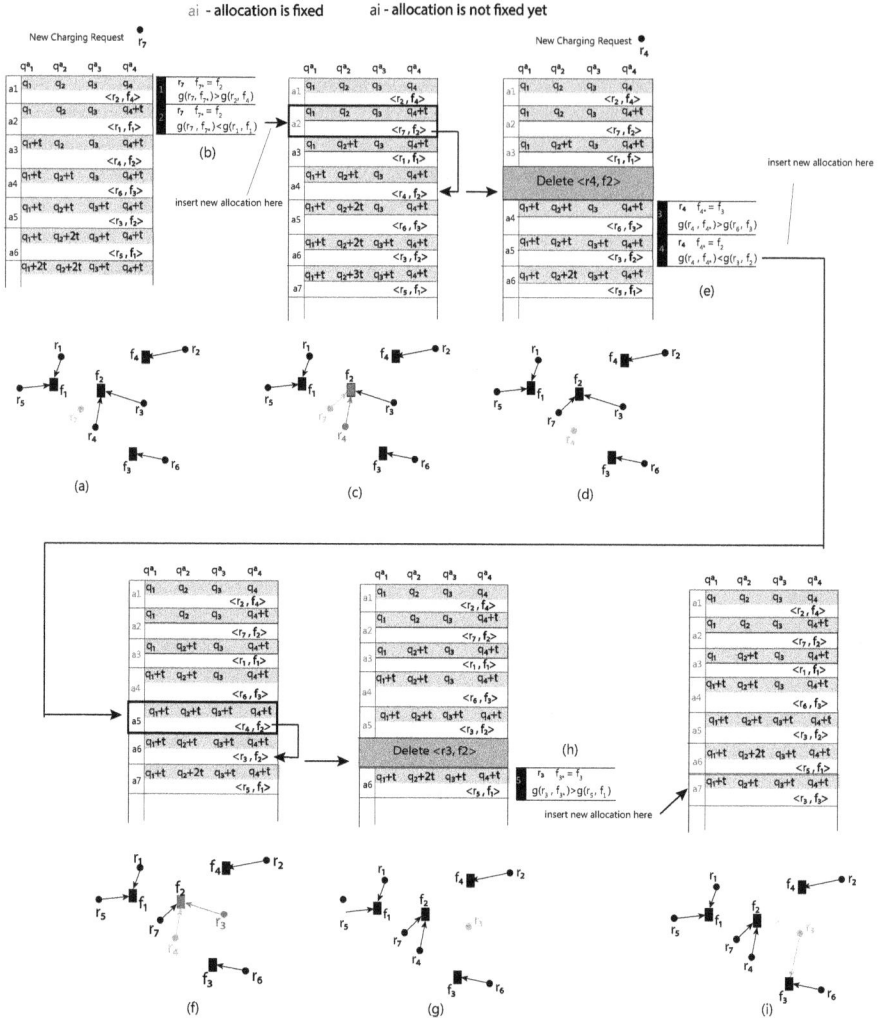

FIGURE 6.2 New charging request—online update.

$< g(r_1, f_1)$, it means the driver discomfort of $< r_7, f_2 >$ is less than $< r_1, f_1 >$; thus $< r_7, f_2 >$ is inserted into A_l before $< r_1, f_1 >$ and allocation $< r_7, f_2 >$ is fixed.

3. Among the allocation(s) not fixed yet, the ones impacted by the new queuing time of f_2 are $a_4 =< r_4, f_2 >$, and $a_6 =< r_3, f\,2 >$. The allocation a_4 is processed by deleting it, as shown in Figure 6.2 (c). a_4 is deleted from A_l.

4. Then, r_4 is treated as a new charging facility as r_7, as shown in Figure 6.2 (d). Different from r_7, only the allocations not fixed yet are examined. The new allocation for r_4 is $< r_4, f_2 >$, and it is inserted into A_l, as shown in Figure 6.2 (f). Among the allocation(s) not fixed yet, the ones impacted by the new queuing time of f_2 is $a_6 =< r_3, f2 >$. The allocation a_6 is processed by deleting it, as shown in Figure 6.2 (g).

5. As shown in Figure 6.2 (i), r_3 is treated as a new charging facility. Since inserting the new allocation $< r_3, f_3 >$ does not impact any allocations not fixed yet, the update completes.

The updated solution is equivalent to the solution computing from scratch using the greedy algorithm. That is, for any fixed allocation, say a_i, given the queuing time at a_i, its driver discomfort must be less than that of following allocations a_j ($j > i$). This can be easily observed at each step of the update process when an allocation is fixed. The update is performed by running function NewRequest($\mathcal{F}$, $\mathcal{R}$, A_l, r_{new}, 0), the pseudocode of which is presented in Figure 6.3. The time complexity for update is $O(nm)$.

The queuing time of a facility f_i may increase for some reason. In this situation, the update of A_l can be performed using function NewRequest($\mathcal{F}$, $\mathcal{R}$, A_l, r_j, j) in Algorithm 3, where r_j is the first charging request to which f_i is allocated.

ALGORITHM 3
New Request Update

```
1:   procedure NEWREQUEST(𝓕,𝓡,Aₗ,rₙₑw,start)
2:      i = start;
3:      while aᵢ ≠ null do
4:          identify f_{new*} where queuing times of facilities at aᵢ;
5:          if g(aᵢ) ≤ g(r_{new}, f_{new*}) then
6:              i = i + 1;
7:          else
8:              insert < r_{new}, f_{new*} > in Al before aᵢ;
9:              delete the first allocation < rⱼ, f_{j*} > (j > i and f_{j*} ≡ f_{new*}); NewRequest(𝓕, 𝓡,Aₗ,rⱼ,j);
10:         end if
11:     end while
12:  end procedure
```

6.3.4.2 Allocation Cancellation

When an existing allocation is canceled, the update of A_l is illustrated using an example shown in Figure 6.3. The current A_l is shown in Figure 6.3 (a) and a_2 is canceled. The queuing time of f_1 has been changed. The update takes the following steps:

1. The next allocation $< r_4, f_2 >$ is examined. If $g < r_4, f_2 >$ remains the same, it is fixed and the next allocation is examined, and so on.
2. $g < r_3, f_2 >$ is impacted by the change of queuing time of f_1, as shown in Figure 1.3 (c). As a result, the allocation to r_3 is updated to $< r_3, f_1 >$. The queuing times of f_2 and f_1 have been updated.
3. Then, the next allocation $g < r_5, f_1 >$ is examined. Since $g < r_5, f_1 >$ remains the same, it is fixed.
4. The allocations are ordered in ascending order of driver discomfort.

The update is performed by running function AllCan($\mathcal{F}$,$\mathcal{R}$,A_l,i), where a_i is the allocation canceled. The pseudocode is presented in Algorithm 4. The updated

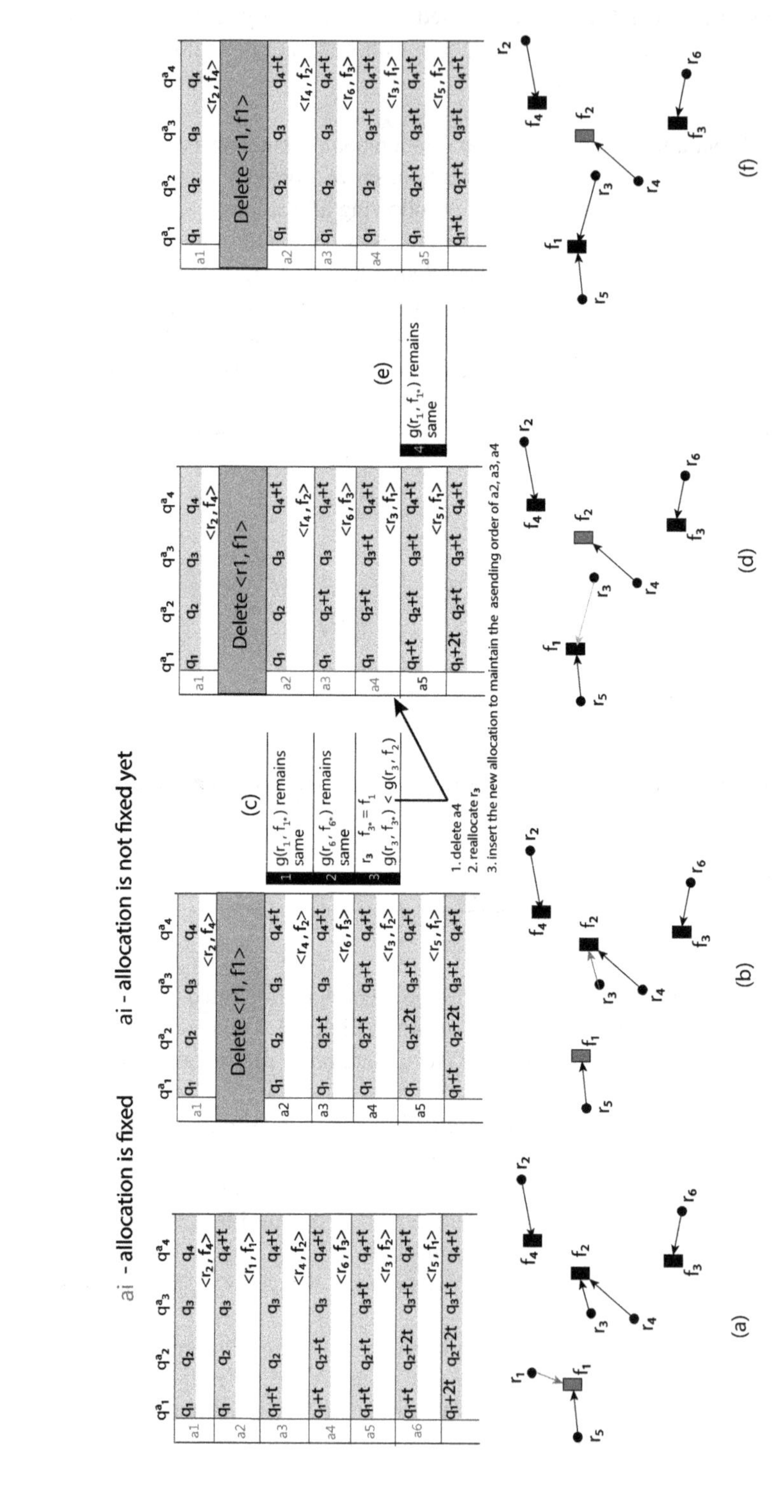

FIGURE 6.3   Allocation cancellation—online update.

solution is equivalent to the solution computing from scratch using the greedy algorithm. The time complexity for the update is $O(nm + m \log m)$, where $m \log m$ is for sorting.

ALGORITHM 4
Allocation Cancellation

1: **procedure** ALLCAN($\mathscr{F}$, $\mathscr{R}$, A_l, i)
2:      $i = i + 1$;
3:      **while** $a_i \neq null$ **do**
4:          identify f_{i*} where queuing times of facilities at a_i;
5:          **if** $g(a_i) \leq g(r_i, f_{i*})$ **then**
6:              $i = i + 1$;
7:          **else**
8:              update a_i to $< r_i, f_{i*} >$; AllCan($\mathscr{F}$, $\mathscr{R}$, A_l, i);
9:          **end if**
10:     **end while**
11:     Sort A_l in ascending order of driver discomfort
12: **end procedure**

In the situation that the queuing time of a facility decreases, the update can be done using function AllCan($\mathscr{F}$, $\mathscr{R}$, A_l, 0).

6.4 EMPIRICAL STUDY

In this section, we use real-world data, such as point of interest, vehicle trajectory, etc., to simulate the proposed bilevel programming and the online charging planning framework. We implement the algorithms by Matlab (R2015b) with YALMIP and CPLEX (v12.6). All simulations are done on a computer with Intel(R) Core(TM) @2.40GHz 4.10GHz, 8GB 2133 LPDDR3 RAM, and the OS Monterey 64 bit operating system.

6.4.1 DATA PREPARATION

Urban Region Data Collection: We simulate the proposed frameworks in Shenzhen City. The candidate EVCS locations are extracted as the roadsides of petroleum stations, and the current building carparks. The charging demand is extracted by gasoline taxi refilling demands distribution [21]. The distribution of refilling demands and the petroleum stations are shown in Figure 6.4. The time slot is set as 1 hour, and 24 hours is considered in the simulation. The average charging time is 4 hours ($acTime = 4$). The Poisson distribution is applied to explain charging demand in each time slot on the road [23]. With the charging demand distribution, traveling distance, maximal EVC capacity at each candidate location, the bilevel framework is simulated and the approximation solution is obtained and shown as the heat map in Figure 6.5.

FIGURE 6.4 The road network and petroleum stations of Shenzhen City.

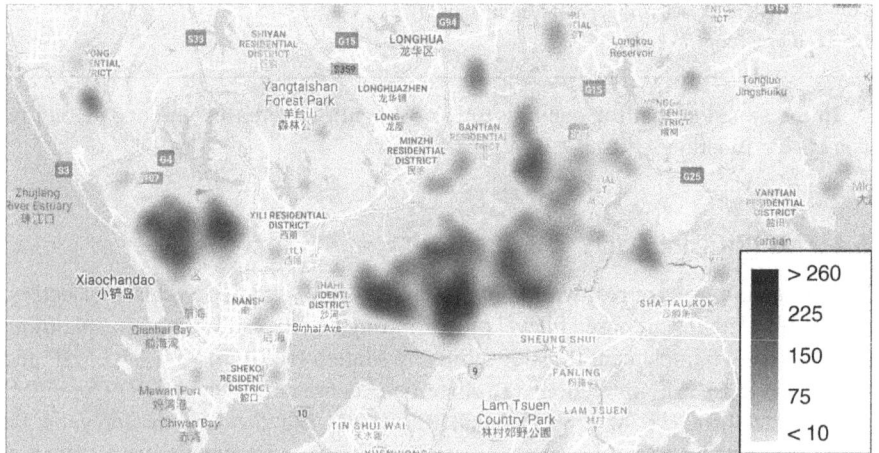

FIGURE 6.5 EV charger distribution in Shenzhen City.

6.4.2 OPTIMAL EVCS PLACEMENT

The result is indicated as *TA-Bilevel*, where *TA* represents *time aware*. The proposed method is compared with two baselines that are *Baseline*-1 and *Baseline*-2, respectively. The objective function of *Baseline*-1 is to maximize the EVCS operation revenue, which is shown in Eq. (6.15). The constraints of EV user charging discomfort are added to the baseline. The objective function of *Baseline*-2 focuses on the EV charging discomfort only and is shown as Eq. (6.16). The constraints of EVCS distribution are added as the same as *TA-Bilevel*. Both baselines are solved by relaxing the integer decision variables to continuous. Then Algorithm 1 is implemented for both baselines.

$$\min_{x,y} -\phi(x,y) = -\sum_{i \in A}\left(\sum_{t \in T}\sum_{j \in J}\pi_i y_{it\,j} - \omega_i x_i\right)$$

$$s.t.\ 0 \le x_i \le x_i^{max}, \forall i \in A,$$

$$x_i \in \mathbb{Z}, \forall i \in A, \quad\quad\quad\quad\quad (6.15)$$

$$\sum_{i \in A}\sum_{r(j) \in \omega} y_{it\,j} = Y_{rt}, \forall t \in T, \forall \omega \in R,$$

$$d\left(y_{it\,j}\right) \le kNN, \forall j \in J, \forall i \in A,$$

$$y_{it\,j} \in \{0,1\}, \forall i \in A, \forall t \in T, \forall j \in J,$$

$$\min_{y} \mathscr{F}(x,y) = \alpha\sum_{i \in A}\sum_{t \in T}\sum_{j \in J}l_{ij}y_{it\,j} + \beta\sum_{i \in A}\sum_{t \in T}\sum_{j \in J}\pi_i y_{it\,j}$$

$$+\gamma\sum_{i \in A}\sum_{t \in T}\sum_{j \in J}P_{it}y_{it\,j}$$

$$+\delta\sum_{i \in A}\sum_{t \in T}\left|\sigma x_i - \sum_{j \in J}y_{it\,j}\right|, \quad\quad\quad (6.16)$$

$$s.t.\ 0 \le x_i \le x_i^{max}, \forall i \in A,$$

$$x_i \in \mathbb{Z}, \forall i \in A,$$

$$\sum_{i \in A}\sum_{r(j) \in \omega} y_{it\,j} = Y_{rt}, \forall t \in T, \forall \omega \in R,$$

$$d\left(y_{it\,j}\right) \le kNN, \forall j \in J, \forall i \in A,$$

$$y_{it\,j} \in \{0,1\}, \forall i \in A, \forall t \in T, \forall j \in J.$$

6.4.3 EFFECTIVENESS

The optimized placement of EVCSs is obtained by *TA-Bilevel*, *Baseline*-1, and *Baseline*-2. The charging demands are satisfied at each time slot. The EVCS operation revenue and EV user charging discomfort are evaluated and compared, respectively. In addition, the proposed online charging planning framework is implemented after the optimization process.

The impact of user discomfort is evaluated as well. As shown in Eq (1.4), α is the weight for traveling distance to charge the EVs at an EVCS, β is charging monetary cost, γ is extra time cost on the EV charging, and δ is the queuing time. By initial, $\alpha = \beta = \gamma = \delta = 1$.

As can be seen from Figure 6.6 to Figure 6.9, the result from *Baseline*-1 receives the most revenue with the most EV charging discomfort (the value of discomfort, as less as better), for *Baseline*-1 does not model the features of EV user charging discomfort. The EVCS distribution would not change based on the charging demand distribution. With the consideration of different weights of charging discomfort, *TA-Bilevel* and *Baseline*-2 obtain similar and better outputs on charging discomfort

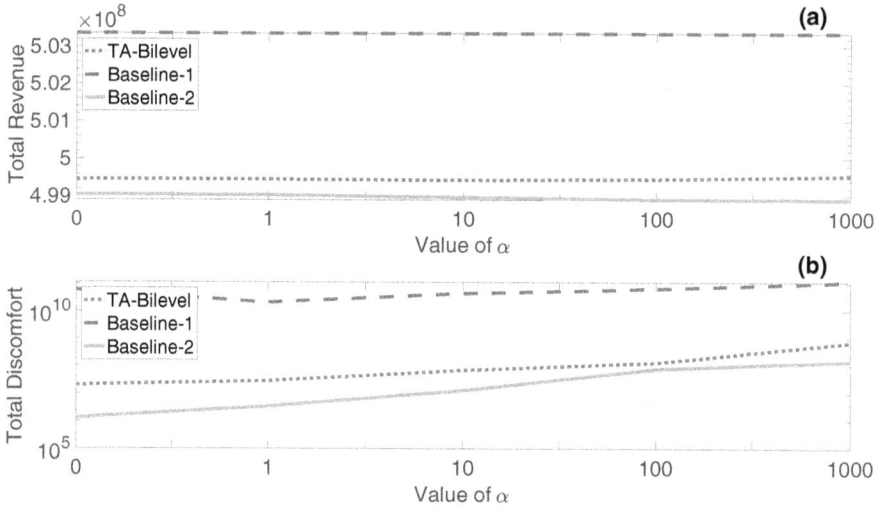

FIGURE 6.6 The impact of α.

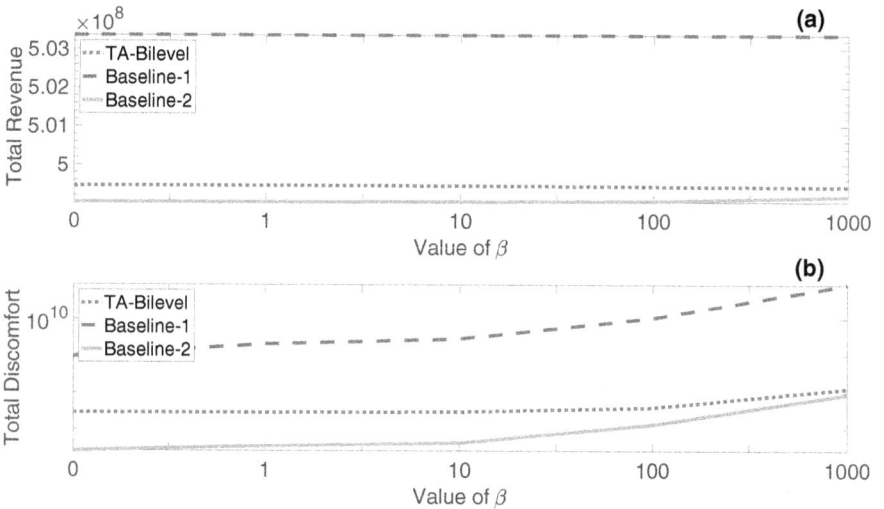

FIGURE 6.7 The impact of β.

than *Baseline*-1. The proposed method always makes more profits than *Baseline*-2. When the queuing time becomes a more significant component, *TA-Bilevel* has a better performance on the charging discomfort part.

6.4.4 IMPACT OF TIME INTERVAL LENGTH

In this work, the distribution of EV charging demand is divided into each time slot rather than an average daily demand. We simulate the proposed bilevel optimization

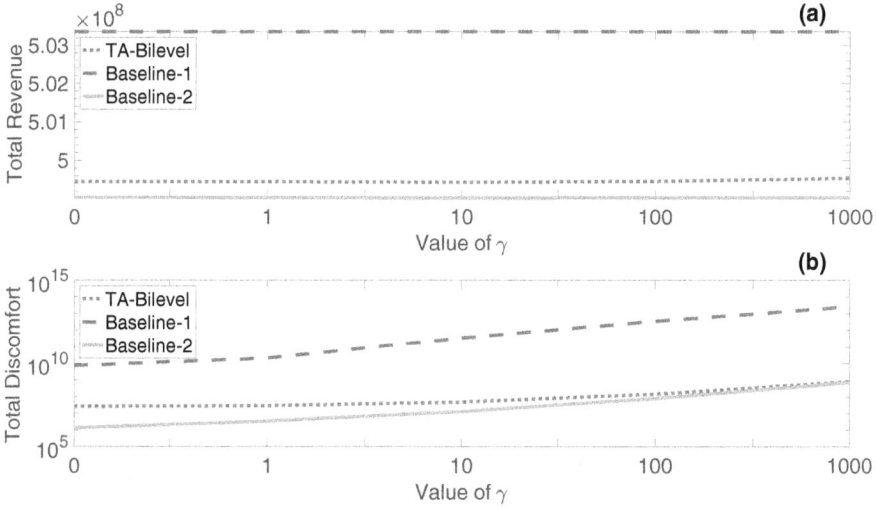

FIGURE 6.8 The impact of γ.

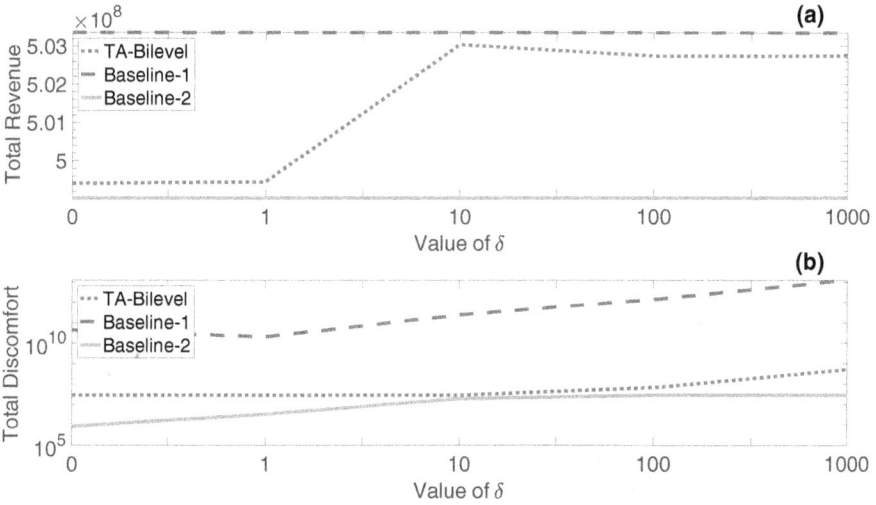

FIGURE 6.9 The impact of δ.

programming and the online charging planning framework in different lengths of time. Five situations are considered, i.e., splitting a day into T intervals, where $T = 8, 6, 4, 3$, and 1, respectively. The interval length is 3, 4, 6, 8, and 24 hours accordingly. The case $T = 1$ is the same charging demand distribution as daily demand. The simulation results are shown in Figure 6.10.

As can be seen from the upper in Figure 6.10, when the charging demand is separated by different time intervals, the revenue is always more than the case with the daily demand. The reason is that EVCS investors satisfy the EV charging demand

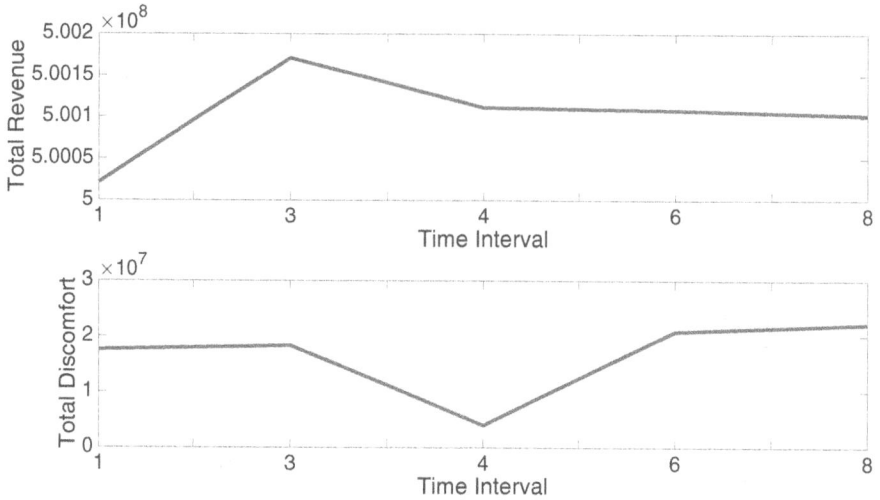

FIGURE 6.10   Impact of time awareness in EVC deployment.

with fewer EVCs. The generated profit by providing charging services is the same as the $T = 1$ case, but the maintenance cost and installation cost is much more than $T = 1$. On the other side, from the lower in Figure 6.10, capacity for more EVCs provide a better service quality that is reflected on EV charging discomfort. The overall queuing time is reduced significantly. The EV user has more opportunity to access an EVC with a lower charging discomfort.

6.5 CONCLUSION

This chapter has proposed a framework to optimize the EVCS placement problem with the consideration of EV user charging discomfort. The proposed method has modeled the number of EV charging demands as fluctuated in different time slots. Also, the EV charging discomfort is considered in four components: traveling distance to charge the EVs at an EVCS, charging monetary cost, extra time cost on the EV charging, and queuing time. After that, in order to achieve the optimization objectives after EVCS installation, an online charging planning framework has been proposed. A series of algorithms have been developed for the online charging request allocating, updating, and cancellation. Both proposed frameworks have been simulated with a large-scale real-world dataset. The result has shown under the optimal placement of EVCS, the EV charging planning helps EV users allocate an EVC with a lower charging discomfort to achieve win-win outcomes. In further work, dynamic traffic congestion will be considered in the online charging planning framework.

REFERENCES

[1] Z. Liu, F. Wen, and G. Ledwich, "Optimal planning of electric-vehicle charging stations in distribution systems," *IEEE Transactions on Power Delivery*, vol. 28, no. 1, pp. 102–110, 2012.

[2] L. Chen, C. Xu, H. Song, and K. Jermsittiparsert, "Optimal sizing and sitting of evcs in the distribution system using metaheuristics: A case study," *Energy Reports*, vol. 7, pp. 208–217, 2021.

[3] P. V. K. Babu and K. Swarnasri, "Multi-objective optimal allocation of electric vehicle charging stations in radial distribution system using teaching learning based optimization," *Int. J. Renew. Energy Res*, vol. 10, no. 1, pp. 366–377, 2020.

[4] W. Kong, Y. Luo, G. Feng, K. Li, and H. Peng, "Optimal location planning method of fast charging station for electric vehicles considering operators, drivers, vehicles, traffic flow and power grid," *Energy*, vol. 186, p. 115826, 2019.

[5] M. M. Islam, H. Shareef, and A. Mohamed, "Optimal location and sizing of fast charging stations for electric vehicles by incorporating traffic and power networks," *IET Intelligent Transport Systems*, vol. 12, no. 8, pp. 947–957, 2018.

[6] S. R. Gampa, K. Jasthi, P. Goli, D. Das, and R. Bansal, "Grasshopper optimization algorithm based two stage fuzzy multiobjective approach for optimum sizing and placement of distributed generations, shunt capacitors and electric vehicle charging stations," *Journal of Energy Storage*, vol. 27, p. 101117, 2020.

[7] X. Luo and R. Qiu, "Electric vehicle charging station location towards sustainable cities," *International Journal of Environmental Research and Public Health*, vol. 17, no. 8, p. 2785, 2020.

[8] P. Sadeghi-Barzani, A. Rajabi-Ghahnavieh, and H. Kazemi-Karegar, "Optimal fast charging station placing and sizing," *Applied Energy*, vol. 125, pp. 289–299, 2014.

[9] G. Battapothula, C. Yammani, and S. Maheswarapu, "Multi-objective simultaneous optimal planning of electrical vehicle fast charging stations and dgs in distribution system," *Journal of Modern Power Systems and Clean Energy*, vol. 7, no. 4, pp. 923–934, 2019.

[10] H. Yang, Y. Deng, J. Qiu, M. Li, M. Lai, and Z. Y. Dong, "Electric vehicle route selection and charging navigation strategy based on crowd sensing," *IEEE Transactions on Industrial Informatics*, vol. 13, no. 5, pp. 2214–2226, 2017.

[11] F. V. Cerna, M. Pourakbari-Kasmaei, R. A. Romero, and M. J. Rider, "Optimal delivery scheduling and charging of evs in the navigation of a city map," *IEEE Transactions on Smart Grid*, vol. 9, no. 5, pp. 4815–4827, 2017.

[12] C. Liu, M. Zhou, J. Wu, C. Long, and Y. Wang, "Electric vehicles en-route charging navigation systems: Joint charging and routing optimization," *IEEE Transactions on Control Systems Technology*, vol. 27, no. 2, pp. 906–914, 2017.

[13] J. Tan and L. Wang, "Real-time charging navigation of electric vehicles to fast charging stations: A hierarchical game approach," *IEEE Transactions on Smart Grid*, vol. 8, no. 2, pp. 846–856, 2015.

[14] X. Shi, Y. Xu, Q. Guo, H. Sun, and W. Gu, "A distributed ev navigation strategy considering the interaction between power system and traffic network," *IEEE Transactions on Smart Grid*, vol. 11, no. 4, pp. 3545–3557, 2020.

[15] T. Qian, C. Shao, X. Wang, and M. Shahidehpour, "Deep reinforcement learning for ev charging navigation by coordinating smart grid and intelligent transportation system," *IEEE Transactions on Smart Grid*, vol. 11, no. 2, pp. 1714–1723, 2019.

[16] X. Li, Y. Xiang, L. Lyu, C. Ji, Q. Zhang, F. Teng, and Y. Liu, "Price incentive-based charging navigation strategy for electric vehicles," *IEEE Transactions on Industry Applications*, vol. 56, no. 5, pp. 5762–5774, 2020.

[17] Z. Li, A. Alsabbagh, Y. Meng, and C. Ma, "User behavior-based spatial charging coordination of ev fleet," in *IECON 2020 The 46th Annual Conference of the IEEE Industrial Electronics Society*. IEEE, 2020, pp. 3635–3640.

[18] Q. Guo, S. Xin, H. Sun, Z. Li, and B. Zhang, "Rapid-charging navigation of electric vehicles based on real-time power systems and traffic data," *IEEE Transactions on Smart Grid*, vol. 5, no. 4, pp. 1969–1979, 2014.

[19] T. Chen, B. Zhang, H. Pourbabak, A. Kavousi-Fard, and W. Su, "Optimal routing and charging of an electric vehicle fleet for high-efficiency dynamic transit systems," *IEEE Transactions on Smart Grid*, vol. 9, no. 4, pp. 3563–3572, 2016.

[20] H. Yang, S. Yang, Y. Xu, E. Cao, M. Lai, and Z. Dong, "Electric vehicle route optimization considering time-of-use electricity price by learnable partheno-genetic algorithm," *IEEE Transactions on Smart Grid*, vol. 6, no. 2, pp. 657–666, 2015.

[21] C. Liu, K. Deng, C. Li, J. Li, Y. Li, and J. Luo, "The optimal distribution of electric-vehicle chargers across a city," in *Proceedings of 2016 IEEE 16th International Conference on Data Mining (ICDM)*, Barcelona, Spain, 2016, pp. 261–270.

[22] G. L. Nemhauser, L. A. Wolsey, and M. L. Fisher, "An analysis of approximations for maximizing submodular set functions—i," *Mathematical Programming*, vol. 14, no. 1, pp. 265–294, 1978.

[23] H. Zhang, S. Moura, Z. Hu, W. Qi, and Y. Song, "A second order cone programming model for pev fast-charging station planning," *arXiv preprint arXiv:1702.01897*, 2017.

7 Smart Control Strategies for AC Switches Used in Electric Vehicle Battery Chargers

Sébastien Jacques

CONTENTS

7.1 INTRODUCTION

At the end of the 26th edition of the Conference of the Parties (COP26) held in Glasgow (Scotland, United Kingdom) in November 2021, the green agenda remains a priority and important decisions were taken in the transportation sector (Laffont et al. 2022). In particular, more than a hundred national governments, cities, states and major companies signed the declaration on zero-emission cars and vans. To achieve this, the preferred solution is electric, but many developments are needed to address not only environmental issues but also autonomy and economic profitability, both for manufacturers and consumers.

DOI: 10.1201/9781003293989-7

The market for electric cars and plug-in hybrids has been booming during the last decade (Xue et al. 2021). In its traditional "Global Electric Vehicle" outlook, the International Energy Agency (IAE), an intergovernmental entity of the Organization for Economic Co-operation and Development (OECD), reports that three million new electric and plug-in hybrid cars registered in 2020, an increase of 41% over 2019 (Heidrich et al. 2022). A figure that is all the more encouraging considering the effect of the Covid-19 pandemic on the global automotive market. Cumulatively, by 2020, the global connected car fleet (battery electric vehicles [BEVs] and plug-in hybrid EVs [PHEVs]) now reaches ten million units. To this is added about one million electric vans, trucks and buses. According to IAE projections, this growth is expected to reach 145–230 million EVs by 2030, if all governments meet their global climate commitments (Dik et al. 2022). If these predictions come true, then we will be able to save between 2 and 3.5 million barrels of oil per day and avoid releasing at least 120 million tons of CO_2 into the atmosphere.

While technology developments have accelerated dramatically to support the expansion of the EV market for BEVs, hybrid EVs (HEVs) and PHEVs, many questions remain unanswered regarding battery management (Naim and Vashist 2022). Batteries play a major role thanks to their cost-effectiveness, energy and power densities, reliability, and charging times, which are all application dependent. In particular, battery life and charging time are highly dependent on the characteristics of the battery charger (Alam et al. 2022).

Two types of battery chargers are currently used: the first is the on-board type, and the second is the stand-alone type that provides a fast charge. Regarding the on-board type, which is particularly considered in this chapter, many experts believe that it is essential to develop chargers with high power and energy efficiency while optimizing their weight, footprint in the EV and the manufacturing cost (Lee et al. 2022).

A mid-range (e.g., 3.7 kW) on-board battery charger converts alternating current (AC) from a widely used 110 V to 230 V/16 A plug to direct current (DC) ranging from 200 V to 430 V/10 A in order to charge batteries by controlling the voltage and current flow. This type of charger usually consists of a large electrolytic bulk capacitor (e.g., 1 μF/W for an RMS voltage of 230 V at a frequency of 50 Hz) (Arif et al. 2021). This output capacitor, also called a DC link capacitor, is responsible for smoothing the ripple of the rectified current before being cut off at high frequency. When the AC-DC power supply is connected, high inrush currents can be generated on the AC grid due to the charging of this output capacitor (Kim et al. 2021). These high currents can easily reach 5–20 times the steady state load current (Jacques et al. 2021). The consequences of these phenomena are numerous (Reymond et al. 2016). First, the equipment itself may be affected (e.g., blown fuses or tripped circuit breakers). Second, individual devices in the AC-DC supply, such as switches, rectifier diodes and smoothing capacitors, can be damaged. Finally, high inrush currents can induce excessive current stress on AC grids (Jacques et al. 2017).

In this chapter, we propose discussing the power component control strategies currently implemented to limit inrush currents during battery charger startup. Their advantages and disadvantages will be discussed later. We will position AC power devices (thyristors and Triacs) in these strategies, while a particularly effective solution will be also proposed, implemented and analyzed.

7.2 INTEREST OF THE ELECTRIC VEHICLE CONNECTED TO THE AC GRID AND CONSEQUENCES ON THE BATTERY CHARGING SYSTEMS

7.2.1 V2H/V2G CONCEPTS REMINDER

Today, the deployment of smart grids is accelerating due to the massive development of renewable energies (e.g. photovoltaic, wind), whose production flow varies according to the daytime and weather conditions (Elfeki et al. 2018). In order to guarantee the balance between electricity production and consumption for all times, management systems require a high level of precision and reactivity, which necessarily involves electricity storage (Bissey et al. 2017). The batteries in electric vehicles can then play a key role by acting as a buffer, i.e., by aligning production with consumption, which is called battery-to-grid (B2G). The development of this connection should eventually allow a high availability rate, resulting in increasing the available gross power and thus an efficient supply/demand balancing (Borge-Diez et al. 2021). As shown in Figure 7.1, one example is the use of EV batteries as a support for the AC grid (Rivera et al. 2021; Aouichak et al. 2022):

- homes equipped with a bidirectional charger: this is called a vehicle-to-home (V2H) connection, and
- parking lots equipped with bidirectional chargers: this is called a vehicle-to-grid (V2G) connection.

These two concepts are similar since they consist in using the energy stored in the batteries of the EV in case it is not consumed. In this way, many car manufacturers have been working on implementation of these two concepts for several years.

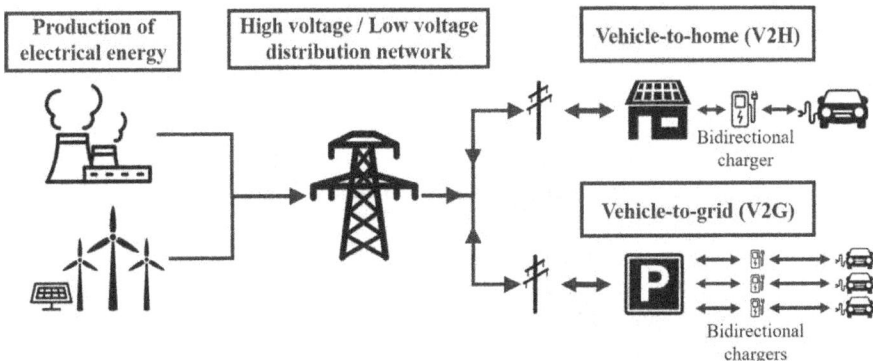

FIGURE 7.1 Illustration of V2H and V2G concepts.

Nissan is particularly aggressive and pioneering in experimenting with V2G and similar architectures producing demonstrators for several years (Schram et al. 2020). For example, in 2020 and driven by not an exactly serene period in terms of climate or geopolitics, the Nissan firm came up with the idea of designing an emergency vehicle concept, based on the compact electric LEAF and capable of supplying electricity for disaster areas. The vehicle, called RE-LEAF, takes advantage of the car's ability to redistribute energy from its 62 kWh lithium-ion battery. Mitsubishi is another major player in the deployment of V2H and V2G concepts (Ravi and Aziz 2022). For instance, in 2021, Mitsubishi Motors Australia Limited introduced bidirectional chargers to the Australian market, which specifically allowed owners of the Outlander PHEV to unlock V2H/V2G capabilities. When it comes to bidirectional chargers, Wallbox, a company from across the Atlantic, recently shook up the calendar for the benefit of connected American motorists by presenting, at the Consumer Electronics Show (CES) 2022 in Las Vegas, a bidirectional charging box that allows you to compose your own small V2G and/or V2H architectures (Von Bonin et al. 2022). Although the first versions of this product were initially intended for the North American market, the firm promised to roll out the concept in Europe and the Asia-Pacific region.

To go further, this technology is already inspiring other sectors. Indeed, the principle of bidirectional charging (V2G) is a trigger for the development of vehicle-to-everything (V2X) technologies (Lin et al. 2022).

7.2.2 Interest in Stationary Storage

The use of EVs in V2G mode is particularly attractive for renewable energy storage, but it also has some important constraints that need to be taken into account (Al Khafaf et al. 2022; Wei et al. 2022):

- the wide availability of charging stations in V2G mode in residential areas and public places (parking),
- the availability of private EVs, such that people should get involved and plug in their cars when they are not using them,
- check the voltage at the battery terminals, and
- check the state of charge and the life of the batteries.

Charging and discharging result in premature wear on the batteries. Once batteries are considered worn out, they can no longer be used in EVs, while it is possible to use the utilized batteries for stationary storage. Assessing the storage potential of the second-life EV batteries requires analysis of two major parameters (Alam et al. 2022):

- the first average life (i.e., use in the vehicle itself), and
- the level of performance degradation of the battery at the end of its first life (indicating its state of health).

The majority of technical studies carried out by car manufacturers and research laboratories give an estimated initial life of between 7 and 15 years for batteries. The first real lifetime of a battery is directly related to its usage factors (e.g., the number of complete cycles, the depth of cycles, operating conditions, temperature). The level of performance degradation of a battery at the end of its first life (energy capacity, power) is mostly estimated to be between 20% and 30%, with the performance between 70% and 80% of that of the same new battery.

With the exception of Nissan and Renault, initiatives to reuse batteries from electric cars are almost non-existent. This is surprising when one considers that the cost of energy storage for the recycled batteries can be up to half as high as that of new specific batteries. By 2025, the difference could even reach 70%. For example, in partnership with the Irish American multinational company Eaton Corporation PLC, Nissan has just developed a device called xStorage Home, controllable from a smartphone (Shahjalal et al. 2022). It uses the utilized second-hand batteries to create an electricity storage device for homes. If the home produces its own renewable energy, including the solar panels, the system recharges, allowing the stored electricity to be redistributed when the home needs it or sold back to the grid. In the absence of photovoltaic panels, xStorage Home is nevertheless interesting, allowing the storage of the electricity during off-peak hours before redistributing it during peak hours. For Renault, the issue of recycling and reuse of batteries is also central, since 80% of them are rented and their replacement is at about the motorist's expense. In addition to a solution similar to xStorage Home, the French carmaker reuses the utilized second-hand batteries to transform them into charging stations or emergency storage systems—in case of a power cut-off, for example.

7.2.3 PROBLEM OF THE INTERCONNECTION OF BATTERY CHARGERS WITH THE AC GRID

During normal operation, i.e., when the electrical distribution system must supply energy to recharge the EV's battery via one of the systems presented in the previous section, overcurrents can occur. If these are not limited, the following two problems can occur (Reymond et al. 2016):

- When the device is plugged into the AC grid, as shown in Figure 7.2, a large inrush current (i.e., 5–20 times the value of the steady-state current) can be produced to charge the storage capacitor up to the peak value of the grid voltage.
- The AC grid can experience voltage dips. These are variable and if they are deep enough, the storage capacitor discharges completely. When the AC grid voltage returns, the storage capacitor must recharge.

IEC 61000–3–3 specifies the limits of permissible voltage variations between the AC mains voltage and the terminal voltage of the device itself connected to the AC mains (Gerber et al. 2022). Specifically, equation (7.1) defines the rate of voltage change (in %), where DU is the voltage change (in volts) and U_{device} is the voltage across the device.

$$d = \Delta U / U_{device} \qquad (7.1)$$

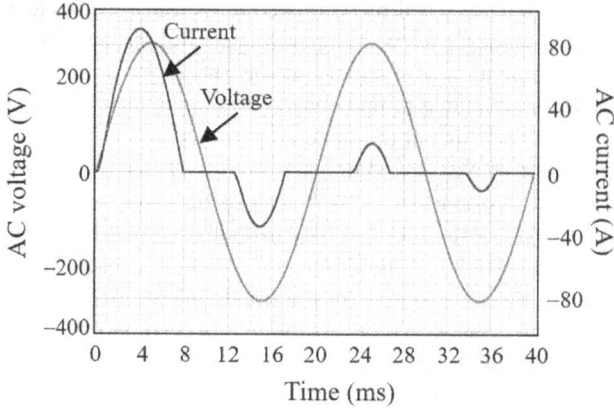

FIGURE 7.2 Example of voltage and current input waveforms during power-up without inrush current limiting.

A model for appliances likely to generate many voltage drops, such as variable load motors, is given in the IEC 61000–3–3 standard. This model expresses the number of variations per minute as a function of the value of the voltage variation rate. Depending on the type of equipment, a classification is provided according to the rate of variation not to be exceeded (this rate is noted d_{max}):

- d_{max} = 4% for devices that switch on automatically and immediately after a power failure.
- d_{max} = 6% for devices that are switched manually or automatically more than twice a day and whose start-up is either manual after a power supply interruption or delayed (the delay being at least several tens of seconds).
- d_{max} = 7% for devices which are either monitored during their usage time, switched on automatically or intended to be switched on manually twice a day at most. Their restart is either delayed (the delay being at least several tens of seconds) or manual after an interruption in the power supply.

The IEC 61000–3–3 standard also specifies that the cumulative time, during which the change rate of the voltage across the device under test can be greater than 3.3%, must be less than 500 ms. In this case, the inrush current (in root mean square [RMS] value) not to be exceeded is given by equation (7.2) where Z_{ref} and U are the reference impedance (in ohms) of the AC grid and the RMS voltage (in volts) of the AC grid, respectively.

$$I_{IN}(\text{ARMS}) = d_{max} \times U / Z_{ref} \qquad (7.2)$$

In order to comply with IEC 61000–3–3, assume that current conduction occurs for 1/6 of the AC voltage period when the charger stages are powered up (excluding the PFC), Figure 7.3 gives the AC voltage and current waveforms for the ideal case.

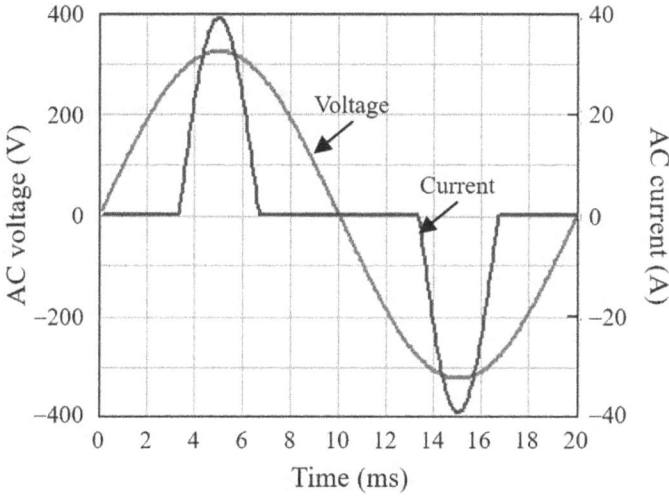

FIGURE 7.3   Waveforms of the AC grid voltage and absorbed current in the ideal case.

Using these waveforms, the maximum current value that should not exceed the standard value, is approximately 39 A.

7.3 REVIEW OF EXISTING SOLUTIONS TO LIMIT INRUSH CURRENT

7.3.1 AS A PREAMBLE

As shown in Figure 7.4, on-board chargers in EVs consist of two high-efficiency stages (Qaisar and Alyamani 2022):

- An AC-DC stage operating as a rectifier coupled with power factor correction (PFC) so that the current waveform corresponds exactly to the sinusoidal shape of the AC voltage meeting the requirements of the industry standards of the AC grid;
- A DC-DC stage supplying energy for the vehicle battery.

For the AC-DC stage, the majority of the market is based on the use of a bridge rectifier coupled to a cascade step-up chopper. In this type of architecture shown in Figure 7.4, the current limiter can be positioned on the:

- AC side, either on line (the case example in Figure 7.4) or on neutral, or the
- DC side, i.e., on the DC bus.

For the DC-DC stage, Figure 7.4 shows an example of a resonant LLC converter that provides high power density and efficiency while minimizing electromagnetic

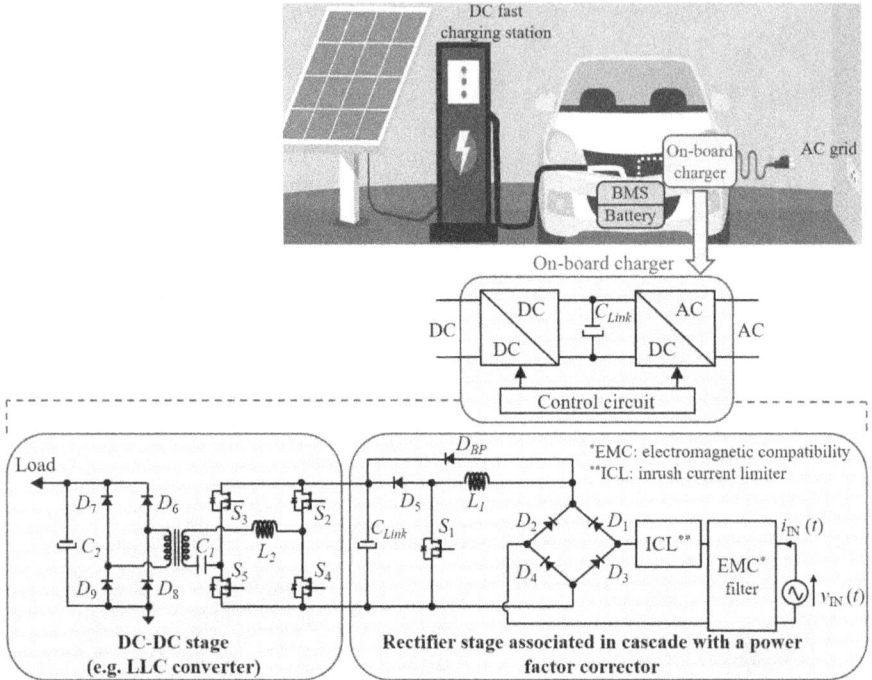

FIGURE 7.4 Example of an electrical diagram of an on-board charger in an electric vehicle.

interferences (Al Attar et al. 2021). Nevertheless, its design may require more optimization efforts compared to pulse width modulation (PWM) converters.

In the rest of this section, we will analyze the different solutions currently implemented regarding the limit inrush currents in AC-DC converters. For the readability of the following diagrams, we will call the cascade association of the power factor corrector and the DC-DC stage (see Figure 7.4) a switched-mode power supply (SMPS). The advantages and disadvantages of each solution will be also explained. For the sake of conciseness, we will not discuss the solutions implemented in low-power converters (i.e., less than 500 W).

7.3.2 PASSIVE INRUSH CURRENT LIMITER BASED ON DIFFERENTIAL INDUCTANCE

As shown in Figure 7.5, one of the simplest passive solutions to limit inrush currents is the use of an inductance on the AC bus side, which has the advantage of filtering currents in differential mode (Lee et al. 2012; Gupta et al. 2021).

However, depending on the targeted power range, its value, which is very important, can easily reach one Henry. Such inductance values are expensive, bulky and obstructive in the compactness of the whole converter. Moreover, this strategy can be limited by the physical properties of the inductor itself and, in particular, the saturation of its magnetic core at high currents.

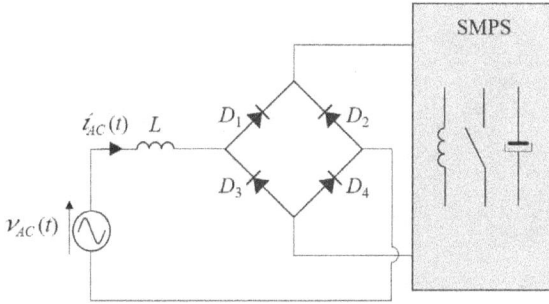

FIGURE 7.5 Passive inrush current limiter based on differential inductance.

7.3.3 INRUSH CURRENT LIMITER BASED ON A RESISTOR OR A THERMISTOR COUPLED TO A SWITCH

Another strategy for limiting inrush currents is to couple a variable resistor to a switch that can be either electromechanical (a relay) or an AC power component (a thyristor or Triac) (Kim et al. 2021). Figure 7.6 illustrates such a strategy where the used variable resistor can be a thermistor with a negative or positive temperature coefficient.

In the case of a negative temperature coefficient (NTC) thermistor, the resistance of the thermistor is high at room temperature (25°C). It heats up when a current flows through it and its Ohm value decreases. This behavior is reversed for the thermistors with a positive temperature coefficient (PTC). The thermistor has a low resistance at room temperature. It heats up when a current flows through it and its Ohm value increases. Although the NTC-type thermistor is much more common in the applications covered in this chapter for inrush current management, the PTC-type thermistor is used in specific applications, such as domestic electrical applications (e.g., disconnecting a motor winding of a refrigerator compressor after the motor starts).

The sizing of an NTC thermistor is similar to that of a resistor (limitation of Joule effect losses), with the exception that the temperature evolution during steady-state operation must be taken into account. The simplified Steinhart–Hart relationship can be used for this, as shown in equation (7.3). In this equation, R_0 is the resistance value (in Ohm) at temperature T_0 given by the manufacturer (298.15 K, or 25°C). The parameter β (in Kelvin) is a temperature coefficient available in the thermistor manufacturer's datasheet. It is important to note that this parameter is only constant in certain temperature ranges (for example, in the 25–50°C or 25–85°C ranges). As shown in Figure 7.7, the higher this coefficient, the better the temperature sensitivity.

$$R(T)(\text{Ohm}) \approx R_0 \times \exp\left(\beta \times \left(1/T - 1/T_0\right)\right) \tag{7.3}$$

However, if in the diagram in Figure 7.6, the NTC thermistor is used alone (i.e., without a parallel switch), some drawbacks appear. At very low temperatures (below

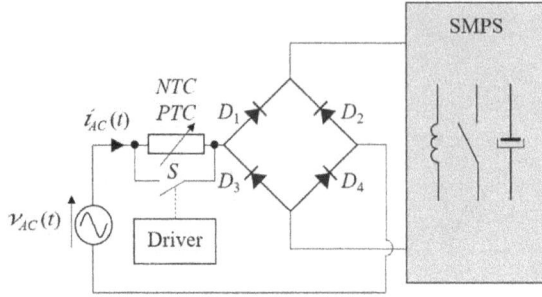

FIGURE 7.6   Example of an inrush current limiter based on a thermistor coupled to a switch.

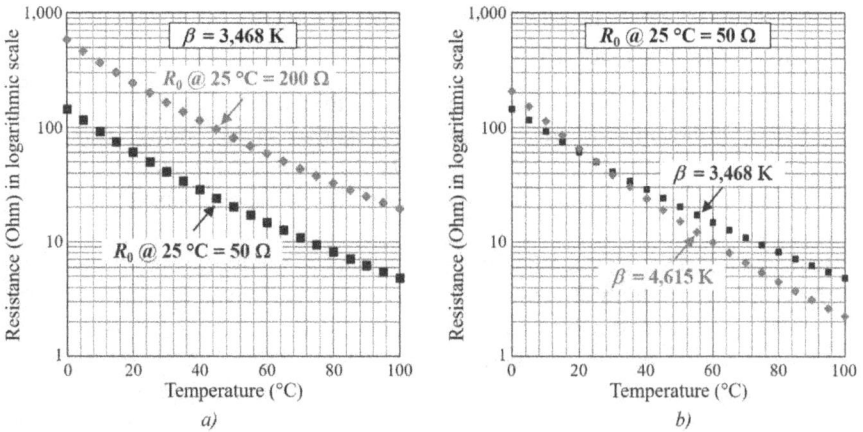

FIGURE 7.7   Impact of a) parameter R_0 and b) parameter b on the resistance of an NTC thermistor (from the RL35/40/45 series datasheet).

0°C), initial problems can occur because the resistance is too high for reaching the rated current of the application. In addition, the current limitation is insufficient at too high temperatures due to the decrease in the Ohm value of the thermistor. Another disadvantage is the possible micro interruption of the power supply. For example, for an interruption of about 100 ms, the electrochemical capacitor (see C_{Link} in Figure 7.4) discharges, and when power is restored the thermistor is still hot. The current limitation is then no longer effective, which finally leads to the Joule effect (about 1% of the device power) caused by the NTC thermistor.

If an electromechanical relay is used in parallel with the resistor or thermistor, the resistor acts as a current limiter when the device is connected to the AC mains. The electromechanical relay, on the other hand, is controlled at the ultimate charging process of the C_{Link} capacitor (see Figure 7.4). Since the contact resistance of the relay (i.e., 50 mΩ) is much lower than the power resistance (i.e., typically 200 times lower) in this case, the conduction losses are extremely reduced compared to using a

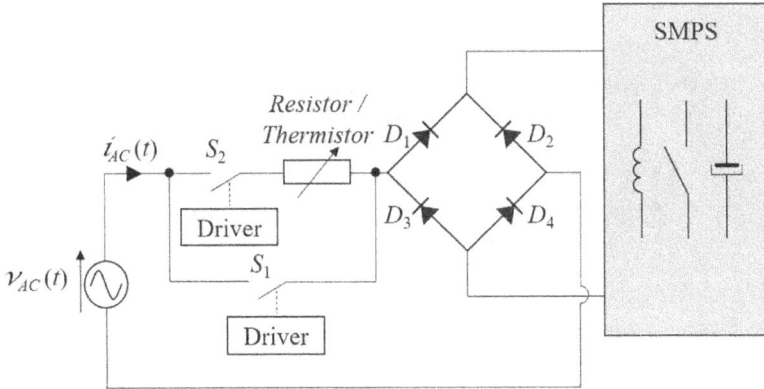

FIGURE 7.8   Example of an inrush current limiter to reduce standby losses.

resistor or thermistor alone. Therefore, for high-power applications, this solution is interesting because the conduction losses are considerably reduced.

Despite these advantages, this solution can be particularly problematic if the volume density and compactness constraints are important. In addition, the use of electromechanical relays can reduce the reliability of the charger itself, particularly for the following two reasons (Jacques et al. 2021):

- The relay is sensitive to vibration and may cause an explosion danger in a flammable environment.
- The relay may remain "blocked" in case of a short interruption of the AC grid. This is because the numerous switches on load reduce the contact quality.

Even though the steady state losses are considerably reduced compared to using a resistor or thermistor alone, there are still losses due to the power consumption of the relay coil (e.g., 500 mW for a 16 A RMS relay) and conduction losses (e.g., 12.8 W for a 16 A RMS relay with a contact resistance of 50 mΩ). In addition, the contact resistance gradually degrades with the number of executed cycles, which will consequently increase the steady-state losses.

Finally, to reduce the standby losses and as shown in Figure 7.8, a second relay can be added in series with the resistor or thermistor to disconnect the DC bus (Van den Bossche et al. 2018).

7.3.4 INTEREST OF AC POWER COMPONENTS IN PREVIOUS SOLUTIONS

To improve the reliability of the battery charger, one solution is to replace the electromechanical relays with thyristors because thyristors are not sensitive to vibrations. Moreover, the possible micro-outages of the electrical systems can be easily managed with such components. The compactness of the charger can also be reduced compared to a solution based on the electromechanical relays (Zhang et al. 2019).

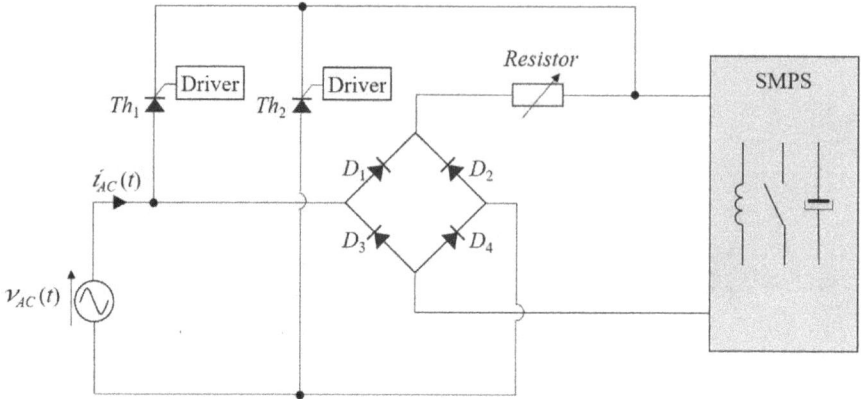

FIGURE 7.9 Example of an inrush current limiter based on two thyristors and a power resistor.

Figure 7.9 shows an example of an inrush current limiter based on two thyristors associated with a power resistor. In this example, the current limiter is positioned on the DC bus side.

The operation of this circuit is as follows. The C_{Link} capacitor (see Figure 7.4) is initially charged via the rectifier bridge and the limiting resistor while the two thyristors Th_1 and Th_2 are blocked. Once the C_{Link} capacitor is charged, the two thyristors are controlled in full wave (i.e., one for the positive alternation of the AC grid voltage, the other for the negative alternation) in order to short-circuit the resistor and thus reduce the losses in steady state.

Since the conduction losses are reduced (e.g., for 16 A RMS, about 9 W with thyristors compared to 12.8 W in the case of the relay-based solution presented in the previous section) in Figure 7.9 making this solution an interesting one, the increase of the cost of the converter (about 150% due to the addition of two thyristors, the associated control circuits and their dedicated power supplies) can be considered as its major drawback.

In the rest of this chapter, we will see how an active inrush current–limiting solution based on thyristors or Triacs can be used in soft start mode. In particular, a solution will be detailed and implemented in an AC-DC converter. Moreover, a validation by experimental measurements will finally be proposed to confirm the operation of the inrush current limiter and thus demonstrate the relevance of the proposed approach.

7.4 PROPOSAL FOR AN ACTIVE INRUSH CURRENT–LIMITING SOLUTION BASED ON THYRISTOR SOFT START

7.4.1 OPERATING PRINCIPLE

As shown in Figure 7.10, one of the simplest ways to generate soft start is to replace the common cathode diodes in the rectifier bridge with thyristors (Jacques et al.

2021). A major challenge is to optimize the phase shift angle of both thyristors (Th_1 and Th_2) to limit inrush currents while optimizing the charge of the C_{Link} capacitor in Figure 7.4.

The turn-on delay, which will be denoted by Dt in the rest of this chapter, can be adjusted by the control circuit of each thyristor. There are then two possibilities: either Dt can be constant or variable to absorb a constant sinusoidal current.

The second solution is particularly interesting because of the sinusoidal current absorption capability. The principle is illustrated in Figure 7.11, and we can see that the two thyristors Th_1 and Th_2 are alternatively controlled according to the polarity of the AC bus via adjusting their activation delay.

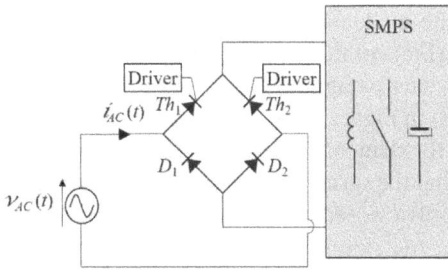

FIGURE 7.10 Example of an inrush current limiter based on thyristor soft start.

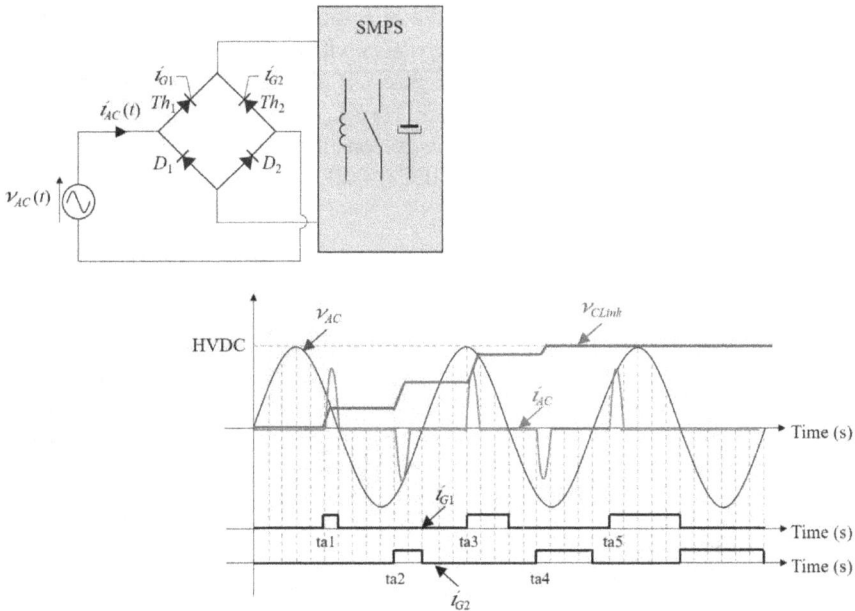

FIGURE 7.11 Soft-start principle with variable turn-on delay to create a sinusoidal absorption of the alternating current.

To implement this method, Reymond et al. have recently proposed a method to calculate the phase shift of each thyristor in order to obtain a constant value of the peak current absorbed on the AC side and comply with the regulations in force (Reymond et al. 2016). A correspondence table is obtained, which can be implemented in the thyristor control circuit. The advantage of this method is that the charging of the DC bus link capacitor (see C_{Link} in Figure 7.4) is accelerated when the battery charger is started. Its charge then goes from nearly a second, when the delay in the turn-on process of the thyristors is constant, to about 100 milliseconds, which is an adjustable variable.

7.4.2 Experimental Validation

Figure 7.12 shows the evaluation board realized in partnership with STMicroelectronics of Tours (France) that implements the inrush current–limiting function presented in the previous paragraph expressing a mixed bridge AC-DC converter (thyristors and diodes) in Figure 7.10.

At the start-up of the demonstrator, the inrush current limitation is based on the soft start by controlling the turn-on delay of the two thyristors of the mixed bridge. As explained in the previous section, this time delay is adjustable to ensure that the inrush current on the AC side is sinusoidal and of constant maximum amplitude, meeting the requirements of IEC 61000–3–3.

The evaluation board is also designed to limit standby losses by completely disconnecting the DC bus from the AC grid when not required to operate. This disconnection is achieved by turning off the thyristors of the mixed rectifier bridge.

Figure 7.13 shows the validation of the operation of the inrush current limiter during soft start by controlling the two thyristors with an adjustable turn-on delay. In this example, the C_{Link} capacitance (see Figure 7.4) is equal to 1 mF, and the AC-DC converter is connected to a 230 V RMS, 50 Hz AC grid. Figure 7.13 also illustrates that the current absorbed on the AC bus is of constant magnitude and its values are well in line with what is expected in the IEC 61000–3–3 standard.

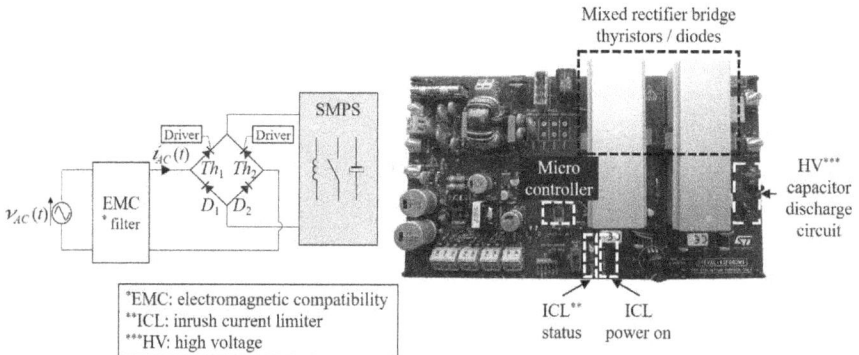

FIGURE 7.12 Evaluation board used to validate the operation of the proposed inrush current limiter.

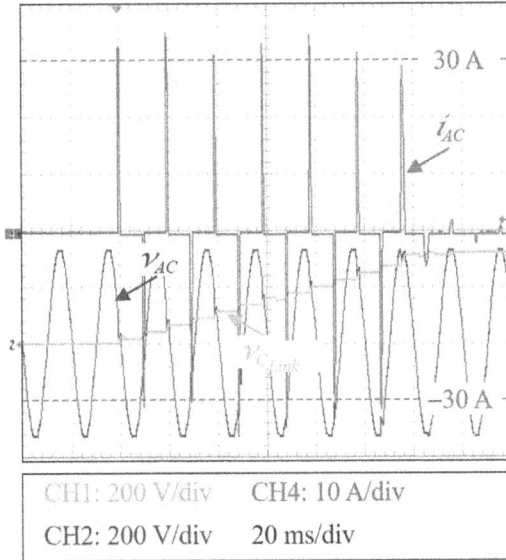

FIGURE 7.13   Experimental validation of the proposed inrush current limiter.

The experimental results show that the charge of the DC link capacitor, of the order of 130 ms, is accelerated by the mentioned thyristor control strategy. If the thyristors had been controlled by a constant turn-on delay, then this duration would have been about 850 ms under the same experimental conditions. To conclude, this solution, which is still interesting because of its cost effectiveness, allows the C_{Link} capacitor to be charged quickly, satisfying the requirements of the IEC 61000–3–3 standard.

7.5 CONCLUSIONS

The EV is still gaining ground dramatically, and its growth must be supported by continuous developments addressing better climate challenges. In particular, the electric car take-off still leaves many technological, environmental and industrial questions unanswered: production costs for the manufacturer and the vehicle purchasing costs for the consumer, battery autonomy, battery technology (lithium-ion, fuel cell), charging stations and the associated chargers. In this way, manufacturers must be ready to provide answers to all these questions.

Among its undeniable environmental benefits, the parked and charged electric car can also provide additional flexibility to the electrical system by importing and exporting some of the electricity contained in the battery to the electrical grid. As part of a smart grid integration approach, the use of second-life batteries for electricity storage is not only another way to optimize the resources needed for manufacturing them but also a complementary storage mode to accelerate an increasingly decentralized energy transition.

Battery chargers, especially those embedded in plug-in electric vehicles (battery electric vehicles and plug-in hybrid electric vehicles), play a major role in both optimizing the charging time and increasing the reliability of batteries. These on-board chargers are AC-DC converters whose performance in terms of efficiency and power quality has been widely discussed in the literature in recent years.

When a charger is connected to the AC grid, high inrush currents can be generated and impact the power system and the charger itself. During the transient phases, the maximum amplitude of the inrush current must meet the limits fixed by the particular requirements in the standard IEC 61000–3–3.

This chapter reviewed the passive and active solutions currently implemented in industrial medium power chargers (3.7 kW) in order to limit the inrush currents. One of the most commonly used solutions is the one that associates a power resistor or a thermistor to a switch, either electromechanical (relay) or electronic (alternative components, such as thyristor or Triac). Some reliability issues may arise if the power grid fails, especially if relays are used, making it a cumbersome but inexpensive solution.

This chapter also provided an opportunity to show the importance of AC switches to overcome the problems listed earlier. In particular, a simple and inexpensive solution was analyzed based on the control of the thyristor turn-on delay in order to limit the inrush currents. This soft-start solution was implemented in an AC-DC converter whose AC stage is a mixed rectifier bridge based on two thyristors and two diodes. The turn-on delay of the thyristor has been set appropriately to absorb a constant maximum amplitude of the AC current. The results of the experimental measurements validated the operation of this solution showing that the link capacitor on the DC bus can be charged quickly meeting the requirements of IEC 61000–3–3.

REFERENCES

Alam, M. S., Pillai, R. K. and Murugesan, N. (Eds.), 2022. *Developing Charging Infrastructure and Technologies for Electric Vehicles*. Hershey, PA: IGI Global. https://doi.org/10.4018/978-1-7998-6858-3

Al Attar, H., Hamida, M.A., Ghanes, M. and Taleb, M., 2021. LLC DC-DC converter performances improvement for bidirectional electric vehicle charger application. *World Electric Vehicle Journal* 13, 2. https://doi.org/10.3390/wevj13010002

Al Khafaf, N., Rezaei, A. A., Amani, A. M., Jalili, M., McGrath, B., Meegahapola, L. and Vahidnia, A., 2022. Impact of battery storage on residential energy consumption: An Australian case study based on smart meter data. *Renewable Energy* 182, 390–400. https://doi.org/10.1016/j.renene.2021.10.005

Aouichak, I., Jacques, S., Bissey, S., Reymond, C., Besson, T. and Le Bunetel, J.-C., 2022. A bidirectional grid-connected DC-AC converter for autonomous and intelligent electricity storage in the residential sector. *Energies* 15, 1194. https://doi.org/10.3390/en15031194

Arif, S. M., Lie, T. T., Seet, B. C., Ayyadi, S. and Jensen, K., 2021. Review of electric vehicle technologies, charging methods, standards and optimization techniques. *Electronics* 10, 1910. https://doi.org/10.3390/electronics10161910

Bissey, S., Jacques, S. and Le Bunetel, J. C., 2017. The Fuzzy logic method to efficiently optimize electricity consumption in individual housing. *Energies* 10, 1701. https://doi.org/10.3390/en10111701

Borge-Diez, D., Icaza, D., Açıkkalp, E. and Amaris, H., 2021. Combined vehicle to building (V2B) and vehicle to home (V2H) strategy to increase electric vehicle market share. *Energy* 237, 121608. https://doi.org/10.1016/j.energy.2021.121608

Dik, A., Omer, S. and Boukhanouf, R., 2022. Electric vehicles: V2G for rapid, safe, and green EV penetration. *Energies* 15, 803. https://doi.org/10.3390/en15030803

Elfeki, I., Jacques, S., Aouichak, I., Doligez, T., Raingeaud, Y. and Le Bunetel, J.-C., 2018. Characterization of narrowband noise and channel capacity for powerline communication in France. *Energies* 11, 3022. https://doi.org/10.3390/en11113022

Gerber, D. L., Ghatpande, O. A., Nazir, M., Heredia, W. G. B., Feng, W. and Brown, R. E., 2022. Energy and power quality measurement for electrical distribution in AC and DC microgrid buildings. *Applied Energy* 308, 118308. https://doi.org/10.1016/j.apenergy.2021.118308

Gupta, J., Maurya, R. and Arya, S. R., 2021. Soft-switching based integrated on-board charger for electric vehicles. *Proceedings of the 47th Annual Conference of the IEEE Industrial Electronics Society (IECON)*, 1–6. https://doi.org/10.1109/IECON48115.2021.9589306

Heidrich, O., Dissanayake, D., Lambert, S. et al., 2022. How cities can drive the electric vehicle revolution. *Nature Electronics* 5, 11–13. https://doi.org/10.1038/s41928-021-00709-3

Jacques, S., Reymond, C., Benabdelaziz, G. and Le Bunetel, J. C., 2017. A relevant inrush current limitation based on SCRs' smart control used in EV battery chargers. *Proceedings of the International Conference on Renewable Energies and Power Quality (ICREPQ)* 15, 49–53. https://doi.org/10.24084/repqj15.21

Jacques, S., Reymond, C., Le Bunetel, J. C. and Benabdelaziz, G., 2021. Comparison of the power balance in a Totem-Pole Bridgeless PFC topology with several inrush current limiting strategies. *Journal of Electrical Engineering* 72, 12–19. https://doi.org/10.2478/jee-2021-0002

Kim, H., Park, J., Kim, S., Hakim, R. M., Kieu, H. P. and Choi, S., 2021. Single-stage EV on-board charger with single- and three-phase grid compatibility. *Proceedings of the IEEE Applied Power Electronics Conference and Exposition (APEC)*, 583–589. https://doi.org/10.1109/APEC42165.2021.9487195

Laffont, P., Waygood, E. O. D. and Patterson, Z., 2022. How many electric vehicles are needed to reach CO_2 emissions goals? A case study from montreal, Canada. *Sustainability* 14, 1441. https://doi.org/10.3390/su14031441

Lee, E. J., Ahn, J. H., Shin, S. M. and Lee, B. K., 2012. Comparative analysis of active inrush current limiter for high-voltage DC power supply system. *Proceedings of the 2012 IEEE Vehicle Power and Propulsion Conference*, 1256–1260. https://doi.org/10.1109/VPPC.2012.6422710

Lee, S. Y., Lee, W. S., Lee, J. Y., et al., 2022. High-efficiency 11 kW bi-directional on-board charger for EVs. *Journal of Power Electronics* 22, 363–376. https://doi.org/10.1007/s43236-021-00344-3

Lin, S.-Y., Huang, C.-M. and Wu, T.-Y, 2022. Multi-access edge computing-based vehicle-vehicle-RSU data offloading over the multi-RSU-overlapped environment. *IEEE Open Journal of Intelligent Transportation Systems* 3, 7–32. https://doi.org/10.1109/OJITS.2022.3142065

Naim, A. and Vashist, D., 2022. Technological advancements for reduced charging time of electric vehicle batteries: A review. In: Khosla, A. and Aggarwal, M. (eds) *Smart*

Structures in Energy Infrastructure. Studies in Infrastructure and Control. Singapore: Springer. https://doi.org/10.1007/978-981-16-4744-4_11

Qaisar, S. M. and Alyamani, N., 2022. A review of charging schemes and machine learning techniques for intelligent management of electric vehicles in smart grid. In: Visvizi, A. and Troisi, O. (eds) *Managing Smart Cities.* Cham: Springer. https://doi.org/10.1007/978-3-030-93585-6_4

Ravi, S. S. and Aziz, M., 2022. Utilization of electric vehicles for vehicle-to-grid services: Progress and perspectives. *Energies* 15, 589. https://doi.org/10.3390/en15020589

Reymond, C., Jacques, S., Benabdelaziz, G. and Le Bunetel, J. C., 2016. An active inrush current limiter based on SCR phase shift control for EV charging systems. *Journal of Energy and Power Engineering* 10, 247–257. https://doi.org/10.17265/1934-8975/2016.04.006

Rivera, S., Kouro, S., Vazquez, S., Goetz, S. M., Lizana, R. and Romero-Cadaval, E., 2021. Electric vehicle charging infrastructure: From grid to battery. *IEEE Industrial Electronics Magazine* 15(2), 37–51. https://doi.org/10.1109/MIE.2020.3039039

Schram, W., Brinkel, N., Smink, G., van Wijk, T. and van Sark, W., 2020. Empirical evaluation of V2G round-trip efficiency. *Proceedings of the 2020 International Conference on Smart Energy Systems and Technologies (SEST)*, 1–6. https://doi.org/10.1109/SEST48500.2020.9203459

Shahjalal, M., Roy, P. B., Shams, T., Fly, A., Chowdhury, J. I., Ahmed, M. R. and Liu, K., 2022. A review on second-life of Li-ion batteries: Prospects, challenges, and issues. *Energy* 241, 122881. https://doi.org/10.1016/j.energy.2021.122881

Van den Bossche, A., Stoyanov, R. S. and Valchev, V. C., 2018. A thyristor- and thermistor-based inrush current limiter for DC-link start-up. *International Journal of Electronics* 105, 1615–1627. https://doi.org/10.1080/00207217.2018.1477194

Von Bonin, M., Dörre, E., Al-Khzouz, H., Braun, M. and Zhou, X., 2022. Impact of dynamic electricity tariff and home PV system incentives on electric vehicle charging behavior: Study on potential grid implications and economic effects for households. *Energies* 15, 1079. https://doi.org/10.3390/en15031079

Wei, H., Zhang, Y., Wang, Y., Hua, W., Jing, R. and Zhou, Y., 2022. Planning integrated energy systems coupling V2G as a flexible storage. *Energy* 239 Part B, 122215. https://doi.org/10.1016/j.energy.2021.122215

Xue, C., Zhou, H., Wu, Q., Wu, X. and Xu, X., 2021. Impact of incentive policies and other socio-economic factors on electric vehicle market share: A panel data analysis from the 20 countries. *Sustainability* 13, 2928. https://doi.org/10.3390/su13052928

Zhang, D., Lin, H., Zhang, Q., Kang, S. and Z. Lu, 2019. Analysis, design, and implementation of a single-stage multipulse flexible-topology thyristor rectifier for battery charging in electric vehicles. *IEEE Transactions on Energy Conversion* 34(1), 47–57. https://doi.org/10.1109/TEC.2018.2856905

8 Smart Grid–Based Control of Modes of Railway Power Supply Systems

Y. Bulatov, A. Kryukov, and K. Suslov

CONTENTS

8.1 INTRODUCTION: DESCRIPTION OF THE CONTROL OBJECT

The AC railway power supply system (RPSS) (Figure 8.1) is a complex technical system of a significantly large dimension. The complexity of the RPSS is also determined by the stochastic set of possible states and interactions with environment. These factors considerably complicate development of adequate RPSS models.

The RPSS $\mathbf{S}_{SERR}$ comprises three complex subsystems:

$$\mathbf{S}_{SERR} = \mathbf{S}_{EES} \cup \mathbf{S}_{STE} \cup \sum_{k=1}^{n} \mathbf{S}_{RES}^{(k)},$$

DOI: 10.1201/9781003293989-8

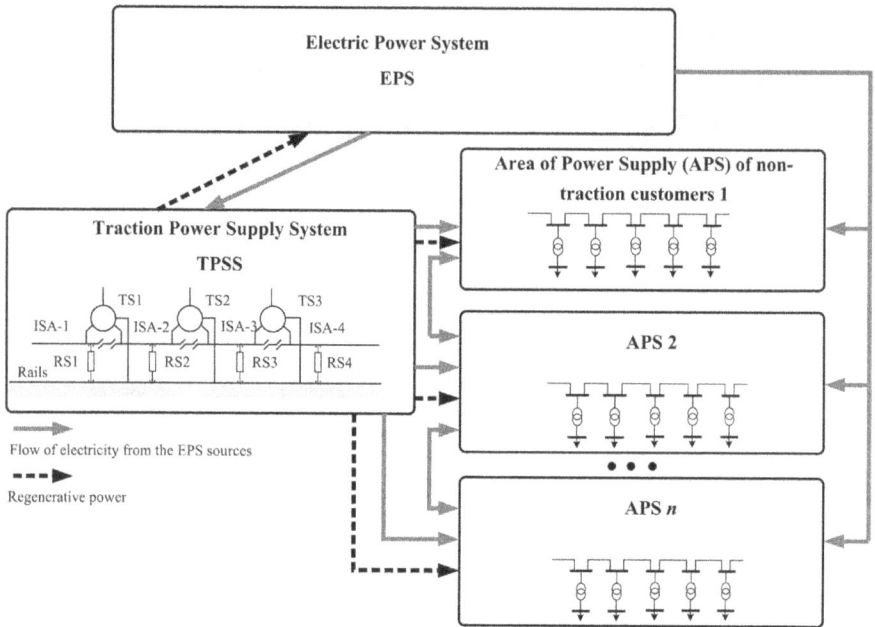

FIGURE 8.1 Diagram of the alternating current RPSS: TS—traction substation; ISA—intersubstation area; RS—rolling stock.

where S_{EES}—supply electric power system (EPS); $S_{STE} = S_{STE}^{(25)} \cup S_{STE}^{(2 \times 25)}$ is a traction power supply system (TPSS); $S_{RES}^{(k)}$ is a kth power supply area for non-traction and non-transport consumers; $S_{STE}^{(25)}$–25 kV TPSS; and $S_{STE}^{(2 \times 25)}$–2×25 kV TPSS.

Traction substations are normally powered from 110 to 220 kV networks of EPS. The traction power supply system includes 25 kV and 2x25 kV single-phase traction networks, which include three segments: traction transformers, contact suspensions and a rail network. Traction transformers are often made with three windings. The third winding can have a voltage of 6kV, 10 kV or 35 kV and can be used to power stationary railway consumers, as well as the facilities not included in the railway infrastructure. The 6–10–35 kV power lines and transformer substations connected to them form the power supply areas of stationary consumers.

Each subsystem can be associated with the following tuple definition:

$$S : \left\{ \{ \mathbf{el} \}, \ \{ \mathbf{Lin} \}, \ F \right\},$$

where $\{ \mathbf{el} \} = \{ \mathbf{el}_E \} \cup \{ \mathbf{el}_I \}$ are elements of the RPSS; $\{ \mathbf{el}_E \}$ are power elements directly participating in power transformations; $\{ \mathbf{el}_I \}$ are information elements that receive, process and transmit information about the RPSS; $\{ \mathbf{Lin} \}$ are connections between elements that determine the power and energy structures of the RPSS; F is the function of the RPSS that determines its main emergent property, which is not inherent in individual elements.

Since the function of the RPSS is the complete and uninterrupted supply of power to transportation and stationary objects of railroads, we can write

$$\left(\forall t_k \in T_H\right) P_\Sigma\left(t_k\right) = P_{\Sigma T}\left(t_k\right) + P_{\Sigma HP}\left(t_k\right) + \Delta P\left(t_k\right);$$
$$Q_\Sigma\left(t_k\right) = Q_{\Sigma T}\left(t_k\right) + Q_{\Sigma HP}\left(t_k\right) + \Delta Q\left(t_k\right), \tag{8.1}$$

where t_k is a time moment; $T_H = T - T_A$ is total time of the system regular operation; T_A is a total time of emergencies in which the balance relations (8.1) may not be observed; T is a lifetime of the system; $P_\Sigma\left(t_k\right)$ is a supplied active power; $P_{\Sigma T}\left(t_k\right)$ is a total active power consumed by the rolling stock; $P_{\Sigma HP}\left(t_k\right)$ is a total power of non-traction consumers; $\Delta P\left(t_k\right)$ are losses; $Q_\Sigma\left(t_k\right), Q_{\Sigma T}\left(t_k\right), Q_{\Sigma HP}\left(t_k\right), \Delta Q\left(t_k\right)$ are similar indicators for reactive power.

Relations (8.1) must be realized under the following conditions:

1. maximum possible efficiency, i.e.

$$Z_\Sigma \to \min,$$

where Z_Σ are power transmission and distribution costs;
2. optimal power supply reliability, i.e.

$$P_{off} \to \min \text{ at } Z_\Sigma \to \text{opt or } T_A \to \min,$$

where P_{off} is the power outages probability;
3. compliance with regulatory requirements for power quality

$$\mathbf{G} \in \mathbf{G}_D,$$

G is a vector of indicators characterizing the power quality; **G**$_D$ is a range of admissible values of power-quality indicators (PQIs);
4. compliance with the admissible levels of electromagnetic fields generated by traction networks

$$\mathbf{E} \in \mathbf{E}_D ; \mathbf{H} \in \mathbf{H}_D,$$

where **E**, **H** are vectors formed by the amplitude values of the strengths of the electric and magnetic fields calculated at controlled points in the area surrounding the traction network; **E**$_D$, **H**$_D$ areas of admissible stress values determined by safety rules and hygienic standards.

Solving the problems of controlling the modes of the RPSS implies difficulties associated with the features of the object. These issues can be divided into two groups: structural and mode. Structural features are as follows:

- Significant spacial distribution; the length of the traction network of the main railroad can reach several thousand kilometers;

- Structural heterogeneity of subsystems due to the fact that traction power supply systems form single-phase traction networks, and external power supply and areas of power supply (APS) to non-traction consumers are built on the basis of three-phase networks.

The following features are characteristic of the RPSS modes:

- Active power pulsation in the single-phase TPSS;
- Moving power consumers;
- Indicators characterizing the non-stationarity of traction loads are significantly higher than similar parameters in general-purpose electrical networks;
- Significant voltage asymmetry in 110–220 kV networks and on windings of traction transformers that power APS caused by single-phase traction loads; during low short-circuit powers on the high-voltage busbars of the transformer substation, the levels of asymmetry can significantly exceed the admissible limits;
- Large harmonic voltage distortions in traction networks (TN) and three-phase networks adjacent to traction substations;
- Noticeable effects of an electromagnetically unbalanced traction network on power lines and communications, as well as extended metal structures located near the railroad route; and
- Increased level of electromagnetic fields (EMF) created by electromagnetically unbalanced traction networks.

The active power pulsation is associated with the presence of single-phase traction loads. In a symmetrical three-phase network, the total instantaneous power does not change with time

$$p(t) = \left[\mathbf{u}(t)\right]^T \mathbf{i}(t) = 3UI\cos\phi = P, \tag{8.2}$$

where $\mathbf{u}(t) = \left[U_{am}\sin\omega t \quad U_{bm}\sin\left(\omega t - \dfrac{2\pi}{3}\right) \quad U_{cm}\sin\left(\omega t - \dfrac{4\pi}{3}\right) \right]^T$; $\mathbf{i}(t) = \left[I_{am}\sin \right.$

$\left. (\omega t - \phi_a) \quad I_{bm}\sin\left(\omega t - \dfrac{2\pi}{3} - \phi_b\right) \quad I_{cm}\sin\left(\omega t - \dfrac{4\pi}{3} - \phi_c\right) \right]^T$ are phase voltages and currents.

Relation (8.2) holds, if $U_{am} = U_{bm} = U_{cm} = U_m$; $I_{am} = I_{bm} = I_{cm} = I_m$; $\phi_a = \phi_b = \phi_c = \phi$. In the presence of a single-phase load, these conditions are not met and the ripple of power $p(t)$ starts (Figure 8.2).

The change in traction loads over time is sharply variable, which is confirmed by Figure 8.3, which shows time dependences of active powers of phases $P_A = P_A(t)$, $P_B = P_B(t)$, $P_C = P_C(t)$ obtained from measurements at the inputs of 220 kV traction substation. For comparison, Figure 8.4 shows graphs of changes in active power of industrial load in time.

(a)

(b)

(c)

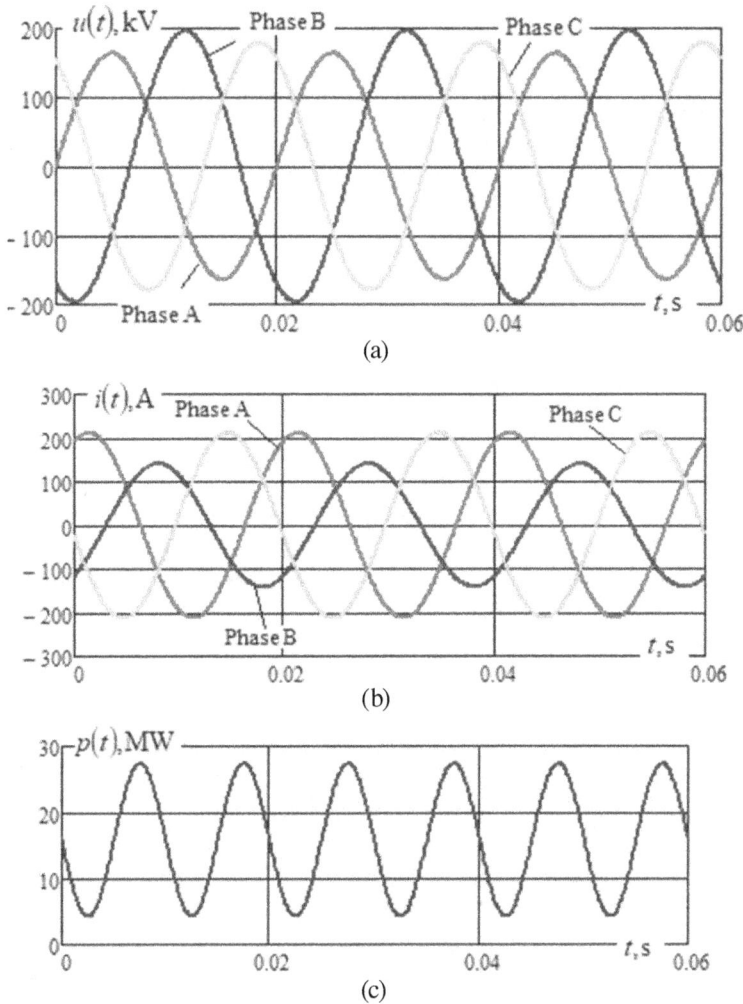

FIGURE 8.2 Active power ripple in the case of unbalance: (a)—time dependence of voltage; (b)—time dependence of current; (c)—time dependence of instantaneous power.

It is worthwhile to note one more factor that distinguishes the traction load from the general industrial one. This is the possibility of changing the sign of active power flows of the traction load in the case of regenerative braking of trains (Figure 8.5). A single-phase traction load leads to a significant voltage unbalance on the buses of traction substations and on the input terminals of consumers that are powered by these buses. With the connection of traction substations to electrical networks with low short-circuit powers at nodal points, the values of the voltage unbalance factor k_2U may exceed the allowable limits (Figure 8.6).

Rectifier units of electric locomotives, being nonlinear loads, generate harmonic currents into the network. The dynamics of the change in the harmonic factor of the

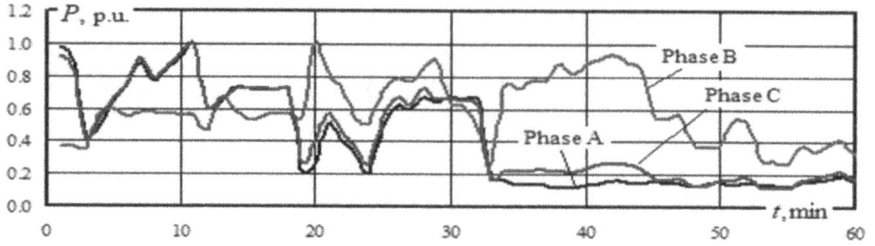

FIGURE 8.3   Change in active power of traction load over time.

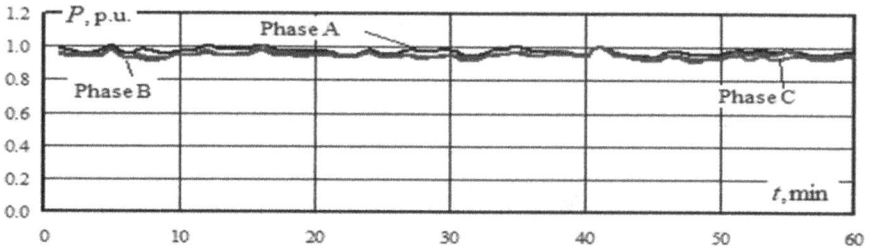

FIGURE 8.4   Change in active power of industrial load over time.

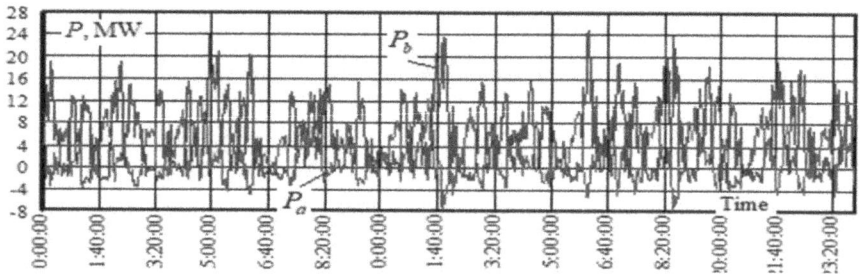

FIGURE 8.5   Relationships $P_a = P_a(t)$, $P_b = P_b(t)$ at brake energy recovery.

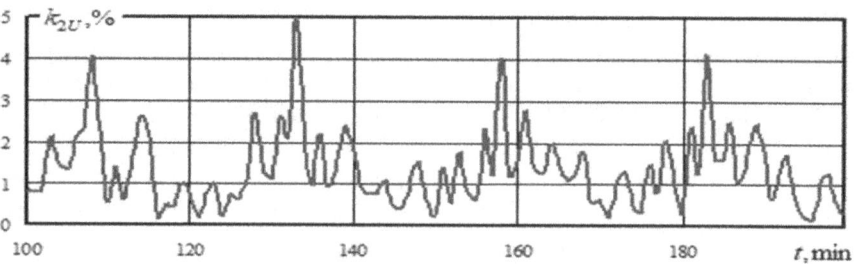

FIGURE 8.6   Buses of 220 kV traction substation. Negative sequence voltage unbalance factor.

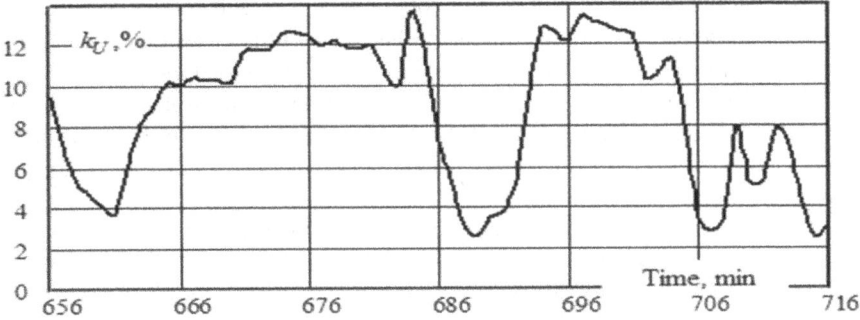

FIGURE 8.7   Buses of 220 kV traction substation. Harmonic factor of line voltage AB.

line voltage AB on the 220 kV traction substation buses are shown in Figure 8.7. As seen in the figure, the harmonic factor k_U exceeds the maximum allowable values.

The specific features of the railway power supply system analyzed earlier should be factored in when developing the control systems of operating conditions using smart grid technologies. Otherwise, there can be a sharp decrease in the efficiency of these technologies, and in some cases, premature failure of expensive devices.

The railway power supply system can be described using the following tuple:

$$SPSR = \langle X, M_k, P_f, P_C \rangle,$$

where X is a vector of continuously changing parameters; M_k is a set of functionally and structurally related components; P_f is a complex process of functioning characterized by the frequent occurrence of emergencies that lead to a shortage of time for decision-making; P_c is a complex control process.

The RPSS model, which takes into account electromagnetic and electromechanical transient processes, can be represented as a non-autonomous system of nonlinear differential equations

$$\frac{dX(t)}{dt} = \Phi_t \left[X(t), C(t), Y(t), S(t) \right], \tag{8.3}$$

where $X(t)$ is an n-dimensional vector of the state parameters characterizing the RPSS mode; t is time; Φ_t is a nonlinear operator depending, generally, on t, and its ith component is defined as $\phi_{ti} = \phi_{ti} \left[S(t), C(t), Y(t) \right]$; $Y(t)$ is a vector of perturbations; $C(t)$ is an l-dimensional vector of controls; $S(t)$ is a p-dimensional vector, which components represent structure parameters of the RPSS.

Formation of $X(t)$ is usually carried out using either Cartesian coordinates $U'_i, U''_i, U'_i + jU''_i, j = \sqrt{-1}$ or polar coordinates $U_i, \delta_i, U_i e^{\delta_i}$ of nodal voltages. The vector $Y(t)$ is formed by active $P_i(t)$ and reactive $Q_i(t)$ traction loads that change in time according to complex laws and move at high speeds. In addition, the components of this vector are the loads of stationary consumers that also change over time but with a much lower intensity. The vector of controls is formed on the basis of the train schedule. The vector $S(t)$ includes the parameters of the elements of the electric power system (EPS), TPSS and APS; at the same time, due to switching in the networks of subsystems of the RPSS, as well as due to the presence of active elements for controlling modes, individual components of this vector can change over time.

Efficient control of the RPSS modes can be implemented using smart grid technologies [1–9]. Next is a description of the results of research aimed at implementing the concept of intelligent RPSS.

8.2 PRINCIPLES OF BUILDING A SMART GRID RAILWAY POWER SUPPLY SYSTEM

Smart grid technologies are applicable to any power supply systems (PSS), including railway power supply systems. A diagram illustrating the conceptual provisions of the intelligent railway power supply system (IRPSS) is shown in Figure 8.8. The structure of the IPSS includes segments that form two units: power unit and information unit. The first unit includes measuring complexes united by the developed data transmission networks. They provide continuous monitoring and operate on-line. The formation of control actions is carried out by means of an adaptive system based on digital devices.

The power unit is formed by active elements capable of realizing the generated control actions and effectively changing the modes of the RPSS in the required direction. Those elements include:

- Controlled reactive power sources (RPS);
- Distributed generation plants [10–16];
- Energy storage devices;
- Special means to improve the power quality.

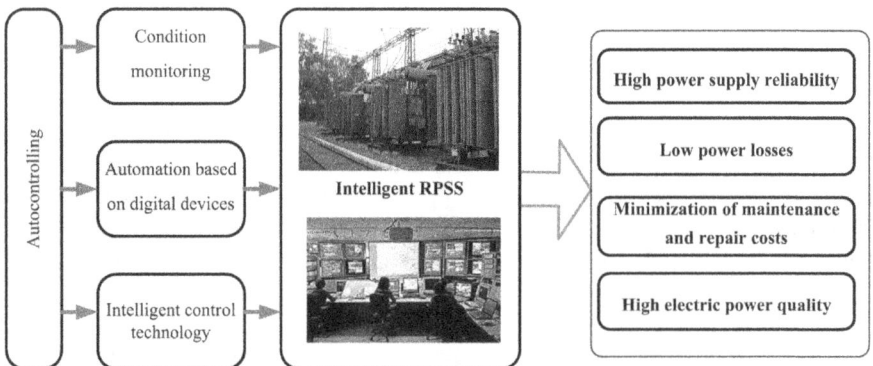

FIGURE 8.8 Conceptual provisions of the IRPSS.

To address the latter problem, a wide range of devices can be used:

- Passive and active filters of higher harmonics;
- Balancing transformers, as well as symmetrizers built on the basis of the Steinmetz scheme; and
- Devices for controlling circulating currents in the traction networks.

It should be highlighted that some devices are multifunctional. In particular, phase-controlled RPS can effectively stabilize voltage levels, as well as reduce asymmetry in supply networks and APS. Electricity storage devices help reduce losses, deviations and voltage fluctuations.

The methods and algorithms for modeling the modes of the RPSS [17–25] developed at Irkutsk State Transport University using smart grid technologies make it possible to solve the problems of controlling the modes and quality of electricity.

When solving control problems in the IRPSS, it is necessary to take into account the following:

- The multilevel RPSS, which is a nonlinear dynamic object of large dimension characterized by a continuous change not only in the numerical values of the parameters but also in the structure of the model; the latter factor is associated with the movement of traction loads; and
- Active elements behind the smart grid technologies further complicate building of models.

These difficulties can be resolved through the use of decomposition; in this case, modeling includes two stages that employ different models. At the first stage, simulation modeling is carried out, and the results are used to build dynamic models at the second stage.

When developing the IRPSS that provide an effective solution to the problems of controlling modes and increasing energy efficiency and power quality, the following results obtained at Irkutsk State Transport University can be used:

- Methods and tools for modeling a series of the RPSS modes with a discretization step that provides an adequate display of train movement according to a given schedule;
- Prediction of power consumption by traction substations based on neural network technologies;
- Situational analysis of modes and their management using algorithms of classical cluster analysis and fuzzy clustering;
- Implementation of uncertainty in modeling the modes of the RPSS using interval methods;
- Synthesis of equivalent models of external power supply systems;
- Parametric identification of elements of the RPSS: power lines, transformers, traction networks [20];

- Fuzzy modeling of the modes of three-phase and single-phase networks using phase coordinates [6];
- Optimal tuning of automatic excitation regulators and automatic speed regulators for generators of DG plants implemented on the basis of genetic algorithms, fuzzy logic, and wavelet analysis [3, 5, 26];
- Multi-agent mode control systems [2, 4].

8.3 MICROGRID, NANOGRID AND PICOGRID

Implementation of the intelligent RPSS requires development of a multi-level control system characterized by an advanced automation and increased reliability [1–5, 8, 9]. Intelligent control systems make it possible to obtain the following results:

- Increase of power supply reliability;
- Customer focus, i.e., full consideration of consumer needs;
- High quality of electricity in traction networks, APS and adjacent parts of the EPS;
- Maximum use of bandwidth;
- Expanded use of distributed generation plants, in particular with renewable energy sources; and
- Achievement of optimal costs for the production, transmission and distribution of electricity.

One of the main incentives for building IPSSs is the current transition of the electric power industry to a new technological platform based on technologies of intelligent power systems with active-adaptive networks.

One of the possible structures of an intelligent PSS is shown in Figure 8.9. It is built as a microgrid connected to a network of the intelligent EPS (IEPS) using a DC link. Due to the DC link, the adaptability of the PSS increases. In particular, it becomes possible to work efficiently when connected to networks with a reduced quality of power energy, for example, railroad networks. Equipping the microgrid with power electronic devices allows the use of non-traditional renewable energy sources (NRES), such as wind generators, micro hydroelectric power plants, solar panels, fuel cells. Combining individual microgrids opens the way to the implementation of a larger smart grid shown in Figure 8.10.

Introduction of the IEPS technologies with an active-adaptive network (AAN) can help solve the following tasks:

- Increase reliability of the EPS operation and power supply to consumers;
- Improve the quality of electrical energy and stimulate the energy efficiency growth; and
- Reduce man-made impact on the environment through the use of new types of power transmission lines with lower levels of EMF, as well as through nontraditional energy sources.

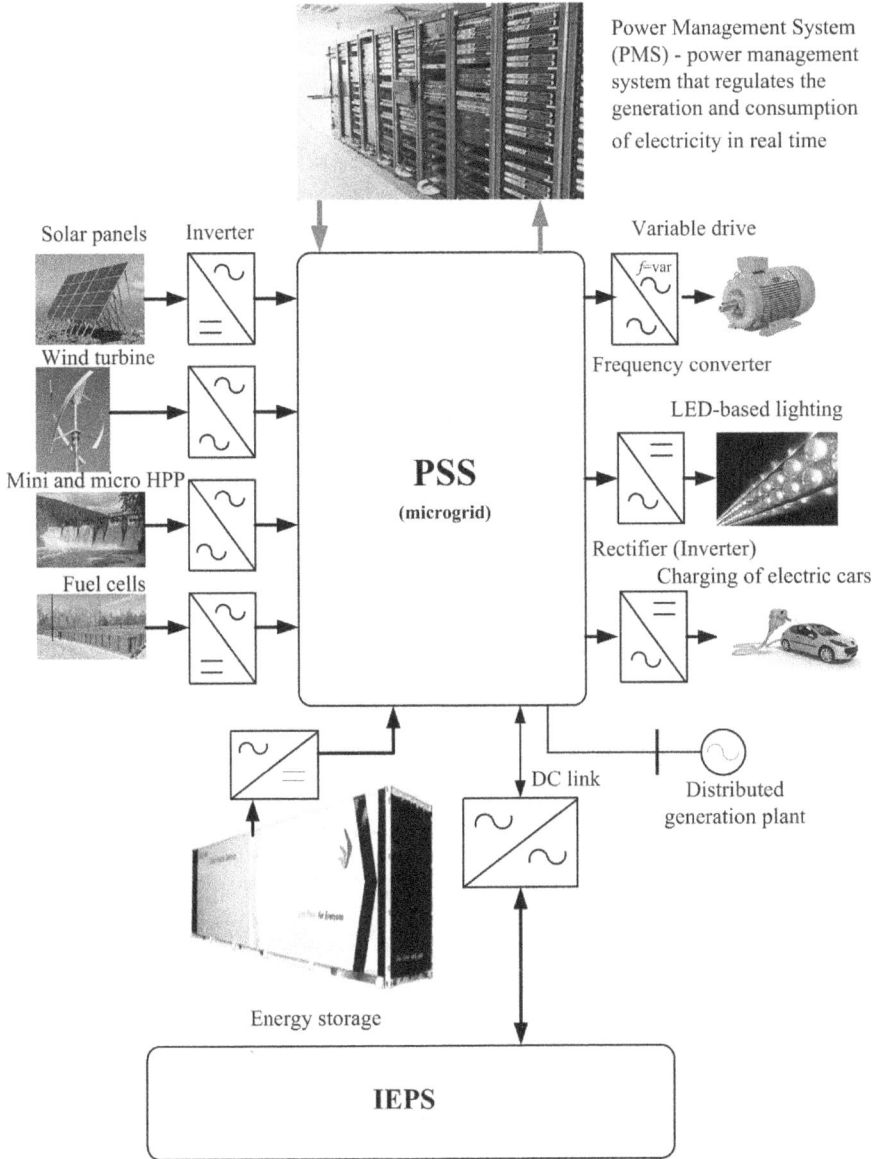

Power Management System (PMS) - power management system that regulates the generation and consumption of electricity in real time

Solar panels   Inverter

Variable drive

Wind turbine

Frequency converter

Mini and micro HPP

LED-based lighting

PSS
(microgrid)

Fuel cells

Rectifier (Inverter)

Charging of electric cars

DC link

Distributed generation plant

Energy storage

IEPS

FIGURE 8.9 A microgrid-based IPSS.

The intelligent EPS is built according to the hierarchical principle and includes three levels:

1. A megasystem covering the entire EPS;
2. Energy facilities, such as power plants and substations; and
3. Complex power equipment, as well as consumer energy complexes: smart houses, streets, etc.

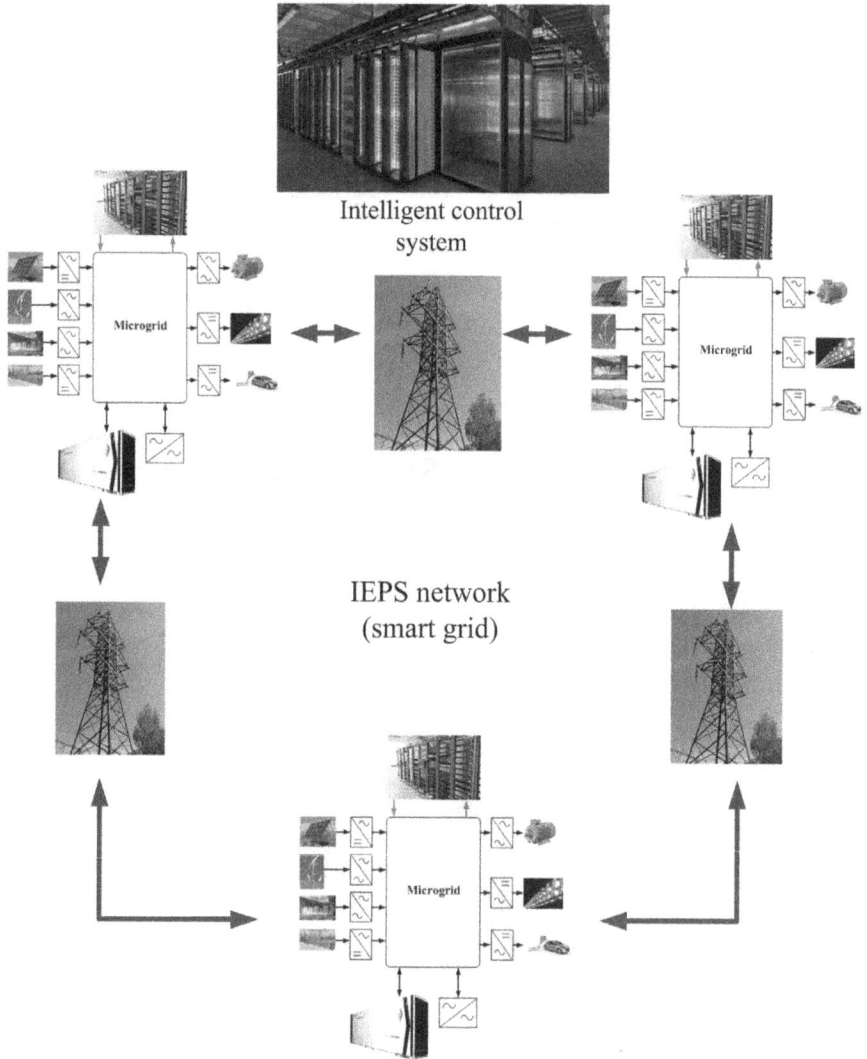

FIGURE 8.10 Combining microgrids into a smart grid.

The IEPS has two interacting layers:

- Power equipment, including active devices that provide adjustment to changes in the environment, for example, flexible alternating current transmission systems (FACTS), active harmonic conditioners (AHCs), distributed generation (DG), etc.; and
- Well-developed information infrastructure implemented on the basis of digital technologies.

One of the important components of the IPES are DG plants which are located near power-receiving facilities and tackle electricity shortages in the provinces. Reliability of the EPS, which has a large number of low-power generators, is not inferior to a traditionally built power system based on large generators; in this case, the following inequality might be fulfilled

$$\sum_{i=1}^{n} P_{Gi}^{(T)} > \sum_{i=1}^{m} P_{Gi}^{(T)} + \sum_{i=1}^{s} P_{Gi}^{(DG)}; \, n \ll m + s,$$

where $P_{Gi}^{(T)}$ are capacities of large conventional generators; $P_{Gi}^{(DG)}$ are capacities of small-scale DG generators.

Thus, an EPS with a DG may have a lower total power compared to a traditional power system. Further development of the smart grid concept is based on the ideas of EnergyNet (the Internet of Energy) that implies the use of intelligent technologies in the network facilities of medium (6–20 kV) and low (0.4 kV) voltage. The EnergyNet strategy includes development of nanogrid (6–20 kV) and picogrid (0.4 kV) systems. Figures 8.11–8.14 illustrate possible implementation of these systems in APS of non-traction consumers.

To build a nanogrid (Figure 8.15), it is necessary to use intelligent technologies for controlling the modes of power supply systems, as well as interfaces that provide connection of renewable energy sources. Energy routers can be effectively used as such interfaces [27–30]. The previously conducted studies [29, 30] show that the use of energy routers improves the quality of electricity in low-voltage networks connected to district windings of traction transformers. Energy routers, implemented on the basis of solid-state transformers, provide the connection of electrical receivers of both AC and DC voltage. Thus, energy routers can be effectively used in power supply systems of AC trunk railroads to provide energy flow control, connection of distributed generation plants, including those based on renewable energy sources.

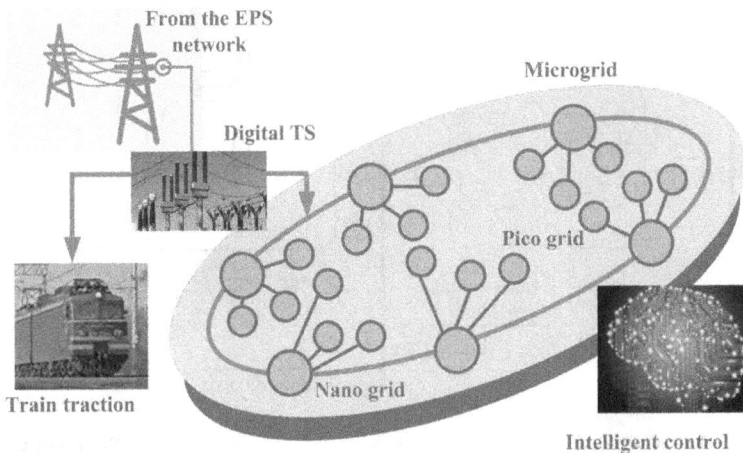

FIGURE 8.11 Microgrid structure.

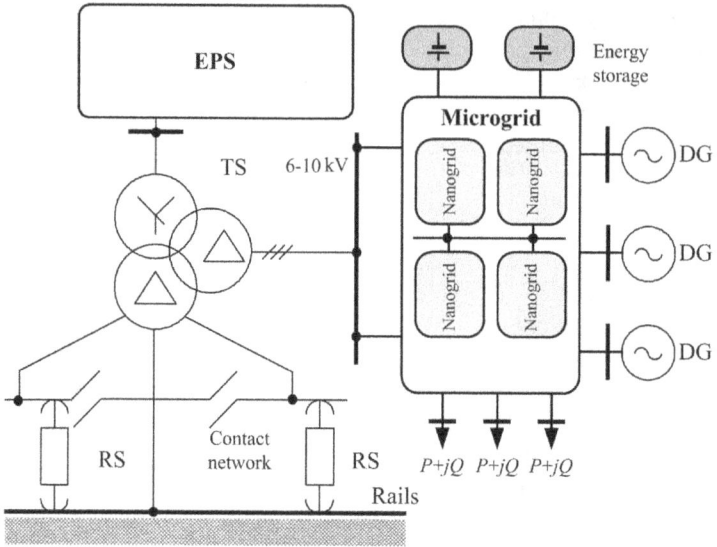

FIGURE 8.12 Microgrid for critical non-traction consumers.

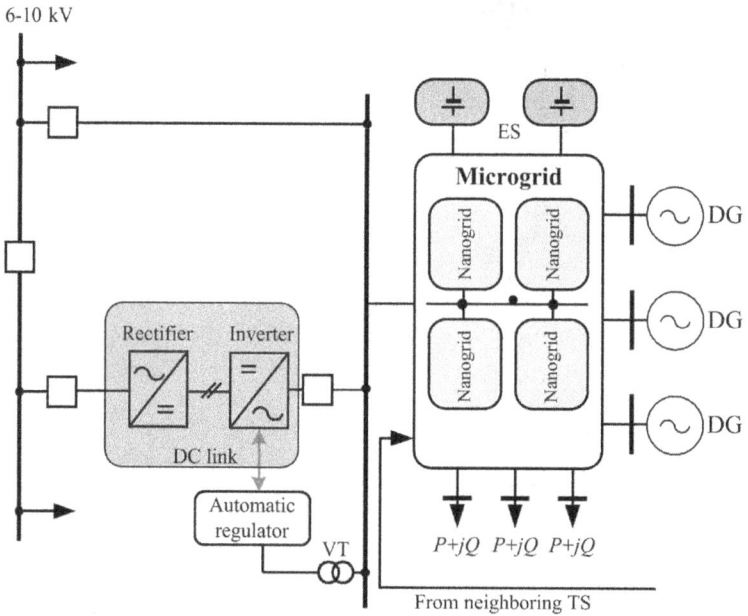

FIGURE 8.13 Connection through DC link: VT—voltage transformer; ES—energy storage.

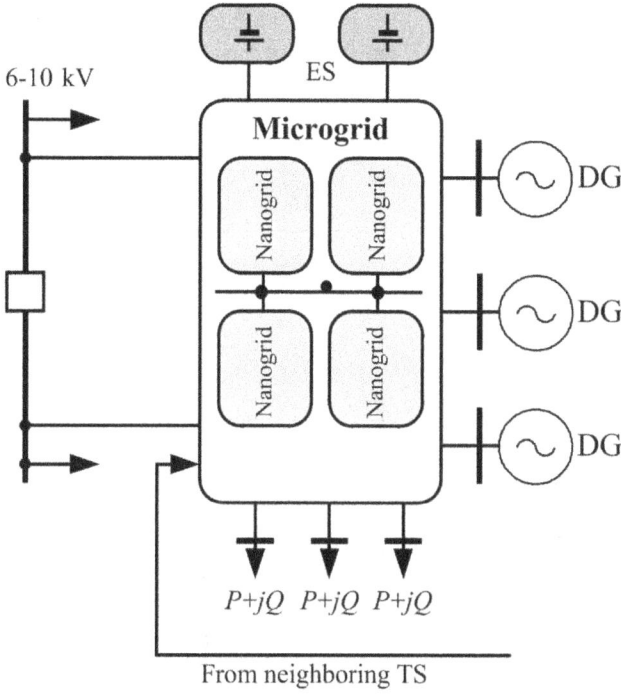

FIGURE 8.14   Direct connection to 6–10 kV barbuses.

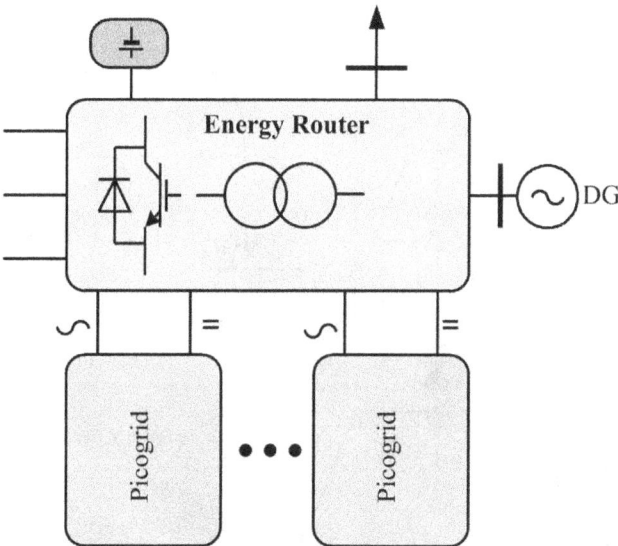

FIGURE 8.15   Nanogrid diagram.

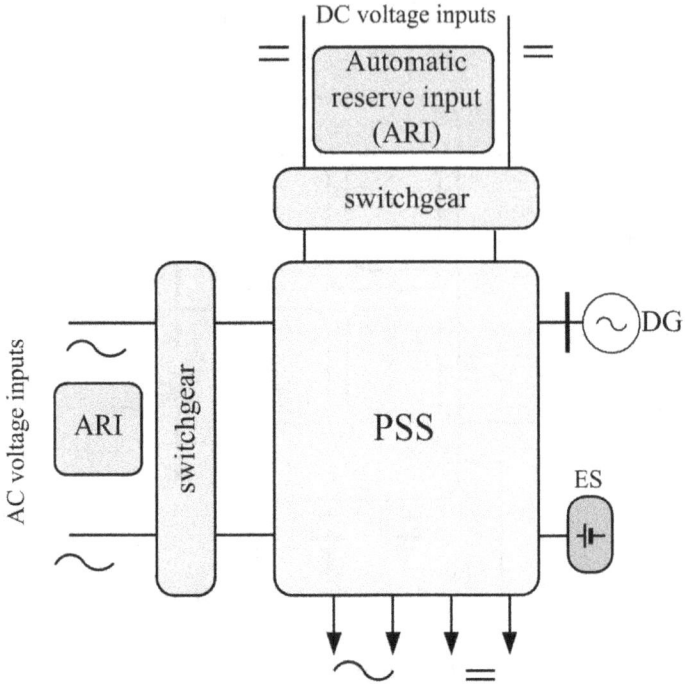

FIGURE 8.16   Picogrid diagram.

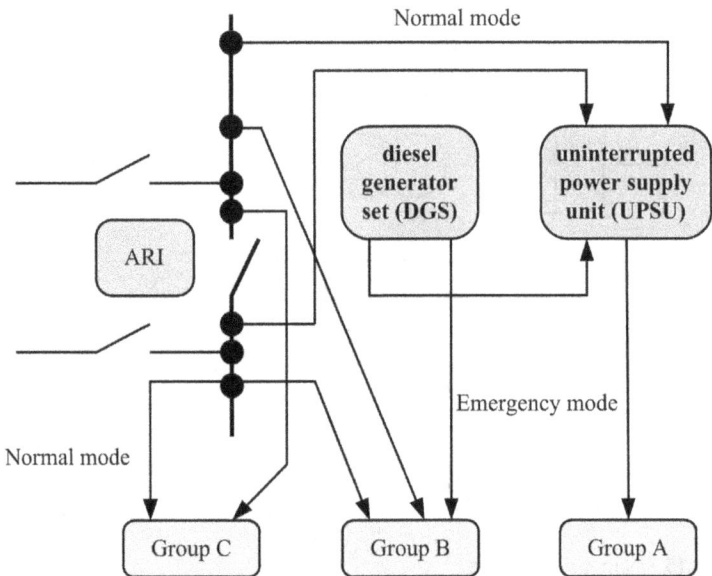

FIGURE 8.17   Picogrid with UPSU and DGS.

A generalized diagram of a picogrid that presents an intelligent (low-voltage) power supply system is shown in Figure 8.16. A picogrid for supplying essential consumers can be implemented on the basis of uninterruptible and guaranteed power supply systems (Figure 8.17).

8.4 MULTIAGENT TECHNOLOGIES FOR CONTROLLING MODES OF IPSS

New approaches to addressing the issues of intelligent control of the RPSS modes can be implemented on the basis of multi-agent technologies. The search for optimal solutions in a multi-agent control system (MACS) is carried out as a result of interaction between the software and hardware units (agents). This factor significantly distinguishes multi-agent methods from classical ones, which, as a rule, use deterministic algorithms.

The multi-agent system is decentralized, and its intelligent agents have autonomy and limited representation [2, 4]. Decentralization is ensured by the fact that there are no agents in the MACS that control the entire object. Autonomy means that the agents are, at least partially, independent. The limitation of representation is determined by the fact that none of the agents has complete information about the control object and the external environment. This is due to the complexity of the control object. Therefore, full knowledge about it has no practical value for the agent. The listed characteristics provide the realization of the properties of self-organization and complex behavior with simple strategies for the actions of individual agents. The MACS can be considered a system comprising a set of agents acting on control objects. It is convenient to represent the MACS as a black box with vectors of input and output parameters, $\mathbf{X}$ and $\mathbf{Y}$ correspondingly (Figure 8.18).

The multi-agent system receives information about the parameters of an object and its state with the help of sensors (receivers), processes this information, and activates the regulators and executive elements of the system (effectors), thereby changing the operating conditions of the object defined by the system of equations (8.3). The set of agents included in this structure, as a rule, is ranked into clusters formed by the main agents that determine the optimal parameters of the regulators and the

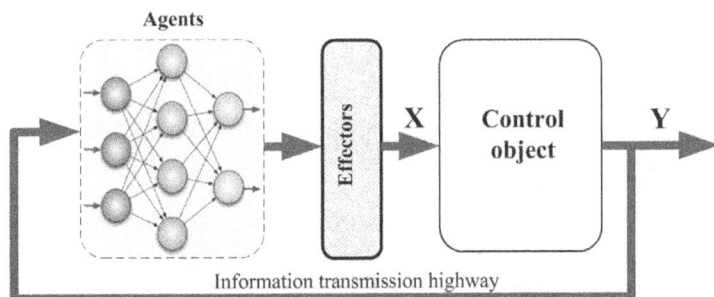

FIGURE 8.18 Functional diagram of the MACS.

system as a whole, and by the auxiliary agents that solve local problems and are distributed over individual objects.

A formal description of the MACS can be represented as a tuple

$$MASC = \{ AF, E, P, SV \},$$

where **AF** is a family of agents; **E** is a vector of states of control objects and the environment; **P** is a vector function of perception; **SV** is a vector function of state (behavior) change.

The global goal of controlling modes of the RPSS is to ensure an uninterrupted and energy-efficient power supply for transportation and non-traction consumers. The implementation of this goal requires the solution of the following tasks:

- Maintenance of voltage levels at pantographs of the rolling stock required by regulatory documents;
- Reduction to optimal values of surge currents and power losses in traction networks and areas of electricity supply; and
- Compliance with the requirements for power-quality indicators at the points of common connection of the RPSS and the feeding EPS.

A diagram of the MACS for the RPSS is given in Figure 8.19. Development of the MACS for a railway power supply system is a complex scientific and technical problem. In what follows, we present the results of studies that help implement individual segments of the MACS for the RPSS.

- Receptors that perceive the external environment; in accordance with Figure 8.19, such devices are measuring instruments that determine the complex values of currents and power flows, as well as voltage synchrophasors; and
- Effectors, which are the executive bodies that affect the RPSS: FACTS, AHCs, on-load tap-changers, energy storage devices, reactive power sources, regulators of DG plants, etc.

The scheme of interaction between agents of the MACS for the RPSS is shown in Figure 8.20. The first stage of developing the MACS requires tools that determine parameters of effectors and their rational placement in the RPSS. This problem can be solved using methods for the simulation modeling of the RPSS in phase coordinates [6, 21–23].

Next are the results of modeling a multi-agent control system for DG plants operating on the basis of synchronous generators (Figure 8.21), where agents had the following basic set of properties:

- Activity: the ability to organize and implement impacts on the control object, as well as on other agents;
- Reactivity: perception of the object state through sensors and messages from other agents;

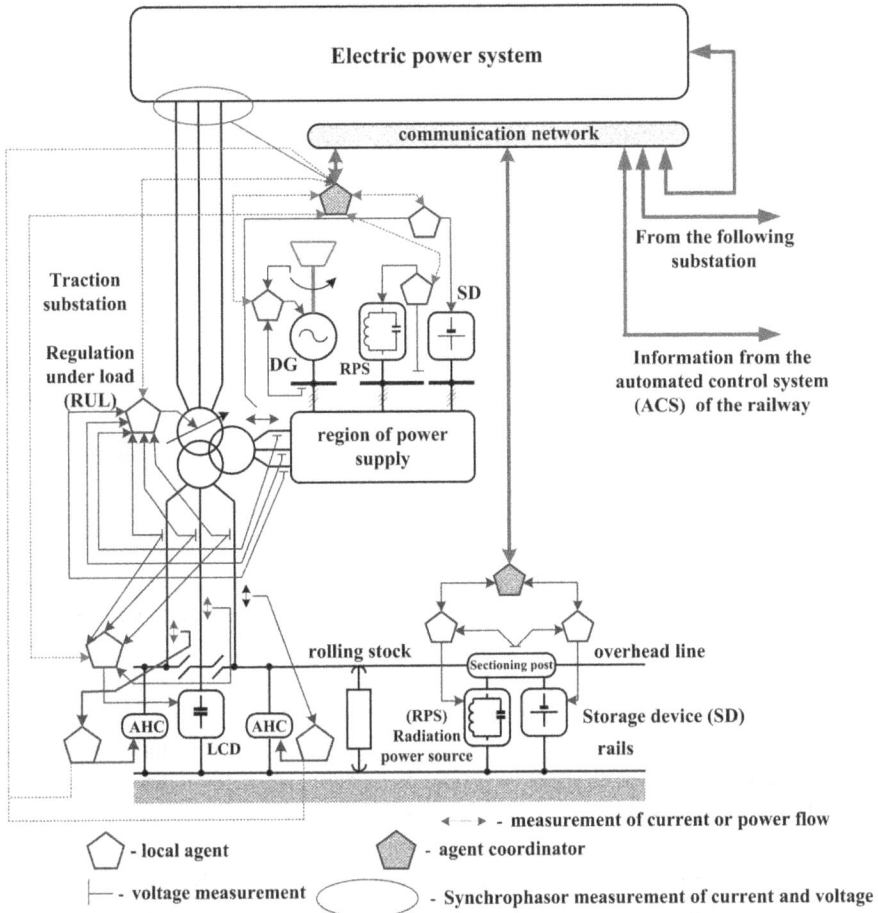

FIGURE 8.19 Diagram of the MACS: LCD—longitudinal compensation device.

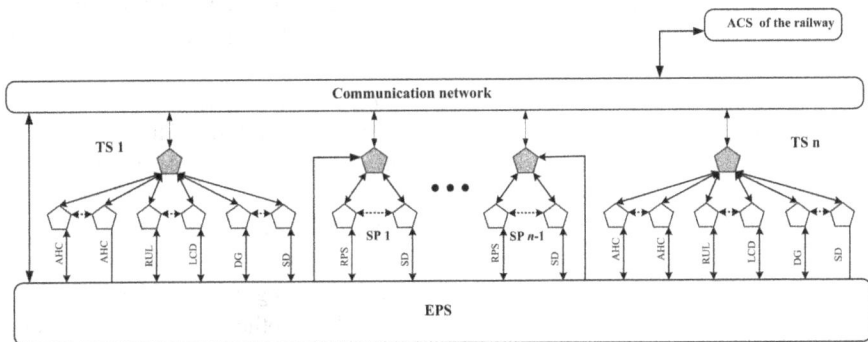

FIGURE 8.20 Scheme of interaction between agents: SP—sectioning post.

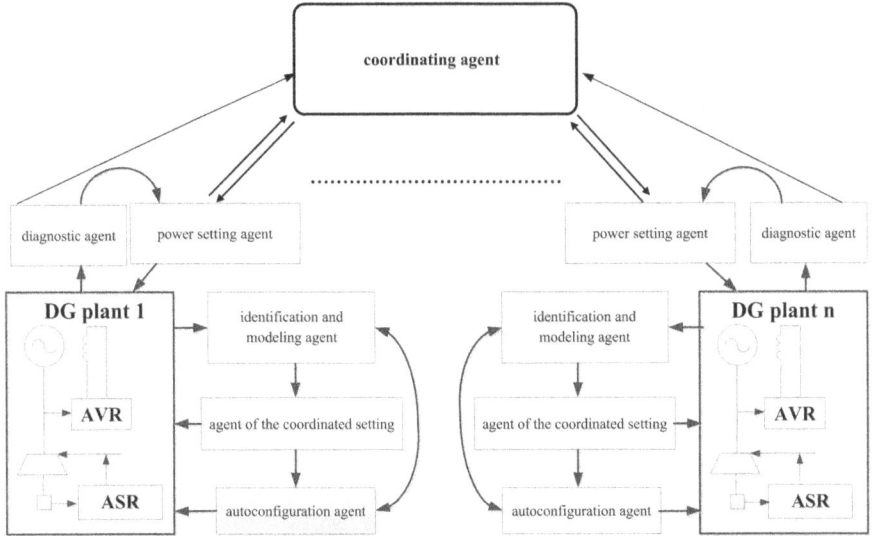

FIGURE 8.21 Diagram of the multi-agent control system for DG plants located in different nodes of the network: AVR—automatic voltage regulator; ASR—automatic speed regulator.

- Autonomy: relative independence from the environment, as well as some agents' free will that determines their own behavior;
- Sociability provided by developed communication protocols and allowing an individual agent to solve its problems together with others; and
- Purposefulness: own sources of motivation.

Through communication with other agents, each agent in the system, having its own local goals and objectives, participates in achieving the global goal proposed by the MACS, which can be formulated as an increase in the reliability, efficiency and survivability of the power supply system. In the proposed MACS, sensors, as well as meters and analyzers of power quality indicators, are used as j receptor. Effectors are programmable regulators that implement automatic excitation and speed for generators that are part of the DG plants. The MACS processor is a server for collecting, processing and storing information[1] that supplies a specific agent with the necessary data in case of failure.

The main agents are interchangeable and, in case of failure, send the corresponding message to other agents, thereby ensuring system survivability. The communication of agents can be represented by an event model (Figure 8.22) described via the Joiner network [2].

Figure 8.22 shows the Joiner network of the MACS event model for two DG plants. The nodes of this network are processes of agents and events that agents generate after their work is done. Processes communicate using input and output events. The output events of one process can be the input for another, in other words, they initiate the launch of another agent. The main elements of the description of agent-based scenarios are presented in Table 8.1.

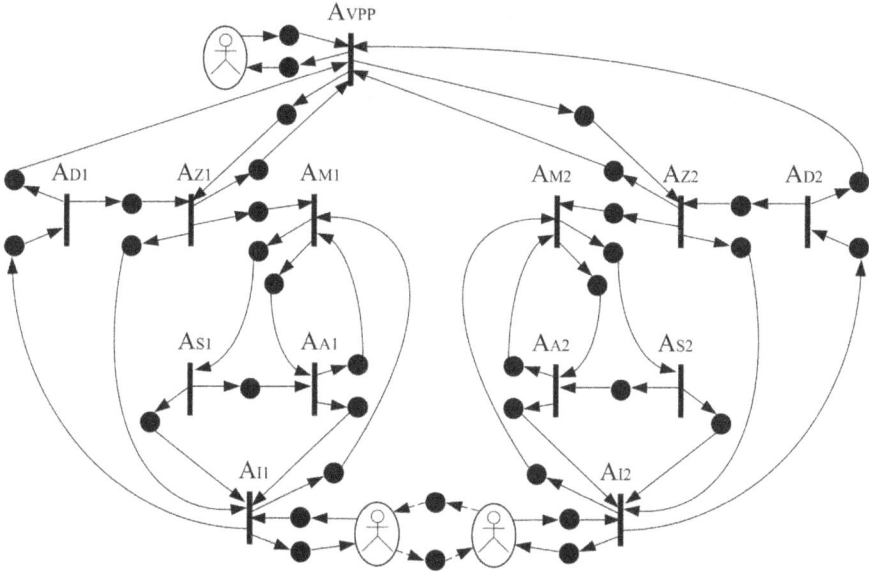

FIGURE 8.22   Joiner network of the event scenario model for the multi-agent control system of DG plants: A_{VPP}—coordinating agent; A_D—diagnostic agent; A_Z—power setting agent.

TABLE 8.1
Joiner Elements

Symbol	Meaning
$\vert$	Agent work process
●	The event that the agent generates after finishing its work
人	User interacting with the program
→	The direction of the event

A_M—identification and modeling agent; A_S—agent of the coordinated setting; A_A—autoconfiguration agent; A_I—agent of executive elements (regulators).

The basic functions of the agents of the proposed MACS can be formulated as follows:

1. The *coordinating agent* A_{VPP} sends commands to the power setting agents (PSAs) in the form of unit load graphs and also receives messages from PSAs and diagnostic agents of the DG plants, in accordance with which it

can adjust its tasks. The coordinating agent may include internal segments that can be interpreted as agents, for example:

- A *prediction agent* that predicts power consumption and power losses in the network;
- A *generation optimization agent* that selects the optimal load of DG plants according to the criteria for ensuring the specified voltage and frequency levels; and
- An *electricity market agent* involved in tariff settings.

2. The *diagnostic agent* A_D performs continuous unit diagnostics and sends messages to A_{VPP} and PSAs if the parameters under diagnostics are out of range.

3. The *power setting agent* A_Z receives commands from A_{VPP} and affects the turbine control mechanism. If necessary, it also corrects the task informing the coordinating agent about this: "I am unable to complete the task, I am making an adjustment." The PSA also receives messages from A_D and limits the unit power.

4. When the operating mode of the unit changes, the *identification and modeling agent* A_M performs the identification procedure, using wavelet transform to extract the regulator noise [26] and build a virtual model for the current mode, as well as to determine the optimal settings of the regulators and the need to change the current settings. The results obtained by agent A_M are transferred to agents A_S and A_A.

5. The *coordinated configuration agent* A_S uses the model obtained by A_M to determine the optimal coordinated configuration of the AVR and ASR of the generator for the current mode using a genetic algorithm [26] and transmits the obtained values of the configuration coefficients to A_A; if the latter fails—to the agent A_I.

6. The *autoconfiguration agent* A_A uses fuzzy control system technologies, accumulates a knowledge base of the AVR and ASR settings for various modes of operation of the unit [3, 5], checks these settings on a virtual model using the agent A_M and, if a positive effect is obtained, changes the current settings through the agent A_I.

7. The *executive element agent* A_I is the main control unit for the DG plant that provides an interface for the operator to select the mode of its operation and plan the load schedule. This agent can also change the settings of the AVR and ASR.

The proposed MACS is positioned as a self-organizing system with the ability of agents to initiate a dialogue by analyzing a situation not prescribed in advance while working under conditions of uncertainty.

The diagnostic agent has receptors that measure the generator parameters, and the PSA is equipped with an effector, according to which the generation of the active power of the generator changes. Agents are connected by information channels through ports and connectors, creating a network through which they can send messages to each other.

Next we consider algorithms for the functioning of complex intelligent agents: identification and modeling, coordinated configuration and autoconfiguration.

8.4.1 Functioning Algorithm of the Identification and Simulation Agent

The identification and modeling agent, using its receptors, captures the control signals coming from the AVR and ASR, as well as the voltage and frequency, performing the procedure for identifying the model of the DG plant for the current mode of operation. The block diagram of the algorithm for the method of passive identification of the mathematical model of the DG plant with automatic regulators used by the agent A_M is shown in Figure 8.23. Identification of the DG plant model involves obtaining a mathematical description of the system, taking into account automatic regulators in the form of a characteristic polynomial. To do this, it is necessary to present the regulated DG plant as a closed structure shown in Figure 8.24. In this case, to identify the mathematical model, two inputs from the regulators and two adjustable values will be used: the rotor speed ω_g and the generator voltage U_g (Figure 8.24). In the proposed structure, a PID regulator can be used as an ASR, and a microprocessor-based powerful AVR, which also implements the PID law, can be used as an AVR. These regulators can be represented by the following complex transfer coefficient:

$$W_{ASR} = \left(k_p + \frac{k_i}{0.1 j\omega} + \frac{k_d j\omega}{j\omega+1} \right) \cdot \frac{1}{0.01 j\omega+1},$$

where W_{ASR} is a complex transfer coefficient of the ASR; k_p, k_i, k_d are configuration coefficients of the ASR; ω is frequency;

$$W_{AVR}^{\omega} = \frac{1+0.5 j\omega}{0.5 j\omega} \cdot \left[\frac{2 k_{0\omega} j\omega}{(2 j\omega+1)(0.02 j\omega+1)} + \frac{0.05 k_{1\omega} j\omega}{0.05 j\omega+1} \right];$$

$$W_{AVR}^{U} = \frac{1+0.5 j\omega}{0.5 j\omega} \cdot \left(k_{0u} - \frac{0.02 k_{1u} j\omega}{0.06 j\omega+1} \right),$$

where W_{AVR}^{ω} is a complex transfer frequency coefficient of the AVR; W_{AVR}^{U} is a complex transfer voltage coefficient of the AVR; k_{0u}, k_{1u}, k_{0w} and k_{1w} are configuration coefficients of the AVR.

To obtain the complex transfer coefficients of the generator of the DG plant, we propose to use a noise, which was isolated by the wavelet transform [26], as a test action. The block diagram of the proposed regulator noise extraction algorithm is shown in Figure 8.25.

The isolation of the regulator noise used for identification is performed by the expression

$$f_v(t) = f(t) - f_w(t),$$

```
┌─────────────────────────────────────────────────────────────┐
│  Reading of experimental values of the input-output          │
│  parameters " of the closed regulated turbine-generator       │
│  system for various modes of operation of the DG plant        │
│  1                                                            │
└─────────────────────────────────────────────────────────────┘
                              │
                              ▼
┌─────────────────────────────────────────────────────────────┐
│  Using wavelet transform to Isolate regulator noise from      │
│  all observed input-output parameters                         │
│  2                                                            │
└─────────────────────────────────────────────────────────────┘
                              │
                              ▼
┌─────────────────────────────────────────────────────────────┐
│  Using the Fourier transform to get spectra of model input    │
│  and output signals                                           │
│  3                                                            │
└─────────────────────────────────────────────────────────────┘
                              │
                              ▼
┌─────────────────────────────────────────────────────────────┐
│  Formation of a matrix transfer function based on the         │
│  complex transfer coefficients of the main channels and       │
│  cross-links of the turbine-generator system                  │
│  4                                                            │
└─────────────────────────────────────────────────────────────┘
                              │
                              ▼
┌─────────────────────────────────────────────────────────────┐
│  Smoothing of empirical estimates of complex transfer         │
│  coefficients of the turbine-generator system model           │
│  5                                                            │
└─────────────────────────────────────────────────────────────┘
                              │
                              ▼
┌─────────────────────────────────────────────────────────────┐
│  Adequacy estimation obtained experimental model for the      │
│  DG plant                                                     │
│  6                                                            │
└─────────────────────────────────────────────────────────────┘
                              │
                              ▼
                         ╱─────────╲          No
                    ╱─────────────────╲───────────
                   ╲  7  Is model       ╱
                    ╲   adequate?      ╱
                     ╲───────────────╱
                              │ Yes
                              ▼
┌─────────────────────────────────────────────────────────────┐
│  Compilation of the characteristic polynomial of the closed   │
│  controlled turbine-generator system of the DG plant          │
│  8                                                            │
└─────────────────────────────────────────────────────────────┘
```

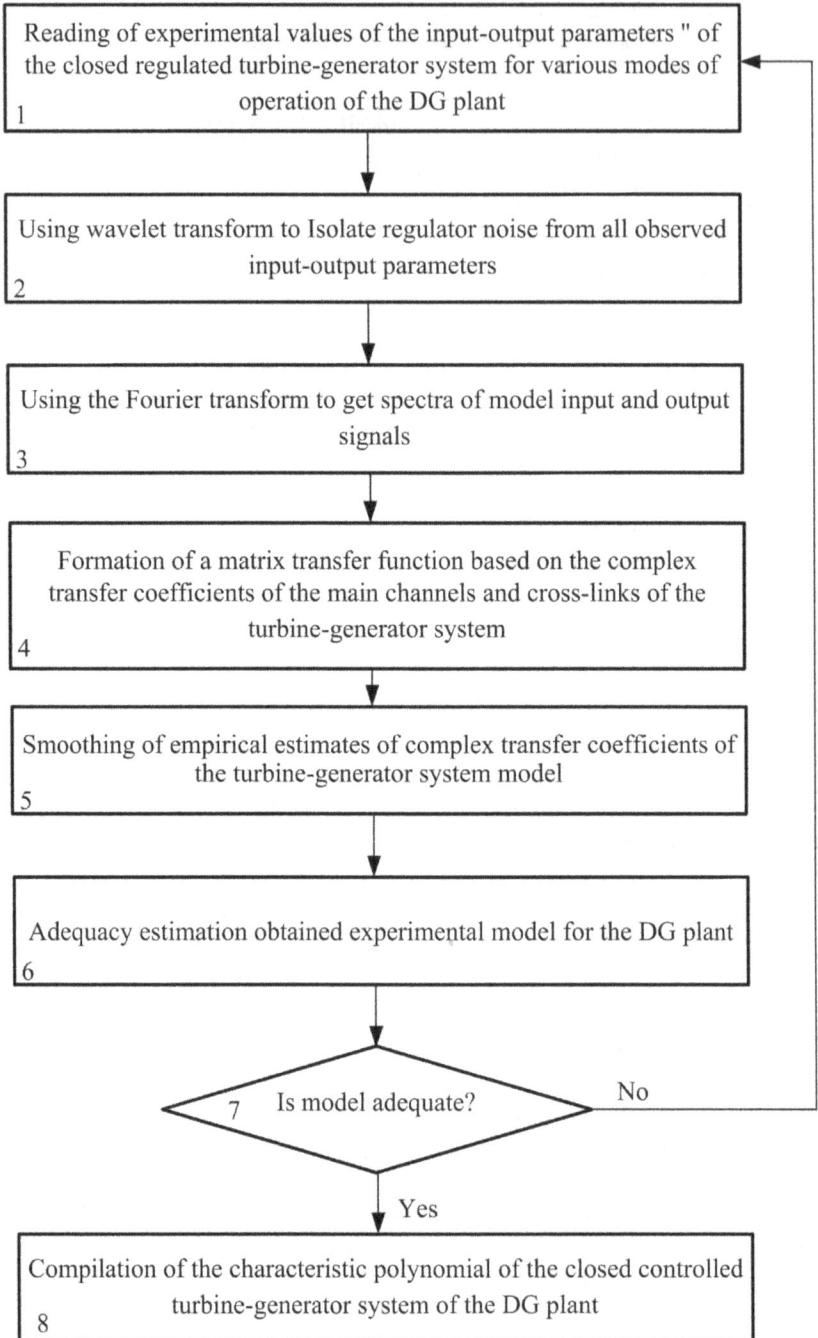

FIGURE 8.23 Block diagram of the algorithm for identifying a model of the distributed generation plant.

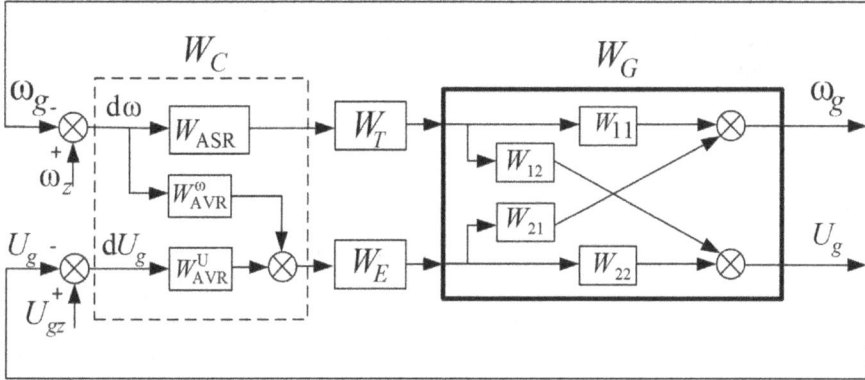

FIGURE 8.24 Diagram of a controlled distributed generation plant: W_G, W_T, W_E, W_C are transfer functions of the generator, turbine, exciter and controller.

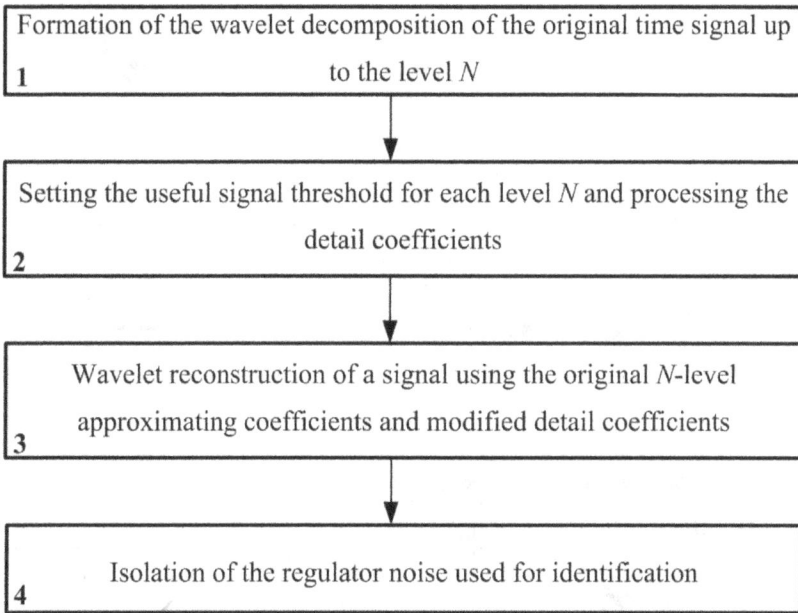

FIGURE 8.25 Block diagram of the regulator noise extraction algorithm.

where $f(t)$ is a noisy input; $f_w(t)$ is a useful component of the input obtained as a result of the wavelet transform; $f_V(t)$ is noise.

The performance of the regulator noise extraction algorithm used in identification is shown in Figure 8.26. Irregularities in the contour lines of the scaling diagram (Figure 8.26b) indicate the noise.

The transfer functions of the AVR, the ASR, turbine and the exciter are usually known. In this case, the main problem of model identification is to determine the matrix transfer function of the generator

$$W_G = \begin{vmatrix} W_{11} & W_{12} \\ W_{21} & W_{22} \end{vmatrix}.$$

Denote the noise spectra of the input signals by U_1, U_2 and the noise spectra of the output signals—by U_g, ω_g. Then the ratio of the spectral characteristics of the noise of the output and input signals will determine the necessary complex transfer coefficients. For a separate DG plant, it is necessary to determine the noise spectra of the input and output signals for two different load modes and obtain two systems of equations, from which the complex transfer coefficients of the main channels W_{11}, W_{22} and cross-couplings W_{12}, W_{21} are determined:

$$\begin{cases} Y_1 = W_{11} \cdot U_1 + W_{12} \cdot U_2, \\ Y_2 = W_{21} \cdot U_1 + W_{22} \cdot U_2 \end{cases};$$

$$\begin{cases} Y_1' = W_{11} \cdot U_1' + W_{12} \cdot U_2', \\ Y_2' = W_{21} \cdot U_1' + W_{22} \cdot U_2' \end{cases}.$$

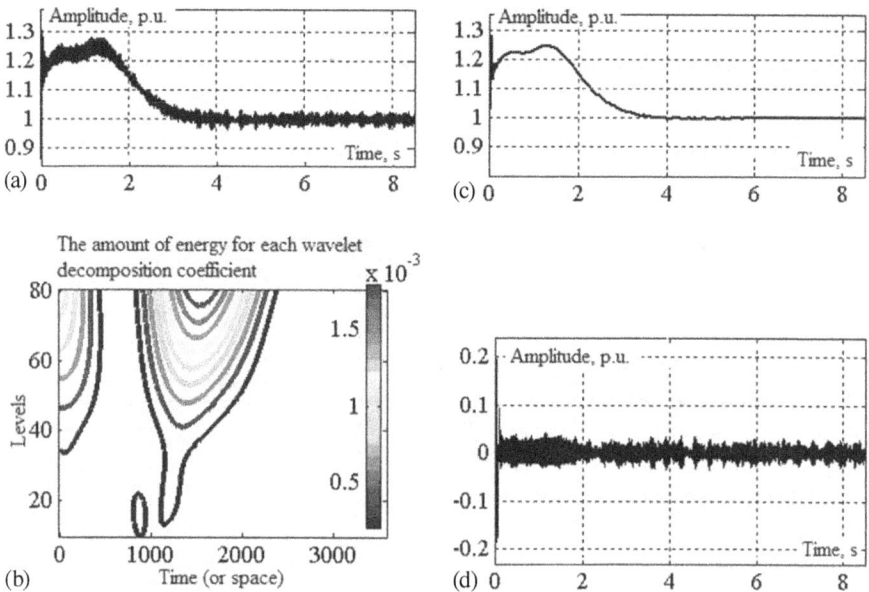

FIGURE 8.26 Performance results of the regulator noise extraction algorithm: (a)—initial noisy regulator signal; (b)—scaling diagram of the regulator signal; (c)—isolated useful signal; (d)—noise.

The noise spectra of the input and output signals are calculated using the discrete Fourier transform. If we construct the frequency responses from the complex transfer coefficients obtained with the help of regulator noise, they will be not "smooth" enough. Therefore, it is necessary to carry out additional processing of the obtained transfer coefficients. To do this, we propose to use digital signal processing based on weight windows, which implies the use of an estimate based on successive averaging of each spectrum of a discrete sample of the system signals under study instead of the complex transfer coefficient obtained during identification.

$$W'(j\omega) = \frac{\sum\limits_{i=1}^{\Omega} W_{\lambda}(\omega) \cdot |U(i\Delta\omega)|^2 \cdot W(i\Delta\omega)}{\sum\limits_{i=1}^{\Omega} W_{\lambda}(\omega) \cdot |U(i\Delta\omega)|^2}, \tag{8.4}$$

where $W(i\Delta\omega) = \dfrac{Y(i\Delta\omega)}{U(i\Delta\omega)}$ is an experimental complex transfer coefficient obtained during identification through the regulator noise; $U(i\Delta\omega)$ is the input noise spectrum; $Y(i\Delta\omega)$ is the output noise spectrum; $W_{\lambda}(\omega)$ is a weight function or a weight window.

The algorithm for smoothing the complex transfer coefficient was implemented as a code that allows the identification and modeling agent to obtain a reliable mathematical model of the regulated object and build frequency responses.

To select the range of frequency change when identifying the mathematical model of the DG plant, we propose to use the method based on the wave approach [31] that determines the system bandwidth as the total frequency of natural oscillations of units of the DG plants:

$$\Omega_i = 2\omega_{pi} \cdot \sin\left(\frac{\varphi_i}{2}\right) \cdot \sqrt{\frac{K_i}{1 + 4K_i \cdot \sin^2\left(\dfrac{\varphi_i}{2}\right)}}, \tag{8.5}$$

where $\omega_{pi} = \dfrac{\sqrt{\dfrac{\partial P_{Gi}}{\partial \delta_{Gi}} \cdot \dfrac{\omega_0}{T_{je}}}}{2\pi}$ is a partial frequency (frequency of natural oscillations of the rotor of the ith generator of the DG plant) [Hz]; $\dfrac{\partial P_{Gi}}{\partial \delta_{Gi}} = \dfrac{E_{qi} \cdot U_{gi}}{X_{d\Sigma i}} \cdot \cos\delta_i$ is a synchronizing power of the ith generator of the DG plant [p.u.]; $\omega_0 = 314$ [rad/s] is a nominal angular frequency of rotation of the generator rotor; T_{je} is an equivalent mechanical inertia constant of the unit [s]; φ_i is a wavenumber (change in the phase of the spatial harmonic between two adjacent nodes of the chain-to-chain circuit) [rad]; $K_i = \dfrac{U_S \cdot X_{di}}{E_{qi} \cdot X_{d\Sigma i}}$ is a coefficient of relative rigidity of connections of the ith generator with the system [p.u.]; X_{di} is an inductive reactance of the generator along the longitudinal axis [p.u.]; $X_{d\Sigma i} = X_{di} + X_{BH}$ is the total resistance of the connection between the generator and the external network; U_S is a system bus voltage.

For a chain homogeneous circuit consisting of N generators (if it is connected to the system) the wavenumber is determined as follows [31]

$$\varphi_i = \frac{(i-0.5)\cdot\pi}{N+0.5};$$

The time sampling step of the initial samples of the input and output signals of the DG plant in accordance with the Kotelnikov theorem depends on the maximum frequency of the system bandwidth:

$$\Delta t = \frac{1}{2}\Omega_{max}.$$

The length of a fixed sample of input and output signals of the system is determined by the number of points that fix the accuracy of the estimate

$$T = n\cdot\Delta t.$$

Experimental research [31] revealed that for the study of electromechanical oscillations in the EPS, the values $n = 256$ or 512 and $\Delta t = 0.05,0.1$ s are most suitable.

To analyze the consistency of the experimental model in the form of complex transfer coefficients, we propose to use the spectrum function of the squared coherence

$$\gamma^2_{YU}(\omega) = \frac{|S_{YU}(j\omega)|^2}{S_Y(\omega)S_U(\omega)}, \tag{8.6}$$

where $S_U(\omega)$, $S_Y(\omega)$ are density functions of the power spectra of the input and output noises of the observed signals of the system; $S_{YU}(j\omega)$ is a cross-spectrum density function taking real values from the interval $0 \le \gamma^2_{YU}(\omega) \le 1$.

The densities of the noise power spectra of the observed signals of the system can be found using digital averaged complex spectra of the input and output signals

$$S_U(\omega) = \frac{1}{n\cdot\Delta t}\cdot|U(j\omega)|^2,$$

$$S_Y(\omega) = \frac{1}{n\cdot\Delta t}\cdot|Y(j\omega)|^2, \tag{8.7}$$

$$S_{YU}(j\omega) = \frac{1}{n\cdot\Delta t}\cdot Y(j\omega)\cdot U(j\omega)$$

For an ideal linear input-output connection, the following equality takes place in the entire frequency range

$$\gamma^2_{YU}(\omega) = 1,$$

i.e., the spectrum function of the squared coherence (8.6) reaches its maximum value at those frequencies where the relationship between the input and output signals is linear.

Using $\gamma^2_{YU}(\omega)$, we can calculate a mean square random error in determining the absolute value of the complex transfer coefficient

$$\varepsilon\left[|W(j\omega)|\right] = \frac{\sqrt{1-\gamma^2_{YU}(\omega)}}{\left|\gamma^2_{YU}(\omega)\right| \cdot \varepsilon_0},\qquad(8.8)$$

where ε_0 is a statistical error in determining frequency spectra, $\varepsilon_0 = \sqrt{\dfrac{1}{n}}$; n is a number of points in the frequency range.

It can be seen from the expression (8.8) that in the case of an "ideal" experiment, the identification values are close to unity and the computational identification error tends to zero. It should be noted that the accuracy of the identified mathematical model of the turbine-generator system can be achieved by obtaining complex transfer coefficients of the main channels and cross-links for all possible modes of operation of the DG plant.

In case of a satisfactory assessment of the model, the completion of the identification procedure will be the compilation of the characteristic polynomial of the DG plant for a specific operating mode according to the following expression:

$$D^M(j\omega) = \det\left[\mathbf{E} + \mathbf{W}_{ob} \cdot \mathbf{W}_p\right],\qquad(8.9)$$

where $\mathbf{E}$ is an identity matrix; $\mathbf{W}_{ob}$ is a matrix transfer function of the regulated object that includes complex transfer coefficients of main channels and cross-links of the generator connections, as well as complex transfer coefficients of the turbine and the exciter; $\mathbf{W}_p$ is a matrix transfer function of the regulator that takes into account the relationship between the AVR and the ASR:

$$\mathbf{W}_p = \begin{bmatrix} W_{ASR}(j\omega) & W^\omega_{AVR}(j\omega) \\ 0 & W^U_{AVR}(j\omega) \end{bmatrix}.$$

The characteristic polynomial of the DG plants obtained during identification makes it possible to determine the optimal configuration coefficients for the AVR and the ASR of the generator, taking into account the mutual influence of the regulators on electromechanical and electromagnetic transient processes [26].

The presented identification algorithm was implemented in the specialized ARE&ARR software package that serves as the basis for building an identification and modeling agent.

8.4.2 Functioning Algorithm of the Coordinated Configuration Agent

To optimize settings of the regulators, it is necessary to set an objective function, the minimization of which will provide the necessary parameters of the transient process and the stability margin. To do this, we can use the following functional [5, 26]:

$$J = \int_0^\Omega e^2 (j\omega) d\omega \rightarrow \min, \qquad (8.10)$$

where $e(j\omega) = D^D (j\omega) - D^M (j\omega)$ is a mismatch between the desired and model sets of coefficients of characteristic polynomials; ω is a current frequency value from the range [0; Ω], which determines the "bandwidth" of the system.

Since the mismatch value

$$e(j\omega) = Re_e (\omega) + jIm_e (\omega)$$

is complex, minimization of the functional (8.9) is quite challenging. Therefore, we can try and use a linear convolution

$$J = \frac{1}{2} J_{Re} + \frac{1}{2} J_{Im} \rightarrow \min, \qquad (8.11)$$

where J_{Re}, J_{Im} are the criteria corresponding to the proximity of hodographs in the areas of real and imaginary values.

The criteria J_{Re} and J_{Im} are formed as

$$J_{Re} = \int_0^\Omega \left(ReD^D (\omega) - ReD^M (\omega) \right)^2 d\omega,$$

$$J_{Im} = \int_0^\Omega \left(ImD^D (\omega) - ImD^M (\omega) \right)^2 d\omega. \qquad (8.12)$$

The functional to be minimized (8.10) has a large number of local extrema, and it is difficult to search for a global minimum by classical optimization methods. Therefore, following the recommendations of [26], the genetic algorithm (GA) introduced by J. Holland is used to search for the global extremum.

The genetic algorithm represents an optimum search procedure based on the mechanisms of natural selection and inheritance. The main differences between the GA and traditional algorithms are:

1. Solution of the problem involves processing of an encoded form of parameter values;
2. Search for a solution is performed from a certain set of points called population;
3. Only the objective function is used, derivatives or other additional relations are not involved; and
4. Probabilistic selection rules are applied.

The most important concept in the GA theory is the fitness function. This function helps assess the fitness degree of individuals and select the most adapted ones that correspond to the highest values of this function.

The classic GA includes the following steps:

1. Selection of the initial population of chromosomes;
2. Assessment of their fitness;
3. Verification of the algorithm termination condition;
4. Selection of chromosomes;
5. Use of crossover and mutation operators;
6. Formation of a new population; and
7. Selection of the best chromosome.

The initial population is randomly generated. Next, an iterative process is carried out, which ends when the stopping criterion is reached or a given number of generations is formed. In each generation, individuals are selected for further processing using the value

$$Ps(i) = \frac{f(i)}{\sum\limits_{i=1}^{k} f(i)}, \qquad (8.13)$$

where k is a number of individuals and $f(i)$ is a fitness function of the ith individual.

As a result of selection, a parental population of the size equal to that of the current population is formed. After selection, the selected individuals undergo crossover; in this case, a given probability Pc is used. At the first stage of the crossover, pairs of chromosomes are randomly selected from the parental population. The crossover probability lies in the range [0.5–1]. Then, for each pair of parents, the position of the gene in the chromosome is drawn and the crossing point is determined. When parental chromosomes are crossed, the next pair of offspring appears. After the crossover, a mutation operation is performed with the probability Pm, which is taken to be smaller than at the previous stage. Information about the population obtained after the mutation is overwritten by the old one. The next generations are processesed similarly.

The choice of coding method is an important step in the GA when finding the optimal coefficients for configuration of regulators and can affect the speed and accuracy of the result. When using the real coding method, an individual gene is one of the desired parameters in the form of a real number, and their combination is a chromosome, i.e., a possible set of regulator configuration coefficients (see

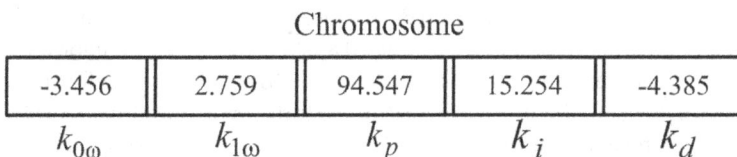

Chromosome

-3.456	2.759	94.547	15.254	-4.385
$k_{0\omega}$	$k_{1\omega}$	k_p	k_i	k_d

FIGURE 8.27   Example of a chromosome with real coding of the configuration coefficients for the AVR and ASR.

Figure 8.27). Individuals are randomly formed from a chain of chromosomes, and the initial population is determined.

Figure 8.27 shows a chromosome when encoding two configuration coefficients of the AVR for a stabilizing control channel for the frequency deviation $k_{0\omega}$ and its rate $k_{1\omega}$, as well as three configuration coefficients of the ASR operating according to the PID law: k_p, k_i, k_d.

Currently, researchers of GAs offer various methods of selection, crossover and mutation, including the following: tournament selection, elite methods that guarantee the "survival" of the best members of the population, two-point and uniform crossover, etc.

As studies have shown [26], the final result is greatly influenced by the following factors: the method of determining the initial population, crossover share probabilities (crossover probabilities) and mutation algorithm.

Their incorrect choice can lead to an ambiguous solution and to a large cost of processor time. In addition, since the GA is stochastic, its application to the same optimization problem may yield different results.

To obtain the best results of the GA operation and automatically determine the family of the initial population with the most appropriate crossover and mutation probability, we propose to use an adaptive algorithm that provides optimal GA settings for a specific objective function [26]. This algorithm essentially implies that we perform the GA twice. At the first stage, for a given objective function, a search range and an initial reference point are formed. In this case, it is enough to carry out 10–100 iterations of the GA, depending on the complexity of the function. Then, using the settings and additional procedures obtained at the first stage, a global solution is formed. For example, the final family of individuals determined at the first stage is taken as the initial population, and the adaptive feasible algorithm coded in the MATLAB Flex Tool is used as the mutation function. The optimal value of the probability of the crossover fraction is determined at the first stage of the algorithm.

The proposed adaptive GA uses a tournament selection method and a two-point crossover, since these methods are quite effective. In addition, to reduce the number of GA iterations and refine the global solution after the GA procedure, the Nelder–Mead method coded in MATLAB is used. Studies have shown that this method is the most effective for finding optimal settings for the AVR and ASR.

Thus, the adaptability of the proposed algorithm is stipulated by the automatic process of selecting the initial population, as well as by setting the optimal value for the probability of crossing and mutation. The procedure of the adaptive genetic algorithm was coded in MATLAB using a specialized software package that quickly finds optimal settings for the AVR and ASR taking into account their connectivity.

Figure 8.28 shows the performance results both of the proposed adaptive algorithm and the GA coded in MATLAB with default settings (random formation of the initial population, crossing probability 0.8, standard mutation algorithm with a proportion of selected chromosome components 0.01). We searched for the optimal control coefficients of the AVR and ASR of the turbogenerator of the DG plant.

Figure 8.28 reveals that the adaptive GA yields a solution during a smaller number of iterations and with a more accurate result (see Figure 8.28b), which shows the effectiveness of the algorithm proposed.

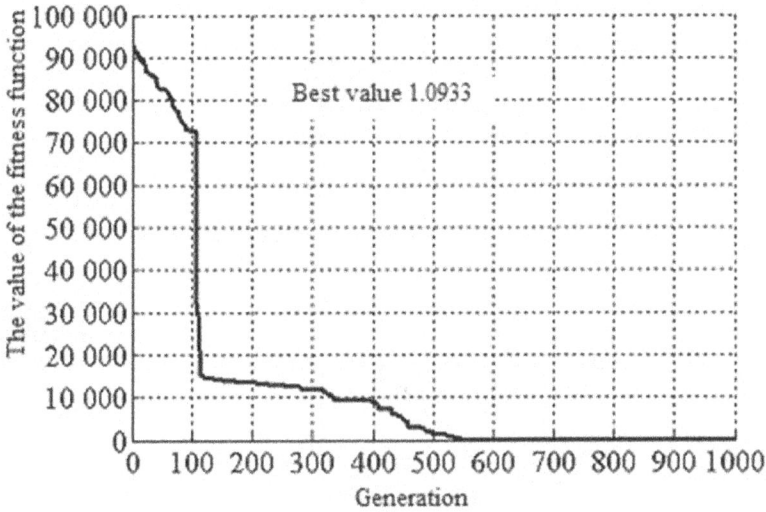

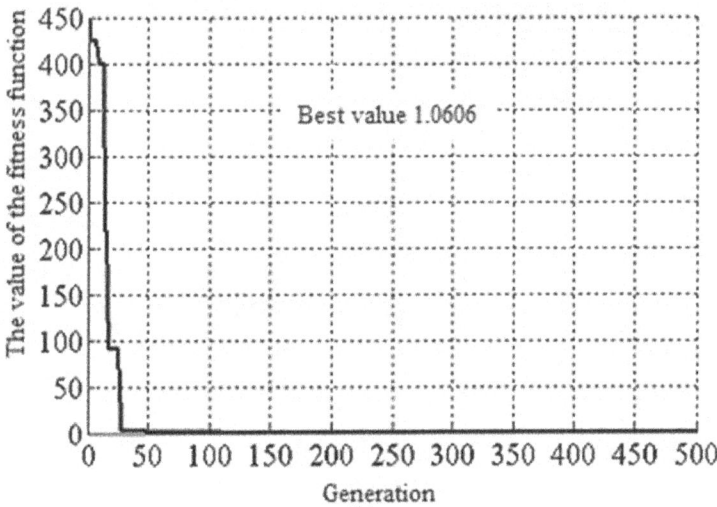

FIGURE 8.28 Dependencies of the fitness function best value on generation with default GA settings (a) and optimal settings (b).

The coordinated configuration agent can use a classical GA scheme or the adaptive GA, which automatically determines the initial population, the optimal values of the probabilities of crossing, and mutation. The adaptive GA quickly and accurately solves the problem of optimizing the regular configuration coefficients, which is an important quality when the MACS operates in real time.

The GA can be used to set up automatic regulators of DG plants using multi-agent technologies, since MACSs are characterized by survivability, complex behavior and self-organization properties. The same properties are manifested in the survival of the fittest individuals in genetic algorithms. Therefore, each individual can be represented by an agent seeking to reach the extremum of the objective function by choosing the best breeding partner.

Figure 8.29 shows the structure of the proposed multi-agent GA implementation for solving the problems of determining the optimal configuration coefficients for the AVR and ASR. Each individual is represented by a separate agent, characterized by a certain set of chromosomes and a fitness function. Since selection is usually centralized, there exists an agent responsible for this process. Each agent exchanges information with other agents and the selection agent.

The activity of individual agents is to calculate the objective function and notify the selection agent about its value and the coefficients of regulator configuration, as well as to determine the properties of the representative agent of the next generation. The selection agent collects data on the fitness function of each individual agent and selects the best individual.

The modified GA proposed for the multi-agent implementation realizes an iterative process consisting of identical steps. For each individual, the initial set of chromosomes is formed randomly and may coincide for some individuals. Individual agents, communicating with each other, determine the objective function and choose a partner for crossing. The latter is possible only for different individuals. As a result of crossing, each pair of individual agents produces one offspring that differs from the parents in the set of chromosomes. The breeder selects one of the best offspring for the next generation. For the optimal values of the configuration coefficients of

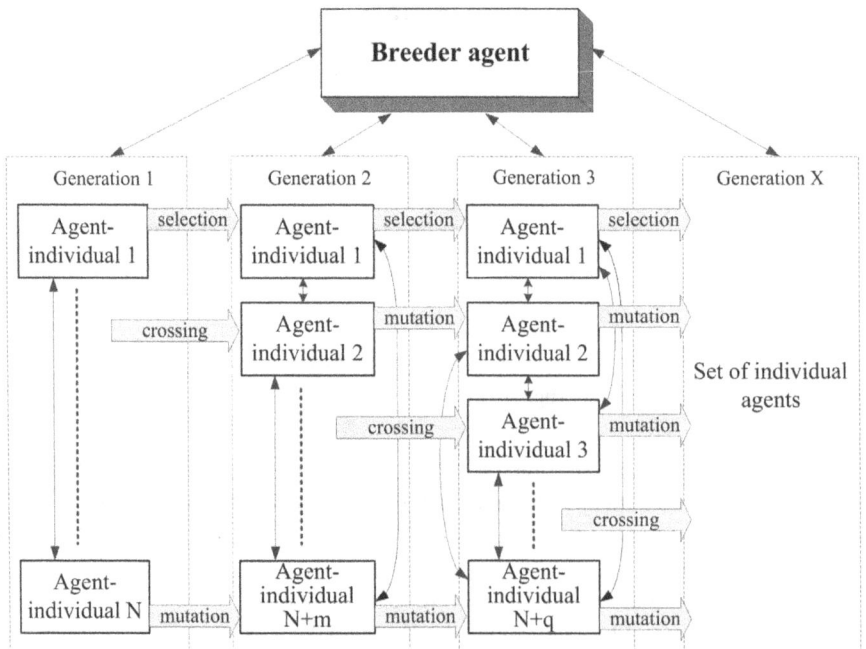

FIGURE 8.29   The structure of the multi-agent implementation of the genetic algorithm.

the AVR and ASR to be determined quickly, it is necessary to have many solutions. Therefore, the number of individual agents must increase with each new generation. To do this, all individuals are mutated, with the exception of the one selected by the selection agent. Mutants pass into the next generation as new individual agents (Figure 8.29). When a given number of generations that do not change the fitness function is reached, the iterative process stops.

8.4.3 Functioning Algorithm of the Autoconfiguration Agent

For optimal control, adjustment of the AVR and ASR settings is required in case of significant changes in the operating modes of the DG and EPS plants. By changes we mean the following situations:

- Noticeable changes in the average voltage levels at the connection points of the DG plants;
- A large variation in the load of consumers leading to large deviations in voltage and frequency;
- Connection of DG plants and their synchronization with the network;
- The occurrence of emergency modes and the subsequent reconfiguration of the topology of the electrical network caused by disconnections of damaged elements; and
- Change in power supply schemes.

The principle of operation of the autoconfiguration agent is to identify the operating mode of the generator of the DG plant and correct the current settings of the AVR and ASR when the operating mode is changed. To solve the identification problem, an adaptive network based on a fuzzy inference system (ANFIS) is used. It is a variant of hybrid neuro-fuzzy networks with direct propagation of a special type of signal. Its architecture is isomorphic to a fuzzy knowledge base. Such networks use smooth membership functions and differentiable implementations of triangular norms.[2] Therefore, when configuring them, one can use fast-learning algorithms based on the backpropagation method. When identifying, it is advisable to use a two-stage procedure for constructing fuzzy models of the Sugeno type. At the first stage, fuzzy rules are determined by subtractive clustering using experimental data on voltage (U_g), frequency (ω_g), power (P_g, Q_g) and power quality indicators in the connection node of the DG plant. The following operating modes the generator of the DG plant are distinguished:

1. Unit start mode;
2. Idle mode;
3. Autonomous operating mode for the assigned load;
4. Mode of parallel operation with the network;
5. Emergency mode;
6. Asynchronous running mode;
7. Overload mode; and
8. Mode of parallel operation with a network of reduced power quality, etc.

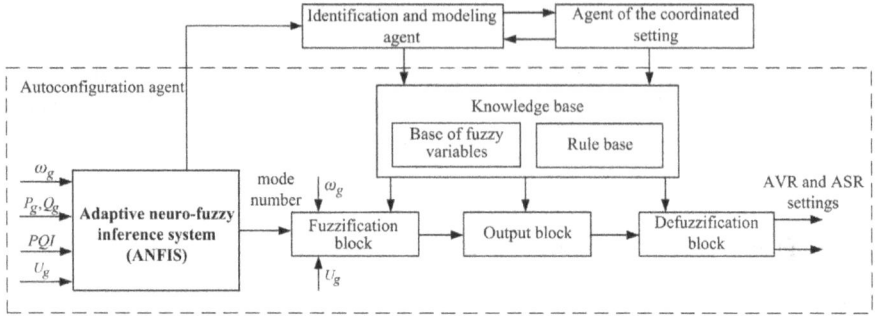

FIGURE 8.30   Block diagram for the autoconfiguration agent.

At the second stage, the parameters of the fuzzy model are tuned using the ANFIS algorithm. Subtractive clustering is used here to quickly synthesize fuzzy rules from data. As a result, a fuzzy model of the operating modes of the DG plant generator is obtained, which is a starting point for learning. The basis of the autoconfiguration agent is a fuzzy inference system that has a knowledge base about all possible modes of operation of the DG plant and the corresponding optimal configuration coefficients for each mode. The block diagram of the fuzzy inference system of the autoconfiguration agent is shown in Figure 8.30.

The type of the current mode of operation of the DG plant is determined by the autoconfiguration agent using the ANFIS unit, whose interaction with the identification and modeling agent makes it possible to form the knowledge base of the autoconfiguration agent and simplify its rule base. At the same time, the identification and modeling agent compares the obtained mathematical model with the current mode of operation and transfers this information to the knowledge base of the autoconfiguration agent.

8.4.4 Modeling of the MACS for DG Plants

Over the past decade, more than a hundred software platforms have been developed that implement agent-based modeling, among which the AnyLogic system can be distinguished. It combines all main types of simulation modeling. We used it to model the proposed MACS for DG plants equipped with turbogenerators that have the AVR and ASR. We considered the most difficult situation of autonomous operation of the APS, i.e., when communications with the traction substation were disconnected. The model is based on the following system of equations in per units [2, 4], which is a special case of system (8.3):

1. The turbine generator rotor equation

$$\frac{d\Delta\omega(t)}{dt} = \frac{1}{T_J}\left[P_T\left(\omega_0 + \Delta\omega(t)\right) - \frac{E_q\left(i_f\right)U_G}{x_G}\sin\left(\delta_0 + \Delta\delta\right)\right],$$

where T_J is the inertia constant of the turbogenerator; P_T is the mechanical power of the turbine; E_q is the EMF of the generator; U_G is generator voltage; x_G is generator resistance; ω_0 is a synchronous frequency of the rotor rotation; δ_0 is the load angle (the angle between the voltage and the EMF of the generator) in the steady state;

2. The generator bus voltage equation

$$\dot{U}_G = \dot{E}_q - j\dot{I}_G x_G,$$

where I_G is the generator current;

3. The generator exciter equation

$$\frac{dE_q}{dt} = \frac{1}{T_f}\left(U_{AVR} - E_q\right),$$

where T_f is a field winding time constant; U_{AVR} is a control signal of the AVR;

4. The turbine equation

$$\frac{dP_T}{dt} = \frac{1}{T_T}\left(U_{ASR} - T_T\right),$$

where T_T is a turbine time constant; U_{ASR} is a control signal if the ASR;

5. Equations defining deviation of the load angle

$$\frac{d^2\Delta\delta}{dt^2} = \frac{\omega_0}{E_G \cdot T_J}\left[-\Delta E_q U_G \sin(\delta_0) - \frac{d\Delta\delta}{dt} x_G D - \Delta\delta\right];$$

$$\frac{d\Delta\delta}{dt} = \Delta\omega,$$

where $\Delta\delta$ is a small change in the load angle; D is a damper moment;

6. The AVR equation

$$U_{AVR} = \Delta U_G \cdot \left(K_{p1} + u_{i1}\right) + u_{d1}; \quad \frac{du_{i1}}{dt} = \frac{K_{i1}}{T_{i1}}; \quad \frac{d\Delta\omega}{dt} = \frac{u_{d1}}{K_{d1}},$$

where ΔU_G is a generator voltage deviation from the set value; K_{p1}, K_{i1}, K_{d1} are the AVR configuration coefficients;

7. The ASR equation

$$U_{ASR} = \Delta\omega \cdot \left(K_{p2} + u_{i2}\right), \quad \frac{du_{i2}}{dt} = \frac{K_{i2}}{T_{i2}},$$

where K_{p2}, K_{i2} are the ASR configuration coefficients.

The AVR equation describes a PID regulator with stabilization in terms of the rotational speed of the generator rotor. This equation is a simplified analogue of the model of a unified powerful AVR. The ASR equation describes a proportional-integral regulator.

We assumed the following values of the turbogenerator parameters, the AVR and the ASR: $T_j = 5$s, $x_G = 2.43$ p.u., $T_f = 0.025$s, $T_T = 0.2$s, $T_{i1} = T_{i2} = 0.1$s. Configuration coefficients of the AVR and ASR for the initial load mode were determined by the coordinated setting method [26]: for the AVR $K_{p1} = 35, K_{i1} = 5, K_{d1} = 1$; for the ASR $K_{p2} = 100, K_{i2} = 1$.

Models of agents were developed following the description of their main functions, given earlier. As an example, Figure 8.31 shows the structure of the model for the coordinating agent that determines the algorithm of its work.

This agent is characterized by dynamic variables (*P1* and *P2*) that determine the power of the DG plants. The algorithm for calculating *P1* and *P2* works in accordance with its state chart through the agents of prediction, generation optimization and electricity market. The coordinator can change its state and power assignments for DG plants upon receiving messages from other agents, and it is equipped with communication ports (port, port1) and a network (connections, agentLink). The event element imitates the appearance of an event that critically changes the behavior of the MACS agents (see Figure 8.31). Such an event may be a trip of protective relays or a significant deterioration of power-quality indicators. The structure of agent connections in the model is shown in Figure 8.32. The diagnostic agent has receptors that measure the parameters of the turbogenerator, and the PSA is equipped with an effector according to which the active power generation is changed.

The identification and modeling agent, using its receptors, captures the control signals coming from the AVR and ASR, as well as the voltage and frequency, performing the identification procedure described earlier. The autoconfiguration and coordinated configuration agents use the corresponding effectors to change the configuration coefficients of the AVR and ASR when altering the operating mode of

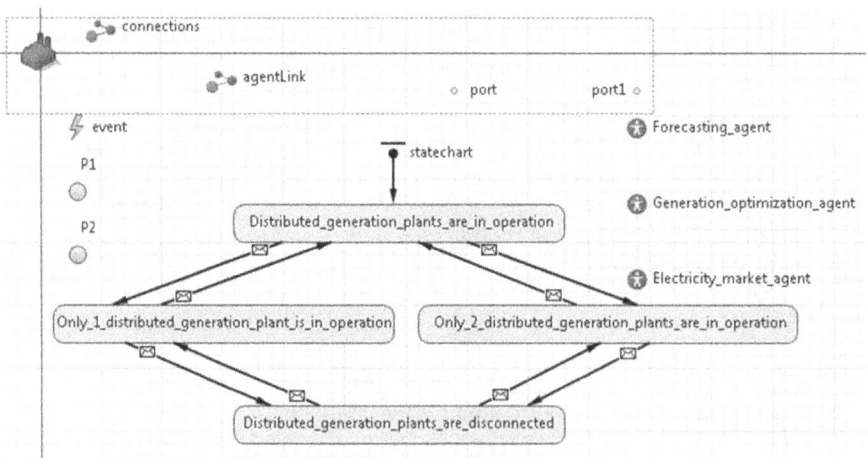

FIGURE 8.31 Structure of the model for the coordinating agent.

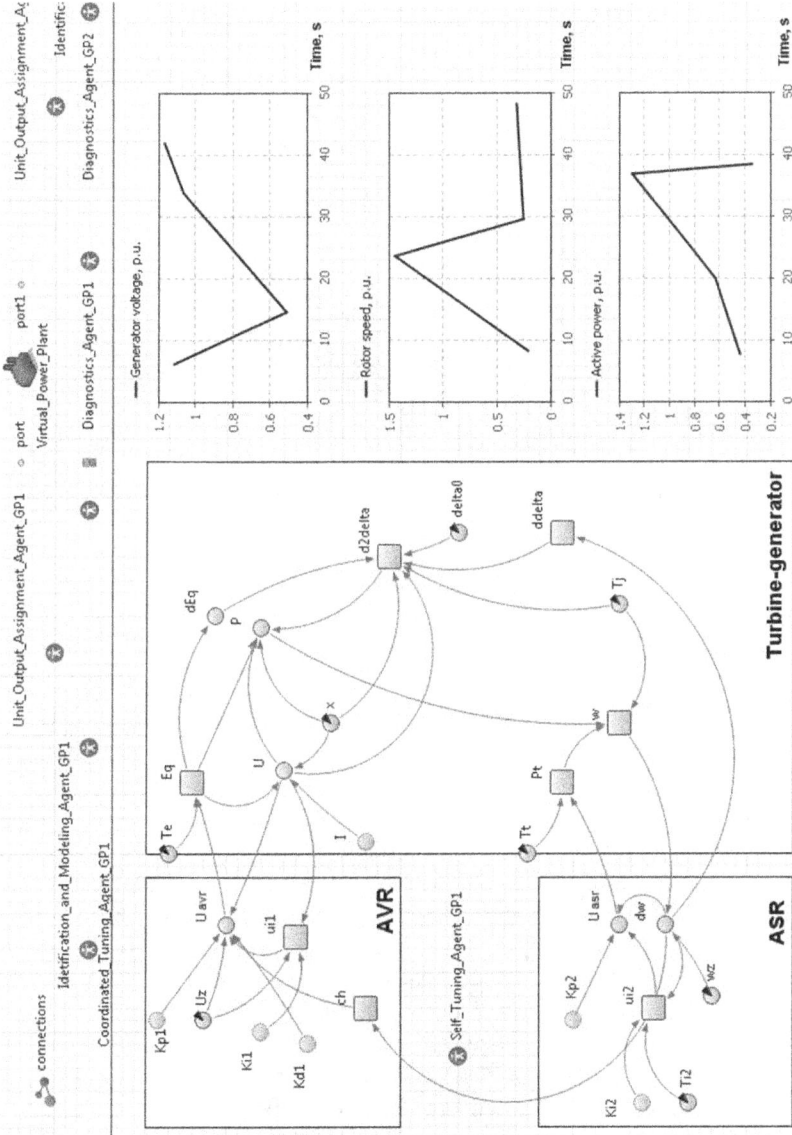

FIGURE 8.32 Fragment of the model for the MACS of DG plants.

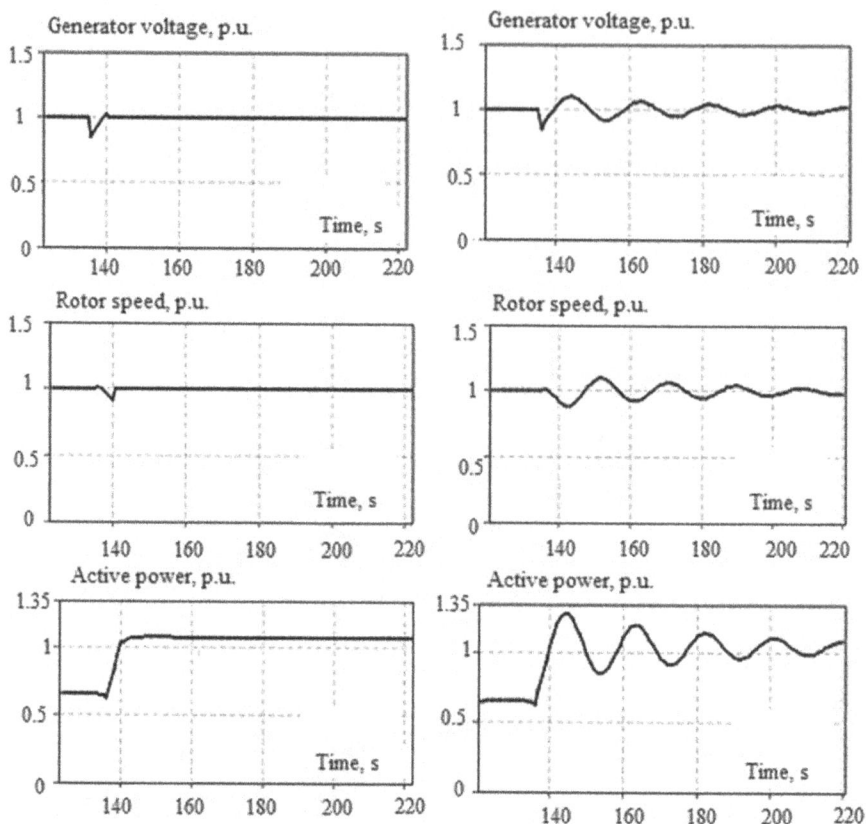

FIGURE 8.33 Changes in generator voltage, rotor speed and active power when the load changes: (a)—during the MACS operation; (b)—when the MACS is out of operation.

the turbogenerator. Agents are connected by information transfer channels through ports and connectors that build a network through which they can send messages to each other. Using the developed model, we ran an experiment to assess the effectiveness of the interaction between agents when the power of one of the DG plants was changed. At the same time, the coordinating agent sent a message to the PSA about a change in the generated active power. As a result of a positive response from the diagnostic agent, the value of the active power of the turbogenerator changed. Based on this, the autoconfiguration agent refined the configuration coefficients of the AVR and ASR. Figure 8.33 shows the results of the performance in the form of time dependencies for voltage, rotor speed and generated active power.

The simulation results show a positive effect during the operation of the MACS: we can observe a significant reduction in the control time and overshoot value, as well as an improvement in the quality of the transient process. Figure 8.33 reveals that the change in the settings of regulators is carried out with some delay due to the expectation of a noticeable change in the power of the DG plant embedded in the MACS algorithm.

8.5 ADAPTIVE CONTROL SYSTEM FOR REGULATORS OF DISTRIBUTED GENERATION PLANTS

To ensure the stability of the parallel operation of the DG plants based on synchronous generators, they can be equipped with automatic control systems (AVR and ASR). To optimize the control, it is necessary to adjust the settings of the AVR and the ASR in case of significant changes in the operating modes of the DG plants and the RPSS units. Improved stability and reliable integration of DG plants in the EPS can be achieved using an adaptive system that adjusts the settings of the AVR and ASR and employs the concept of a fuzzy logical inference (Figure 8.34).

The adaptive system identifies the operating modes of generators of the DG plants and adjusts the current settings of the AVR and the ASR when modes are changing. To solve the problem of identification based on the experimental data derived for voltage U_g, frequency ω_g, power P_g, Q_g and power-quality indicators, we propose to use an adaptive ANFIS-based network. A fuzzy regulator with an autotuning unit is a fuzzy inference system with elements for identification and coordinated configuration of the AVR and the ASR. The input vector is formed by the current values of voltage U_g and rotational speed ω_g of the generator rotor, as well as by the parameters of the identified mode. The output vector of the regulator includes the optimal for the current mode configuration coefficients of the AVR and the ASR. The block diagram of the fuzzy regulator with the autotuning unit is similar to the block diagram of the autoconfiguration agent shown in Figure 8.30. At the same time, the autotuning unit comprises identification and modeling units, as well as a coordinated configuration unit, and allows the formation of knowledge base of the fuzzy regulator. This block uses the algorithms for the coordinated configuration of the AVR and the ASR of the generators of DG plants described in [5, 26].

Identification and coordinated configuration of the AVR and the ASR should be carried out for different modes. This helps create a rule base that determines the functioning of the adaptive regulator. The base is made of production rules of the IF-THEN type using the knowledge of an expert or an autotuning unit.

Approbation of the control system was carried out on the RPSS model (Figure 8.35). The DG plant included two synchronous generators with a capacity of 2.5 MV×A each. The load of the APS consumers was assumed to be 5 MV×A. The APS was powered by TS1 or TS2 via a 6 kV transmission line (Figure 8.35). The AVR

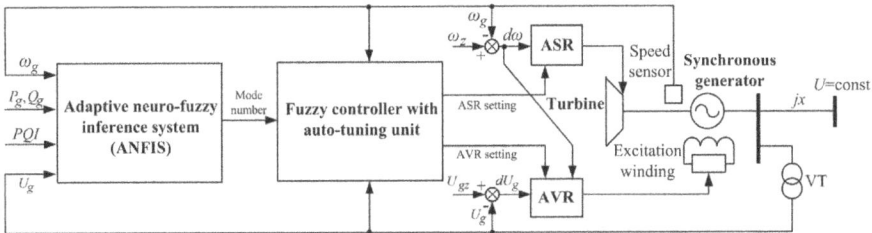

FIGURE 8.34 Diagram of the adaptive control system for the AVR and the ASR of the DG plant.

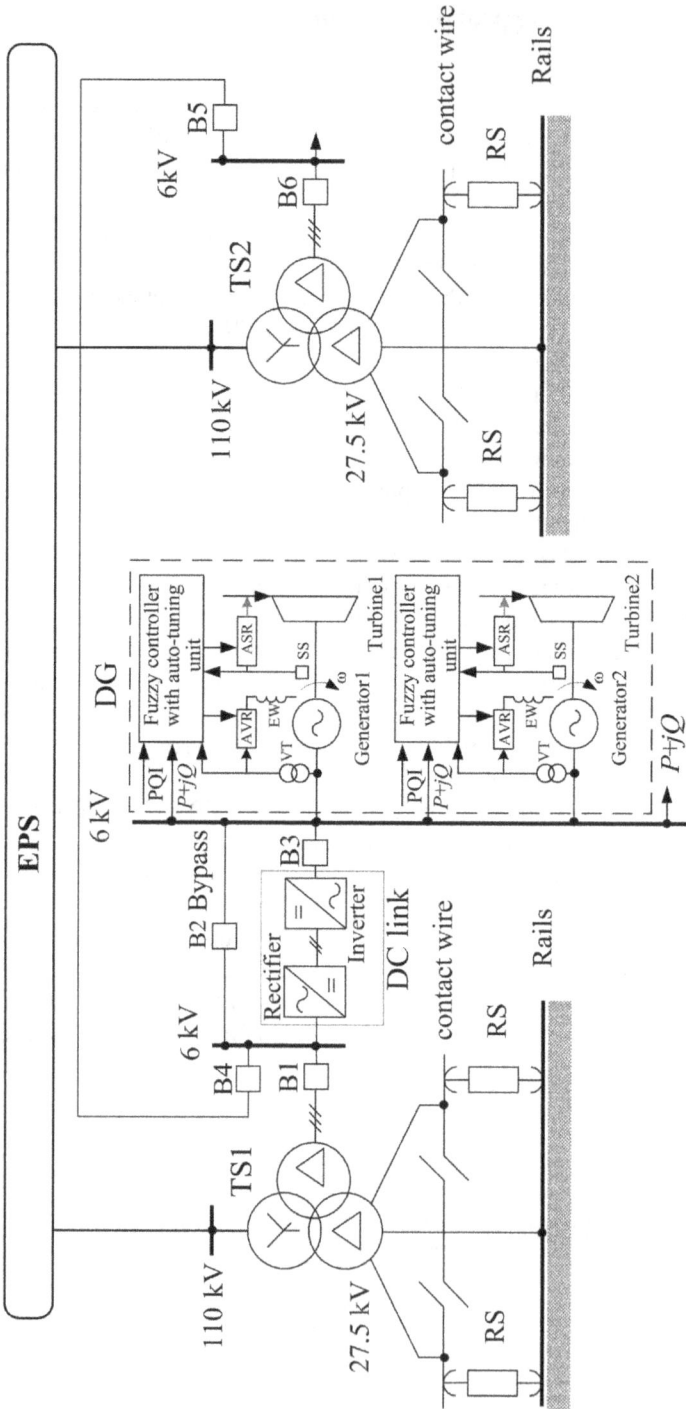

FIGURE 8.35 Fragment of the RPSS.

and the ASR were used to control the DG plant. Their configuration was adjusted by a fuzzy regulator that was equipped with an autotuning unit.

The optimal settings for the AVR and the ASR were determined for the following main modes:

- Autonomous operation of the DG plant for the maximum allocated load;
- Parallel operation of the DG plant and the EPS through the DC link when powered by TS1;
- Parallel operation of the DG plant and the EPS through the DC link when powered by TS2;
- Parallel operation of the DG plant and the EPS through the bypass when powered by TS1; and
- Parallel operation of the DG plant and the EPS through the bypass when powered by TS2.

We studied the following modes:

1. Connection of an additional load in the mode of parallel operation of the DG plant and the EPS through the DC link when powered by TS1;
2. The occurrence of a short circuit on the 6 kV buses and its shutdown by relay protection after 0.5 s during parallel operation of the DG plant and the EPS through the DC link when powered by TS1; and
3. Disconnection of the DC link and automatic supply (after 0.5 s) of power from the EPS through the bypass. In this mode, there was a significant deterioration of the PQI, which changed the frequency range of natural oscillations and required adjustment of the AVR and ASR settings.

Dependences of voltage and rotor speed on time are shown in Figures 8.36–8.38.

1—without changing the configuration coefficients of regulators;
2—using a fuzzy regulator that changes settings.

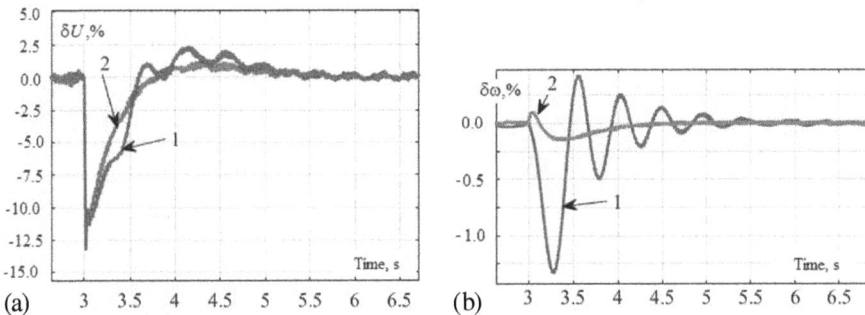

FIGURE 8.36 Deviations of voltage (a) and frequency (b) when connecting an additional load.

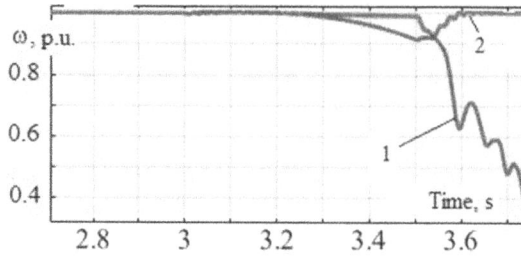

FIGURE 8.37 The rotor speed during a 0.5 s short circuit: 1—without changing the configuration coefficients of regulators; 2—using a fuzzy regulator that changes settings.

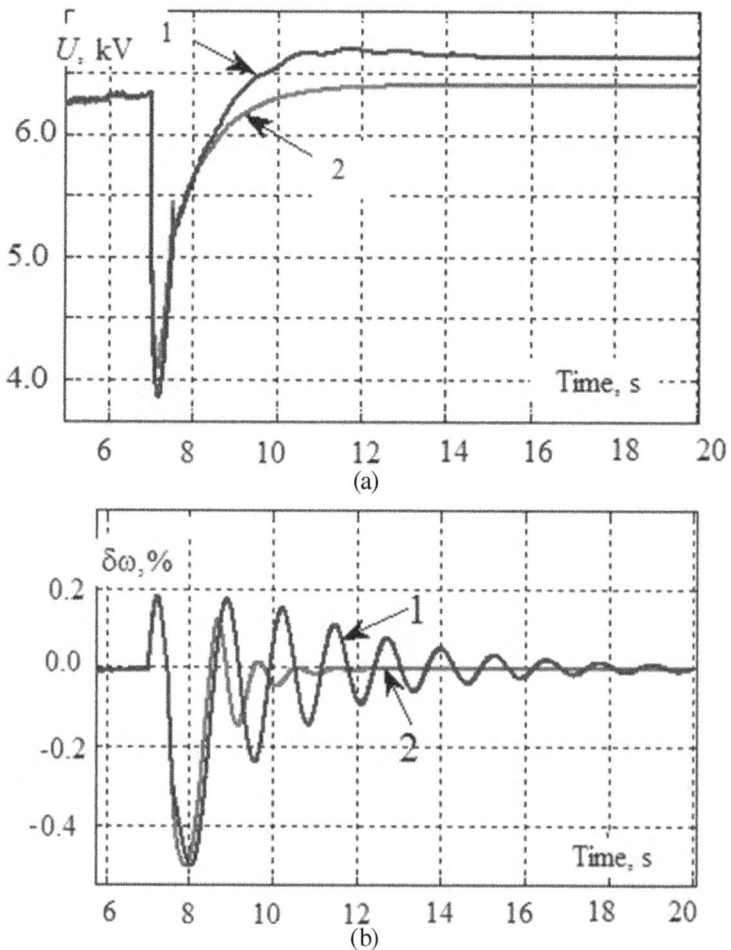

FIGURE 8.38 Voltage on the consumer's buses (a) and the frequency deviation (b) when the DC link is deactivated and the bypass is activated after 0.5 s: 1—without changing the configuration coefficients of the regulators; 2—using a fuzzy regulator that changes settings.

The results of computer simulation allow us to draw a conclusion about the effectiveness of the application of adaptive neuro-fuzzy control of the coordinated configuration of the AVR and ASR of distributed generation. The use of the intelligent control system makes it possible to fully implement the RPSS function (8.1); reduce the transient time, voltage and frequency overshoot; and provide the necessary stability margin, survivability and adaptability of the system to possible operating conditions.

Thus, the intelligent technologies used in the identification of modes and the control of regulators together with the method of coordinated configuration allowed creating a knowledge base for an adaptive control system of a distributed generation plant. The intelligence of the proposed system can be increased due to a more detailed ranking of the possible modes of operation of the DG plant when forming the knowledge base of the fuzzy regulator.

8.6 FUZZY WIND TURBINE CONTROL SYSTEM AND THE METHOD FOR ITS CONFIGURATION

Wind power plants (WPPs) operate on a random schedule, and attempts to control their modes lead to the problem of uncertainty of the initial dates. Therefore, it is necessary to take into account random variations in the speed of the air flow.

The mechanical power developed by a horizontal-axis wind turbine is defined by the formula

$$P_m = \frac{1}{8}\rho\pi D^2 V^3 C_p(\lambda,\beta),$$

where ρ is air density, kg/m³; D is a diameter of the area swept by the wind wheel, m; V is the wind speed, m/s; and C_p is the wind power utilization factor.

The utilization factor depends on the design features of the wind turbine, in particular, on the rotation angle of blades β and the speed λ defined as

$$\lambda = \frac{\omega \cdot r}{V},$$

where ω is a wind turbine rotor speed and r is a wind wheel radius.

The power of a wind turbine can be adjusted by changing the length of the blades and their rotation angle. Next, we consider a method for regulating the output power and rotational speed of a horizontal-axis WPP using an electric drive[3] to rotate the blades of a wind wheel (the so-called pitch-regulation).

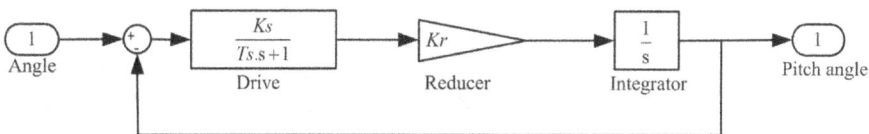

FIGURE 8.39   MATLAB model of the mechanism for rotating the blades.

A servo drive with a gearbox was used as a mechanism for rotating the blades (Figure 8.39). The drive was set by an aperiodic link with a gain K_s and a time constant T_s. The reducer was presented as an amplifier with a coefficient K_r. To determine the deviation angle, the signal was integrated.

The WPP was controlled via the Mamdani and Sugeno fuzzy inference algorithms. They involve the following steps:

1. Creation of a rule base presented in the form IF "Condition" THEN "Conclusion" (F), where F is understood as a weight coefficient that takes values from the interval [0, 1].
2. Fuzzification of input variables to establish a correspondence between a specific value of an individual variable and the value of the membership function of the term corresponding to it.
3. Aggregation of subconditions in fuzzy production rules, representing the procedure for determining the truth degree of conditions for each of the rules.
4. Activation or composition of subconclusions that implements the procedure or process of finding the degree of truth of each of them. In the proposed fuzzy regulator, the following min-activation method was used

$$\mu'(y) = \min\{c_i, \mu(y)\},$$

where c_i are values of the truth degrees of subconclusions and $\mu(y)$ is a term membership function.
5. Accumulation of conclusions by the fuzzy set max-union method,[4] the goal of which is to combine all degrees of truth to obtain a membership function.
6. Defuzzification of output variables by the gravity center method to obtain quantitative values for each of the output variables. They can be used by devices that are not part of the fuzzy inference system.

The fuzzy regulator of the WPP (Figure 8.40) comprises the following units:

• Units that perform fuzzification and defuzzification;

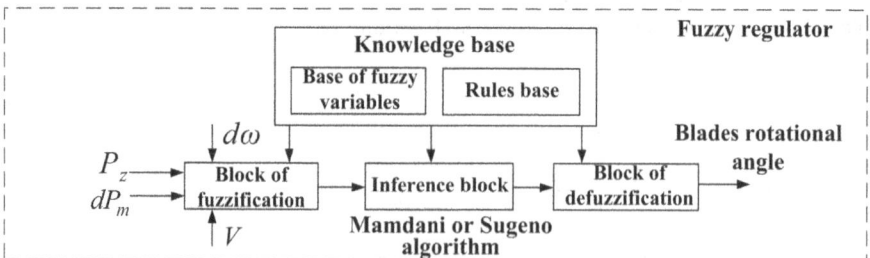

FIGURE 8.40 Diagram of the fuzzy regulator for the blade rotation angle.

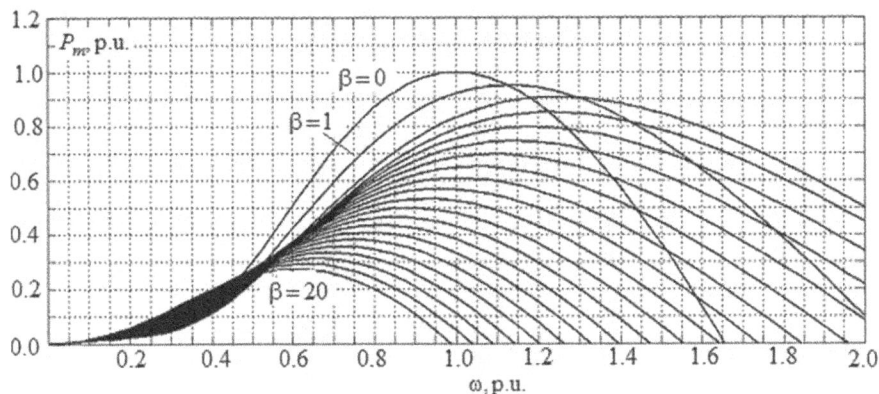

FIGURE 8.41 Wind turbine power characteristics.

TABLE 8.2
Linguistic Variables and Term Sets of the Fuzzy Control System of the WPP

Linguistic variable	Term set
Rotor speed deviation $d\omega$	NB—negative big; NS—negative small; Z—zero; PS—positive small; PB—positive big
Given WPP power P_z	VS—very small; S—small; A—average; B—big; M—maximal
Wind speed V	W—weak; B—basic; S—strong
Mechanical power gain dP_m	N—negative; Z—zero; P—positive
Blade rotation angle *Angle*	Z—zero; VS—very small; S—small; A—average; B—big; VB—very big; L—limit

- Knowledge base that contains bases of fuzzy variables and rules; and
- Output unit implemented on the basis of the Mamdani or Sugeno algorithms.

To configure a fuzzy regulator, it is enough to have an idea about the behavior of the object, and its mathematical description is not required. The input of the proposed fuzzy regulator is fed with signals and names of linguistic variables with the corresponding term-sets presented in Table 8.2. To form an expert knowledge base of fuzzy rules, it is necessary to obtain experimental dependencies of the mechanical power of the WPP on the rotor speed at different angles of rotation of the blades and wind speeds. The dependencies used are shown in Figure 8.41.

The obtained dependencies are divided into intervals corresponding to the term sets of linguistic variables presented in Table 8.2. They are used to build a base of rules taking into account the main criterion—achieving the maximum power of the WPP under given conditions.

The MATLAB FIS Editor and the Sugeno algorithm were used to develop and configure a fuzzy logical inference system for controlling the WPP. The tuned fuzzy

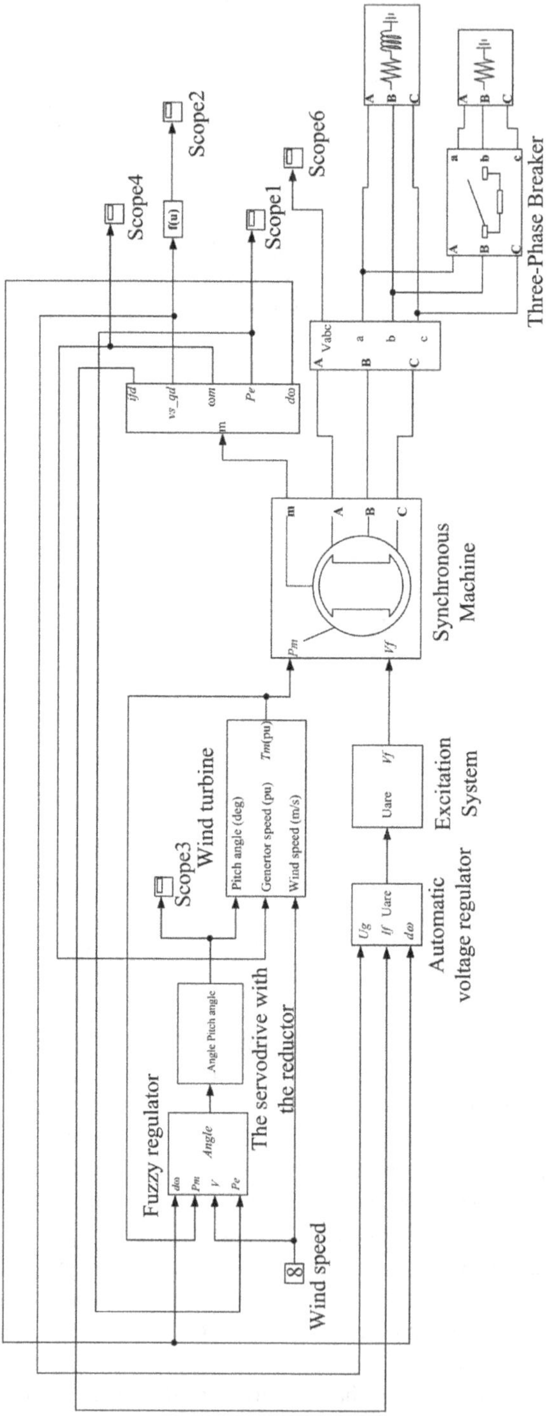

FIGURE 8.42 The MATLAB model.

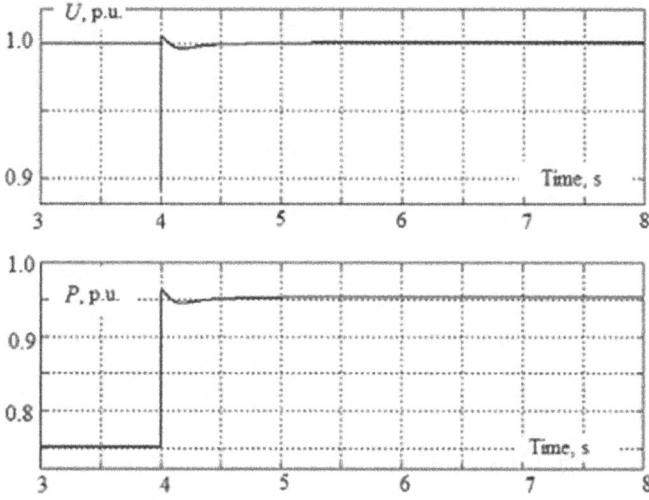

FIGURE 8.43 The WPP operates with a fuzzy rotation angle of blades: P, U are the generator power and voltage, respectively.

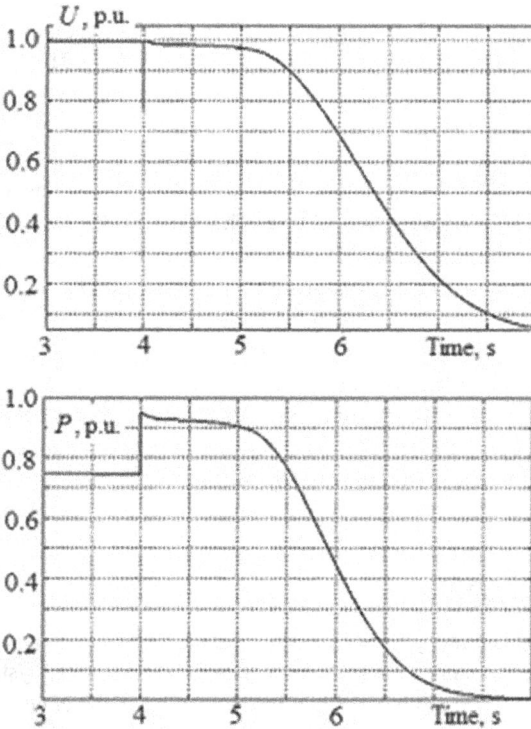

FIGURE 8.44 The WPP operates with the regulator for the rotation angle of blades (deviation angle is $\beta = 20°$): P, U are the generator power and voltage, respectively.

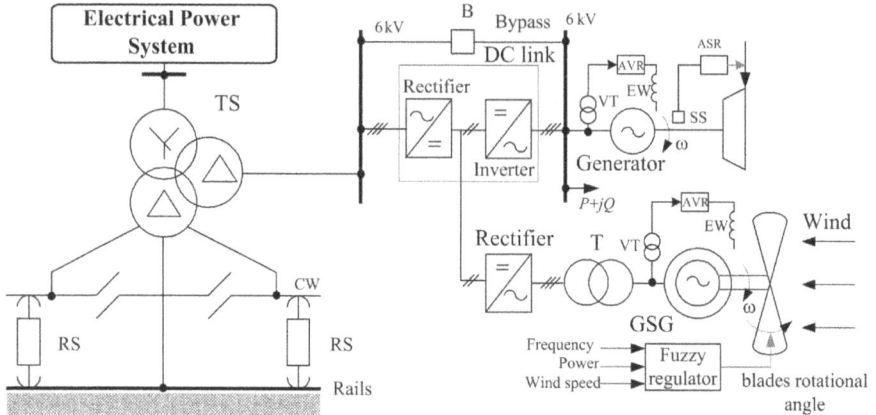

FIGURE 8.45 Fragment of the RPSS: GSG—gearless synchronous generator; CW—contact wire.

controller is efficient in maintaining the consumer's set power value, frequency and voltage when an additional load is connected. This was confirmed by the results of studies [32] on the model shown in Figure 8.42.

The time dependencies shown in Figure 8.43 reveal that when an additional load is connected at 4 s, the generated power and voltage stabilize. If the rotation angle of the blades is not regulated, then the system is unstable (Figure 8.44). Simulation was also performed for the RPSS shown in Figure 8.45.

In Figure 8.45, the DG plant that reached a power of 2.5 MW feeds loads with a total power of 5 MW. At the same time, the WPP had a power of 1 MW and was connected to the DC buses of the DC link. The AVR model [26] was used to control the voltage of the WPP and the DG plant. Modeling was performed when the main power was switched off for 0.8 s. During this period, the DG and WPP with a total capacity of 3.5 MW operated under a load with a capacity of 5 MW, which caused a decrease in voltage. The use of the WPP with a fuzzy regulator maintained the voltage at the level close to the nominal one. Without the WPP, the voltage stabilization becomes impossible and the power quality significantly deteriorates (Figure 8.46).

The results of computer simulation show that the power control of the WPP using a fuzzy controller helps maintain stability of the plant's operation not only when the wind speed changes but also when the consumer load also varies.

8.7 CONCLUSIONS

The AC railway power supply system is a complex technical system characterized by a significant dimensionality. Complexity of the AC power supply system is determined by a stochastic set of possible states and interactions with the environment. These factors significantly complicate the construction of adequate models and control of the RPSS. Effective solutions to this problem can be based on intelligent technologies.

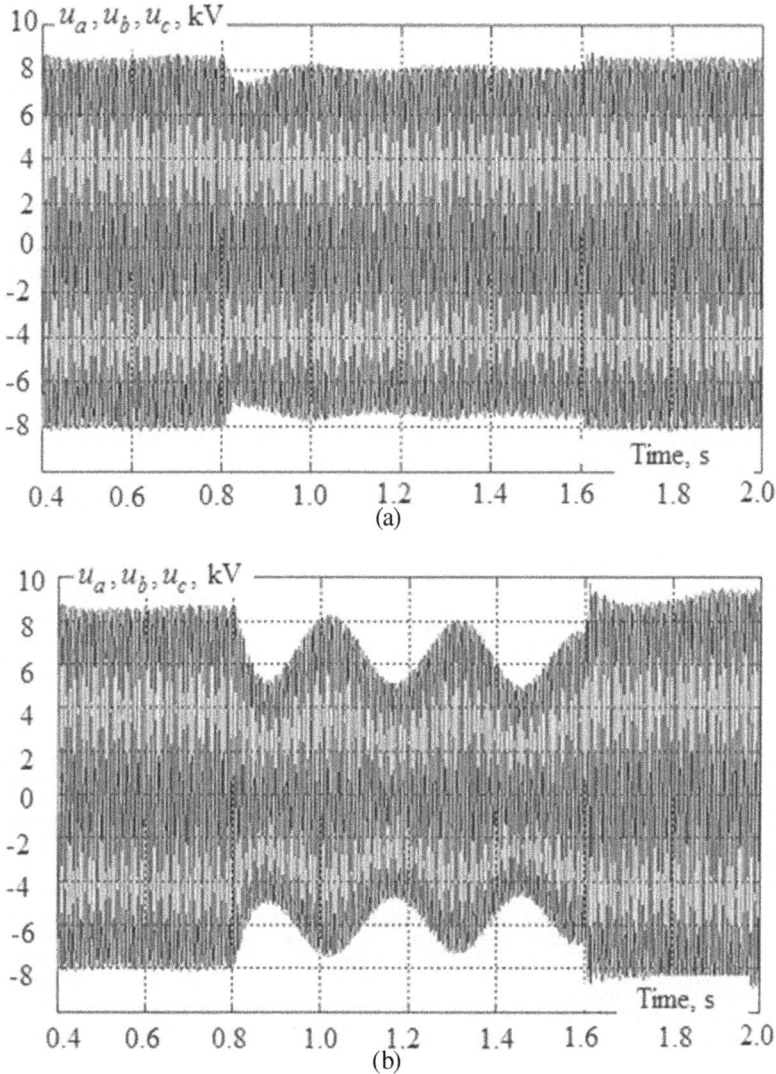

FIGURE 8.46   Voltage on the busbars of the non-traction consumer when the main power supply is switched off: (a)—the DG and WPP operate as a part of the grid cluster; (b)—the WPP is switched off.

The power unit consists of active elements that can implement the generated control actions and effectively change the RPSS modes in the right way. Such elements include regulated reactive power sources, distributed generation plants, power storage units and special facilities for power quality improvement.

The results of modeling have shown that the complex issue of controlling the modes of the AC trunk railway power supply system can be addressed through the use of multiagent technologies. The multiagent control system of the distributed generation plants of the APS significantly reduced the regulation time and the overshoot of the regulated values and improved the transient process quality.

Intelligent technologies help develop and implement an adaptive control system for the parameters of the regulators of distributed generation. The adaptive system knowledge base can be generated by the autotuning unit. The adaptive network based on the fuzzy inference system and the method of subtractive clustering provide real-time identification of the operation mode of the generator and greatly simplify formation of the fuzzy regulator rule base. The results of computer simulation revealed the effectiveness of the adaptive neuro-fuzzy control unit for the coordinated configuration of the AVR and the ASR. The intellectual control system reduces the time of transient process, voltage and frequency overshoot and ensures the necessary stability margin, survivability and adaptivity of the system to all possible modes of operation.

The control of the capacity of the WPP using fuzzy regulator helps maintain the stability of its operation not only under wind speed variations but also in the case of abrupt changes in the load of consumers.

Thus, the research demonstrated that the control of modes of railway power supply systems and the PSS of non-traction consumers yields the best result when we employ intelligent technologies and algorithms: neural networks, fuzzy logic inference system, genetic algorithms and multiagent control systems.

The presented research can evolve toward the development of cyber-physical power supply (CPPS) systems of railways. Unlike conventional control systems, those of CPPS provide closer communication and coordination between computing and physical resources. Energy processes are monitored and controlled using a large amount of feedback. The operation of algorithms affects the physical components, while information about the operating conditions of RPPS and actual parameters of its individual components is used to modify control algorithms and reset automatic controllers.

The establishment of CPPS systems can involve the most advanced information and computer technologies: artificial intelligence, big data, the Internet of Things, quantum computing and others. The core of the virtual part of the CPPS, however, should consist of digital models based on the algorithms for solving conventional power engineering problems. These include the calculation of normal and emergency operating conditions, determination of power quality by voltage deviations and oscillations, and by the levels of unbalance and harmonic distortion. In addition, it is very important for operational practice to ensure personnel safety; plan ice melting conditions; determine wire heating temperatures, thermal wear of transformers and others.

NOTES

1. This information includes the following segments: structure of the MACS, the measured parameters of the DG plants, functions and tasks of agents, knowledge base, etc.
2. Multiplication and probabilistic OR.
3. It was assumed that the turbine was oriented toward the wind, so the option of rotating the nacelle was not considered.
4. The Sugeno algorithm does not have this step since the calculations are carried out with real numbers.

REFERENCES

[1] B.M. Buchholz, Z. Styczynski, "Smart grids: Fundamentals and technologies in electricity networks," Springer, Heidelberg, New York, Dordrecht, London, 2014, 396 p.

[2] Y.N. Bulatov, A.V. Kryukov, "A multi-agent control system of distributed generation plants," *Industrial Engineering, Applications and Manufacturing (ICIEAM)*, 2017, pp. 1–7.

[3] Y. Bulatov, A. Kryukov, "Neuro Fuzzy control system for distributed generation plants," *Advances in Intelligent Systems Research*, 2018, Vol. 158, *Vth International workshop "Critical infrastructures: contingency management, intelligent, agent-based, cloud computing and cyber security" (IWCI 2018)*, pp. 13–19.

[4] Y.N. Bulatov, A.V. Kryukov, K.V. Suslov, "Multi-agent technologies for control of distributed generation plants in the isolated power systems," *Far East Journal of Electronics and Communications*, 2017, Vol. 17, No. 5, pp. 1197–1212.

[5] A.V. Kryukov, S.K. Kargapol'cev, Y.N. Bulatov, O.N. Skrypnik, B.F. Kuznetsov, "Intelligent control of the regulators adjustment of the distributed generation installation," *Far East Journal of Electronics and Communications*, 2017, Vol. 17, No. 5, pp. 1127–1140.

[6] A.V. Kryukov, V.P. Zakaryukin, V.S. Alekseenko, "Modeling of smart grid active elements based on phase coordinates," *Smart Grid for Efficient Energy Power System for the Future, Proceeding*. 2012, Vol. 1, Otto—von—Guericke University Magdeburg, pp. 12–17.

[7] V. Zakaryukin, A. Kryukov, A. Cherepanov, "Intelligent traction power supply system," *International Scientific Conference Energy Management of Municipal Transportation Facilities and Transport (EMMFT)*, Advances in Intelligent Systems and Computing, 2017, Vol. 692, pp. 91–99.

[8] V.P. Zakaryukin, A.V. Kryukov, "Intelligent traction power supply system," *The Power Grid of the Future, Proceeding № 2, Otto—von—Guericke University Magdeburg*, 2013, pp. 44–48.

[9] V.P. Zakaryukin, A.V. Kryukov, V.A. Alekseenko, "Use of smart grid technologies for optimal operation of railroad power supply system," *The Power Grid of the Future, Proceeding № 3, Otto—von—Guericke University Magdeburg*, 2013, pp. 22–26.

[10] Y.N. Bulatov, A.V. Kryukov, K.V. Suslov, "Solving the flicker noise origin problem by optimally controlled units of distributed generation," *18th International Conference on Harmonics and Quality of Power (ICHQP)*, Ljubljana, Slovenia, 13–16 May 2018, pp. 1–4.

[11] Y.N. Bulatov, A.V. Kryukov, Van Huan Nguyen, "Effect of distributed generation plants' automatic controllers on power quality factors," *E3S Web of Conferences, 114, 04001 Energy Systems Researh*, 2019, pp. 1–5.

[12] Y. Bulatov, A. Kryukov, "Enhancing reliability of power supply systems with distributed generation plants," *E3S Web of Conferences, 58, 01006, RSES*, 2018. pp. 1–6.

[13] Y. Bulatov, A. Kryukov, K. Suslov, N. Shamarova, "Ensuring postemergency modes stability in power supply systems equipped with distributed generation plants," *Proceedings of the 10th International Scientific Symposium on Electrical Power Engineering, Elektroenergetika*, 2019, pp. 38–42.

[14] Y.N. Bulatov, A.V. Kryukov, "Control of distributed generators in Island mode," *International Conference on Industrial Engineering, Applications and Manufacturing (ICIEAM)*, 2019, pp. 1–6.

[15] Y.N. Bulatov, A.V. Kryukov, "Emergency control in power supply systems with distributed generation plants," *International Russian Automation Conference (RusAutoCon)*, Sochi, Russia, 9–16 Sept. 2018, pp. 1–5.

[16] A.V. Kryukov, V.P. Zakaryukin, M.O. Arsentiev, "Distributed generations systems for transport electric power industry," *Journal of East China Jiaotong University, Nanchang China*, 2009, Vol. 26, pp. 216–223.

[17] A. Cherepanov, V. Zakaryukin, A. Kryukov, "Modelling of traction power supply systems for movement of high-speed trains," *MATEC Web of Conference, 216, 02006, Polytransport Systems*, 2018, pp. 1–7.

[18] Y.N. Bulatov, A.V. Kryukov, A.V. Cherepanov, I.M. Avdienko, A.P. Kutsyi, "Electric energy quality management in railroads power supply systems," *Proceedings of the 6th International Symposium on Innovation and Sustainability of Modern Railroad ISMR 2018*, China Railroad Publishing House, Beijing, 2018, pp. 314–318.

[19] A.V. Kryukov, A.V. Cherepanov, "Static models for active harmonics conditioners," *Smart Grid for Efficient Energy Power System for the Future, Proceeding*, 2012, Vol. 1, Otto—von—Guericke University Magdeburg, 2012, pp. 18–22.

[20] V.P. Zakaryukin, A.V. Kryukov, A.V. Kushov, A.V. Cherepanov, "Parametrical identification of railroad traction power supply systems of alternating current," *Proceedings of the 6th International Symposium on Innovation and Sustainability of Modern Railroad ISMR 2018*, China Railroad Publishing House, Beijing, 2018, pp. 333–338.

[21] V.P. Zakaryukin, A.V. Kryukov, "Mathematical model of multiphase power transmission line," *The Power Grid of the Future, Proceeding № 3, Otto—von—Guericke University Magdeburg*, 2013, pp. 70–74.

[22] V.P. Zakaryukin, A.V. Kryukov, "Multifunctional mathematical models of railroad electric systems," *Innovation and Sustainability of Modern Railroad Proceeding of ISMR 2008*, China Railroad Publishing House, Beijing, 2008, pp. 504–508.

[23] V.P. Zakaryukin, A.V. Kryukov, "Multifunzionale modellazione di sistemi di energia elettrica—energia," *Italian Science Review*, 2014, 3(12), pp. 267–272.

[24] V.P. Zakaryukin, A.V. Kryukov, N.A. Abramov, "Electro energetic technological control in Eastern Siberia Railroad," *JEPE Journal of Energy and Power Engineering*, 2012, Vol. 6, No. 2, pp. 293–299.

[25] V.P. Zakaryukin, A.V. Kryukov, E.A. Alekseenko, "Simulation of emergency conditions in rail electric traction networks," *Proceedings of the 6th International Symposium on Innovation and Sustainability of Modern Railroad ISMR 2018*, China Railroad Publishing House, Beijing, 2018, pp. 285–291.

[26] Y.N. Bulatov, A.V. Kryukov, "Optimization of automatic regulator settings of the distributed generation plants on the basis of genetic algorithm," *2nd International Conference on Industrial Engineering, Applications and Manufacturing (ICIEAM)*, 2016, pp. 1–6.

[27] A. Juneja, S. Bhattacharya, "Energy router: Architectures and functionalities toward Energy Internet," *2011 IEEE International Conference on Smart Grid Communications (SmartGridComm)*, 2011, pp. 31–36.

[28] Y.N. Bulatov, A.V. Kryukov, G.O. Arsentiev, "Use of power routers and renewable energy resources in smart power supply systems," *International Ural Conference on Green Energy (UralCon)*, 2018, pp. 143–148.

[29] Y.N. Bulatov, A. Kryukov, G. Arsentyev, "Application of energy routers in railroad power supply systems," *MATEC Web of Conferences, 239, 01047, TransSiberia*, 2018, pp. 1–6.

[30] Y.N. Bulatov, A.V. Kryukov, G.O. Arsentiev, "Intelligent electrical networks based on controlled energy routers," *2018 International Conference on Industrial Engineering, Applications and Manufacturing (ICIEAM)*, 2018, pp. 1–5.

[31] V.V. Bushuev, N.N. Lizalek, N.L. Novikov, "Dynamic properties of power systems," Moscow: Energoatomizdat, 1995, 320 p.

[32] Y.N Bulatov, A. Kryukov, Nguyen Van Huan, Tran Duy Hung, "Fuzzy controller of rotation angle of blades of horizontal-axial wind power generation plant," *Advances in Intelligent Systems and Computing*, 2019, Vol. 983, pp. 892–901.

9 Sustainable Electrified Transportation Systems

Integration of EV and E-bus Charging Infrastructures to Electric Railway Systems

Hamed Jafari Kaleybar, Morris Brenna, Federica Foiadelli, and Francesco Castelli Dezza

ABBREVIATIONS

DER	Distributed energy resource
E-bus	Electric bus
ERS	Electric railway system
ESS	Energy storage system
EV	Electric vehicle
EVCI	Electric vehicle charging infrastructure
EVCS	Electric vehicle charging station
G2V	Grid-to-vehicle
HFT	High-frequency transformer
HSR	High-speed railway
LRT	Light rail transit
LVDC	Low-voltage direct current
MPPT	Maximum power point tracking
MVAC	Medium-voltage alternative current
MVDC	Medium-voltage direct current
OCS	Overhead catenary system
OPCS	Opportunity charging system
PE	Power electronics
PQ	Power quality
PMSG	Permanent magnet synchronous generator
PSFB	Phase shift full bridge
RBE	Regenerative braking energy
RES	Renewable energy source
RPC	Railway power conditioner
SST	Solid state transformer
T2V	Train-to-vehicle

DOI: 10.1201/9781003293989-9

TE	Traction energy
TPSS	Traction power substation
V2T	Vehicle-to-train
V2G	Vehicle-to-grid
ZVS	Zero voltage switching

CONTENTS

9.1 INTRODUCTION

During the last several years, investment in developing electric vehicles (EVs) and E-bus charging infrastructures have expanded as the main step by countries to decrease carbon emissions and the use of fossil fuels. In addition, the majority of power supply EV charging stations (EVCSs) can overload the main grid and lead to some indirect emissions. However, the significant amount of energy that is generated by applying regenerative braking in electric railway systems (ERSs) can be adopted as an auxiliary supply. Considering that EVs are parked most hours of the day either at home or offices, their internal batteries can be exploited as promising energy storage systems (ESSs) to save regenerative braking energy (RBE) of trains or even operate as complementary supply ERSs. Integration of ERSs and EV/E-bus charging stations at strategic points, like parking areas close to ERS stations or rail freight intermodal terminals with picked up using of trains RBE as an ancillary supply, can ameliorate the system efficiency and decrease the cost. Meanwhile, the current situation of charging infrastructures can be further improved based on smart grid technology, considering the interaction between these two means of transportation. However, according to the diversity of ERSs, different integration architectures can be defined utilizing the DC or AC energy hub concept. Although several research projects have been conducted on the integration of different transportation charging infrastructures, there is still a long way to achieve the goals regarding the smart grid concept and energy hubs. This chapter, as one of the pioneering studies in this field, proposes different DC and AC hub-based integration architectures and analyzes the incorporation of EV/E-bus charging infrastructures with ERSs. Meanwhile, the concept of train-to-vehicle (T2V) and vehicle-to-train (V2T) technologies, together with the challenges of the architecture of future supplying systems in sustainable transportation, will be explained.

9.2 BACKGROUND OF ELECTRIC RAILWAY SYSTEMS

Electric railway systems (ERSs) have experienced substantial development and evolution in the last several decades due to the increased demand for transportation and the dramatic growth of technology. The historical overview and comprehensive classification of ERS configurations have been addressed in [1]. However, the current ERSs, depending on the main supplying system, can be divided into two main groups of DC and AC systems.

9.2.1 DC Railway Systems

The DC ERSs were introduced as the first systems with lower requirements and power ratings in transportation networks. Urban ERSs containing subways, trams, and light rails are the most popular types. Depending on the supplying voltage level they can be classified into four groups, as shown in Table 9.1.

The allowable voltage limits are determined according to BS EN 50163 and IEC 60850 standards. These values are determined according to the number of wagons and the distance of the locomotive from the feeding substation. Choosing the right

TABLE 9.1
Different Voltage Levels of DC ERSs

Maximum instantaneous voltage	Maximum permanent voltage	Nominal voltage	Minimum permanent voltage	Minimum instantaneous voltage	ERS voltage
800 V	720 V	600 V	400 V	400 V	600 V DC
1 kV	900 V	750 V	500 V	500 V	750 V DC
1950 V	1800 V	1500 V	1000 V	1000 V	1500 V DC
3 kV	3 kV	3 kV	2 kV	2 kV	3 kV DC

voltage level in a project depends on several structural and utilization conditions. Meanwhile depending on the operation type and power range DC ERSs can be classified into four groups, as shown in Figure 9.1. Urban low-voltage low-power systems known as trams, streetcars, or light rail transit (LRT) are the first category with tracks and trains running along the streets and operating with another road traffic. These systems mostly are supplied by overhead wires with a nominal voltage of 450–750 V DC in the power range of 0.5–1 MW. The second category is dedicated to the urban low-voltage medium-power systems known as subways, metros, or rapid transit systems with the often-underground railway track. These trains mostly are supplied by both third rail systems or overhead wires with a nominal voltage of 750–1500 V DC in the power range of 2–4 MW. Frequently braking and stopping diffused existing DC lines in the city area and similar supply voltage ranges are encouraging features of urban ERSs for T2V integration. Suburban or regional ERSs are the third category that connects suburbs or commuter towns to the center of a city. They mostly are supplied by overhead wires with a nominal voltage of 1500–3000 V DC up to 10 MW. The next category is related to the high-speed railway (HSR) systems with medium-voltage DC (MVDC), which operate at intercity distances and connect inland cities and sometimes international routes. HSR MVDC trains are faster than other groups of trains, adopting special rolling stock and dedicated tracks. The operating speed of HSR MVDC trains usually is more than 250 km/h. The supply system is based on overhead catenary wires with a nominal voltage of 1500–3000 V DC and a power range of up to 12 MW for each train. The high train power and restriction on the number of operating trains because of the high currents taken from the overhead catenary system (OCS) made the experts think about increasing the voltage between 7.5 and 24 kV in the future. However, such systems are still under study and development.

9.2.2 AC RAILWAY SYSTEMS

With the high-power needs of HSR systems and the fact that MVDC supply still has a long way to be established, the medium voltage AC (MVAC) grid is a mature and popular technology adopted in these systems. MVAC-based HSR systems are

TABLE 9.2
Different Voltage Levels of AC ERSs

Maximum instantaneous voltage	Maximum permanent voltage	Nominal voltage	Minimum permanent voltage	Minimum instantaneous voltage	ERS voltage
18 kV	17.25 kV	15 kV	12 kV	11 kV	15 kV AC, 16.7 Hz
29 kV	27.5 kV	25 kV	19 kV	17.5 kV	25 kV AC, 50 Hz

the final category of ERS types, which have been implemented by 15 kV-16.67 Hz or 25 kV-50/60 Hz OCS and a power range up to 15 MW for each train. Suburban and HSR ERSs usually are closed to remote areas and motorways or highway lines. Therefore, the opportunity of utilizing the existing ERS installations in such areas for power supplying of establishing EVCSs can be a prominent advantage and feature that will be mentioned in the following sections. Depending on the supplying voltage level, AC ERSs can be classified into two groups, as shown in Table 9.2.

9.3 BACKGROUND OF ELECTRIC VEHICLE CHARGING INFRASTRUCTURES

Against the diversity of ERSs, which is summarized in Figure 9.1, EV charging infrastructures (EVCIs) also contain various levels that can be classified based on the power range, power distribution type, and standards. The five categorized groups of EVCIs illustrated in Figure 9.1 can be divided into two groups of low-power AC EVCIs and high-power DC EVCIs [2].

9.3.1 Low-Power AC EVCIs

This group of EVCIs provides EVs to be connected directly to a single-phase/three-phase AC grid due to the internal battery charger. Accordingly, they are utilized for an enormous installation of charging lots covering a wide spread of EVs. They are commonly used nowadays and equipped with a household socket (NEMA 5–15) with 120/240 V and around 32A with the standard. However, this category is characterized by low charging power (generally up to 22 kW). Thus, it is the basic level of EVCS and is known as a slow charging method (2–8 hours). They are appropriate for household applications or while the vehicle can stay stopped for a long time, at least a few hours.

9.3.2 High-Power AC EVCIs

These types of EVCIs are characterized by a high charging power due to the transfer of battery chargers from the vehicle inside to the charging station itself. In this

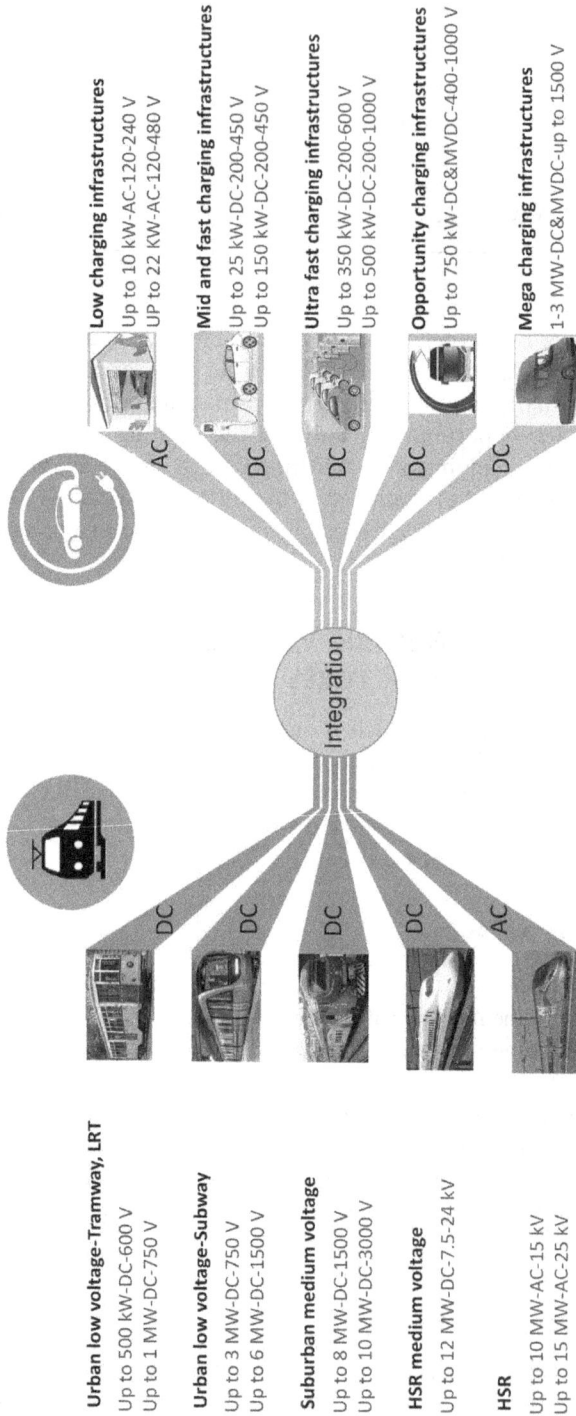

FIGURE 9.1 Different structures of ERSs and EVCIs.

Low charging infrastructures
Up to 10 kW-AC-120-240 V
UP to 22 kW-AC-120-480 V

Mid and fast charging infrastructures
Up to 25 kW-DC-200-450 V
Up to 150 kW-DC-200-450 V

Ultra fast charging infrastructures
Up to 350 kW-DC-200-600 V
Up to 500 kW-DC-200-1000 V

Opportunity charging infrastructures
Up to 750 kW-DC&MVDC-400-1000 V

Mega charging infrastructures
1-3 MW-DC&MVDC-up to 1500 V

Urban low voltage-Tramway, LRT
Up to 500 kW-DC-600 V
Up to 1 MW-DC-750 V

Urban low voltage-Subway
Up to 3 MW-DC-750 V
Up to 6 MW-DC-1500 V

Suburban medium voltage
Up to 8 MW-DC-1500 V
Up to 10 MW-DC-3000 V

HSR medium voltage
Up to 12 MW-DC-7.5-24 kV

HSR
Up to 10 MW-AC-15 kV
Up to 15 MW-AC-25 kV

context, the power electronic (PE) converter rating inside the charger can be enhanced because there is no limitation in terms of volume and weight. Nowadays, the EVCIs contain not only the control and protection units but also all operations related to the power appliances and converting and regulating the power from an AC three-phase system to the onboard DC batteries. The expense of these EVCIs is much higher than the AC types. Therefore, DC EVCIs are often used for reducing the charging time and are known as fast chargers. Furthermore, they are substantially established along the motorways to actualize the so-called electric corridors, such as electrified roads, which provide an EV to travel for long distances.

DC EVCIs can be utilized for various types of vehicles like cars, buses, trucks, and motorbikes. Accordingly, high-power EVCIs also can be divided into four categories.

9.3.2.1 Mid and Fast Charging Infrastructure

The next category is dedicated to the mid (up to 25 kW) and fast (up to 150 kW) charging infrastructure commonly used nowadays equipped with Combo 2 and/ or ChAdeMO plug. These are able to charge most of the current EVs and all the upcoming ones, preferably with the Combo 2 connection as chosen by the European Union as the charging standard. Due to the low power value, it is desirable to consider their connection in the low-voltage DC section. These systems have to be able to charge vehicles equipped with batteries up to 1000 V. Mid charging infrastructures are suitable for a charging time of 1–2 hours, while fast charging is mainly used to reduce the charging time to half an hour.

9.3.2.2 Ultrafast Charging Infrastructure

Ultrafast charging infrastructure, typically up to 350–500 kW, is under development even if some real installations are already active. The main problems related to ultrafast charging infrastructures are their huge impact on the AC mains and the need for a water-cooled cable and connectors to handle high currents (up to 500 A). They are designed to reduce the charging time to the order of ten minutes, even if not all the EVs are able to accept so high a charging power. The presence of a high-voltage DC grid allows ultrafast charging infrastructure to reduce their impact on the mains and to directly use the connected renewable sources and the braking energy coming from the trains.

9.3.2.3 Opportunity Infrastructure

These systems are nowadays used for E-buses, even it is possible to use them for electric trucks. Their duty is to provide an intermittent short charge to the E-buses to extend their range in a working day. Since the connection to the vehicle is through a pantograph, the contact bars are able to handle a high amount of current, to reach a maximum charging power up to 750 kW. These systems can benefit from an MVDC grid for their supply due to the high amount of power required. Moreover, they contribute to promoting the use of public transport since they can be implemented in correspondence with the train stations facilitating intermodal mobility (train + E-bus).

9.3.2.4 Mega Charging Infrastructure

The next introduction of full-electric trucks and tractor-trailers from the main vehicle manufacturers will need the definition of new ultra-high-power charging infrastructures named mega chargers. These kinds of vehicles that are under development can be equipped with a 1000 kWh onboard battery. Accordingly, chargers in the order of 1–3 MW will be needed to reduce their charging time to an acceptable value.

9.4 INTEGRATION OF EV/E-BUS CHARGING INFRASTRUCTURES TO ERSS

The association of ERSs and EVCIs is an important action to reach sustainable transportation systems. Two perspectives can be considered regarding this integration. One is attaining intermodal transportation which regards the assessment of flexible connection of roads and railways considering appropriate integration in strategic points. It is mostly related to geographical and geopolitical investigations and is beyond the scope of this chapter. The other one, which is going to be discussed in the following sections, is the integration of power supply systems and charging infrastructures for both ERSs and EVs. Based on the ERSs' intrinsic characteristics, T2V/V2T integration perspectives can be evaluated and accomplished based on the following frameworks:

- Employing regenerative braking energy (RBE)
- Assuming EVs as stationary ESSs and distributed energy resources
- Utilizing the unemployed capacity of existing power installations
- Utilizing ERS lines as energy hubs preparing suitable connection areas for renewable energy sources (RESs)

9.4.1 Integration Concept Employing Regenerative Braking Energy

RBE as an indivisible feature of ERSs is a popular topic due to the significant effects on energy consumption depletion together with increasing system efficiency and sustainability. Today, most trains are equipped with an electrical braking system to not waste energy on rheostats. In this section, the potential of RBE to supply EVCIs and realize T2V technology will be discussed.

RBE rating and recovery methods in DC and AC ERSs can be varied. In DC systems, especially low-voltage types due to the lower voltage level, higher losses, and the proliferation of TPSSs, there are substantial impediments to recuperating. It is worth mentioning that, due to the line limitations and protection issues, all the recoverable RBE can't be recuperated. The recoverable RBE range for a typical metro line in Milan is measured as 32–36% and the total daily traction energy and RBE of the line for a typical timetable are calculated as 320 MWh and 145 MWh, respectively. The energy figure according to the headway of trains is shown in Figure 9.2a. It can be seen that by increasing headway time, the RBE rating is decreased because of the low number of trains and power losses during transfer between trains. In fact, with non-reversible TPSSs, the first priority to utilize RBE is to supply the other adjacent train in motoring mode. Even adopting the best optimization methods, a significant portion of RBE can't be utilized. The field measurements have demonstrated more than 25% of wasting RBE as heat in onboard rheostats in worse cases and more than 10% on average.

For AC ERSs, RBE production rate is also linked to train speed, train mass, total inertia, and braking duration. Furthermore, the total energy range in AC ERSs is much higher than DC ERSs. Accordingly, the production of RBE also in AC and high-speed ERSs is important and based on real measurements (as shown in Figure 9.2b) that can be in the range of 4–13%. For example, the total daily traction energy and RBE of an AC ERS line for a typical timetable are calculated as 1800 MWh and 138 MWh, respectively. It is obvious that even with the low percentage of RBE contribution, its quantity is high and almost equal to the daily amount of a DC metro line. Due to the neutral zones and phase differences for each section of TPSSs, transferring RBE from one section to the other is not possible and therefore in AC ERSs RBE is fed back to the primary side of TPSS. This returned RBE has many issues in terms of PQ, and

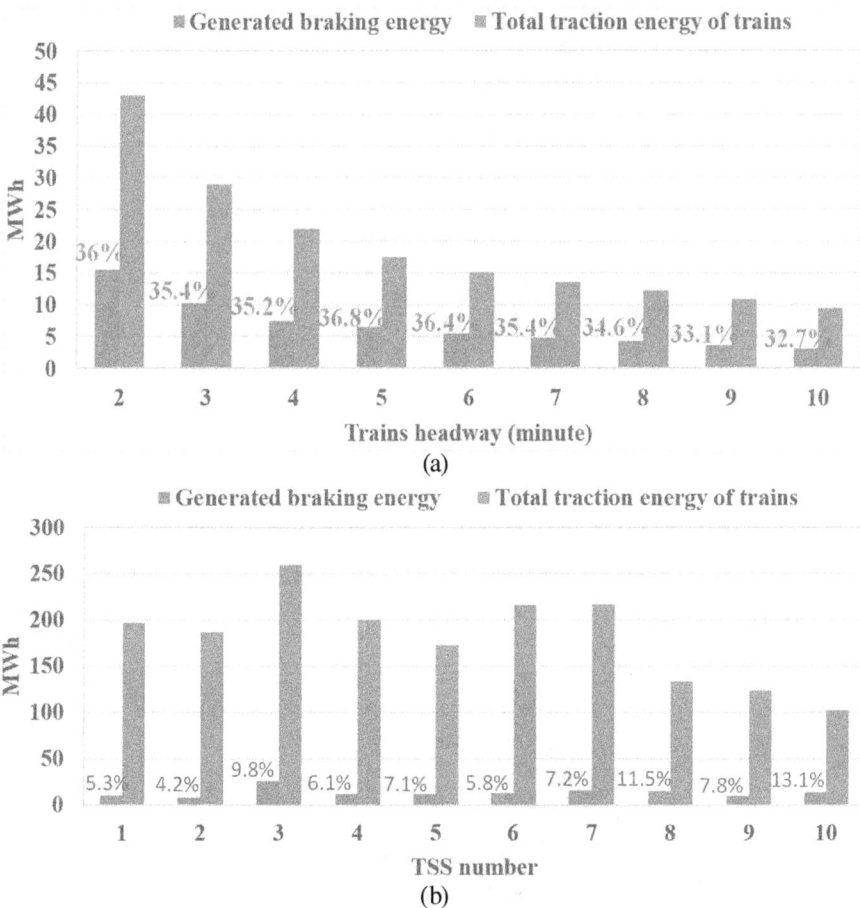

FIGURE 9.2 Trains' total required traction energy and recoverable braking energy. (a) In a typical 1500 V DC metro line. (b) In a 25 kV AC high-speed line.

the railway companies are reluctant to use it for internal consumption. Consequently, a suitable way to use RBE in AC ERSs has not been applied, and experts are looking for innovative technologies.

In addition to the ratings of RBE, evaluating its recuperating methods is also significant. The different solutions can be defined for RBE utilization in both DC and AC ERSs [3], which can be classified into meantime and storing-based methods. These approaches are listed in Table 9.3, and the detailed evaluation together with advantages and disadvantages are explained in [3]. Considering only the unused portion of daily RBE, which is lost as heat, and assuming the average battery capacity of each EV as 50 kWh, more than 280 EVs can be fully charged in a day.

It is obvious from the table that ESSs are popular for RBE management and can be utilized in larger sizes, providing high energy storage volume. However, the cost of implementing ESSs for each TPSS is too much. Accordingly, utilizing available dedicated batteries of EVs instead of establishing new stations can substantially decrease the costs and facilitate achieving sustainable transportation. To supply EVCIs by RBE, the features of demanded load and production must be evaluated. From the ERS side, the main issue of RBE is its impulsive and discontinuity production features. In fact, the electric braking of trains has high power in a very short duration. Depending on the ERS type and train specifications, the generated energy

TABLE 9.3
RBE Management Techniques Classification

		Method & Application	ERS type	
			DC	AC
Meantime methods		Supply other vicinity trains	✓	†
		Supply onboard auxiliary services	✓	–
		Send back to the primary grid/mains	‡	✓
		Waste by onboard rheostats	✓	–
Storing-based methods	Onboard	Supply onboard auxiliary services	✓	–
		Supply train in autonomous operation mode	✓	–
	Offboard	Supply stationary auxiliary services	✓	✓
		Supply accelerating trains	✓	–
		Supply track—emergency situation	✓	✓
		Supply track—peak shave/load shift	✓	✓
	Wayside	Supply track—emergency situation	✓	✓
		Supply track—voltage regulation	✓	–
		Supply track—peak shave/load shift	✓	✓
		Local distributed energy resource	✓	–

† In case of utilizing RPC in TSS or sub-sectioning posts

‡ In case of possible reversible substation

flux parameters can be varied. For instance, for low-voltage ERSs, the speed and inertia of the trains are lower than suburban and high-speed trains.

Therefore, braking time, peak power, and the value of RBE are smaller too. However, the frequency of braking is much higher due to the short headways between trains, which causes an RBE profile more like the continuous mode. Accordingly, management of RBE is more comfortable in low-voltage DC ERSs with respect to the other types. Based on what has been discussed so far, two scenarios as the meantime and ESS-based integration can be defined to supply EVCIs by the RBE of ERSs.

9.4.1.1 Meantime Integration of ERSs and EVCIs

The first mode of integration is the meantime type in which RBE will be transferred to EVCSs uninterruptedly. Figure 9.3 shows the proposed scheme related to this scenario. As mentioned before, one of the main issues in this mode is the mismatch between produced and consumed powers. The power of trains during the braking is in the range of megawatts, while the power required by EVs is in the range of several kilowatts. Thus, it is mandatory to have many numbers of EVs in the parking area or locate multiple EVCSs connected along the ERS line and work together to absorb and consume RBE in a high range. This method is suitable to be implemented around ERSs, which are nearby to the large EVCSs, such as business buildings and commercial centers with hundreds of available parking areas. According to the reports, the majority of vehicles approximately are parked for around 8 hours in such places [4]. Furthermore, the TPSSs close to the metropolis and motorways with a high volume of parking space are suitable for such an integration. The main challenge of the meantime scenario is the discontinuity of the braking energy production. It means that when trains are arriving at the station RBE will be available. Thus, in such a model, there should be an additional connection between the grid and EVCS (as shown in Figure 9.3). Considering a long time without RBE production, EVs can be charged directly through the grid via

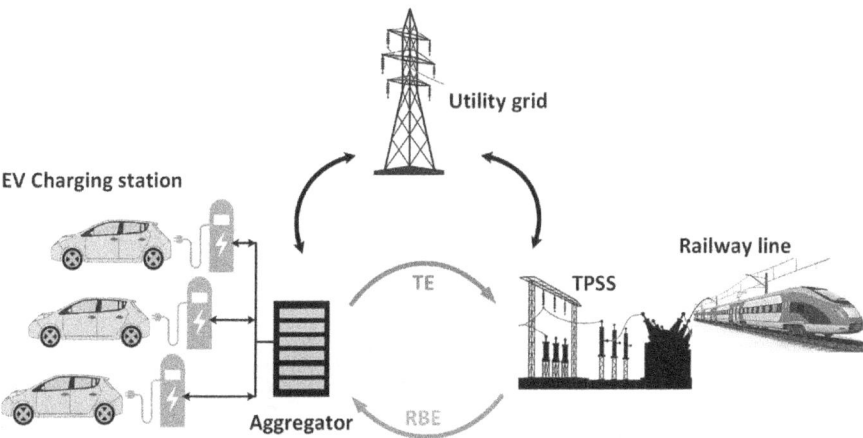

FIGURE 9.3 Scheme of meantime integration between ERSs and EVCIs.

grid-to-vehicle (G2V) technology. A precise aggregator should be implemented to evaluate the EVs' condition, with the available maximum charging rate to realize integration based on calculated prioritization [5].

9.4.1.2 ESS-Based Integration of ERSs and EVCIs

Given the fact that the majority of present EVCSs are in the small/medium range, the meantime integration scenario is not a suitable choice for implementation. Therefore, another indirect method can be introduced by utilizing ESSs as an auxiliary device between ERSs and EVs. Thus, the additional energy during braking can be stored on ESSs and when there is no RBE, EVCSs can be fed from the grid. This scheme allows the continuity of the power transferring, and absorbing RBE with the low number of EVs [6]. In comparison to the meantime method, there is no need for an additional connection with the grid. Figure 9.4 illustrates the scheme of ESS-based integration.

9.4.2 Integration Concept Assuming EVs as Stationary ESS and DER

In the last section, transferring RBE from ERS to EVCI is discussed. Similarly, power flow from EVCI to ERS supplying traction energy (TE) of trains also can be realized by taking advantage of dedicated batteries inside EVs. The meantime power flow can be executed when trains are accelerating or during peak hours. In this way, by management of the aggregator, those batteries of EVs that are fully charged are participated to operate together as distributed energy resources and auxiliary supply ERS. Assuming the high number of participated EVs, a general power threshold (P_g) for EVCSs can be specified that is matchable with the accelerating

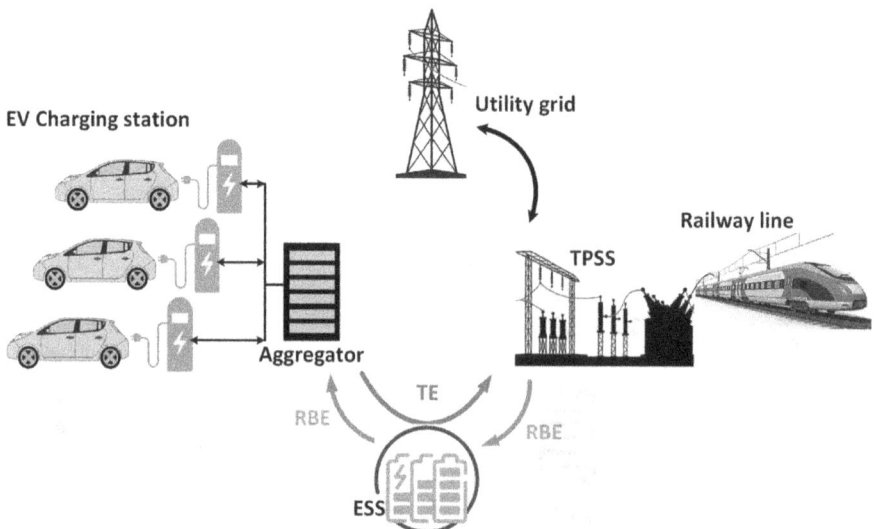

FIGURE 9.4 Scheme of ESS-based integration between ERSs and EVCIs.

traction power (P_{at}) of trains. In this mode, to completely supply the accelerating train, the following equation must be valid.

$$\begin{cases} \sum_{m=1}^{N} P_{d,m} \geq P_g \\ P_g \geq P_{at} \end{cases}, \tag{9.1}$$

where $P_{d,m}$ is discharging power rate and N is the total number of participated EVs. The amounts of P_{at} for each TPSS should be evaluated based on the average threshold powers of departure trains. From an EVCI point of view, the implementation of such integration depends on the number of EVs, the charging/discharging power rate of EVs, and their SOC. In case of insufficient EV number, ESS-based connection can be proposed, taking advantage of hybrid ESSs as an auxiliary device. Thus, during acceleration of trains or peak shaving intervals, ESSs collaborating with EV batteries can supply trains. Literally, ERSs can be supplied by cheaper energy through this integration since EVs are often charged by lower tariffs or RESs.

9.4.3 INTEGRATION CONCEPT UTILIZING THE UNEMPLOYED CAPACITY OF EXISTING POWER INSTALLATION

The power rating of ERSs and equipment is determined based on the line information and the peak power related to the rush hours. Generally, the designed and installed capacity of TPSSs are assigned more than the required operating capacity. Therefore, the significant capacity of ERSs and dedicated equipment are not employed and are available for long intervals during the day, which can be utilized to charge EVs. Using existing ERSs and installations to supply the EVCIs will substantially decrease the operating costs. In order to explain it more clearly, the power profile related to some of the Milan tram substations is evaluated. The daily power profile of six TPSSs that supply the tramway line is illustrated in Figure 9.5. It must

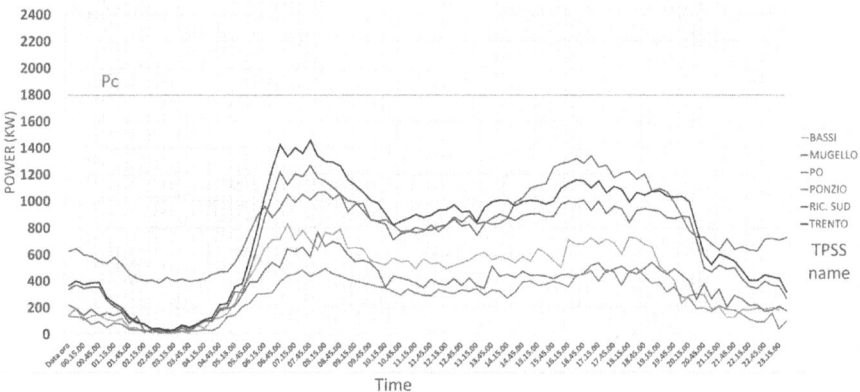

FIGURE 9.5 Daily power profile for Milan tramway substations.

be noted that each TPSS possesses 2500 kW maximum power capacity. Considering the yearly power profile, the suitable contracted power (P_c) can be assigned as 1800 kW, which is indicated by the dashed red line in the figure. Comparing the power profiles with P_c, it can be concluded that a large amount of the existing capacity of the tramway system and its installations are unemployed, which can be used to supply EVCIs. It is worth mentioning that a suitable optimized threshold for P_c is important and will lead to the optimum energy consumption invoice. Adopting the energy stored in EVs to operate the peak-shaving function of traction substations can significantly decrease P_c for ERSs.

One practical example of the pioneering projects regarding this T2V framework has been implemented on Zara Street in Milan, Italy [7], where E-bus opportunity chargers are integrated and supplied by tramway charging lines. The proximity of both an E-bus charging infrastructure and tramway lines at this location (as shown in Figure 9.6) has increased its operational potential.

9.4.4 INTEGRATION CONCEPT UTILIZING ERS LINES AS ENERGY HUB PROVIDING SUITABLE CONNECTION AREAS OF RESs

Regarding the EU climate commitment assigned purpose, attaining a zero-carbon eco-friendly transport system by 2050 has become the main objective in the field of transportation. In this way, the integration of renewable energy sources (RESs)

(a)

FIGURE 9.6 Integration of E-bus opportunity charging station to tramway's DC grid in Milan, Zara Street.

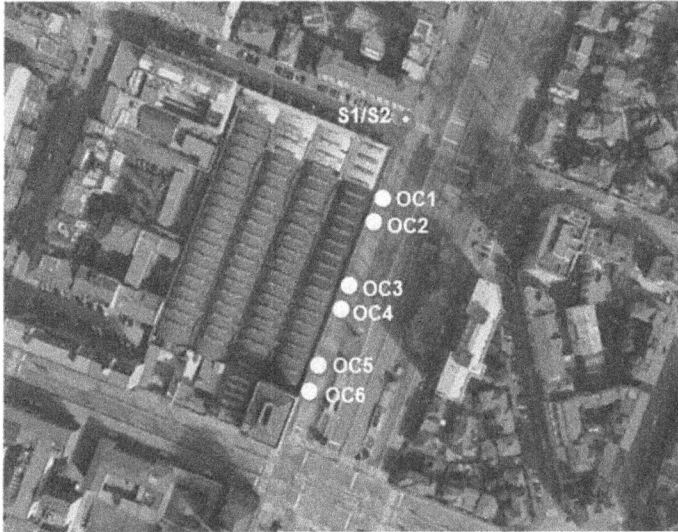

(b)

FIGURE 9.6   (Continued)

with existing electric transportation, can be an important step. In recent years, numerous research has been accomplished regarding the integration of EVCIs and RESs, but the development of such an integration in ERSs is insufficient. However, ERSs have a high potential to be integrated with RESs because of widely distributed lines all over the regions that are closed to wind farms or PV plants. Figure 9.7 demonstrates two real locations in which ERSs are close to the wind farms [8,9]. In other words, using the existing installations and supply lines of the ERSs, the power flow between RESs and loads can be facilitated [10,11]. This is not limited to the lines, but also at railway stations. Due to the large area of stations, the roofs are the best place to install PV panels in addition to trackside PV panels. Based on the report of the Indian Ministry of Railways, 111 MW of solar power capacity has already been installed on rooftops of different stations in January 2021 [12]. Figure 9.8a demonstrates a picture of an Indian railway substation with PV panels. This country is a pioneer moving toward net-zero carbon emission railway systems. Meanwhile, Japan Railway East plans to replace about 20% of the total electricity used in railway networks supplied by RESs, including wind and solar power [13]. Figure 9.8b shows a Tokyo Station Tokaido line platform where PV panels with 453 kW capacity have been implemented. Overall, the concept of ERS and EVCI integration in this framework can be addressed by acting as an energy hub. In other words, for the realization of such an integration, the distributed railway lines act as an energy hub and transfer energy between traction substations and EVCSs. Consequently, both systems can take advantage of RESs.

(a)

(b)

FIGURE 9.7 The proximity of ERSs and wind farms. (a) One of the largest wind farms in California called San Gorgonio Pass [8]. (b) H1S railway line in the UK [9].

(a)

FIGURE 9.8 Rooftop PV implementation in railway substation. (a) India substation [12] (b) 200 and 400 kW PV panels on the roof of Takasaki-Tokyo stations in Japan in 2004 [13].

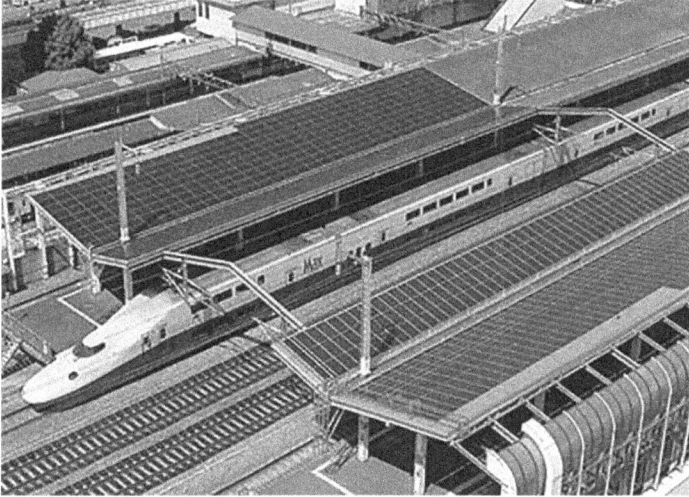

(b)

FIGURE 9.8 (Continued)

9.5 DIFFERENT ARCHITECTURES OF ERS AND EVCI INTEGRATION

In the previous sections, various frameworks for ERSs and EVCIs were explained. The outstanding features and benefits of such an integration motivated the authors to study and analyze the different aspects of incorporating RESs. In this section, different structures of integration considering DC and AC systems and the concept of an energy hub will be explained.

9.5.1 LOW-VOLTAGE DC HUB-BASED INTEGRATION

DC ERSs are mostly low-power systems localized in metropolitan and urban zones with a large population. The dedicated supplying DC lines for these systems are distributed throughout the city. On the other side, the majority of DERs are based on DC networks, such as PV and ESSs, or they contain a DC part inside, like wind generators [2]. Meanwhile, mid and fast EVCIs with DC-type charging systems are distributed places close to railway lines or stations. This similarity can increase the interconnection possibility. The general concept of T2V/V2T integration in low-voltage ERSs is shown in Figure 9.9. LVDC ERSs (0.6–1.5 kV), are usually used in urban rails, trams, light rails, and subways in low/medium distances and are known as low-power transportation systems. Accordingly, low-power elements can be integrated with such a system. In this architecture the shared DC busbar as an energy hub and integration place can be implemented in two positions: substation location and along the line. Substations are the most appropriate place since there are existing switching and protection devices together with buildings that can be shared for new connections. Furthermore, the proximity of the AC mains provides greater power transfer. For the along line position, the integrations can be more complex but provide supplying EVCIs in strategic points, such as parking lots near railway stations.

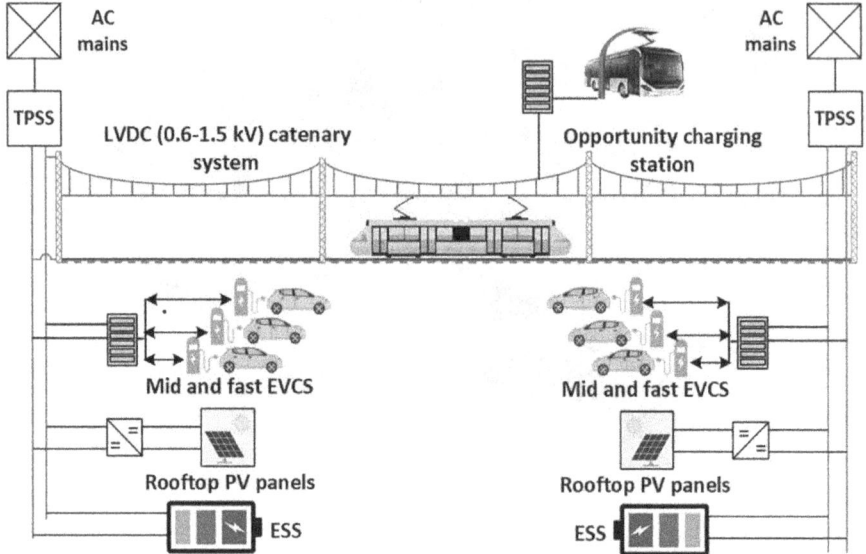

FIGURE 9.9 T2V/V2T integration architecture in LVDC ERSs.

RESs that can be connected to LVDC architecture are in the class of low-power generators, such as small-scale PV plants or low-power wind generation. However, due to the lack of empty space or its high cost in cities, the suitable choice seems to be rooftop PV panels implemented in substations. The outstanding specification of this architecture is the direct connection of EVCIs to DC bus due to the proximity of the voltage levels. Despite the fact LVDC architecture is not appropriate for mega chargers and ultrafast charging systems, it will be effective for mid/fast charging systems, which mostly use and are equipped with Combo or Chademo plugs. In addition, as mentioned in previous sections, opportunity charging systems for E-buses can be easily integrated without an interface in this architecture.

In order to realize T2V/V2T technology more efficiently, hybrid ESSs also can be implemented in DC bus, which can collaborate with EVs to handle the power peaks of railway substations.

9.5.2 MEDIUM-VOLTAGE DC HUB-BASED INTEGRATION

As mentioned before, LVDC ERSs suffer from a low power rating, which can affect the number of operating trains. Therefore, for suburban and HSR lines with high-power demand, a higher voltage is required. Motivated by the PQ-based issues of AC ERSs, the experts are working toward modern medium-voltage DC (MVDC) ERSs with high-power ability. The high-power capability of MVDC ERSs simplifies the direct connection of RESs, such as high-power wind and PV generation. From an EVCI point of the view, MVDC architecture can provide the realization of ultrafast (up to 500 kW) and mega charging stations (in the order of 1–3 MW) which could be troublesome if they are connected to the AC utility grid for their power consumption range. On the other side, OPCSs that provide an intermittent short charge to

E-buses can take advantage of MVDC architecture, reaching a maximum charging power up to 750 kW. Figure 9.10 demonstrates the architecture of MVDC ERSs integrated with RESs and EVCIs. In this architecture, the high-power elements can be connected directly, but low-power elements, such as small PV generators and mid or fast EVCIs, require some DC/DC converter. In some suitable nearby positions, MVDC ERSs can also supply LVDC ERSs independently or even supportively. In fact, MVDC ERSs can incorporate LVDC ERSs via suitable DC/DC converters. Similar to the LVDC ERSs, the MVDC shared bus can be implemented in two various positions of substation and along the line.

9.5.3 AC Hub-Based Integration

Despite the many advantages of MVDC ERSs, they require a lot of time to be developed and expanded. Therefore, the alternative choice to have high-power railway lines is utilizing AC ERSs, which now are mature technology.

However, these systems are implicated by significant PQ issues that necessitate the existence of additional compensators. On the other side, in such a system, transferring RBE between trains is less likely because of the neutral sections and unlike the LVDC ERSs, it is possible to return it to the utility grid. Such a system refers to AC ERSs. Accordingly, the implementation of T2V/V2T technologies in such a system required some limitations. The general concept of integration in AC ERS is demonstrated in Figure 9.11. This architecture contains traction transformer-based substations with

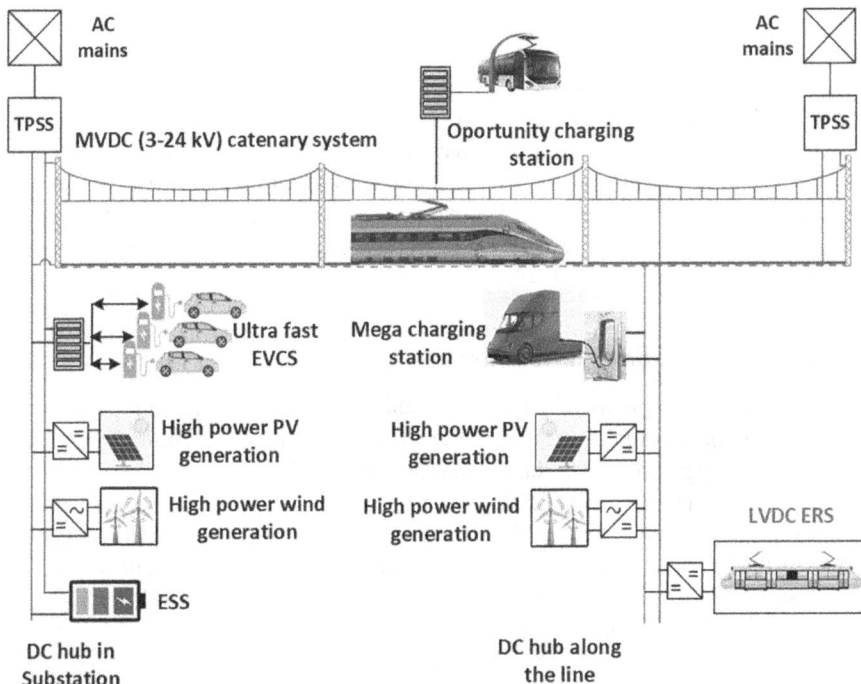

FIGURE 9.10   T2V/V2T integration architecture in MVDC ERSs.

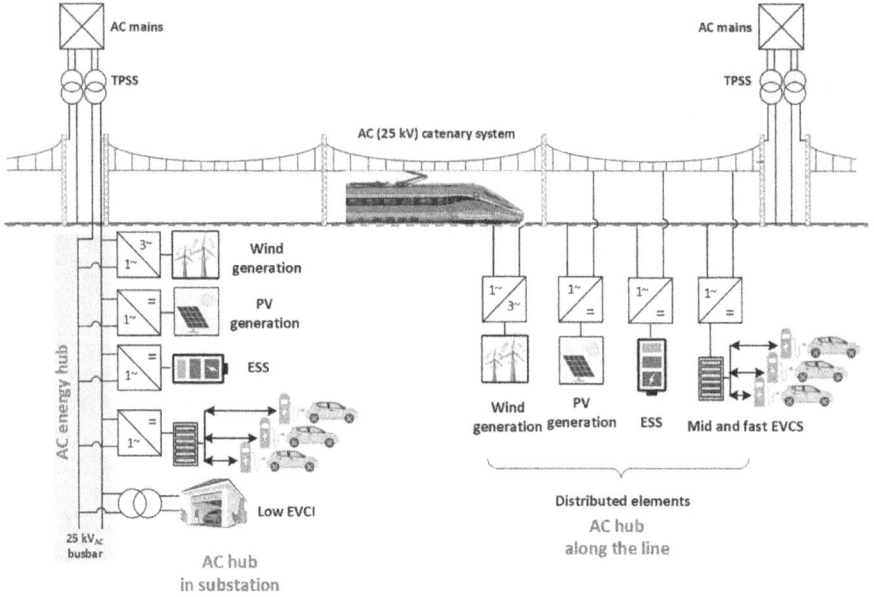

FIGURE 9.11   T2V/V2T integration architecture in AC ERSs.

25 kV OCS separated in an insulated area. The joint AC bus as an energy hub can be implemented in two various formats: concentrated in the substation or distributed along the line. In this architecture, elements are mostly integrated into the AC hub through DC/AC converters. Avoiding PQ issues relating to the switching frequencies of such converters, the power flow capacity must be lower. Consequently, RESs connected to AC ERSs are in the low-power range. From an EVCI point of view, the low-power features of such a structure may not be suitable for mega chargers or even ultrafast charging systems. However, the integration of mid and fast EVCIs through special rectifiers and low EVCIs through step-down transformers can be realized. It is worth mentioning that the integration of elements through the rectification process can intensify PQ issues in AC ERSs.

9.5.4 Hybrid Hub-Based Integration

To overcome the problems of AC-based architecture and increase the power-flow capacity, a hybrid configuration can be proposed. In hybrid architecture, taking advantage of a high-power interfacing device (different kinds of interface converters or compensators [14]), a joint DC bus can be established to facilitate the connections of elements. Depending on the interface device, various configurations can be proposed for hybrid architecture, which is explained in detail in [2]. The scheme of hybrid architecture for T2V/V2T integration is shown in Figure 9.11. Interface converter can create an increased bidirectional power flow between AC and DC buses. The high-power capability of hybrid architecture allows the direct connection of high-power RESs. Meanwhile, from an EVCI point of the view, this system can provide the

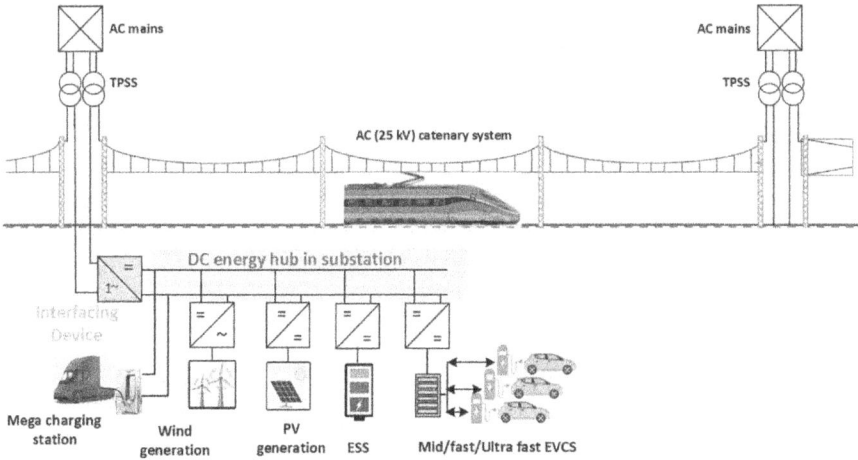

FIGURE 9.12 T2V/V2T integration architecture in hybrid ERSs.

realization of mega EVCIs, which could be problematic in the AC architecture. It must be noted that during busy times with high consumed loads if all the elements work at their full rates, the suitable control interface device, and consequently PQ controlling, will be very complicated. Due to the high cost of implementing DC bus along the line, this architecture it is better to be realized in the substation position.

9.6 MODELING AND INTEGRATION IN MVDC-BASED ARCHITECTURE

In order to realize the integration architecture mentioned in previous sections, modeling of such an integration in MVDC ERSs is discussed in the present section. As mentioned, RESs will play a crucial role in the power supply of ERSs and the feasibility of this integration to ERSs, especially for high-speed railways, is under scrutiny in order to evaluate the technical and economic aspects of such an idea. This section presents an example modeling on the integration of wind turbines (WT) and photovoltaic (PV) as auxiliary power supply to an MVDC railway microgrid together with the concept of T2V and V2T. Thus, power electronic converters are the key factor for the integration. Accordingly, the layout of the proposed MVDC ERS will include the following:

- TPSS connected to the main high-voltage AC through step-down transformer and AC/DC converter in order to convert the AC to DC;
- Catenary system that connects all the units, including TPSS, ESS, RES, and charging infrastructure;
- RESs, including WT and PV, connected to the catenary system through AC/DC and DC/DC power converters;
- ESSs, including batteries and ultra-capacitors, connected to the catenary system through DC/DC converter; and

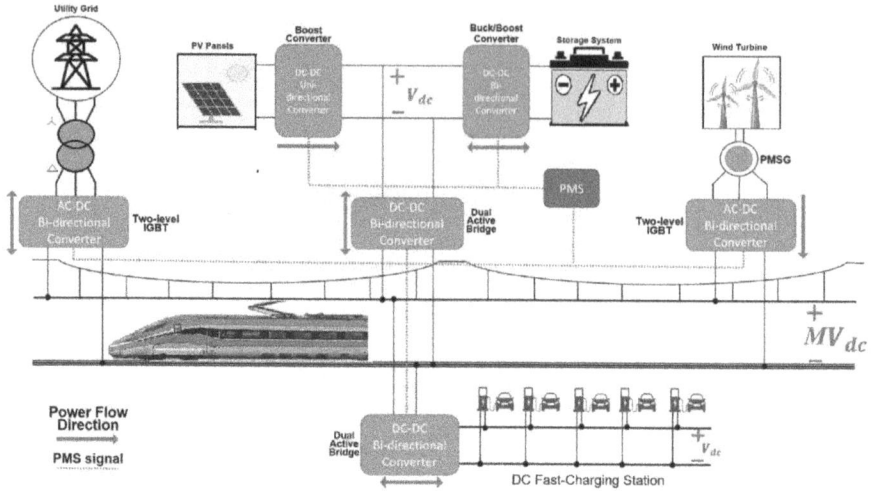

FIGURE 9.13 Proposed MVDC railway microgrid system architecture.

- Charging infrastructure, including mid/fast charging stations, ultrafast chargers for EVs, opportunity charging, and mega chargers for trucks, connected to the catenary system through DC/DC converters.

A simple example of the proposed MVDC system is shown in Figure 9.13. It includes two RESs (one PV plant and one wind farm) and an ESS unit. The grid supplies the catenary system through TPSS, and the EV fast charging station is integrated through a DAB converter along the line. PV panels and ESS are integrated using dedicated DAB-based power electronic converters. DAB converters support bidirectional power flow between the MVDC catenary system and the other connected components. Meanwhile, EV fast charging infrastructure connection via a DAB converter can provide the system bidirectional power flow to realize the G2V and V2G technology.

9.6.1 INTEGRATION OF PV PANELS

The PV system includes panels and a boost converter for maximum power point tracking (MPPT) purposes. Generally speaking, a large range of DC/DC power converters can be implemented to connect the PV farms to the MVDC railway system. For instance, non-isolated soft switched interlocked boost (SSIB) DC/DC converter, modular cascaded DC/DC converter, hybrid resonant PWM, full-bridge DC/DC converter, or phase shift full bridge (PSFB) DC/DC converter can be applied. Nevertheless, the choice for the final step DC/DC converter is a DAB converter due to its galvanic isolation, soft switching, etc. However, for simplicity, a boost converter has been modeled to connect PV to the MVDC system in this chapter. Because of the changing features of solar radiation and temperature, the output power of PV is variable. Consequently, the MPPT algorithm is adopted to guarantee the maximal power extraction. Different MPPT algorithms, including perturb and observe (P&O),

incremental current, artificial intelligence (AI), etc., are presented in the research. The selected method in this chapter is P&O due to its simplicity and feasibility. The concept of the P&O method is shown in Figure 9.14. It is clear that in this method the PV output voltage changes with the small variation of irradiance. It modifies the output power of the PV system illustrated with ΔP. Its principles are described as: If $\Delta P > 0$, it reveals that it is getting close to MPP, and therefore, any increment in the same direction will move the operating point toward MPP; If $\Delta P < 0$, it presents that the operating point moves away from MPP, thus its direction must be reversed.

9.6.2 Integration of Wind Turbines

Wind speed and wind turbine (WT) diameters are two major determining factors of output power; as the turbine's blade diameter increases (larger turbines), the turbine will capture a higher level of wind power that will consequently be more efficient compared to turbines that are connected in a different manner. Different types of WT can be addressed and elaborated traditionally and at the commercial level; AC WT can be connected in various configurations that can be then connected to an MVDC railway system through AC/DC power electronic converters. However, due to the development of direct current (DC) power transmission, implementing DC WT can be advantageous in comparison to traditional AC WT. The principal requirement of such an idea is developing DC WT, which is just at the incubator of the research phase, but all details of both technologies will be elaborated to have a better insight into the future of the MVDC

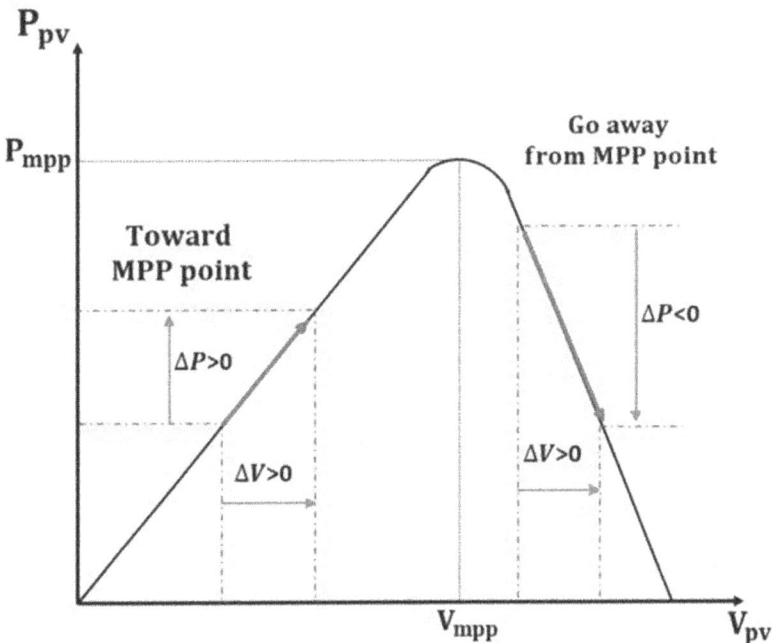

FIGURE 9.14 MPPT working principle.

railway system inherited RESs. The other main objective is to select an optimum point that has the highest mean wind speed during the year and is close to the railway in order to avoid any cable losses. Accordingly, the designed parameter is not considered in this chapter. In this context, a permanent magnet synchronous generator (PMSG) is adopted connected with the DC catenary system using an AC-DC power converter.

For low-power generation, a diode rectifier plus DC-DC converter is utilized, whereas in high-power purposes an active PWM rectifier is utilized. The dynamic relationship of a PMSC is as follows:

$$\begin{cases} V_{sd} = -L_q p \omega_r i_{sq} + L_d \dfrac{di_{sd}}{dt} \\ V_{sq} = p \omega_r \varphi_v + L_d p \omega_r i_{sd} + L_q \dfrac{di_{sq}}{dt} \end{cases}, \tag{9.2}$$

where (i_{sd}, i_{sq}) and (V_{sd}, V_{sq}) are the DQ coordinates and (i_{sa}, i_{sb}, i_{sc}) and (V_{sa}, V_{sb}, V_{sc}) are of the ABC coordinates of stator current and voltage, respectively. p is the number of pole pairs, φ_v shows the magnetic flux linkage, and ω_r is the rotor angular speed. The WT generator with its control diagram is illustrated in Figure 9.15.

9.6.3 INTEGRATION OF ESSs

In the proposed MVDC ERS, storage systems are mainly needed to handle the power peaks that come from trains, both in traction and in the braking phase, and from RESs, especially from wind generators. In this way, it is possible to minimize the impact on the AC mains and the sizing of the AC/DC substations. The development of energy

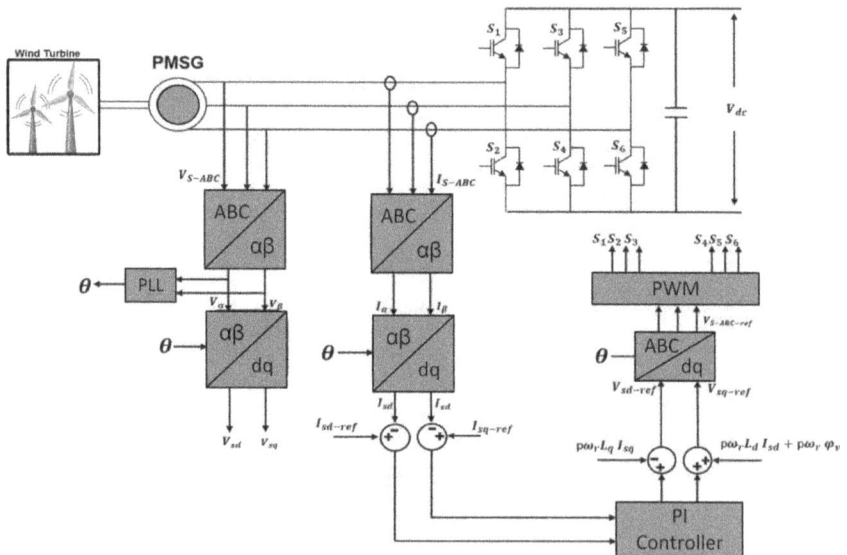

FIGURE 9.15   WT generator with dedicated control system.

consumption and efficiency includes extra advantages for the ERS's operator and for the electric system in general. In designing ESSs and determining their suitable size for smart ERSs, two main subjects needed to be studied: the RBE level in relation to the railway traffic and the amount of installed renewables. In the ERSs, storage systems can save train RBE and handle the power peaks that come from trains and renewable sources. Therefore, the high-density ESSs are applied as supercapacitors and batteries. The ESSs in MVDC systems can be divided into three types: onboard, offboard, and wayside [3]. The onboard ESS is applied inside trains or even on the roof. Despite the advantages of lower loss and better saving of energy, the higher weight of onboard ESS can enhance the consumed power of trains and decrease speed and efficiency. The offboard ESS is commonly employed in substations. Besides the advantages of onboard ESS, the offboard ESS can supply the auxiliary and internal loads of substations, i.e., the lighting, air conditioning, and escalator. As mentioned, the limitations of size and weight have presented an obstacle to the application of onboard ESSs in ERSs. Moreover, the intersections of railway networks and RESs are propitious to the utilization of local renewable energy. Therefore, increasing the utilization rate of RBE and RES via ESS helps achieve the energy-saving goals more effectively.

9.6.4 INTEGRATION OF EV CHARGING INFRASTRUCTURES

To integrate the charging infrastructures into the DC catenary system of the ERPS, the following two architectures applicable for both low- and high-power charging stations can be proposed.

9.6.4.1 Two-Stage DC Charging Infrastructures
As it is shown in Figure 9.16a, the two-stage chargers are applicable for low-power purposes, where the first-stage solid-state transformer (SST) as a DC-DC converter

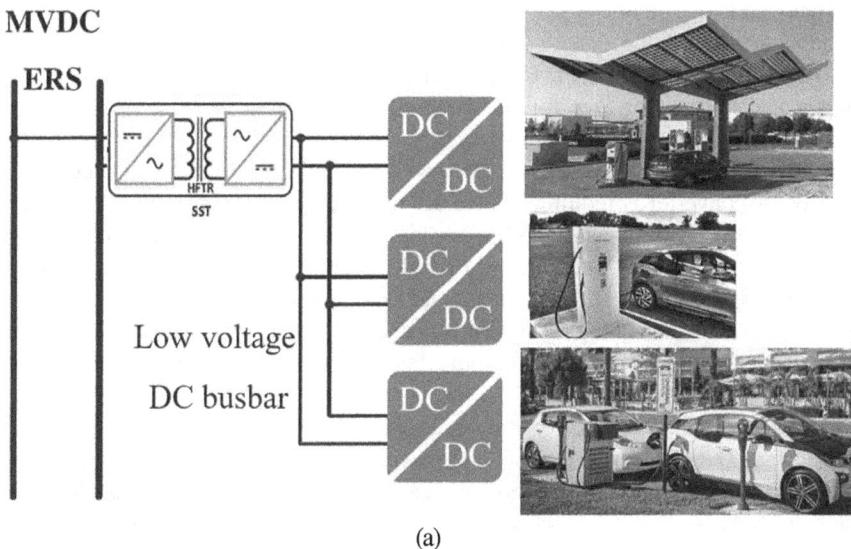

(a)

FIGURE 9.16 Integration configurations. (a) Two-stage charging station. (b) Single-stage charging station.

MVDC

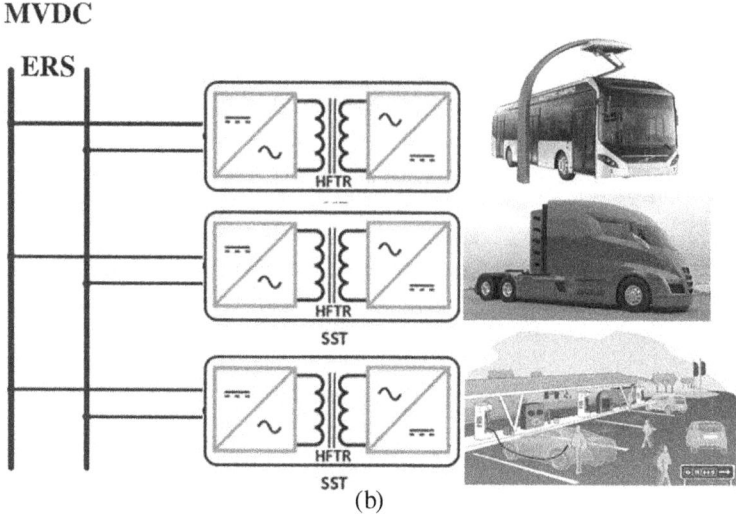

FIGURE 9.16 (Continued)

is isolated and interlinks the DC catenary system with the charging station. The second-stage DC-DC converter can be chosen to be isolated or non-isolated, and it is mainly installed to be able to charge various types of EVs with different voltage requirements.

9.6.4.2 Single-Stage DC Charging Infrastructures

The architecture of a single-stage charging station can be seen in Figure 9.20b in which one power electronic converter is directly interlinking the DC catenary system with various types of chargers, including mega chargers for E-buses and E-trucks. In this type of charger, the single power electronics converters are isolated in order to provide the system with galvanic isolation for meeting the safety factor. As mentioned, most EVs, E-buses, and E-trucks are compatible with various ranges of voltages. However, for the proposed integration, a charging infrastructure is connected to the MVDC catenary system. Also as mentioned, for connecting the charging infrastructure to the DC catenary system of the ERPS, an isolated power electronics converter is required. There are many choices for selecting the isolated DC-DC converter by considering its advantages and disadvantages. A dual active bridge (DAB) converter and phase shift full bridge (PSFB) converter are the most promising options due to their maturity and availability. For modeling the proposed system, a DC EV fast charging station is considered to be integrated with the smart DC catenary of the ERS via a DAB converter. The structure together with its control structure is shown in Figure 9.17.

A DAB converter consists of two single-phase full bridges and a high-frequency transformer (HFT). It transfers the energy from the DC catenary line to the EV's internal battery (and contrariwise in the case of V2G) using phase shifting. Changing the rate of phase-shift between the primary and secondary side converter, the DAB manages the power flow. The control function can be expressed as (9.3)

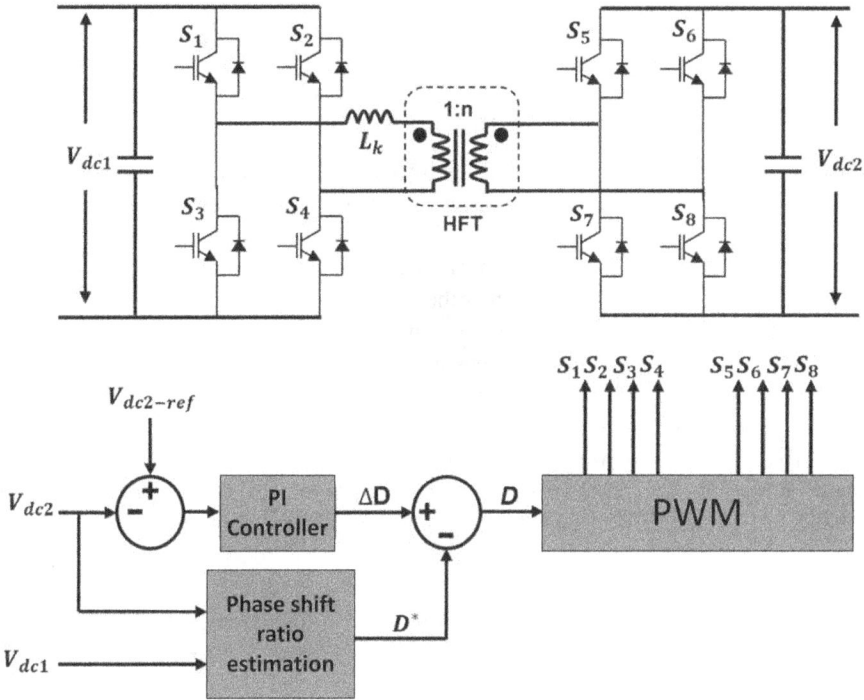

FIGURE 9.17 EVCI integration through a DAB converter and its control system.

$$P = \frac{V_{dc1}V_{dc2}\varphi(1-\varphi)}{n\pi L_k \omega}, \qquad (9.3)$$

where V_{dc1} and V_{dc2} are the primary and secondary DC voltages, n is the turns ratio of primary and secondary windings of the HFT, ω is the angular frequency, L_k is the leakage inductance of the HFT, and φ is the phase-shift amount. The amount of phase shift and leakage inductance can be calculated as follows:

$$\varphi = \frac{\pi}{2}\left(1 - \sqrt{1 - \frac{8f_s L_k P_{out}}{nV_{dc1}V_{dc2}}}\right) \qquad (9.4)$$

$$L_k = \frac{V_{dc1}V_{dc2}d(1-d)}{2f_s nP} \qquad (9.5)$$

In (9.4) and (9.5), fs is the switching frequency and d is the duty cycle percentage. This type of converter can provide the system with galvanic isolation and the zero-voltage switching (ZVS), which decreases the losses and enhances the efficiency. To switch DAB, there are multiple phase-shifting methods, such as single-phase shift, dual-phase shift, triple-phase shift, extended phase shift, and hybrid phase shift. In this scrutiny, a single phase-shifting angle is chosen.

9.6.5 CONTROL MANAGEMENT SYSTEM AND RESULTS

The proposed MVDC system is simulated by MATLAB/Simulink software with real daily profiles as input data and assuming various working situations. To control the proposed system an energy management system or power management system (PMS) can be applied to manage the power flow between loads and sources. PMS is a centralized system that collaborates with the other decentralized units. As inputs, 24-hour real profiles of electric railway power demand at a specific TPSS, wind speed, solar radiation, temperature, and EV charging station power demand are applied.

The EV charging station is assumed in the suburban area where the EVs get charged during working hours and the peak hours are during the morning when the EVs arrive. The daily power profile of the EVCS is illustrated in Figure 9.18, which confirms that the peak hour happens from 7 a.m. to 11 a.m. To simulate the exact behavior of MVDC ERS, generating/demanding powers at one 30 MW TPSS is assumed. Figure 9.19 shows the daily power profile of a typical and middle TPSS in a bilateral MVDC line.

The uncertainty and time-varying specifications of ERS are obvious in this figure. During the night and early in the morning the passenger trains are not working and only freight trains are in operation. Therefore, the amount of required power is low.

The nominal power of the PV system is considered 3 MW, and the nominal output power of WT is 6 MW. The maximum output power demand of the EVCS is 500 kW,

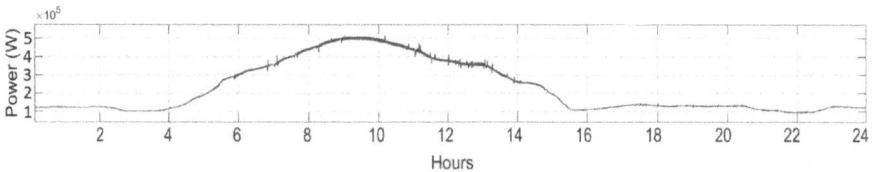

FIGURE 9.18 Fast charging daily power demand [10].

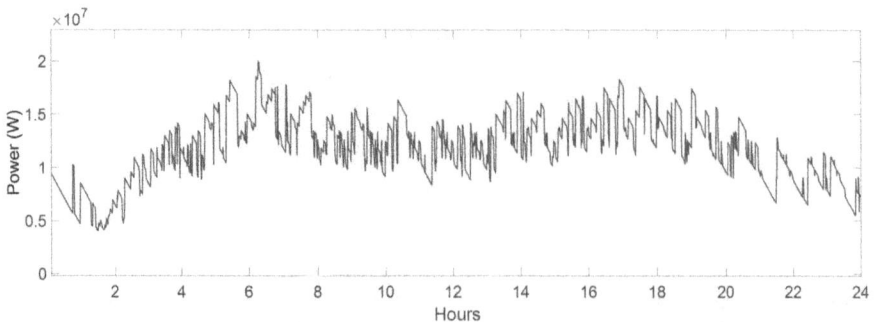

FIGURE 9.19 TPSS daily power profile.

which usually happens during peak hours. However, as is shown in Figure 9.19, the power demand of TPSS differs from 4 MW during off-peak hours to about 20 MW during peak hours. The features of the different parts of the proposed MVDC system are mentioned in Table 9.4.

Figure 9.20 demonstrates the daily voltage profile of the proposed catenary system, which remains on 9 kV DC with a +10% and −10% threshold, which is the acceptable range in ERSs.

For normal working condition, the power grid is the main supplier of the proposed MVDC ERS system. However, for emergency or some other situations, stand-alone mode can be executed too. Figure 9.21 illustrates the power profile of various elements in the proposed MVDC micro-grid, including the main grid, EVCS, RESs, and the ESSs. As it is clear, the proposed system is functional in the multiple working modes specified in the dedicated PMS. During night, the ERS's power demand is low and therefore both grid and WTs (if available) supply the loads. This mode is called feeding mode (FM). However, during the mornings and peak hours, thanks to RESs, it can be seen that peak-shaving mode (PSM) has happened. Therefore, the power supplied from the main grid is lower than the required power of loads by about 30%. For the rest of the day (from 7 a.m. to 8 p.m.) the proposed system is mostly working in the FM and the wind/PV/grid system can feed the loads (EVCS and trains). In some situations, if the SOC of the storage reaches 100%, and RESs

TABLE 9.4
Proposed MVDC System Parameters

System type	Feature
PV	14250 modules-1Soltech 1STH-215-P, 25 series+5750 parallel, 3 MW
WT	PMSC, 6 MW, 8 m/s (basic speed)
ESS	500 V, 4500 Ah Lithium-ion, 1s response time
EVCS-DAB	800 kW, 10 kHz switching frequency
TPSS	50 MVA, 4500 V AC, 50 Hz, 9 kV DC

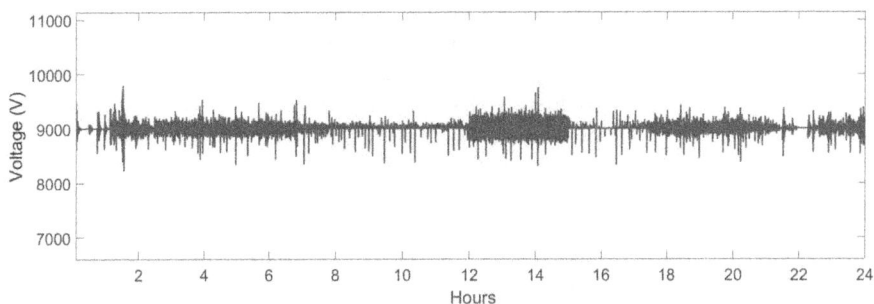

FIGURE 9.20 Voltage profile of the 9 kV catenary system [10].

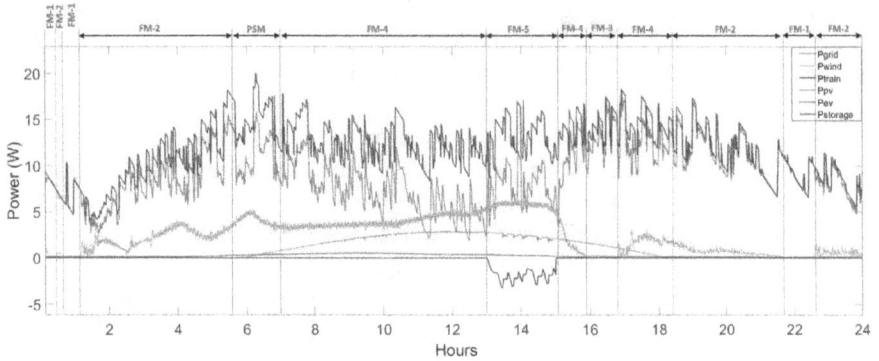

FIGURE 9.21   Power values for different sources and loads [10].

produce power, the additional power will be transferred into the main power grid, which illustrates recovery mode (RM). The results reveal that the proposed MVDC micro-grid system can be a promising solution to reinforce the ERSs feeding line for future and high-traffic issues. Consequently, introducing RESs, ESSs, and EVCIs and integrating them with existing ERSs can enhance system power capacity and efficiency. Meanwhile, it can compensate voltage drops and make smoother voltage profile on the overhead catenary.

9.7 CONCLUSION

Expanding and increasing EV/E-bus charging infrastructures in recent years may overload the utility grid and require new energy production and installations. The integration of ERSs as existing high-power networks with EVCIs can be a good solution to overcome the mentioned problem. In this chapter, the new concept of V2T/T2V technologies are introduced and different integration features are discussed, utilizing intrinsic features of RBE in ERSs. In addition, exploiting ERSs' dedicated lines as an energy hub to provide suitable connection areas with RESs is explained and different integration architectures for DC and AC ERSs are demonstrated. It is revealed that an MVDC-based architecture can provide both the advantages of the AC and LVDC types. A simple model is simulated to reveal the effectiveness of such a system. However, in the route of development of this architecture and despite the numerous advantages stated for the T2V/V2T integration, some challenges can be posed, especially the complexity of the smart control system and the high cost of PE-based modules and converters to evaluate its feasibility attentively for future executions.

REFERENCES

[1] Kaleybar, H.J., Brenna, M., Foiadelli, F., Fazel, S.S. and Zaninelli, D., 2020. Power quality phenomena in electric railway power supply systems: An exhaustive framework and classification. *Energies, 13*(24), p. 6662.

[2] Brenna, M., Foiadelli, F. and Kaleybar, H.J., March 2020. The evolution of railway power supply systems toward smart microgrids: The concept of the energy hub and integration of distributed energy resources. *IEEE Electrification Magazine, 8*(1), pp. 12–23.

[3] Khodaparastan, M., Mohamed, A.A. and Brandauer, W., 2019. Recuperation of regenerative braking energy in electric rail transit systems. *IEEE Transactions on Intelligent Transportation Systems, 20*(8), pp. 2831–2847.

[4] Saber, A.Y. and Venayagamoorthy, G.K., 2010. Plug-in vehicles and renewable energy sources for cost and emission reductions. *IEEE Transactions on Industrial Electronics, 58*(4), pp. 1229–1238.

[5] Krueger, H., Fletcher, D. and Cruden, A. (2021). Vehicle-to-grid (V2G) as line-side energy storage for support of DC-powered electric railway systems. *Journal of Rail Transport Planning & Management, 19*, p. 100263.

[6] Fernandez-Rodriguez, A., *et al.*, 2017. Charging electric vehicles using regenerated energy from urban railways. *2017 IEEE Vehicle Power and Propulsion Conference (VPPC)*, pp. 1–6.

[7] Zaninelli, D., Brenna, M. and Jafari Kaleybar, H., 2021. Tecnologie per le nuove flotte di autobus elettrici. *AEIT Journal*, 30–38.

[8] https://ogrforum.ogaugerr.com/topic/windfarm-question?reply=21703120928990720

[9] http://ecotrains.co.uk/

[10] Ahmadi, M., Jafari Kaleybar, H., Brenna, M., Castelli-Dezza, F. and Carmeli, M.S., 2021. Integration of distributed energy resources and EV fast-charging infrastructure in high-speed railway systems. *Electronics, 10*, 2555.

[11] Ahmadi, M., Kaleybar, H.J., Brenna, M., Castelli-Dezza, F. and Carmeli, M.S., 2021. DC railway micro grid adopting renewable energy and EV fast charging station. *2021 IEEE International Conference on Environment and Electrical Engineering and 2021 IEEE Industrial and Commercial Power Systems Europe (EEEIC/I&CPS Europe)*, pp. 1–6.

[12] https://indianrailways.gov.in/railwayboard/uploads/directorate/secretary_branches/IR_Reforms/Green%20Railways%20(use%20of%20renewable%20energy).pdf

[13] www.pvdatabase.org/projects_view_details.php?ID=227

[14] Kaleybar, H.J., Brenna, M. and Foiadelli, F., 2020. EV charging station integrated with electric railway system powering by train regenerative braking energy. *2020 IEEE Vehicle Power and Propulsion Conference (VPPC)*, pp. 1–6.

10 Modeling of Electric Vehicle DC Fast Charger

Viswanathan Ganesh, V. M. Ajay Krishna, S. Senthilmurugan, and S. Hemavathi

CONTENTS

10.1 INTRODUCTION

Modes of transportation have vastly improved and come a long way in terms of efficiency, but due to population growth, carbon emissions have not been reduced. The emission of greenhouse gases (GHG) has been dominant in developed and oil-exporting countries [1], as shown in Figure 10.1. For enhanced integration of renewable energy sources and the impact of EVs on a distribution grid, see [1]. The mode of charging can also be classified as direct current (DC) and alternating current (AC), which is used for fast charging. A DC charging infrastructure is preferred over an AC charging infrastructure [2]. Since EV has become a global necessity, several policies and standards have been established to ensure that the required standards are met uniformly across all countries [3]. The cloud system can be used for selecting the mode of charging based on the needs of the owner and vehicle with the help of autonomous charging of EVs in charging stations [4].

Currently, the cost of manufacturing and acquiring an EV is significantly higher than expected, but this issue can be resolved with the help of mass production, which meets the need-demand curve for reasonable pricing of various EVs [5]. It can be used as a battery energy storage system (BESS), which can be used to implement a peak shaving algorithm and reduce the impact of peak demand through vehicle-to-grid (V2G) operations. An EV can be used this way. A bidirectional converter has been described in the literature for enhanced operation [6][7]. The location of EV fast charging stations is critical as they can consume a significant amount of energy when a number of EVs are charging at a high rate, upon which the system must not face an imbalance [8]. Consequently, it is important to maintain the dead band voltage and power buffer to ensure continuous functioning of DC fast chargers [9]. In a microgrid, the power can be provided through BESS connected to an off-grid system

DOI: 10.1201/9781003293989-10

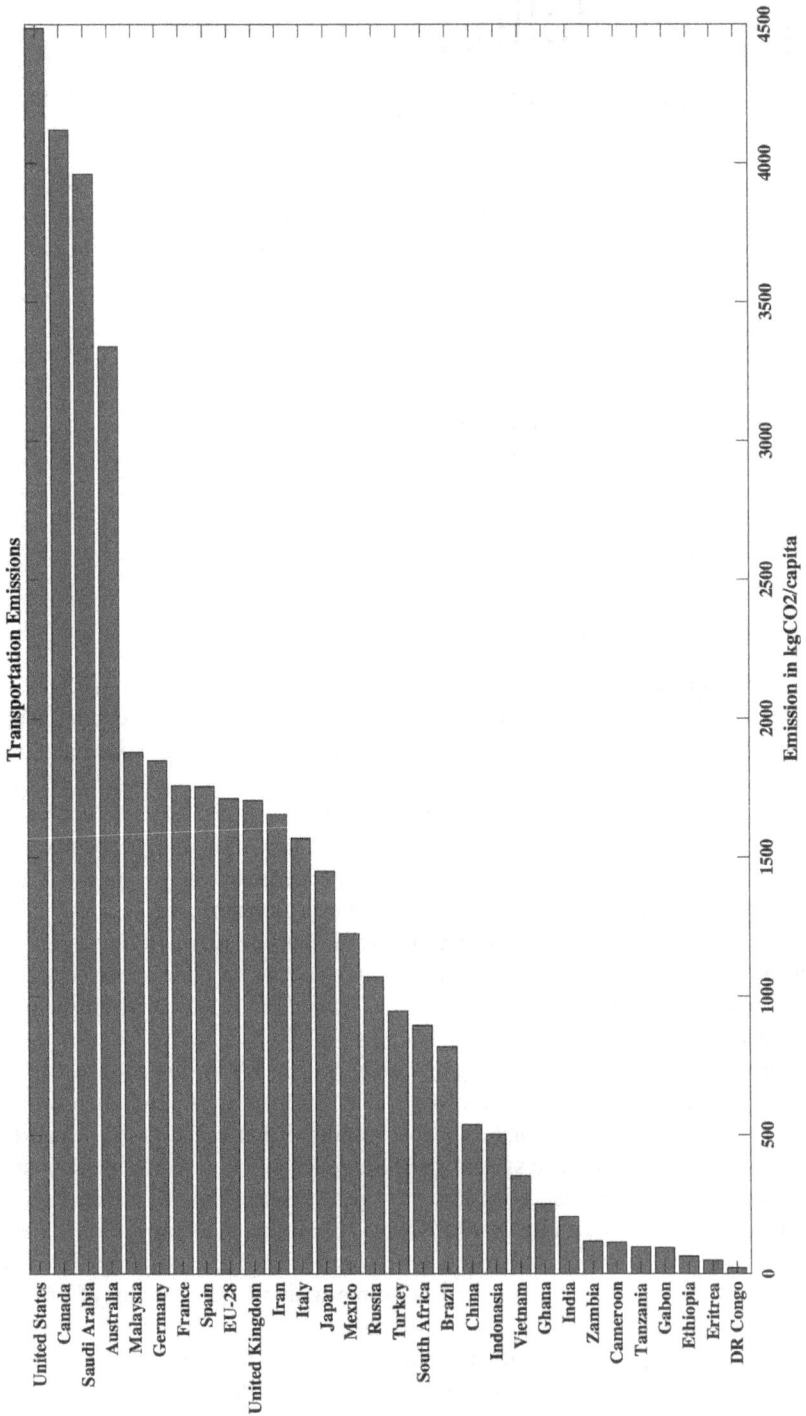

FIGURE 10.1   Per Capita Transportation Carbon Emission.

voltage and power buffer. Power quality can be affected if the system is unstable and one of the primary factors, such as frequency, voltage, phase angle, or current limit, is violated [11]. Hence, the use of a voltage source controller (VSC) is used to ensure that the EV charger is in sequence with the main grid. On the other hand, power electronic devices can cause harmonics in the system, and apart from the fundamental component, odd harmonic components can be dangerous [12]. An alternate solution was proposed in the literature [13], discussing the possibilities of having a combination of photovoltaic (PV) and BESS in the distribution network to enhance the power availability and encourage more individuals to convert to EVs. In [14], an introduction to the modular approach to developing an ultra-fast charging solution for EVs was provided. A review of the current available fast charging solutions for EVs was provided [15]. EVs can accommodate various kinds of electric motors, varying from brushless DC motors (BLDC) and permanent magnet synchronous motors (PMSM), depending on the needs and battery capacity [22]. Currently, there has been a shift from IC vehicles to plug-in hybrid vehicles, which have low battery capacity and can aid the starting and initial acceleration of internal combustion (IC) engine vehicles [24].

In the event of unforeseen fault conditions, we require a safety mechanism operating in the charging station to ensure the EVs are safe and to provide owners with confidence [15] [17]. Internet of Things (IoT) can come hand in hand for live monitoring of battery state at an EV charging station [16]. The placement of the charging station must be planned with the electrical bus system, and proper optimization must be ensured for efficient operation [18] and to be sure there is no impact on the utility sector [19–23]. In the present-day scenario, the charging solution can be broadly classified into two categories: AC charging and DC charging. The detailed classification of AC and DC charging is shown in Table 10.1. EV chargers play a crucial role in microgrid and renewable energy utilization, as they are capable of drawing a significant amount of energy [25] [26]. Hence, a detailed analysis of each component must be completed before designing the desired charger rating. In this chapter, a fast DC charger is selected for brief study and modeling.

10.2 MODELING OF DC FAST CHARGERS

The rating of a DC fast charger depends on the type and capacity of battery being utilized. Most used battery cells are Panasonic NCR18650PF or Samsung 21700 [24]. The experimental data of both cells are used in the modeling of the overall battery pack. The open-circuit voltage (OCV) and internal resistance are shown in Figure 10.2 and Figure 10.3, respectively.

$$R_{in} = \frac{\Delta V}{\Delta I} \qquad (10.1)$$

The internal resistance of a battery cell can be calculated with the ratio shown in equation (10.1), where DV and DI are voltage drop and current drop due to the internal resistance of a cell. On the other hand, the value of internal resistance is

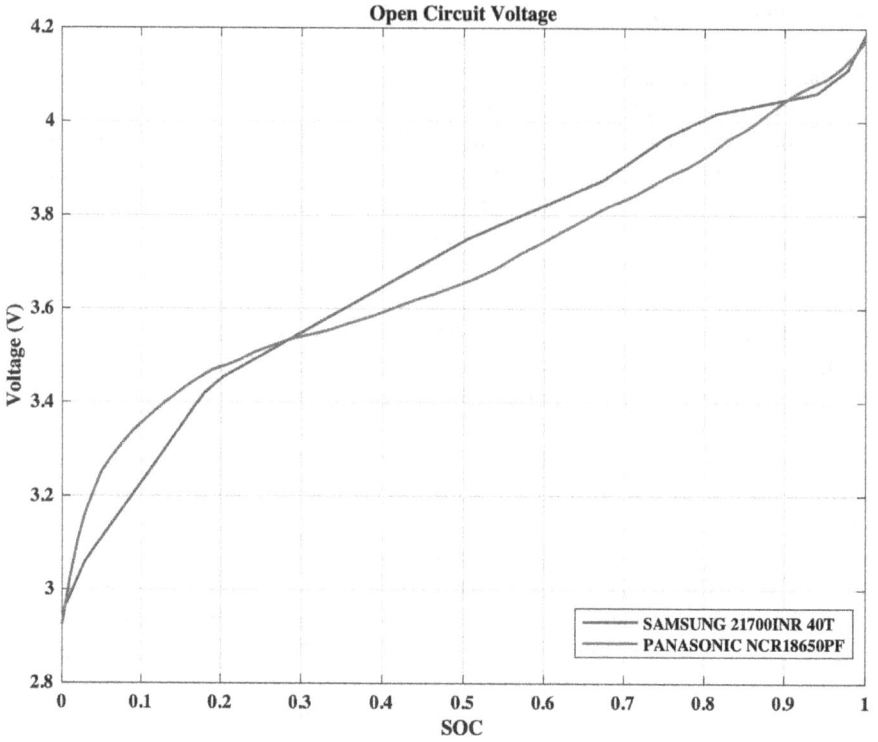

FIGURE 10.2   Open-Circuit Voltage of Predominantly Known Batteries.

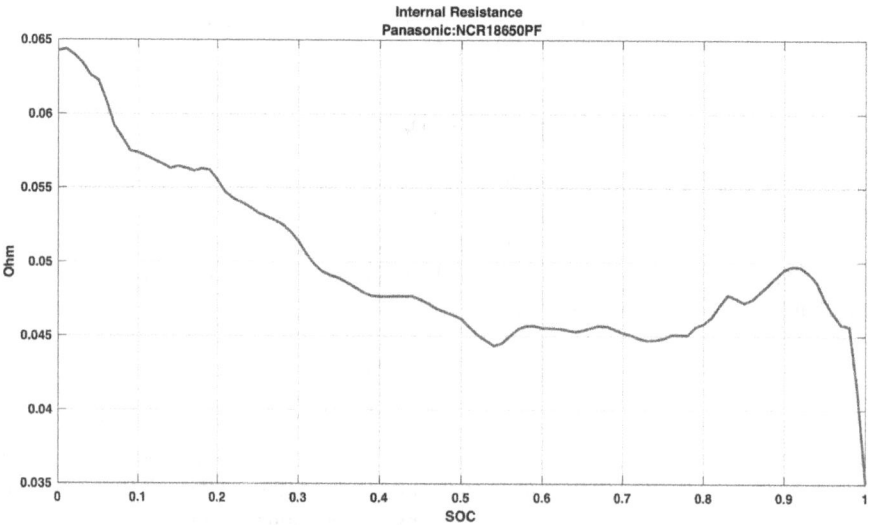

FIGURE 10.3   Internal Resistance of NCR18650PF.

dependent on temperature, the amount of current passing through the cell, the number of cycles, etc.

$$P = \left(OCV - \left(I \times R_{in}\right)\right) \times I \tag{10.2}$$

$$Loss = I^2 \times R_{in} \tag{10.3}$$

$$\eta = \frac{P}{P + Loss} \tag{10.4}$$

The modeling of a DC fast charger must begin with the approximate rating of the battery in an EV. The options for selecting a battery in EV is wide because different manufacturers consider different parameters. Select batteries have been listed in Table 10.1. The current market has provided a few long mileage vehicles, as stated in Table 10.2.

The selection of battery technology is dependent on the factors of specifications, such as overall range expected, total weight of the battery pack, and the rate of current that can be charged and discharged by the battery. Similarly, we can see

TABLE 10.1
Widely Known Battery Technology

Battery model	21700	18650	4680	Unit
Chemistry	Li-ion NCA	Li-ion NCA	Silicon-based	
Nominal voltage	3.7	3.7	3.61	V
Operating voltage	2.5–4.2	2.5–4.2	2.8–4.2	V
Cut off voltage	2.0–2.5	2.0–2.5	2.0–2.8	V
Capacity	4.1	3.6	26	Ah
Weight	65	50	355	g
Charge density	12.5	9.3	380	Wh
Dimension	21x70	18x65	46x80	mm

TABLE 10.2
Examples of EV

EV model	Battery capacity (kWh)	Range (km)
Mercedes EQC 400	80	370
Hyundai Ioniq	28	190
Nissan LEAF	30	170
Tesla Model S 85	80.8	402
Hyundai Kona	64	390
Audi e-tron 55	86.5	365
BMW iX3	74	385
Volkswagen ID 4	77	410

that there is a large possible combination of terminal voltages for a battery pack. Hence, the DC fast charger must be able to provide variable DC voltage (DC Link) to enhance the range of operation of the charger and ensure one charger can be used for several EV models.

There are several charging speeds for an EV categorized as slow and fast chargers, as shown in Table 10.3. In the following sections an introduction in modeling a DC fast charger is discussed.

It is also observed that DC fast chargers provide a high charging current and provide reliable fast charging support for an easy commute. Several companies prefer off-board chargers due to the weight reduction of the overall vehicle, and off-board chargers are capable of charging the batteries at high charging currents, which is not possible with on-board charging.

10.3 OVERVIEW OF DC FAST CHARGERS

There are several possible combinations in construction of a DC fast charger; they can be broadly classified as depicted in Figure 10.4. In Figure 10.4a, the 3-f voltage source can be directly converted to DC with the help of rectification and a power factor correction unit to ensure that the current and voltage waveforms are in phase with each other and ensuring the power factor (PF) is united for highest efficiency and to reduce the possibility of losses in the power electronics component. In Figure 10.4b, the same can be achieved with an inclusion of transformers, but the downside of using a transformer is the losses accounting for stationary/idle condition, hence by reducing the overall efficiency of the system. In Figure 10.4c, the front end and PFC can be split into different circuits to ensure the reliability of the system in case of fault. Similarly in Figure 10.4d, an additional DC-DC converter is used to ensure the isolation component and increase the safety and reliability of the system.

In Table 10.4, a simple conclusion can be made for increased reliability and safety considering the future possible implementation of renewable energy into the EV charging system; hence a combination of Figure 10.4a and Figure 10.4b is being considered to provide variable DC link voltage with isolation and the absence of

TABLE 10.3
EV Charging Classification

Charging mode	Mode 1		Mode 2		Mode 3	Mode 4
Label rating	Level 1		Level 2		Level 3	Level 4
Activity method	On-board		On-board		On-board	Off-board
Location	Residential area		Residential area		Public station	Highway
Electrical Characteristic	1-ϕ	3-ϕ	1-ϕ	3-ϕ	3-ϕ	DC
Voltage (V)	120/240	400	240	400	400	50–900
Current (A)	16	16	32	32	63	100–1250
Power (kW)	3.3	10	7	24	43	50–300
Charging time (h)	7	2.5	3.5	1.5	0.5	0.25

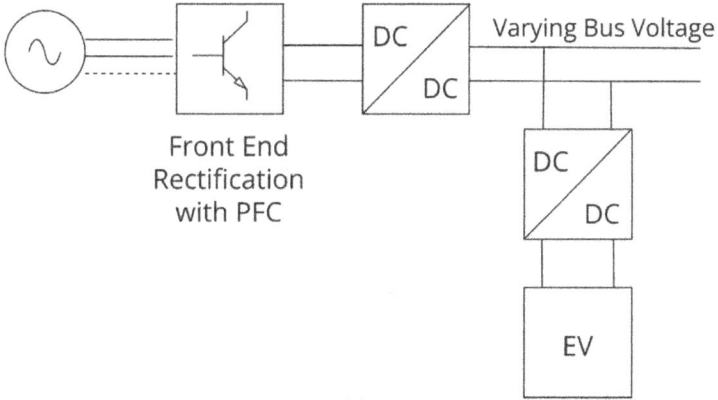

(a)

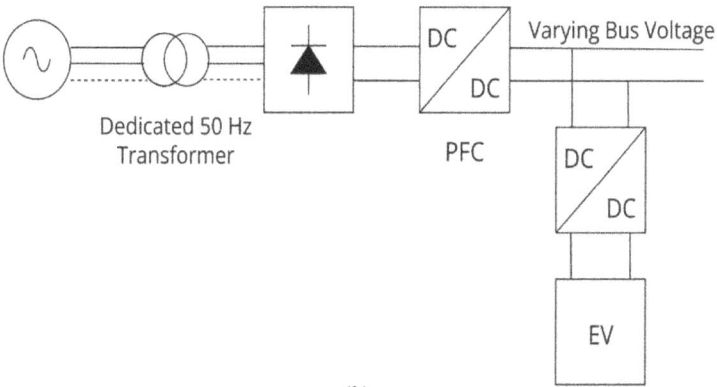

(b)

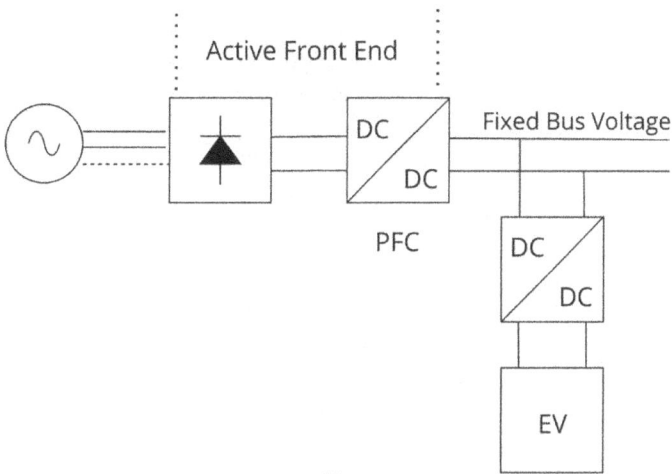

(c)

FIGURE 10.4a   Possible Combinations of a DC Fast Charger.

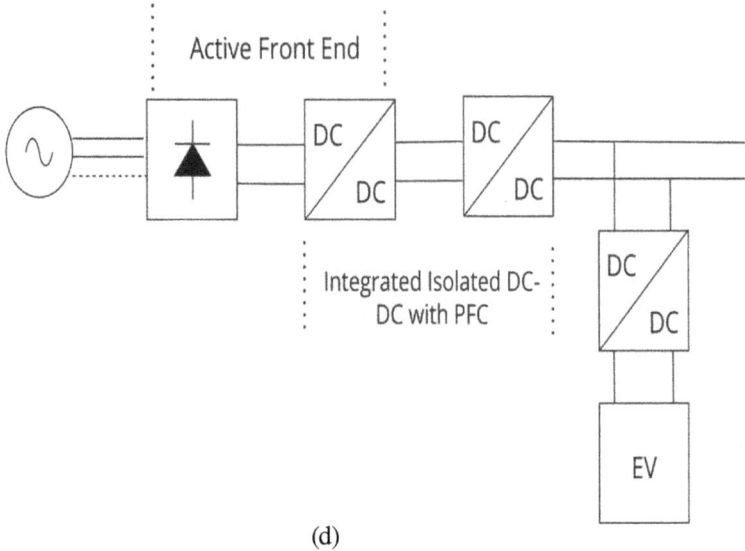

(d)

FIGURE 10.4   (Continued)

TABLE 10.4
Combinations of the EV Charging Model

Combination	Transformer	PFC	FEC	Isolation	DC link voltage
Figure 4 (a)	No	Yes	Yes	No	Variable
Figure 4 (b)	Yes	Yes	No	No	Variable
Figure 4 (c)	No	Yes	Yes	No	Fixed
Figure 4 (d)	No	Yes	Yes	Yes	Fixed

transformers. The variable DC link voltage provides a wide range of operation for EVs, with different battery ratings and maximum charging current supported by the BESS. The circuit considered for DC fast charging is as depicted in Figure 10.5. The grid is considered a 3-f supply provided the required AC power for the system. This is followed by the RL filtering component, which removes the harmonics from the circuit and feeds it into a power factor correction (PFC) and front-end converter (FEC), where AC current and voltages are in phase with each other to enable the circuit as purely resistive and convert AC to DC with the help of FEC. In continuation from FEC, the power is fed into the DC-DC converter with an isolated transformer to resemble a real-life charger and provide safety and reliability factors.

As the final receivers end, a battery is modeled with the help of experimental results, such as OCV, R_{in}, and the capacity of each cell. The simulation parameters are as stated in Table 10.5. Figure 10.6 provides an idea of the flow of the DC fast charger circuit.

The FEC and DC-DC converter control systems play an important role in the optimization and optimal operation of the charging circuit. The output of converter control circuits provides pulse width modulation (PWM) gate signals for the efficient

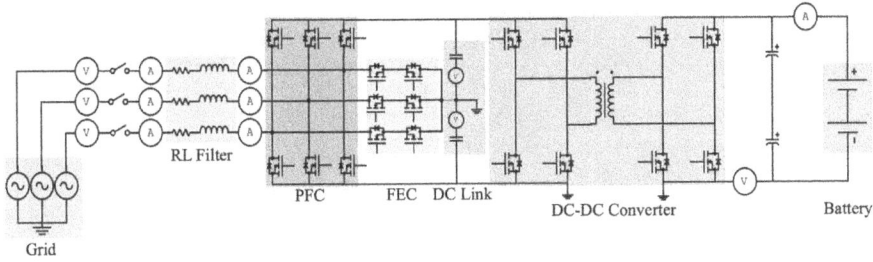

FIGURE 10.5   Electrical Circuit of a DC Fast Charger.

TABLE 10.5
Summary of Controllers

Controller	Gain in time domain	Gain in frequency domain
P	K_p	K_p
PI	$K_p + K_i \int dt$	$K_p + \dfrac{K_i}{S}$
PD	$K_p + K_d \dfrac{d}{dt}$	$K_p + K_d S$
PID	$K_p + K_i \int dt + K_d \dfrac{d}{dt}$	$K_p + \dfrac{K_i}{S} + K_d S$

operation of MOSFET switches. The FEC works in the area of average inverter fidelity with the help of gate signals that are changed.

On the other hand, the DC-DC converter is controlled with the help of gate signals produced with help of proportional integral (PI) controller, as depicted in Figure 10.7, to ensure the voltage and current are as specified to the maximum rating. Commonly known algorithms in DC-DC charger circuits are phase shifting and resonant LLC. In the control algorithm of phase shifting, the duty cycle of switching devices like MOSFET and IGBT's is fixed at 50%, and in the resonant LLC control algorithms, the use of inductors coupled with capacitor banks is used to generate smooth, constant DC output.

In Figure 10.8, we can witness the presence of oscillations in the integral controller, comparatively that the PI, P, and PID have no consecutive oscillations. The proportional controller has a steady state error. On the other hand, I, PI, and PID have no steady state error. The PI controller was selected as the most suitable controller for the operation of FEC and DC-DC converter instead of the P and PID controllers because having only a proportional controller will not gain the required output and will have a steady state error. On the other hand, even though the PID controller has no steady state error, it has an addition of derivative control, which makes the system much more complex compared to the PI controller. Another important selection criteria of the PI controller is that it can act as a low-pass filter to remove any higher-order frequencies.

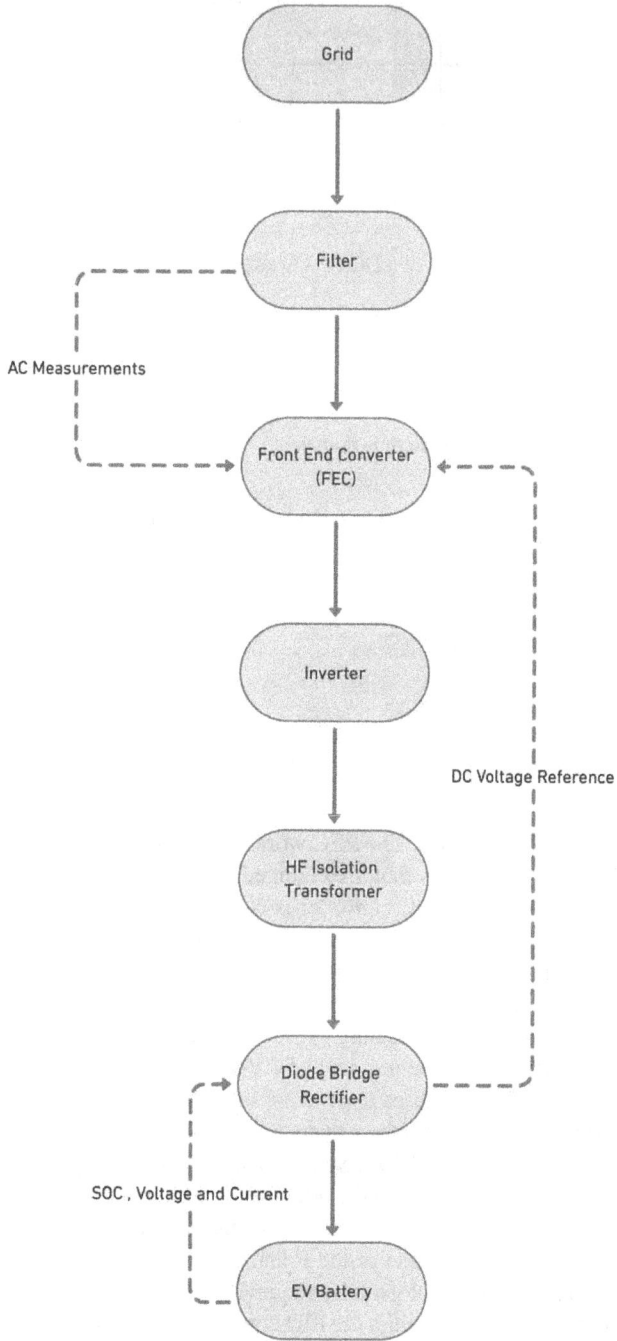

FIGURE 10.6   Modeling Procedure of a DC Fast Charger.

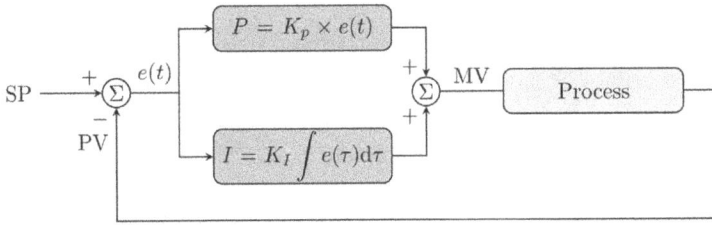

FIGURE 10.7 PI Controller Loop.

TABLE 10.6
Parameters of Various Controllers

Controller	Minimum value	Maximum overshoot	Error	Time to reach steady state (s)
P	−0.105	0.345	0.681	11.2
PI	−0.145	1.12	0	33.5
I	−0.06	1.448	0	80
PID	−0.17	1.092	0	33.5

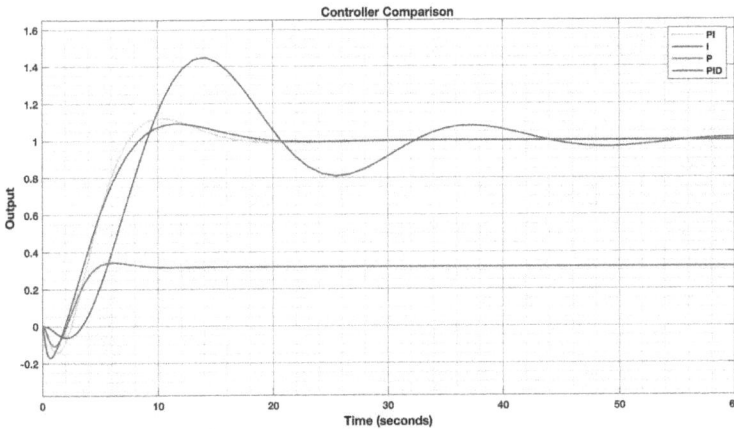

FIGURE 10.8 Comparison of Various Controllers.

10.4 FRONT-END CONVERTER (FEC) MODELING

The FEC is responsible for the initial conversion of AC grid supply to DC bus voltage, hence it is responsible for maintaining key aspects known as voltage and current through the circuit, as depicted in Figures 10.9 and 10.10. The voltage control and current control are handled by individual PI converters.

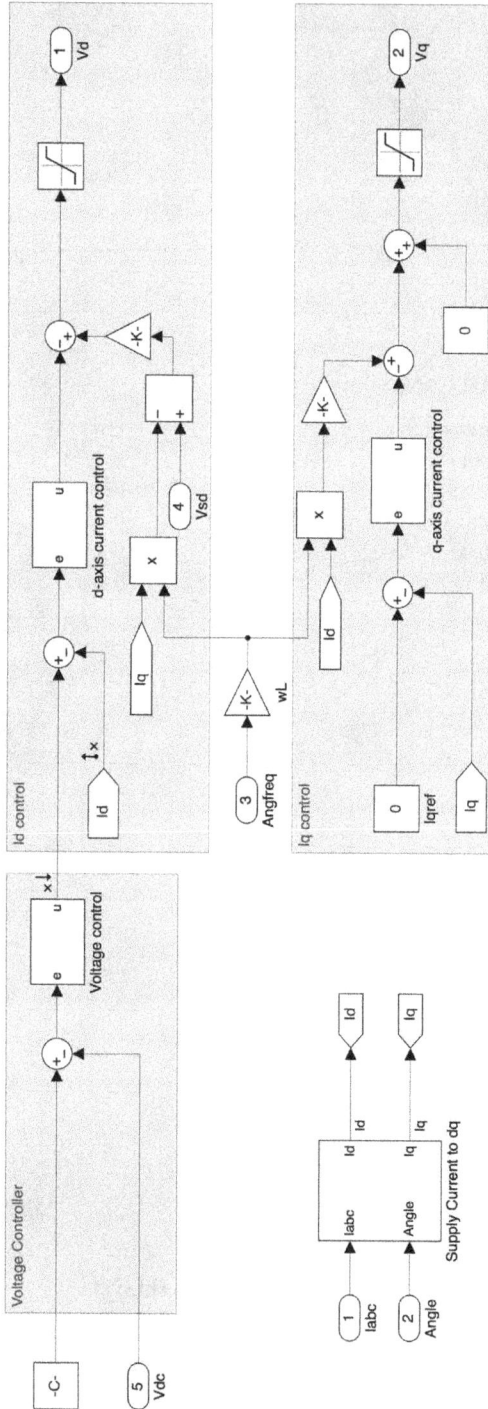

FIGURE 10.9 Front-End Converter Controller.

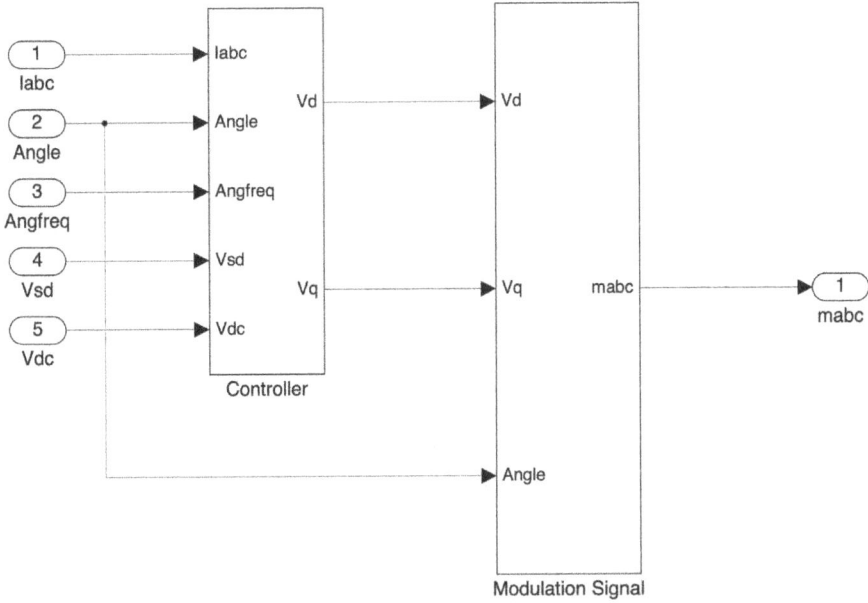

FIGURE 10.10 Front-End Converter Control Circuit.

The modulation signal is converted from the DQ axis to the ABC axis, as in Figure 10.11, with the help of inverse parks transformation. The inverse park transformation is used to convert a rotating reference frame to the component in a 3f system, as depicted in Figure 10.12.

$\widehat{U}_D$ & $\widehat{U}_Q$ are denoted as unit vectors of the new framework in reference to the old reference framework with an arbitrary vector $\widehat{U}_{XY}$.

Where

$$\widehat{U}_D = \cos(\theta)\widehat{U}_X + \sin(\theta)\widehat{U}_Y \tag{10.5}$$

$$\widehat{U}_Q = -\sin(\theta)\widehat{U}_X + \cos(\theta)\widehat{U}_Y \tag{10.6}$$

$$\widehat{U}_{XY} = U_X\widehat{U}_X + U_Y\widehat{U}_Y \tag{10.7}$$

$\widehat{U}_X$ and $\widehat{U}_Y$ represent the unit vectors of reference coordinate system, and θ is the angle between $\widehat{U}_X$ & $\widehat{U}_D$ or the angle between the original reference to the transformed reference.

Hence

$$U_D = \widehat{U}_D \cdot \widehat{U}_{XY} \rightarrow \cos(\theta)U_X + \sin(\theta)U_Y \tag{10.8}$$

$$U_Q = \widehat{U}_Q.\widehat{U}_{XY} \rightarrow -\sin(\theta)U_X + \cos(\theta)U_Y \tag{10.9}$$

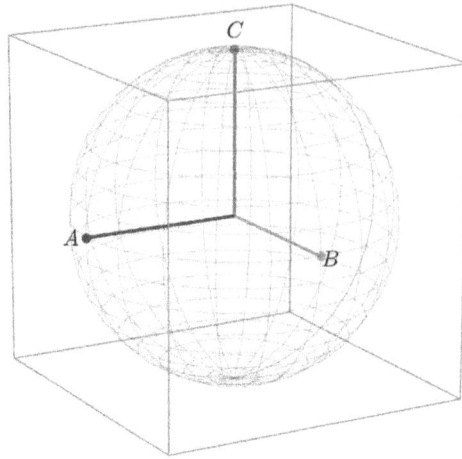

(a)

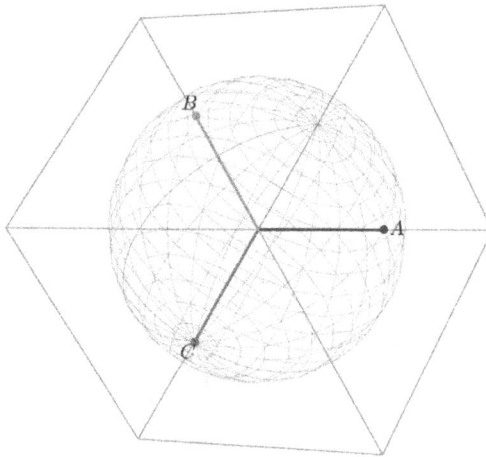

(b)

FIGURE 10.11   (a) ABC Vectors in Sphere Framework; (b) ABC Vector in 2D Framework.

U_D is a transformed projection of $\widehat{U}_{XY}$ in the $\widehat{U}_D$ axis, and similarly U_Q is a projection of $\widehat{U}_{XY}$ in the $\widehat{U}_Q$ axis. The resultant vector is known as $\vec{U}_{DQ}$.

$$\vec{U}_{DQ} = \begin{bmatrix} \cos(\theta) & \sin(\theta) \\ -\sin(\theta) & \cos(\theta) \end{bmatrix} \cdot \vec{U}_{XY} \qquad (10.10)$$

If the rotation is considered on the z-axis the tensor can be equated as

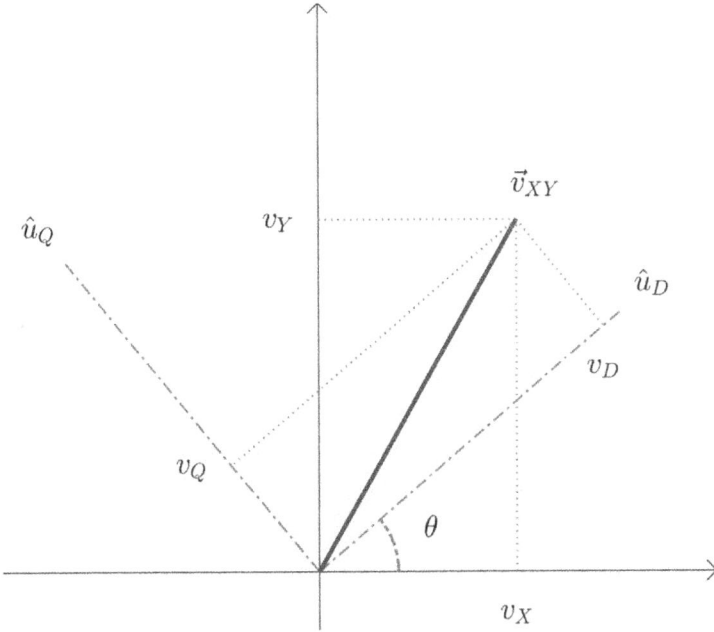

FIGURE 10.12 Resultant Vector in DQ Frame.

$$K_P = \begin{bmatrix} \cos(\theta) & \sin(\theta) & 0 \\ -\sin(\theta) & \cos(\theta) & 0 \\ 0 & 0 & 1 \end{bmatrix} \quad (10.11)$$

As per the formulae

$$\begin{bmatrix} a \\ b \\ c \end{bmatrix} = \sqrt{\frac{2}{3}} \begin{bmatrix} \sin(\theta) & \cos(\theta) & \sqrt{\frac{1}{2}} \\ \sin\left(\theta - \frac{2\pi}{3}\right) & \cos\left(\theta - \frac{2\pi}{3}\right) & \sqrt{\frac{1}{2}} \\ \sin\left(\theta + \frac{2\pi}{3}\right) & \cos\left(\theta + \frac{2\pi}{3}\right) & \sqrt{\frac{1}{2}} \end{bmatrix} \begin{bmatrix} d \\ q \\ 0 \end{bmatrix}. \quad (10.12)$$

$$\begin{bmatrix} a \\ b \\ c \end{bmatrix} = \sqrt{\frac{2}{3}} \begin{bmatrix} \cos(\theta) & -\sin(\theta) & \sqrt{\frac{1}{2}} \\ \cos\left(\theta - \frac{2\pi}{3}\right) & -\sin\left(\theta - \frac{2\pi}{3}\right) & \sqrt{\frac{1}{2}} \\ \cos\left(\theta + \frac{2\pi}{3}\right) & -\sin\left(\theta + \frac{2\pi}{3}\right) & \sqrt{\frac{1}{2}} \end{bmatrix} \begin{bmatrix} d \\ q \\ 0 \end{bmatrix}. \quad (10.13)$$

TABLE 10.7

FEC System Parameters in the Simulation

Controller	K_p	K_i	$\frac{1}{s}\int$	
			Lower limit	Higher limit
Voltage	181.61	1923.07	−500	500
D-axis current	0.00208	200	−10	10
Q-axis current	0.00208	200	−10	10

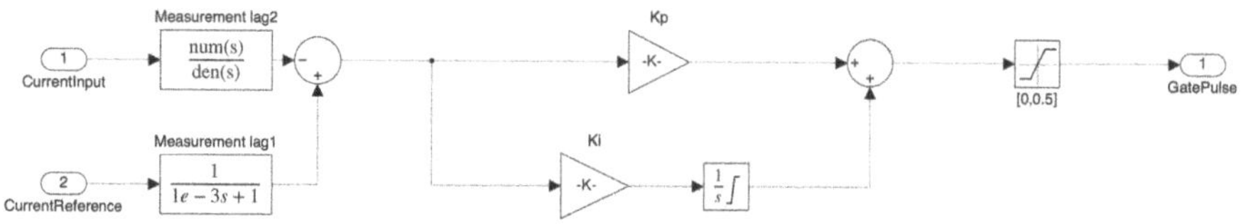

FIGURE 10.13   DC-DC Converter Control Circuit.

TABLE 10.8

DC-DC Converter System Parameters in the Simulation

Parameter	Value
Input current	Measurement from system
Reference current	100
K_p	2
K_i	1
$\frac{1}{s}\int$	−3 to 3

Equation (10.12) is the basic formulae used for having the conversion aligned on the q-axis, and equation (10.13) is formulae used for conversions aligned to the d-axis. In this case an inverse parks transformation with respect to the d-axis is considered. The parameters used for control methodology is tabulated in Table 10.7.

The same principle of the PI controller has been utilized for a simple DC-DC controller circuit in Figure 10.13, which provides the required gate pulses in the form of PWM signals and maintains the current on par with the reference current of 100A.

10.5 SIMULATION AND RESULTS

The simulation and result evaluation has been done using Simulink software; Figure 10.14 depicts the modeled block with respect to the basic circuit shown in Figure 10.5 and system parameters as per Table 10.9, respectively.

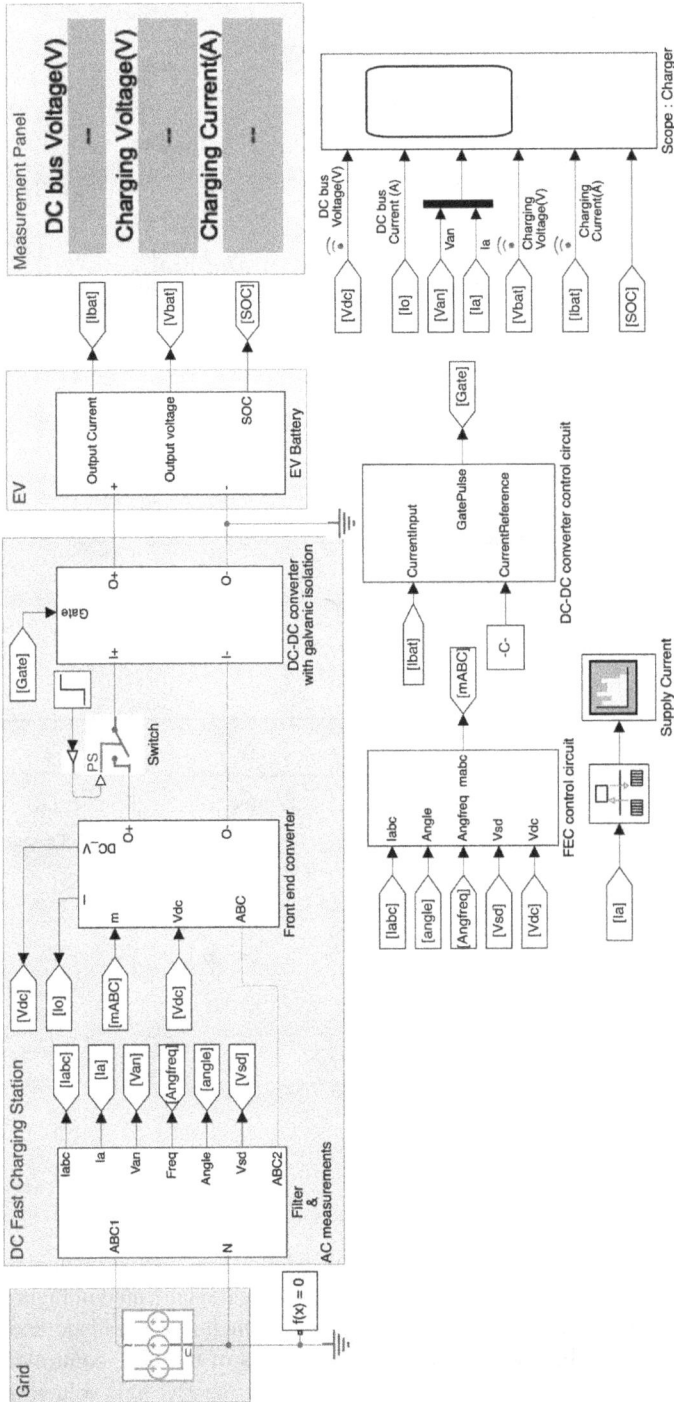

FIGURE 10.14 Overall Model of DC Fast Charger.

TABLE 10.9

System Parameters in the Simulation

System	Parameter	Value	Unit
Grid	Voltage phase-phase	415	V
	Voltage phase-neutral	240	V
	Frequency	50	Hz
Filter	R	20	mW
	L	0.1	mH
Inverter	Switching frequency	10	kHz
	Inductance	10	mH
Transformer	Magnetizing inductance	1	H
	Winding factor	0.5	
Battery	Reference current	100	A
	Initial SOC	20	%
	Capacity	50	Ah
	Inductance	5	mH
	Cell voltage	4.2	V
	Cell capacity	2.73	Ah
	Cells in series	100	
	Cells in parallel	1	

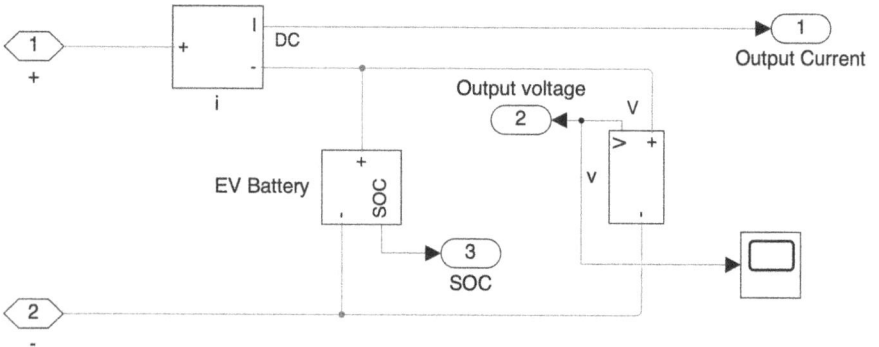

FIGURE 10.15   EV Battery.

The modeling of the battery has been achieved with the of experimental practical readings of OCV, internal resistance, and temperatures implemented in Figure 10.15. This methodology is used to make the simulation much more realistic and practical. As explained in the previous sections, the results of the FEC controller circuit have been plotted in Figure 10.16. It is evident that the DC link voltage is being

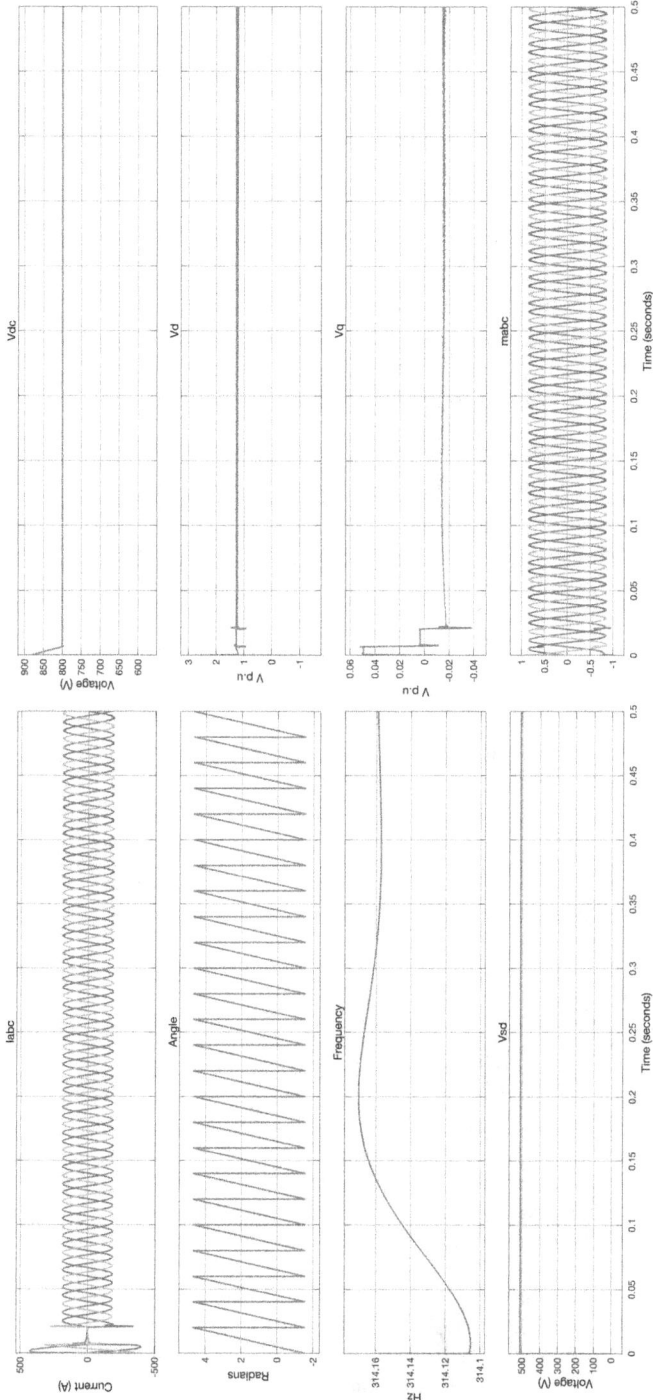

FIGURE 10.16 Results of FEC Controller Circuit.

maintained successfully at rated parameter and the modulation signal has been tuned accurately with the help of PI controller and inverse parks transformations.

In Figure 10.17 the simulation output results of the DC-DC controller circuit have been plotted, with current reference to 100 A, the PI controller is able to handle the simulation at an efficient manner with the help of modulation driving gate signals for the switches. In Figure 10.18 the triggering signals of the inverter present in the DC-DC converter have been plotted to clearly witness the role of PWM in handling the rated voltage and current outputs. In Figure 10.19 the overall input and output waveforms have been plotted depicting the grid phase voltage and phase current, output DC voltage and DC current, battery terminal voltage and charging current, followed by the SOC parameter. Since the phase voltage and phase current are in phase with each other, we can evaluate that the effective working of the PFC module followed by the maximum allowed current passing through the charger is limited to 100 A, which indicates the current control implementation is also working as expected.

10.6 CONCLUSION

This chapter introduced the modeling of DC-DC off-board fast charger technology, by comparing the existing charging rates and battery technologies. Even though the term DC-DC charger is coined, this chapter makes sure the internal components are introduced followed by effective modeling and working of each module is represented through figures and equations. On the other hand, since this charger contains more power electronics components, it explains the role of having voltage and current in phase with each other with the help of the PFC module. The control methodologies used in FEC and the DC-DC converter have been chosen to be simple PI control algorithms to reduce the overall complexity of the simulation. The conversion of the three-phase reference to a two-phase rotating phase and the method of recovering three phases from two have been explained through the inverse park transformation technique. Future simulation work could include a comparative study of multiple control methodologies, such as P, PID, PD, and so on, which can be tested after fault conditions in a DC fast charger module.

REFERENCES

[1] Umar, Muhammad, Xiangfeng Ji, Dervis Kirikkaleli, and Andrew Adewale Alola. "The imperativeness of environmental quality in the United States transportation sector amidst biomass-fossil energy consumption and growth." *Journal of Cleaner Production* 285 (2021): 124863. doi: 10.1016/j.jclepro.2020.124863

[2] LaMonaca, Sarah, and Lisa Ryan. "The state of play in electric vehicle charging services: A review of infrastructure provision, players, and policies." *Renewable and Sustainable Energy Reviews* 154 (2022): 111733. doi: 10.1016/j.rser.2021.111733

[3] Rajendran, Gowthamraj, Chockalingam Aravind Vaithilingam, Norhisam Misron, Kanendra Naidu, and Md Rishad Ahmed. "A comprehensive review on system architecture and international standards for electric vehicle charging stations." *Journal of Energy Storage* 42 (2021): 103099. doi: 10.1016/j.est.2021.103099

[4] Mahfouz, Mostafa, and Reza Iravani. "Autonomous operation of the DC fast-charging station." *IEEE Transactions on Industrial Electronics* (2021). doi: 10.1109/TIE.2021.3076722

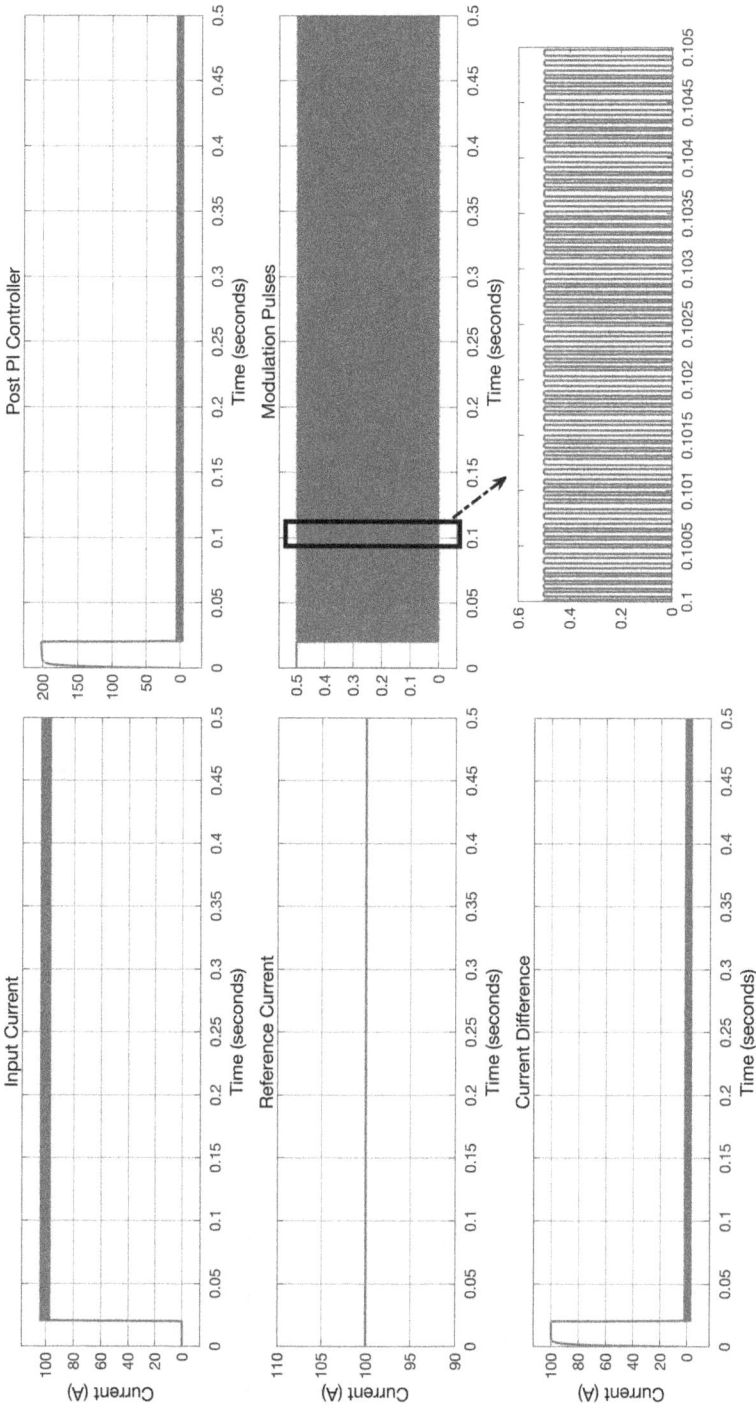

FIGURE 10.17 Results of DC-DC Controller Circuit.

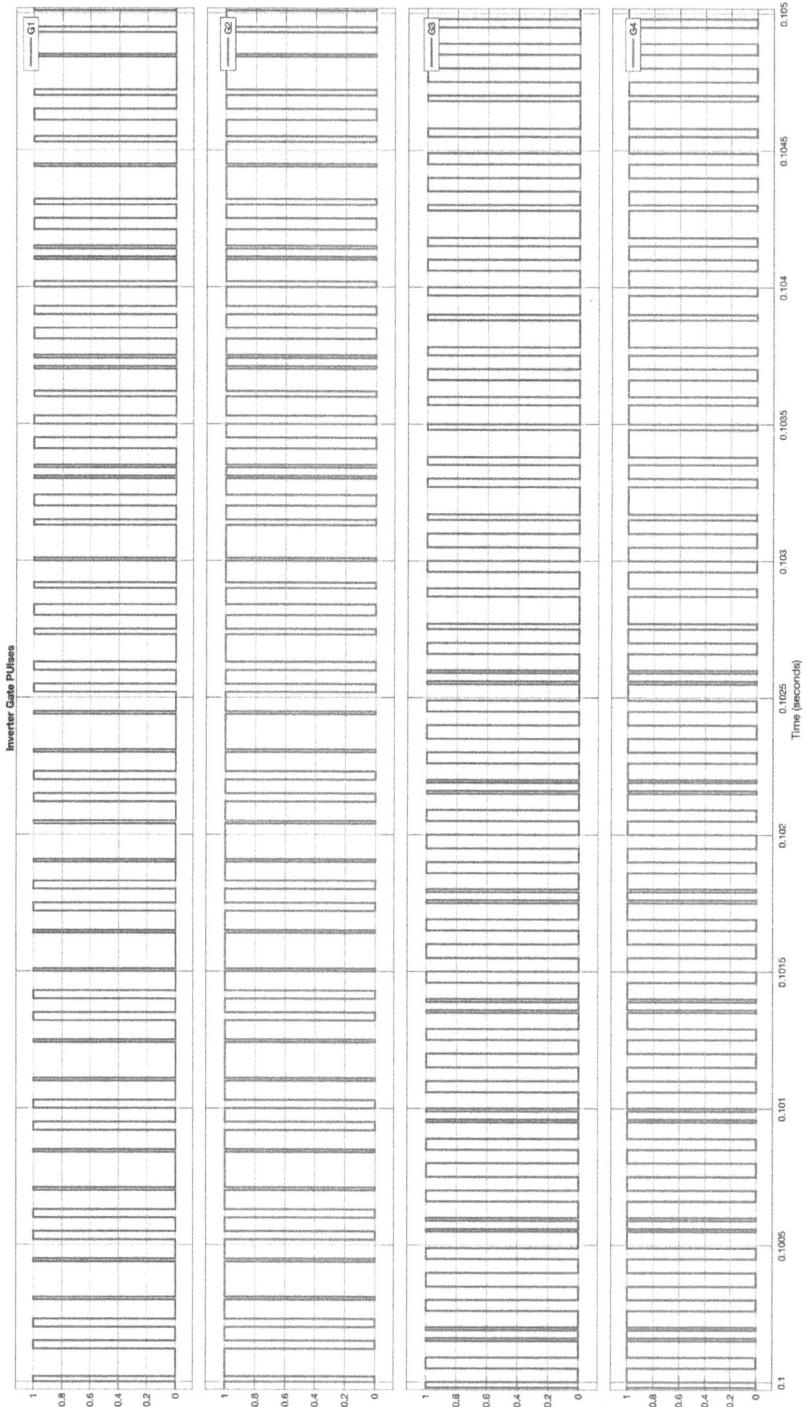

FIGURE 10.18 Inverter Gate Pulses.

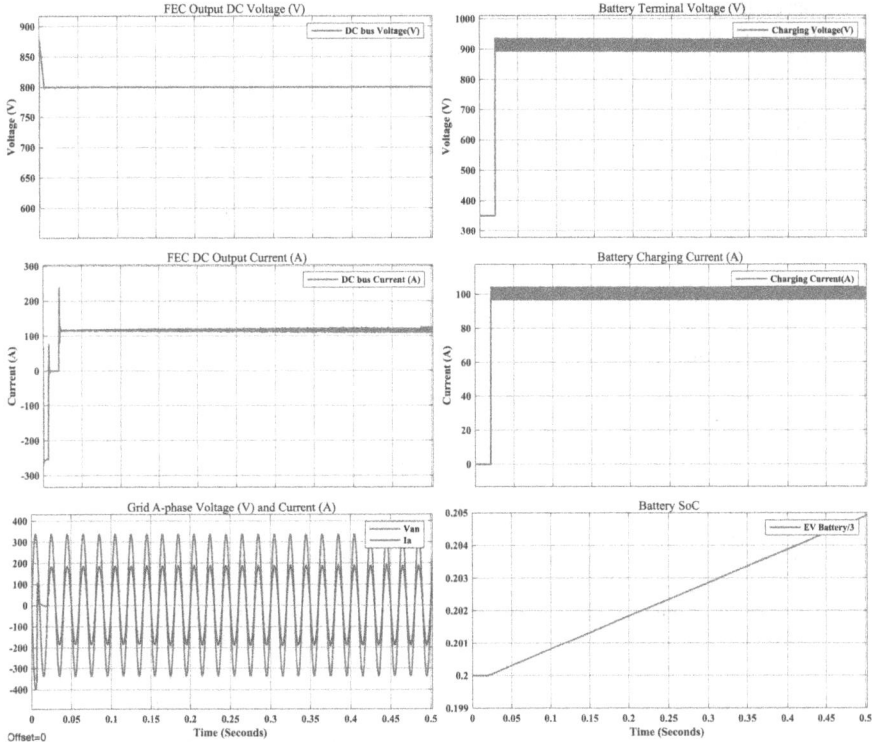

FIGURE 10.19 Results of DC Fast Charger.

[5] Lutsey, Nic, Hongyang Cui, and Rujie Yu. "Evaluating electric vehicle costs and benefits in China in the 2020–2035 time frame." (2021).

[6] Balasundar, C., C.K. Sundarabalan, Jayant Sharma, N.S. Srinath, and Josep M. Guerrero. "Design of power quality enhanced sustainable bidirectional electric vehicle charging station in distribution grid." *Sustainable Cities and Society* 74 (2021): 103242. doi: 10.1016/j.scs.2021.103242

[7] Ma, Tai-Yu, and Simin Xie. "Optimal fast charging station locations for electric ride-sharing with vehicle-charging station assignment." *Transportation Research Part D: Transport and Environment* 90 (2021): 102682. doi: 10.1016/j.trd.2020.102682

[8] Rehman, Waqas Ur, Amirhossein Moeini, Oroghene Oboreh-Snapps, Rui Bo, and Jonathan Kimball. "Deadband voltage control and power buffering for extreme fast charging station." *2021 IEEE Madrid PowerTech*, pp. 1–6. IEEE (2021). doi: 10.1109/PowerTech46648.2021.9494994

[9] Elibol, Burak, Gokturk Poyrazoglu, Bahadır Can Çalışkan, Hatice Kaya, Çiğdem Armağan, Hülya Erdener Akınç, and Alp Kaymaz. "Battery integrated off-grid DC fast charging: Optimised system design case for California." *2021 10th International Conference on Renewable Energy Research and Application (ICRERA)*, pp. 327–332. IEEE (2021). doi: 10.1109/ICRERA52334.2021.9598644

[10] Sawan, Wael, Nirav Karelia, Siddharth Joshi, and Bhinal Mehta. "Power quality improvement in EV charging station based on three-leg VSC D-STATCOM." In *Advances in Electrical and Computer Technologies: Select Proceedings of ICAECT 2020*, pp. 915–924. Springer, Singapore, 2021. doi: 10.1007/978-981-15-9019-1_77

[11] Basta, B., and W.G. Morsi. "Low and high order harmonic distortion in the presence of fast charging stations." *International Journal of Electrical Power & Energy Systems* 126 (2021): 106557. doi: 10.1016/j.ijepes.2020.106557

[12] Pal, Arnab, Aniruddha Bhattacharya, and Ajoy Kumar Chakraborty. "Placement of public fast-charging station and solar distributed generation with battery energy storage in distribution network considering uncertainties and traffic congestion." *Journal of Energy Storage* 41 (2021): 102939. doi: 10.1016/j.est.2021.102939

[13] Leone, Carola, and Michela Longo. "Modular approach to ultra-fast charging stations." *Journal of Electrical Engineering & Technology* (2021): 1–14. doi: 10.1007/s42835-021-00757-x

[14] Deb, Naireeta, Rajendra Singh, Richard R. Brooks, and Kevin Bai. "A review of extremely fast charging stations for electric vehicles." *Energies* 14(22) (2021): 7566. doi: 10.3390/en14227566

[15] Ganesh, Viswanathan, V.M. Ajay Krishna, and R.R. Ajit Ram. "Safety feature in electric vehicle at public charging station." *2021 7th International Conference on Electrical Energy Systems (ICEES)*, pp. 156–161. IEEE (2021). doi: 10.1109/ICEES51510.2021.9383722

[16] Doss, Arun Noyal, M., R. Brindha, A. Ananthi Christy, and Viswanathan Ganesh. "IOT-monitored EV charging stations using DC-DC converter with integrated split battery energy system." In *Proceedings of International Conference on Power Electronics and Renewable Energy Systems*, pp. 659–670. Springer, Singapore, 2022. doi: 10.1007/978-981-16-4943-1_62

[17] Ganesh, Viswanathan, and S. Senthilmurugan. "Role and improvements of battery energy storage systems in energy sector." *SPAST Abstracts* 1(01) (2021).

[18] Perumal, Shyam S.G., Richard M. Lusby, and Jesper Larsen. "Electric bus planning & scheduling: A review of related problems and methodologies." *European Journal of Operational Research* (2021). doi: 10.1016/j.ejor.2021.10.058

[19] Shariff, Samir M., Essam A. Al-Ammar, Ibrahim Al Saidan, Hasan Al Rajhi, Mohammad Saad Alam, Mohd Rizwan Khalid, and Aqueel Ahmad. "A state-of-the-art review on the impact of fast EV charging on the utility sector." *Energy Storage*: e300. doi: 10.1002/est2.300

[20] Hemavathi, S. "Overview of cell balancing methods for Li-ion battery technology." *Energy storage* 3(2) (2021): e203. doi: 10.1002/est2.203

[21] Hemavathi, S. "Modeling of analog battery management system for single cell lithium ion battery." *Energy Storage* 3(4) (2021): e208. doi: 10.1002/est2.208

[22] Srinivasan, Shiva Srenivasan, N.R.S. Lakshanasri, Sahana M. Setty, Piyush Dubey, Anant Singh, and S. Senthilmurugan. "BLDC drive for EV application." *Journal of Physics: Conference Series* 2007(1) (2021): 012059. IOP Publishing. doi: 10.1088/1742–6596/2007/1/012059

[23] Aswin, A., and S. Senthilmurugan. "A survey on power levels of battery charging and infrastructure for plug-in electric and hybrid vehicles." *IOP Conference Series: Materials Science and Engineering* 402(1) (2018): 012154. IOP Publishing. doi: 10.1088/1757–899X/402/1/012154

[24] Nayak, Panugothu, Srinivasa Rao, and Peddanna Gundugallu. "Investigation of MI and performance analysis of SS resonant IPT system for EV battery charging application." *Australian Journal of Electrical and Electronics Engineering* (2021): 1–1 doi: doi.org/10.1080/1448837X.2021.1981588

[25] Ganesh, V., S. Senthilmurugan, R. Ananthanarayanan, S.S. Srinivasan, and N.R.S. Lakshanasri. "Integration strategies of renewable energy sources in a conventional community. In: Fathi M., Zio E., Pardalos P.M. (eds) *Handbook of Smart Energy Systems*. Springer, Cham, 2021. https://doi.org/10.1007/978-3-030-72322-4_120-1

[26] Arunmozhi, M., S. Senthilmurugan, and V. Ganesh. "Design and operational strategies for grid-connected smart home." In: Fathi M., Zio E., Pardalos P.M. (eds) *Handbook of Smart Energy Systems*. Springer, Cham, 2021. https://doi.org/10.1007/978-3-030-72322-4_78-1

11 Advanced Converter Topologies for EV Fast Charging

S. Hemavathi, R. Uthirasamy,
C. Sharmeela, and S. Senthilmurugan

CONTENTS

11.1 INTRODUCTION TO ELECTRIC VEHICLE CHARGING

Fast charging or rapid charging methods are used to recharge electric vehicle (EV) batteries within the minimum time and effort period based on the type of charging methods adopted [1–4]. At least 80% of battery packs have tentatively charged within 20 minutes when fast charging is employed. The rapid development of the EV market has led to the enhancement of new infrastructure for charging stations and charging networks [5–7]. Chargers that rated slow naturally at 3.2 kW are used in domestic applications. This type of charger is not suitable for other applications, such as EVs, testing, etc., because the charges are too slow. Fast chargers are used in car parks and shopping centers and may charge at rates of 7 kW or 22 kW for EV applications. A 22 kW charger might charge 62 kWh EV from zero to a full charge in approximately two and a half to three hours [8–10]. In general, quick charging occurs frequently using a DC charging method and inbuilt rectifiers are in EVs for AC-DC conversion within the charger rather than relying on the EV's onboard rectifier [11–14].

Typically, quick DC chargers are used for charging up to 52 kW. Figure 11.1 shows the general block diagram of the charging infrastructure.

EV charging is classified into two categories: AC bus charging and DC bus charging. The AC household charging system uses a wall box to charge 3–4 times faster, and faster charging for EVs can be accomplished at public DC fast-charging stations that are 50 kW or higher. Using this method the battery is charged in approximately 20 minutes from 20% to 80%. The DC charging infrastructure is classified into unipolar DC bus charging and bipolar DC bus charging. The unipolar DC bus uses two-level voltage source converters, and the bipolar DC bus uses three-level neutral-point clamped converters. Compared with the unipolar architecture, the bipolar architecture provides increased power capacity, the ability to withstand higher voltages, and provides better power quality and flexible means of connecting loads.

Consider our common EV battery recharged in one to three hours. EV owners charge their cars at home to save time and money. Even though more charging facilities are

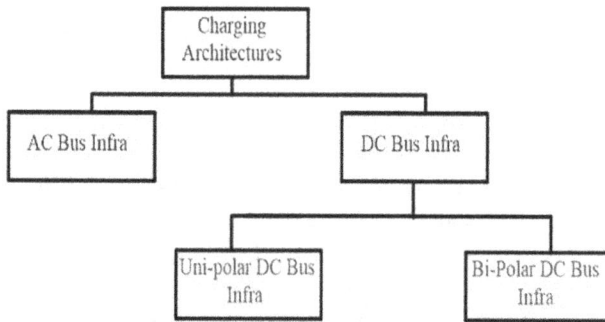

FIGURE 11.1   Block Diagram of Charging Infrastructure.

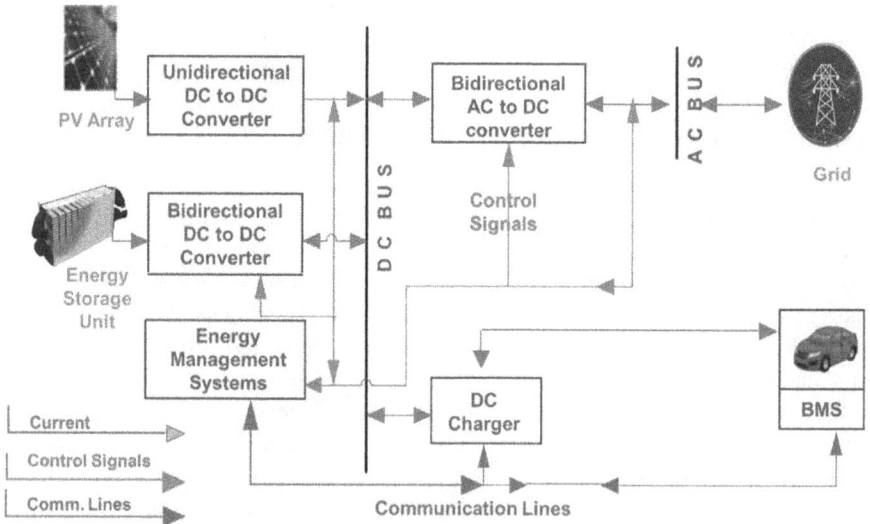

FIGURE 11.2   Triple-Input Single-Output System for EV Application.

available in the residential system, consumers are not satisfied with the provided system due to instability in the grid system. At a 52 kW charging rate, EV cars plan to charge at public charging stations. Generally, based on the power rating, the charging system follows Level 1, Level 2, and Level 3 classified. A block diagram representation of a triple-input single-output system for EV charging is shown in Figure 11.2.

In this, control signal lines indicate power control lines, which is performed in the PV array/energy storage unit/AC grid hybrid power source by the energy management system as follows:

Case 1: If ($P_{pv} \geq P_{EV}$), then the energy management system sets charging of EV by PV and remaining PV power stored in energy storage unit.

Case 2: If ($P_{pv} < P_{EV}$), then the energy management system sets charging of EV by PV and energy storage unit.

Case 3: If ($P_{pv} < P_{EV}$ and $SOC_{ESU} = 20\%$), then the energy management system sets charging of EV by PV and AC grid. The equivalent circuit representation of the dual-output system is shown in Figure 11.3 [15–17].

11.2 DC-DC CONVERTERS

Widely, DC-DC converters are used in industrial, domestic, and EV applications. In many realistic applications, it is essential to convert a fixed DC source input voltage into a variable DC output voltage. DC-DC converters offer soft acceleration control, high efficiency, and a quick response. DC-DC converters are used for UPS, footing motors in aquatic/marine pumps, elevators, EVs, electric chargers, and others. A large body of literature points out the remarkable topologies of DC-DC converters. Fundamentally, DC-DC converters classify into step-up (boost) converters, step-up and step-down (buck-boost) converters, step-down (buck) converters, and isolated converters.

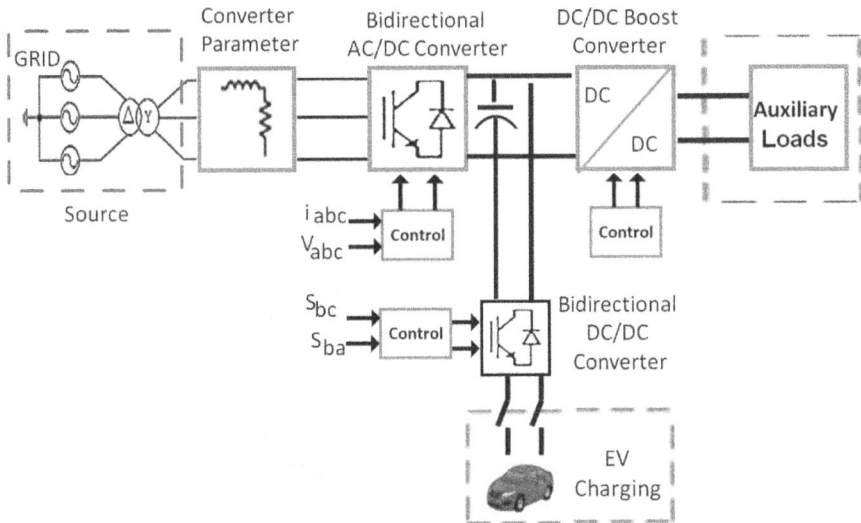

FIGURE 11.3   Dual-Output Voltage for EV System.

11.2.1 BOOST CONVERTER

The step-up converter involves boosting a level of input voltage in accordance with the desired voltage level. Whenever a step-up converter Vo is used, the output voltage is higher than the input voltage [18, 19]. A circuit equivalent to the boost converter is illustrated in Figure 11.4 (A), output waveform of boost converters in Figure 11.4 (B), and control parameters of boost converters in Table 11.1.

11.2.2 BUCK-BOOST CONVERTER

It is possible for buck-boost converters to produce DC output voltages that are either higher or lower than the DC input voltages. This circuit represents the equivalent of a buck converter in Figure 11.4 (C) [20–21] output waveform of buck-boost converter in Figure 11.4 (D), and control parameters of buck-boost converters in Table 11.1.

11.3 DC-DC CONVERTER FOR DC FAST CHARGING

In partial-power converters, the difference between the bus and ΔV voltages is processed, instead of the full voltage. The converter will process a small amount of power and minimize the losses if the voltage buses are small (ΔV small). As opposed to full-power processing converters, Ploss is determined by the voltage of the entire battery bus. With sufficiently low ΔV, Ploss decreases, and the efficiency increases as a result. If the ΔV max is sufficiently small, the converter power rating also becomes small [22–25]. However, DC fast charging systems for EVs do not suffer from a deficit in power transfer from input to output, as shown in Figure 11.5.

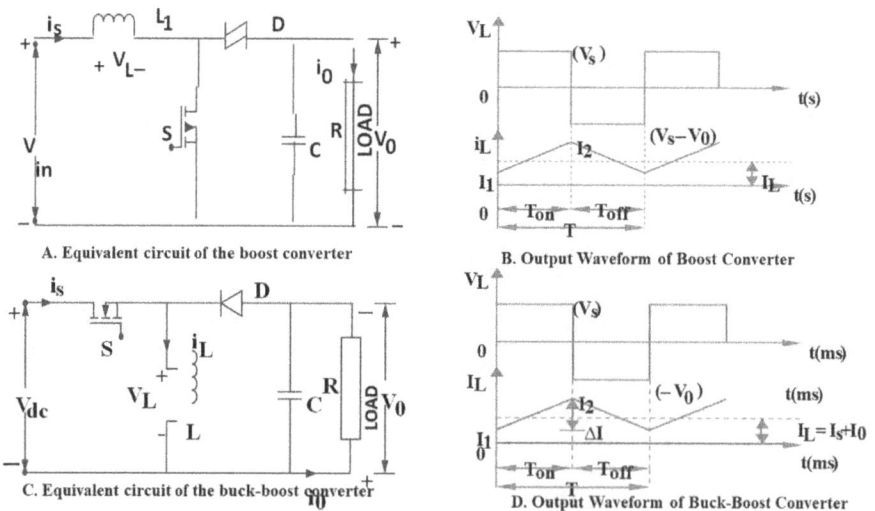

A. Equivalent circuit of the boost converter

B. Output Waveform of Boost Converter

C. Equivalent circuit of the buck-boost converter

D. Output Waveform of Buck-Boost Converter

FIGURE 11.4   Equivalent Circuit and Output of Boost Converter and Buck-Boost Converter.

TABLE 11.1
Modes of Operation of Boost and Buck-Boost Converter

Time period	Control parameters	Boost converter	Buck-boost converter
At t=T_1	Voltage across the inductor	$V_L = L_1 \dfrac{I_b - I_a}{T_1}$	$V_L = L \dfrac{I_b - I_a}{T_i}$
At t=T_1	Input energy to the inductor	$E_i = V_{in} \cdot I_{s1} \cdot T_1$	$E_i = V_{in} \cdot I_{s1} \cdot T_1$
At t=T_2	Output voltage	$V_o = V_{in} + L \dfrac{dI_{s1}}{T_2}$	$V_o = L \dfrac{I_b - I_a}{T_2}$
At t=T_2	Output voltage	$V_o = \dfrac{V_{in}}{1-D}$	$V_o = \left(\dfrac{D * V_{dc}}{1-D} \right)$

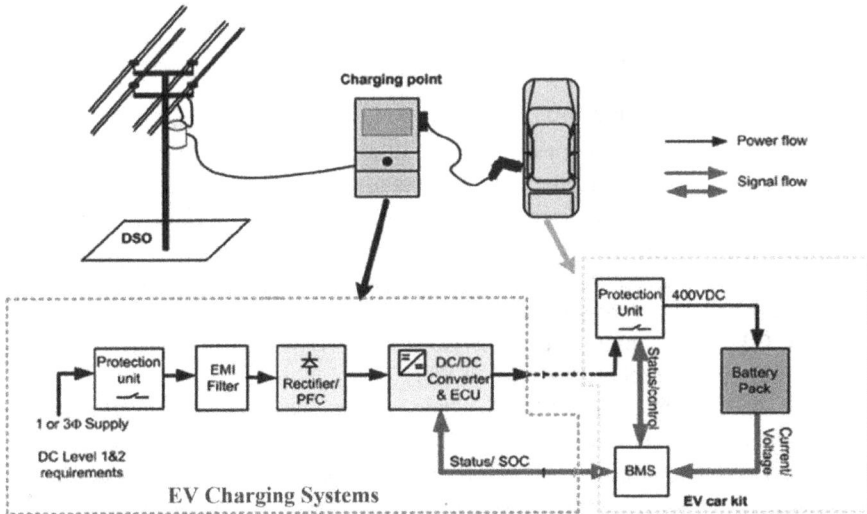

FIGURE 11.5 DC Fast Charging System for EVs.

11.3.1 SNUBBER ASSIST ZERO-VOLTAGE ZERO-CURRENT (SAZZ) CONVERTER FOR EV APPLICATIONS

Normally, SAZZ (snubber assist zero-voltage zero-current) converters are used for fuel cell interfaced electric vehicle (FCEV) or hybrid vehicle applications. The SAZZ pattern operated like a resonant converter and additionally, the designed system was such a way to get zero voltage spikes during the diode operation. The equivalent circuit of the SAZZ Converter is in Figure 11.6.

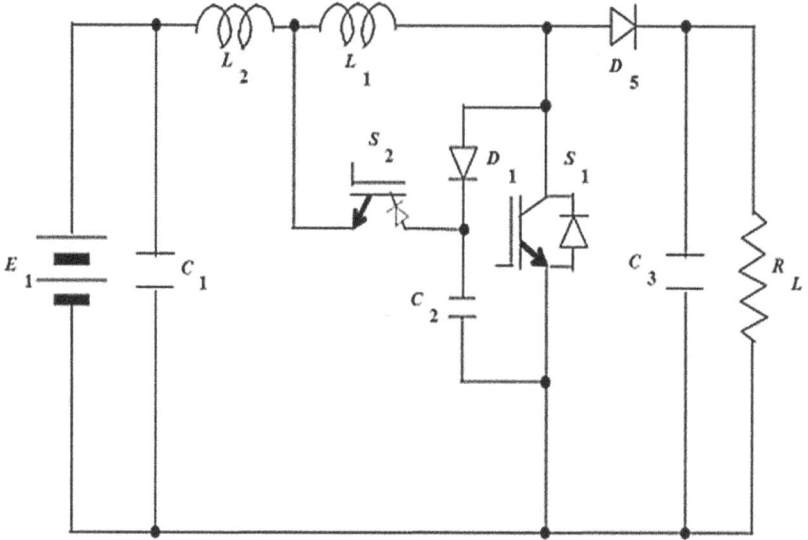

FIGURE 11.6   Equivalent Circuit of SAZZ Converter.

11.3.1.1 Modes of Operation

SAZZ configuration has the operating structure of ZVT and ZCT during on state and ZVS during off state. In the general voltage and current commutation process, capacitor C2 is designed to discharge through auxiliary switch S2 to trigger the on state—main switch S1 reduces the voltage and current stress across the main S1 through the supplementary control S2. The main switch S1 is triggered to the off state, and the voltage across switch S1 begins to ascend as of "0" potential by the capacitor C2. Consequently, it can operate to achieve the quasi-resonant operation, which ensues in the exclusion of reactor SL1.

When switch S anticipates in the designed manner, the proposed converter has the improvements from the quasi-resonant operation scheme. The modes of operation of the proposed converter explain as follows modes of operation 1–6 in Figure 11.7, and Table 11.2 represents the modes of operation of the SAZZ converter.

11.3.2 RE-LIFT CONVERTER

A re-lift converter operates based on the voltage lifting technique. The O/P voltage tends to equate with two times the input voltage. A re-lift converter consists of two power semiconductor switches, S_1 and S_2; three energies transferring inductors concentration, L_{1-3}; four charge transferring capacitors, C_{1-4}; and three unidirectional devices called diodes, D_{1-3}. The proposed re-lift converter is represented in Figure 11.8.

11.3.3 BIDIRECTIONAL DC-DC CONVERTER

The bidirectional converter system attains both the buck and boost operations for EVs with a built regenerative braking system. Whenever the battery is charging,

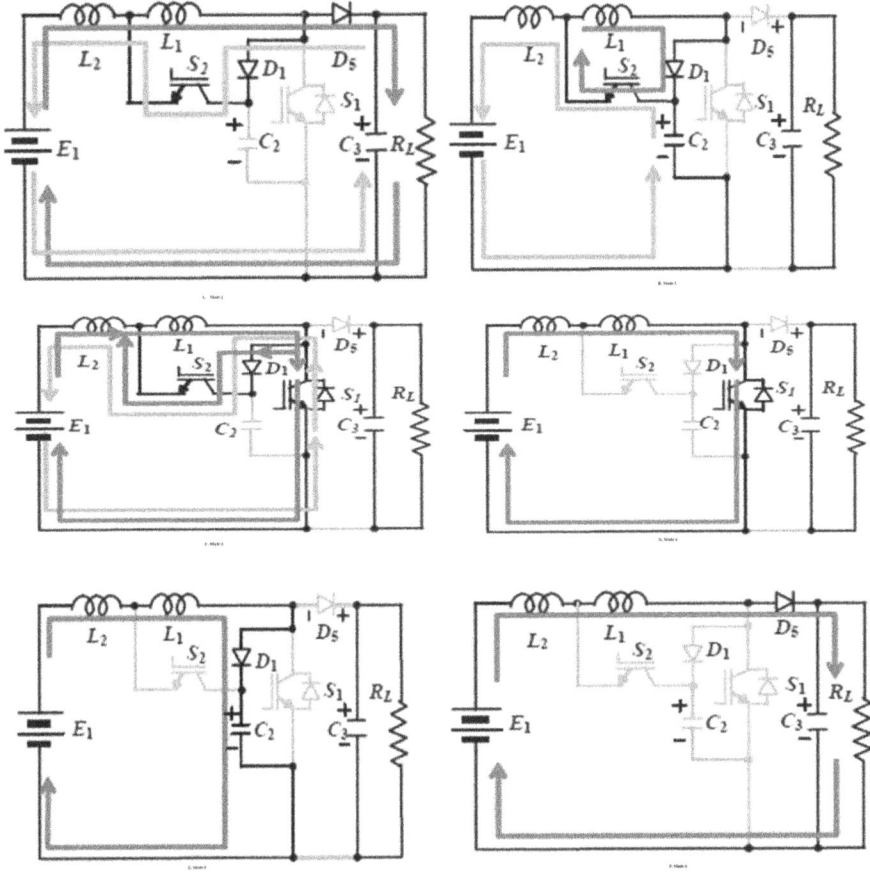

FIGURE 11.7 Modes of Operation of the SAZZ Converter.

TABLE 11.2
Modes of Operation of the SAZZ Converter

Mode of operation	Components	Working
Mode I	S1	Off state
At t = -t_2	S2	On state
	D_1	Off state
	D_s	On state
	C_2	Discharging
	C_3	Charging
	L_1	Energizing
	L_2	Energizing
Mode II	S1	On state
At t = -t_1		
	S2	Off state

(Continued)

TABLE 11.2   *(Continued)*
Modes of Operation of the SAZZ Converter

Mode of operation	Components	Working
	D_1	On state
	D_5	Off state
	C_2	Charging
	C_3	Charging
	L_1	De-energizing
	L_2	De-energizing
Mode III	S1	On state
At t = t_0	S2	On state
	D_1	On state
	D_5	Off state
	C_2	Discharging
	C_3	Charging
	L_1	Energizing
	L_2	Energizing
Mode IV	S1	On state
	S2	Off state
	D_1	Off state
	D_5	Off state
At t = t_1	C_2	Charging
	C_3	Charging
	L_1	Energizing
	L_2	Energizing
Mode V	S1	Off state
At t = t_2	S2	Off state
	D_1	On state
	D_5	Off state
	C_2	Discharging
	C_3	Charging
	L_1	De-energizing
	L_2	De-energizing
Mode VI	S1	Off state
	S2	Off state
	D_1	Off state
	D_5	Off state
	C_2	Charging
	C_3	Charging
	L_1	De-energizing
	L_2	De-energizing

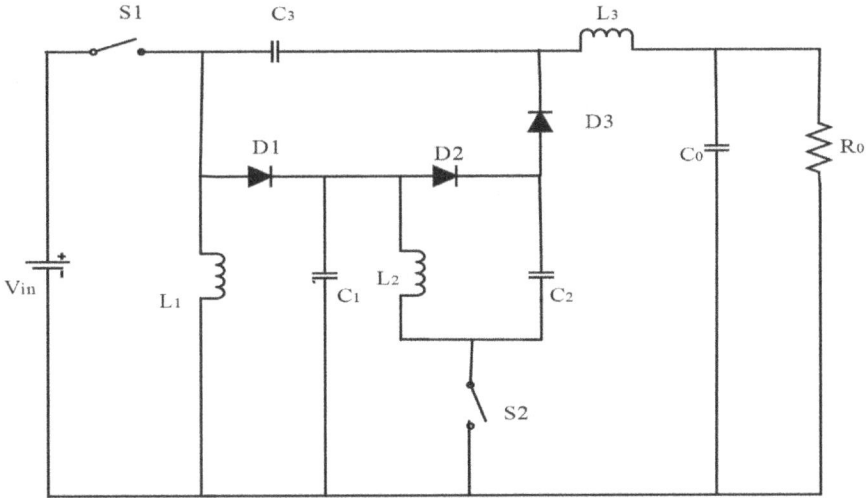

FIGURE 11.8   Circuit Diagram of Re-Lift Converter.

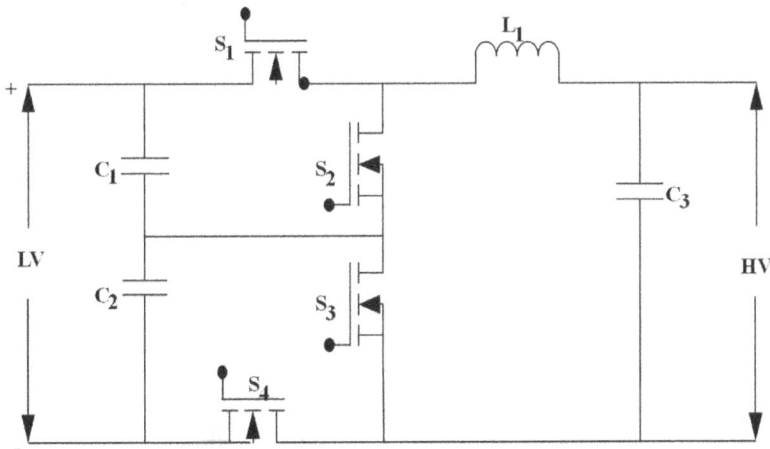

FIGURE 11.9   Circuit Diagram of Bidirectional DC-DC Converter.

the proposed converter should be managed in buck mode, and during discharging it should operate in the boost mode. When the converter consists of both low-voltage V_L and high-voltage V_H sides, the DC bus is connected to the V_H region, and the battery source is connected to the V_L side of the network. The proposed circuit comprises two switches named S_1 and S_2. Though this is simple in structure, it is not suitable for a wide range of applications because the step-up voltage gain is limited because of the effect of the switches and the Equivalent Series Resistance (ESR) of the inductors and capacitors. Thus, the conventional converter is not suitable for a wide voltage-conversion application. The proposed converter has the intrinsic worth of easy configuration and controllability [26–31]. Figure 11.9 illustrates the operation of the circuit diagram of the bidirectional DC-DC converter.

The operation of the bidirectional converter consists of four semiconductor switches named S_1, S_2, S_3, and S_4. It can be operated in two operating modes, like buck mode and boost mode. Each mode has three stages of operations. During the buck mode of operation, the HV side DC bus or the solar PV acts as a source and the battery acts as the load. Battery life is affected by more factors, one of which is battery overcharging, and the battery should be charged at a low voltage [52–54]. So, the bidirectional converter acts as a buck converter during charging. The modes of operation are as follows in stage 1 CCM operation can be obtained within the time and period of (t_0-t_1). In the period in-between, S_3 designs were handled in such a way as to operate as a synchronous AC-DC converter [46–51]. According to the equation, a current that flows through the inductor L_1 is,

$$I_{L1/STEP\ DOWN} = I_{L1}(t_0) + (1/L_1) * ((V_H/2) - V_L) * (t - t_0) \tag{11.1}$$

In stage 2 operations, it can be operated in the time span of $[t_1-t_2]$. Obtaining the current through inductor L_1 is as follows:

$$I_{L1(STEP\ DOWN)} = I_{L1}(t_1) * (V_L/L_1) * (t_1 - t_0) \tag{11.2}$$

In stage 3 converters operated in buck, a mode can be derived at the time interval of $[t_2-t_3]$. Further, the current through the inductor L_1 is obtained as

$$I_{L1(STEPDOWN)} = I_{L1}(t_0) + (1/L_1) * ((V_H/2) - V_L) * (t - t_2) \tag{11.3}$$

Represent the bidirectional converter with switch stats, as shown in Table 11.3, switch on-off state.

11.3.3.1 Boost Mode or Step-Up Mode of Bidirectional DC-DC Converter

Whenever the power from the solar PV is not sufficient for the load, the discharging process of the battery would take place. Normally, the output voltage of the battery is exceptionally low, about 12 V. So, this needs to boost the output voltage from the battery for satisfying the maximum load. During this condition, the bidirectional converter is intending to function in step-up mode or boost mode [40–45]. The operation represented in Figure 11.10 is the boost mode operation in a bidirectional converter.

TABLE 11.3
Switch On-Off State Circuit Diagram of Bidirectional DC-DC Converter

Stage	on-state switches	Off-state switches	Voltage across the inductor
1	S_1 and S_3	S_2 and S_4	$V_{L1(STEP\ DOWN)} = (V_H/2) - V_L$
2	S_2 and S_3	S_1 and S_4	$V_{L1\ (STEPDOWN)} = -V_L$
3	S_2 and S_4	S_1 and S_3	$V_{L1\ (STEPDOWN)} = (V_H/2) - V_L$

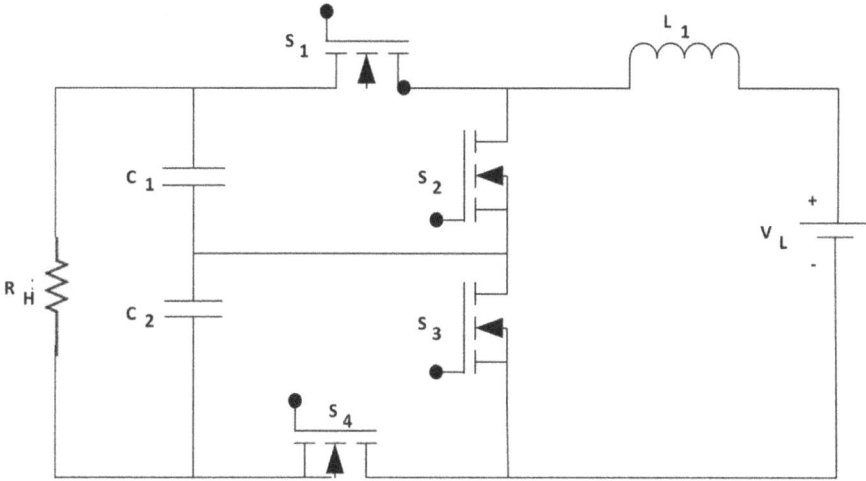

FIGURE 11.10   Boost Mode Operation in a Bidirectional Converter.

In the course of operation for boost mode, the battery resource operated as a supply for a high-voltage DC bus or load. Various operational stages of boost mode are as follows, and the stage 1 operation can take place in the time interval of $(t_0 - t_1)$. The energy of the low-voltage side V_L has transferred to the inductor L_1. The capacitors S_1 and S_2 loaded up to discharge for the load RH [55]. The current through the inductor L_1 follows

$$I_{L1(STEPUP)} = I_{L1}(t_0) + (V_L/L_1) * (t - t_0) \tag{11.4}$$

In stage 2, operation of the step-up mode can take place in the time interval of $(t_1 - t_2)$. The energies of the low-voltage side VL and inductor L_1 are in series to release their energy to capacitor C_1. Capacitors C_1 and C_2 are stacked to discharge to the load RH. The current through the inductor L_1 is given as

$$I_{L1(STEPUP)} = I_{L1}(t_1) + (V_L/L_1) * (t_1) \tag{11.5}$$

The stage 3, operation of step-up mode can take place between the time intervals of $(t_2 - t_3)$. The operation principle is the same as the stage 1. The current through the inductor L_1 follows

$$I_{L1(STEPUP)} = I_{L1}(t_2) + (V_L/L_1) * (t_2) \tag{11.6}$$

The representation of a converter operation with a switch in Table 11.4.

The comparative analysis of different DC-DC converters in Table 11.5; from the analysis, it is inferred that each and every converter has its own merits and demerits.

Voltage stress is zero in the SAZZ converter, but on the other side, the bidirectional property has been restricted. Likewise, re-lift converters are perfectly designed for

TABLE 11.4
Switch On- and Off-State Boost Mode Operation in a Bidirectional Converter

Stage	On-state switches	Off-state switches	Voltage across the inductor
1	S_2 and S_3	S_1 and S_4	$V_{L1(STEPUP)} = V_L$
2	S_1 and S_3	S_2 and S_4	$V_{L1\ (STEPUP)} = V_1 / (V_H/2)$
3	S_1 and S_2	S_3 and S_3	$V_{L1\ (STEPUP)} = V_L$

TABLE 11.5
Comparative Analysis of DC-DC Converters

Parameter	Conventional converters	SAZZ	RelLift	Bidirectional
Device stress	High	Very low	Low	High
Operation	Unidirectional	Unidirectional	Unidirectional	Bidirectional
Conduction loss	High	Low	Low	Medium
Turn on switching loss	Low	Very low	Low	Low
Turn off switching loss	High	Very low	Low	High
Switching frequency	Medium	High	High	High
Efficiency	Medium	High	High	Medium
Boost factor	Medium	High	High	High

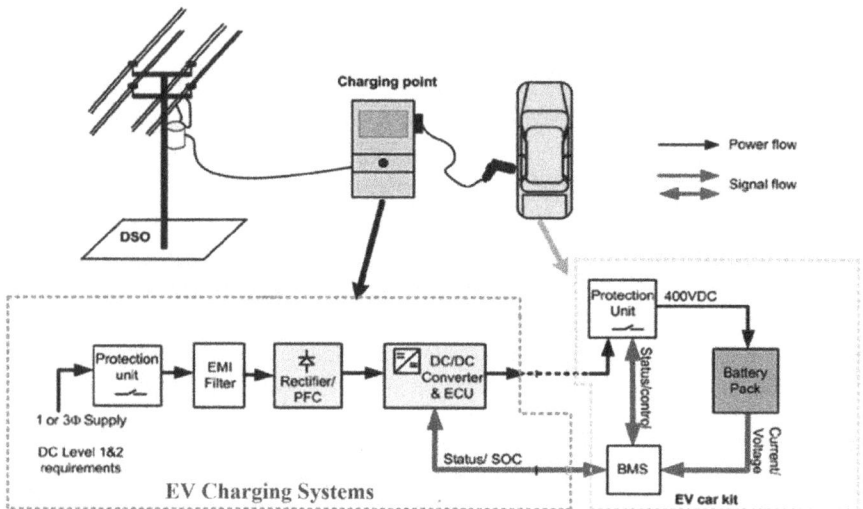

FIGURE 11.11   Bidirectional Converter–Based EV Charging.

EV fast charging, and the bidirectional converter is designed and implemented for EV applications and is represented in Figure 11.11 [32–39].

11.4 WIRELESS CHARGERS WILL HAVE A GREATER POTENTIAL IN THE FUTURE COMPARED TO WIRED CHARGERS

An automatic parking system could be constructed so that the coils are perfectly aligned on the primary coil and the secondary coil at the same time while stationary wireless charging is performed. By dynamic charging, vehicle speed, an automatic adjustment of the vertical and horizontal alignment could be made, as well as the configuration of the charging infrastructure, and as the grid and vehicle control systems interact in real time, the communication system is instrumental in exchanging standardized commands. Parking stations and garages usually use this type of charging system. A transmitter coil is located under the parking pads, and a receiver coil is in the charging vehicle. To charge an EV wirelessly, you need high-frequency AC power, the size of the transmitter, receiver coils, and distance. Various studies investigate how the optimized circuits perform with respect to efficiency, output voltage performance, network size, and power-loss distribution [49–58]. Since onboard chargers require a cable and plug charger, galvanic isolation of the electronics on board, and the weight and size of the charger in adverse weather conditions, they are prone to failure. Wireless charging addresses these problems while offering a convenient alternative to conventional charging methods. To safeguard humans as safely as possible, the electromagnetic field in the active charging zone between primary and secondary needs to be shielded as thoroughly as possible. A wireless charging system can either be dynamic or static, depending on its application. With static wireless charging, a vehicle gets charged while it is stationary. Similarly, its wireless charging for moving electric vehicles can charge dynamically while the vehicle is in motion.

11.5 CONCLUSION

In this chapter, several conventional and advanced DC-DC converters have been reviewed and discussed, and further inferences have been provided for different converters in EV battery-charging operations. Advanced and promising bidirectional DC-DC converters, such as snubber assist zero-voltage zero-current (SAZZ) converters and re-lift converters, have been introduced, along with their topologies, and control strategies have been discussed for EV fast-charging operations, which flows power in a bidirectional way between vehicles and grids, such that it has reduced the impacts on power grids in terms of load leveling, peak load shaving, and improving voltage regulation and the stability of power grid stations.

REFERENCES

[1] Hussain, V.H. Bui, and H.M. Kim, "Optimal sizing of battery energy storage system in a fast EV charging station considering power outages," IEEE Trans. Transp. Electrif., vol. 6, no. 2, pp. 453–463, 2020.

[2] Sbordone, D., I. Bertini, B. Di Pietra, M.C. Falvo, A. Genovese, and L. Martirano, "EV fast charging stations and energy storage technologies: A real implementation in the smart micro grid paradigm," Electr. Power Syst. Res., vol. 120, pp. 96–108, 2015.

[3] Hemavathi, S., "Modeling and energy optimization of hybrid energy storage system." In: Sahoo, U. (ed) Hybrid Renewable Energy Systems. John Wiley & Sons, Ltd, pp. 97–114, 2021. New York: New York, United States. https://doi.org/10.1002/9781119555667.ch4

[4] Shukla, K. Verma, and R. Kumar, "Impact of EV fastcharging station on distribution system embedded with wind generation," J. Eng., vol. 2019, no. 18, pp. 4692–4697, 2019.

[5] Hemavathi, S., and M. Singh, "Microgrid short circuit studies," 2018 IEEE 8th Power India International Conference (PIICON), 2018, pp. 1–6. doi: 10.1109/POWERI.2018.8704389.

[6] Bugatha Ram Vara Prasad, T., Satyavathiv Deepthin, and Hemakumar Satishvarmar, "Solar charging station for electric vehicles," Int.J. Adv. Res. Sci. Commun. Technol., vol. 7, no. 2, pp. 316–325, 2021.

[7] Bugatha Ram Vara Prasad, C. Prasanthi, G. Jyothika Santhoshini, K.J.S.V. Kranti Kumar, and K. Yernaidu, "Smart electrical vehicle," i-manager's J. Digit. Signal Process., vol. 8, no. 1, p. 7, 2020.

[8] S, H., "Modeling of analog battery management system for single cell lithium ion battery," Energy Storage, vol. 3, p. e208, 2021. https://doi.org/10.1002/est2.208

[9] S, H., "Overview of cell balancing methods for Li-ion battery technology," Energy Storage, vol. 3, p. e203, 2021. https://doi.org/10.1002/est2.203

[10] Hemavathi, S., "Li-ion battery health estimation based on battery internal impedance measurement." In: Muthukumar, P., Sarkar, D.K., De, D., De, C.K. (eds) Innovations in Sustainable Energy and Technology: Advances in Sustainability Science and Technology. Springer, Singapore, 2021. https://doi.org/10.1007/978-981-16-1119-3_17.

[11] Bai, W. Gao, J. Li, and H. Liao, "Analyzing the impact ofelectric vehicles on distribution networks," 2012 IEEE PESInnov. Smart Grid Technol. ISGT 2012, 2012.

[12] Shareef, H., M.M. Islam, and A. Mohamed, "A review of thestage-of-the-art charging technologies, placementmethodologies, and impacts of electric vehicles," Renew. Sustain. Energy Rev., vol. 64, pp. 403–420, 2016.

[13] Zheng, Y., Z. Y. Dong, Y. Xu, K. Meng, J. H. Zhao, and J. Qiu, "Electric vehicle battery charging/swap stations in distributionsystems: Comparison study and optimal planning," IEEE Trans. Power Syst., vol. 29, no. 1, pp. 221–229, 2014.

[14] Hemavathi, S. "Modeling and estimation of lithium-ion battery state of charge using intelligent techniques." In: Singh, S., Pandey, R., Panigrahi, B., Kothari, D. (eds) Advances in Power and Control Engineering. Lecture Notes in Electrical Engineering, vol. 609. Springer, Singapore, 2020. https://doi.org/10.1007/978-981-15-0313-9_12

[15] Chen, G., Y.-S. Lee, S. Hui, D. Xu, and Y. Wang, "Actively clamped bidirectional flyback converter," IEEE Transactions on Industrial Electronics, vol. 47, no. 4, pp. 770–779, 2000.

[16] Chen, L., H. Wu, P. Xu, H. Hu, and C. Wan, "A high step-down non-isolated bus converter with partial power conversion based on synchronous LLC resonant converter," IEEE Applied Power Electronics Conference and Exposition (APEC), pp. 1950–1955, 2015.

[17] Duan, R.-Y., and J.-D. Lee, "High-efficiency bidirectional DC-DC converter with coupled inductor," IET Power Electronics, vol. 5, no. 1, pp. 115–123, 2012.

[18] Agamy, M.S., S. Chi, A. Elasser, M. Harfman-Todorovic, Y. Jiang, F. Mueller, and F. Tao, "A highpower-density DC-DC converter for distributed PV architectures," IEEE J. Photovolt., vol. 3, no. 2, pp. 791–798, 2012.

[19] Bellur, D.M., and M.K. Kazimierczuk, "DC-DC converters for electric vehicle applications," 2007 Electrical Insulation Conference and Electrical Manufacturing Expo, pp. 286–293, 2007.

[20] Dolly, R.A., and Bugatha Ram Vara Prasad, "Enhancement of PFC and torque ripple reduction using BL buck-boost converter fed HCC BLDC drive," Int. J. Res., vol. 02, no. 11, pp. 895–901, 2015.

[21] Bugatha Ram Vara Prasad, K.M. Babu, K. Sreekanth, K. Naveen, and C.V. Kumar, "Minimization of torque ripple of brushless DC motor using HCC with DC-DC converter," Int. J. Res. vol. 05, no. 12, pp. 110–117, 2018.

[22] Huang, Y., J.J. Ye, X. Du, and L.Y. Niu, "Simulation study ofsystem operating efficiency of EV charging stations withdifferent power supply topologies," Appl. Mech. Mater., vol. 494–495, pp. 1500–1508, 2014.

[23] Ghahderijani, M.M., M. Castilla, A. Momeneh, J.T. Miret, and L.G. de Vicuna, "Frequencymodulation control of a DC/DC current-source parallel-resonant converter," IEEE Transactions on Industrial Electronics, vol. 64, no. 7, pp. 5392–5402, 2017.

[24] Kimball, J.W., and P.T. Krein, "Singular perturbation theory for DC-DC converters and application to PFC converters," IEEE Transactions on Power Electronics, vol. 23, no. 6, pp. 2970–2981, 2008.

[25] Morrison, A., J.W. Zapata, S. Kouro, M.A. Perez, T.A. Meynard, and H. Renaudineau, "Partial power DC-DC converter for photovoltaic two-stage string inverters," IEEE Energy Conversion Congress and Exposition (ECCE), pp. 1–6, 2016.

[26] Konjedic, T., L. Korošec, M. Truntic, C. Restrepo, M. Rodic, and M. Milanovic, "DCM-based zero-voltage switching control of a bidirectional DC-DC converter with variable switching frequency", IEEE Transactions on Power Electronics, vol. 31, no. 4, pp. 3273–3288, 2015.

[27] Kwon, M., S. Oh, and S. Choi, "High gain soft-switched bidirectional DCDC converter for eco-friendly vehicles," IEEE Trans. Power Electron., vol. 29, no. 4, p. 16591666, Apr. 2014.

[28] Hemavathi, S., "DC to DC energy conversion using novel loaded resonant converter," International Journal of Power Control and Computation (IJPCSC), vol. 7, no. 2, pp. 113–122, 2015.

[29] Aswini, Kotla, Jillidimudi Kamala, Lanka Sriram, Bhasuru Kowshik, Bugatha Ram Vara Prasad, and Damaraju Venkata Sai Bharani, "Design and analysis of bidirectional battery charger for electric vehicle," International Journal of Engineering Research & Technology, vol. 10, no. 07, pp. 410–415, July 2021.

[30] Yu, Du, Xiaohu Zhou, Sanzhong Bai, SrdjanLukic, and Alex Huang, "Review of non-isolated Bi-directional DC-DC converters for plug-in hybrid electric vehicle charge station application at municipal parking decks," Twenty-Fifth Annual IEEE Applied Power Electronics Conference and Exposition (APEC), pp. 1145–1151, 2010.

[31] Seth, Aakash Kumar, and Mukhtiar Singh, "Second-order ripple minimization in single-phase single-stage onboard PEV charger," IEEE Transactions on Transportation Electrification, vol. 7, no. 3, pp. 1186–1195, 2021.

[32] Yudi, Qin, Xuebing Han, Yifan Wei, Languang Lu, Jianqiu Li, Jiuyu Du, Minggao Ouyang, Yujie Sheng, "A vehicle-to-grid frequency regulation framework for fast charging, infrastructures considering power performances of lithium-ion batteries and chargers," In 2021 IEEE 4th International Electrical and Energy Conference (CIEEC), 2021, pp. 1–6. IEEE. doi: 10.1109/CIEEC50170.2021.9510481.

[33] Hemavathi, S., and A. Shinisha, "A study on trends and developments in electric vehicle charging technologies," J. Energy Storage, vol. 52, 105013, 2022. https://doi.org/10.1016/j.est.2022.105013.

[34] Alkasir, S., Ehsan Abdollahi, S. Reza Abdollahi, and P. Wheeler, "A primary side CCSMPC controller for constant current/voltage charging operation of series-series

compensated wireless power transfer systems," 2021 12th Power Electronics, Drive Systems, and Technologies Conference (PEDSTC), 2021, pp. 1–5.

[35] Navaneeth, M., J. Jithin, M. Jasim Ali, and P.V. Swathi, "Improved E-bike charger suitable for advanced charging algorithms," 2021 International Conference on Communication, Control and Information Sciences (ICCISc), 2021, pp. 1–5.

[36] Karthikeyan, S., H. Bragruthshibu, R. Logesh, K. Srinivasan, and S. Tarjanbabu, "Solar based fast tag charger for electrical vehicle," 2021 7th International Conference on Advanced Computing and Communication Systems (ICACCS), 2021, pp. 776–779.

[37] Metwly, M.Y., M.S. Abdel-Majeed, A. Hemeida, A.S. Abdel-Khalik, and S. Ahmed, "NinePhase-based fractional-slot winding layouts for integrated EV on-board battery chargers," 2021 22nd IEEE International Conference on Industrial Technology (ICIT), 2021, pp. 1356–1361.

[38] ElMenshawy, M., and A. Massoud, "Multi-module DC-DC converter-based fast chargers for neighborhood electric vehicles," 2021 IEEE 11th IEEE Symposium on Computer Applications &Industrial Electronics (ISCAIE), 2021, pp. 185–190.

[39] Mohammad, Mostak, Omer C. Onar, Gui-Jia Su, Jason Pries, Veda Prakash Galigekere, Saeed Anwar, Erdem Asa, Jonathan Wilkins, Randy Wiles, Cliff P. White, and Larry E. Seiber, "Bidirectional LCC-LCC-compensated 20-kW wireless power transfer system for medium-duty vehicle charging," IEEE Transactions on Transportation Electrification, vol. 7, no. 3, pp. 1205–1218, Sept. 2021.

[40] Yan, H. Yin, T. Li, and C. Ma, "A two-stage scheme for both power allocation and EV charging coordination in a grid-tied PV: Battery charging station," IEEE Transactions on Industrial Informatics, vol. 17, no. 10, pp. 6994–7004, Oct. 2021.

[41] Mahfouz, M., and R. Iravani, "Autonomous operation of the DC fast-charging station," IEEE Transactions on Industrial Electronics, 2021.

[42] Medina-Garcia, F.J. Romero, D.P. Morales, and N. Rodriguez, "Advanced control methods for asymmetrical half-bridge flyback," IEEE Transactions on Power Electronics, vol. 36, no. 11, pp. 13139–13148, Nov. 2021.

[43] Bi, X., A.J. Chipperfield, and W.K.S. Tang, "Coordinating electric vehicle flow distribution and charger allocation by joint optimization," IEEE Transactions on Industrial Informatics, vol. 17, no. 12, pp. 8112–8121, Dec. 2021.

[44] Dutta, S., S. Gangavarapu, V.R. Vakacharla, A.K. Rathore, V. Khadkikar, and H. Zeineldin, "Small signal analysis and control of single-phase bridgeless cuk-based PFC converter for on-board EV charger," 2021 IEEE Transportation Electrification Conference & Expo (ITEC), 2021, pp. 851–855.

[45] Z.-X. Yang, G. Yu, J. Zhao, P.K. Wong, and X.-B. Wang, "Online equivalent degradation indicator calculation for remaining charging-discharging cycle determination of lithiumIon batteries," IEEE Transactions on Vehicular Technology, vol. 70, no. 7, pp. 6613–6625, July 2021.

[46] Jean-Pierre, G., S. Beheshtaein, N. Altin, and A. Nasiri, "Control and loss analysis of a solid state transformer based DC extreme fast charger," 2021 IEEE Transportation Electrification Conference & Expo (ITEC), 2021, pp. 9–14.

[47] Jafari, H., T.O. Olowu, M. Mahmoudi, and A. Sarwat, "Optimal design of bipolar power pad for dynamic inductive EV charging system application," 2021 IEEE Transportation Electrification Conference & Expo (ITEC), 2021, pp. 333–338.

[48] Kim, J.-W., and P. Barbosa, "PWM-controlled series resonant converter for universal electric vehicle charger," IEEE Transactions on Power Electronics, vol. 36, no. 12, pp. 13578–13588, Dec. 2021.

[49] I. Cortes and W. -J. Kim, "Automated Alignment With Respect to a Moving Inductive Wireless Charger," in *IEEE Transactions on Transportation Electrification*, vol. 8, no. 1, pp. 605–614, March 2022, doi: 10.1109/TTE.2021.3064782.

[50] Luo, Z., S. Nie, M. Pathmanathan, W. Han, and P.W. Lehn, "3-D analytical model of bipolar coils with multiple finite magnetic shields for wireless electric vehicle charging systems," IEEE Transactions on Industrial Electronics, 2021.

[51] Tomar, P.S., M. Srivastava, and A.K. Verma, "Three-port pulse width modulated DC-DC converter for vehicular applications," 2021 International Conference on Sustainable Energy and Future Electric Transportation (SEFET), 2021, pp. 1–6.

[52] Teeneti, R., et al., "System-level approach to designing a smart wireless charging system for power wheelchairs," IEEE Transactions on Industry Applications, vol. 57, no. 5, pp. 5128–5144, Sept.–Oct. 2021.

[53] Kirubadevi, S., N. Hemalatha, E. Sankaran, T. Sathesh Kumar, G. Nageswara Rao, and S. Hemavathi, "Autonomous smart grid management and control by using IoT." Int J Syst Assur Eng Manag, 2022. https://doi.org/10.1007/s13198-021-01565-2

[54] Dilip Borole, Y., M. Kumar Singh, A. Sharma, H. Kumar Singh, A. Poonia, Amarjeet Poonia, and S. Hemavathi, "Artificial neural network-based advanced algorithm for new generation solar PV systems," J Nucl Ene Sci Power Generat Techno, vol. 10, p. 10, 2021

[55] Hemavathi, S., A. Nanoty, M.K. Singh, A. Katiyar, R.P. Manohar, et al., "Power quality issues in autonomous solar photovoltaic utility microgrid employing a STATCOM based intelligent controller," J Nucl Ene Sci Power Generat Techno, vol. 10, p. 10, 2021.

[56] RamaDevi, J., S. Pathur Nisha, S. Karunakaran, S. Hemavathi, S. Majji, and A. Shunmugam, "Machine learning techniques for the energy and performance improvement in Network-on-Chip (NoC)," 2021 4th International Conference on Computing and Communications Technologies (ICCCT), 2021, pp. 590–595. doi: 10.1109/ICCCT53315.2021.9711872

[57] Divya, S., Siva Ramkumar Mathiyalagan, J. Mohana, Vinjamuri S.N. CH Dattu, S. Hemavathi, L. Natrayan, Anand Chakaravarthi, V. Mohanavel, and Ravishankar Sathyamurthy, "Analysing analyzing the performance of combined solar photovoltaic power system with phase change material," Energy Rep., vol. 8, pp. 43–56, 2022. https://doi.org/10.1016/j.egyr.2022.06.109

[58] Bharathi, M.L., Vimal Bhatt, V.V. Ravi Kumar, Rajesh Jagdish Sharma, S. Hemavathi, Bhasker Pant, Rajendra Prasad Arani, T. Sathish, and V. Mohanavel, "Developing a dual axis photoelectric tracking module using a multi quadrant photoelectric device," Energy Rep., vol. 8, pp. 1426–1439, 2022. https://doi.org/10.1016/j.egyr.2022.07.095

12 Systems Optimization of Public Transportation Electrification Using Dynamic Wireless Charging Technology

Yaseen Alwesabi and Nabil Mohammed

CONTENTS

12.1 FUTURE OF PUBLIC TRANSPORTATION ELECTRIFICATION

Public transportation remains one of the main challenges for smart city planning worldwide. Electromobility in public bus systems, such as battery electric buses (EBs), has grown rapidly in the last decade. Transportation systems are currently experiencing a fundamental transformation in their structures and organization, including electrified bus lines, stationary and dynamic charging, as well as static and wireless charging. People will travel together if they are offered competitive quality services and meet their needs, such as travel time, comfortable riding, and reasonable ticket price. Public transportation electrification is one of the bright sides of future smart cities. The major transition to greener public transportation occurs in China,

DOI: 10.1201/9781003293989-12

where over 1,000 EBs have been in service in Tianjin, while Shenzhen has replaced all of its diesel city buses with EBs [1, 2]. Electrifying public buses can reduce deaths by more than 4,000 each year [3]. It is estimated to create 960,000 direct and indirect jobs [3]. Public transportation in the USA consists of 65,000 transit buses, 68,000 demand-response vehicles, and 470,000 school buses [3]. EBs are advantageous compared to fossil fuel–based buses in different aspects, such as smooth riding (low vibration), lower operations cost (maintenance and energy saving), zero-exhaust gas emissions, and contributing toward cleaner and healthier societies [4–8].

12.2 PUBLIC TRANSPORTATION PLANNING

Transit agencies usually start with initial plans to make bus network lines effectively reflect the actual demand. The first stage is bus line planning, which includes collecting data about the demand, capital cost, infrastructure cost (e.g., depot cost, base station cost, etc.), trip frequency, and timetable. Once the timetable is set up, transit agencies move on to stage 2, which involves bus fleet scheduling and crew scheduling (see Figure 12.1).

On the other hand, EBs need additional time in the terminals for charging. Thus, it is recommended to schedule EBs jointly with charging planning (see Figure 12.2). Since the first stage is already set up, the new challenge is to jointly plan the charging infrastructure with EB fleet scheduling to obtain the minimum cost of the system, which adheres to driving range limitation.

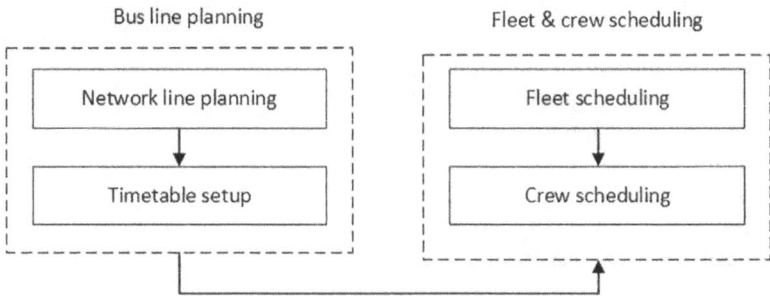

FIGURE 12.1   Public transportation planning stages [9].

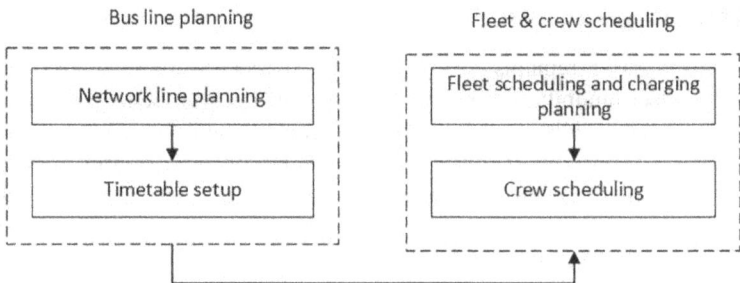

FIGURE 12.2   Electrified public transportation planning stages modified from [9].

12.3 ELECTRIC BUS CHARGING STATIONS

Although there are advantages of EB systems toward the environment and society, EBs are suffering from driving range anxiety because of their limited battery capacity and the need for frequent charging in daily operation. Despite the battery capacity reaching 440 kWh for Catalyst E2 35-foot by Proterra, higher capacity leads to a heavy battery, which reduces bus operation efficiency [10]. Therefore, to obtain high efficiency, a smaller battery size is usually recommended while preserving other operational constraints. Currently, there are three charging technologies to address driving range anxiety, namely, battery swapping, stationary charging, and dynamic wireless charging (DWC) [2, 11, 12] (see Figure 12.3). All three charging technologies are applicable to EBs.

The battery-swapping method is an early step taken by several transit companies due to its similarity to the traditional way of the "refuel and go" method for conventional diesel-based buses. This method allows buses with depleted batteries to be replaced by a fully charged battery [2]. However, it is found that the battery-swapping infrastructure is costly in terms of construction and required land area, which is expensive and sometimes not available in metropolitan cities. The economic benefits of bus battery swapping are discussed by [13], locating the facility [14], and charging schedule [15].

Stationary charging on the other hand consists of three modes, namely, wireless charging, fast plug-in charging, and regular plug-in charging. Wireless charging is convenient and safe but not too much different than plug-in charging in terms of charging time [16]. Regular chargers are usually utilized in the depot for cheaper electricity prices and longer charging time to preserve battery degradation [17]. Fast charging is a widely used technology today due to its fast charging time that can be utilized in terminals or intermediate bus stops [2]. Buses can extend their operating range to 80 kW in less than 10 minutes [2]. The speed of charging reaches 28.67 kW per minute [18]. There are different types of fast chargers with different power, such as 150 kW and 300 kW [19]. Relevant studies on fast charging can be found in [20–24].

The DWC is a new technology that allows buses to charge in motion using electromagnetic pads installed underneath the road [17]. Relevant review studies on wireless charging EV technology can be found in [16, 17, 25, 26].

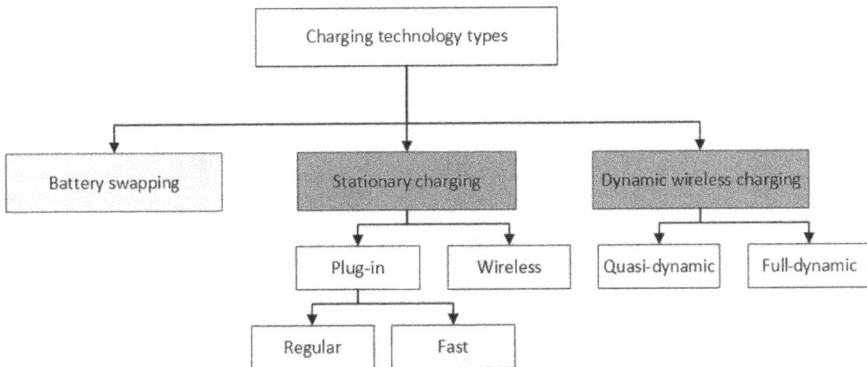

FIGURE 12.3 Different charging technologies.

12.3.1 WIRELESS POWER TRANSFER APPROACHES

Two main challenges encounter the wireless power transfer (WPT): large airgap and coil misalignment [27–31]. The average airgap distance varies from 0.15 to 0.3 meters for small passenger vehicles. Researchers worked on four main wireless charging systems for electric vehicles (EVs), namely, traditional inductive power transfer (IPT), capacitive wireless power transfer (CWPT), magnetic gear wireless power transfer (MGWPT), and resonant inductive power transfer (RIPT) (see [32–38]). The CWPT technology is more fit for low-power appliances, such as laptops and cell phones [39–41]. The IPT and RIPT have higher efficiency than other methods and are more suitable for EB wireless charging systems [42].

12.3.2 DYNAMIC WIRELESS CHARGING MECHANISM

The DWC has two modes, quasi DWC and fully DWC, known simply as DWC. Quasi DWC occurs when an EB moves at a slow speed, deaccelerates to zero or accelerates from zero, while fully DWC occurs when an EB is in motion [17]. The wireless charging system consists of a supply system and a pickup system. The supply system consists of four parts, a power transmission system that transmits the power by electromagnetic field from the grid, a rectifier, an inverter, and a controller. The pickup system is attached to the bottom of the bus, which receives the power from the transmitter and passes it to the battery or the engine. A standard configuration of wireless charging electric bus system is shown in Figure 12.4, while slightly different configurations can be found in [25, 34, 43–46].

12.4 SYSTEMS OPTIMIZATION OF EB CHARGING PLANNING

12.4.1 MATHEMATICAL STRUCTURE

The design of an EB network is more complicated than the conventional bus network due to battery-charging restrictions that pose additional constraints on problem formulation with implications on route design. The optimal deployment of DWC lanes has three forms: continuous, discrete, and network models [17]. Continuous formulation seeks the optimal locations based on continuous decision variables, see

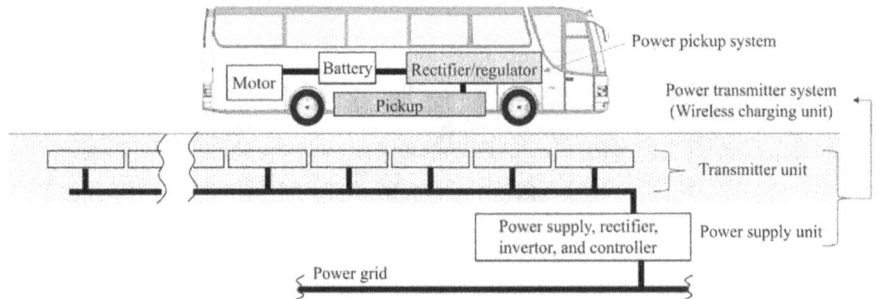

FIGURE 12.4 Wireless charging components [17].

e.g. [8, 47, 48], while the discrete form seeks the optimal locations based on a binary variable that searches for a segmented route, see e.g. [49–55]. The network model divides the routes into nodes and links such that the search decision variable is the link between two nodes [56].

12.4.2 Systems Optimization

Envisioning the maturity of DWC technologies, a handful of studies have investigated the locations of the DWC facilities for EBs. Early studies only consider a single bus route. For example, reference [8] proposed a nonlinear model for a single EB line considering the fixed inverter cost and the variable cable cost. The model has been extended by discretizing the route into multiple small segments, turning the optimization approach by [8] into a discrete problem [50]. Considering the closed environment in which buses operate under regulated velocity and less traffic, [49] developed a mixed-integer linear programming (MILP) model for a single bus line. Reference [47] was the first to incorporate battery-life prediction in the model proposed by [8]. However, in real-world applications, buses operate in a multiple-route environment, which requires modification on the aforementioned model to be adequate in handling multiple-route systems. Some bus routes may overlap with others, which could lead to an overall cost reduction. Reference [57] proposed a nonlinear optimization model to find the optimal locations of transmitters and the uniform optimal battery size for all EBs in a multiple-route environment. As an extension, [56] proposed a novel network-based MILP optimization model to find the optimal locations of transmitters and route-specific optimal battery sizes in a multiple-route environment. Reference [48] proposed a MILP model to solve the DWC facility deployment problem in a multiple-route environment using a continuous search-based approach (see Table 12.1). Most of the studies assume that the charging process is linear, i.e.,

TABLE 12.1
Summary of Recent Studies in DWC Planning for EB Systems

Reference	Vehicle target	Model	Format	Method/software
Single-route approach				
[8]	EB	MINP[1]	Continuous	PSO[2]
[49]	EB	MILP	Discrete	CPLEX
[47]	EB	MILP	Continuous	PSO
[50]	EB	MILP	Discrete	Genetic Algorithm
Multiple-route approach				
[57]	EB	MINP	Discrete	PSO
[56]	EB	MILP	Discrete	CPLEX
[48]	EB	MILP	Continuous	PSO
[54]	EB	MIQCP[3]	Discrete	Gourbi
[53]	EB	MILP	Discrete	Gourbi
[52]	EB	MILP	Discrete	Gourbi

the state of charge increases during charging and is proportional to the charging time. Despite the nonlinear spectrum of the battery charging, it is roughly linear between the maximum and minimum charging points [8, 58]. Also, the bus battery is usually assumed to have a predetermined service life. The battery capacity degeneration is usually ignored for modeling simplicity [49, 59].

The first study investigated the optimal number of EBs required in the system for a planned dynamic wireless charging infrastructure was conducted by Alwesabi et al. [54]. In this study, authors developed a bi-level model to optimize both charging planning decision variables and finding the optimal number of EBs required. It was found for the first time that the optimal EB fleet size could be larger than the current conventional bus size, which used to be ignored in the literature. Later on, reference [53] extended the work in reference [54] by developing a novel model to optimize both charging planning and EB fleet size in an integrated fashion. This model is the first in the literature to incorporate charging time in the EB charging planning, which leads to overall system cost reduction.

12.4.3 ROBUST OPTIMIZATION

Robust optimization (RO) is a powerful risk-management method with high controllable robustness and low computational burden. RO was first introduced by Soyster in the 1970s and required that each uncertain parameter represent the worst scenario [60]. Unlike stochastic optimization models, RO does not require probability distribution [61]. Additionally, the amount of risk can be controlled in RO, which is not possible for other approaches. RO is popular due to its computational tractability for many classes of uncertainty sets and problem types [62]. Unlike RO, stochastic approaches suffer from high execution burdens. In the domain of planning, a deterministic solution could turn out to be infeasible or suboptimal [63].

A handful of references of RO can be found in [63–66]. RO has wide applications in energy systems [61, 67, 68], EV charging [1, 56, 69], bus scheduling [1, 70, 71], marketing [72, 73], supply chain [74, 75], and healthcare [76]. A limited number of studies addressed uncertainty events in EB transportation systems [1, 2, 56]. The first study that investigated a robust model of the EB system that considers DWC planning is [56]. The fleet size proposed by [56] is assumed to be constant regardless of the scheduling part and charging time. Later on, authors in [52] addressed the limitation of the model in [56] through deriving a robust model for EB charging planning with considering fleet size as a decision variable.

12.5 ELECTRIC BUS SCHEDULING

Transit agencies concentrate on two issues, determining the optimal fleet size and assigning buses to their optimal locations on time. The strategy used for vehicle scheduling relies on a connection-based network, each two vertices are connected by one arc [77–79]. Generally, the connection-based model problem is hard to solve by optimization software (e.g., GUROBI) when thousands of trips must be scheduled.

To this end, the connection-based network could lead to an immense number of deadheading trips. As an alternative solution, a space-time network modeling concept has been commonly used [79, 80].

The electric bus scheduling problem (EBSP) is rather sophisticated due to battery-charging restrictions that pose additional constraints on problem formulation. The EBSP refers to the optimal assignment of EBs to cover the whole timetable of the transit system. The chain of trips will be assigned to one bus so that every following trip's time should be later than the arriving time of the previous trip to the destination plus the potential charging time [1]. In addition, no trip can be assigned to different EBs at the same time, and vice versa. The EBSP is categorized into single-depot (SD-EBSP) and multiple-depot (MD-EBSP) (see [1, 20, 22, 81, 82] and [83, 84], respectively).

Considering the electricity price and depth-of-discharge for the battery, [81] proposed a MILP model to address the SD-EBSP with up to 543 trips and 244 kWh of battery capacity. Reference [82] proposed a MILP model to address the SD-EBSP to find the minimal bus fleet size adhering to the charging constraints in the depot and some predefined charging locations. The study was conducted in Zilina City, Slovakia, with 4 datasets and 12 scenarios. The largest number of trips being solved was 160 trips, which took more than 19 hours to get the optimal solution with five chargers and 140 kWh of battery capacity. Table 12.2 illustrates the recent studies on the EBSP.

12.6 DATA COLLECTIONS

The off-campus college transportation (OCCT) case study is considered. The first step is to determine the roads of the case study, which are six roads in this case. The second step is to segmentize the routes into arcs and segments. Google Maps is used for computing the distance for each arc. Also, the speed profile for each route is

TABLE 12.2
Summary of Recent Studies in the EBSP

Reference	Vehicle target	Model	DWC	Method
Single-depot approach				
[81]	EB	MILP	No	CPLEX solver
[1]	EB	MILP	No	Branch & Price/CPLEX
[82]	EB	MILP	No	Standard MILP solver
[85]	Mixed	MILP	No	Heuristic
[54]	EB	MIQCP	Yes	Gourbi solver
[53]	EB	MILP	Yes	Gourbi solver
[52]	EB	MILP	Yes	Gourbi solver
Multiple-depot approach				
[83]	EB	MISOCP[4]	No	Matlab CPLEX solver
[84]	Mixed	MILP	No	IBM ILOG CPLEX
[86–88]	EB (different sizes)	MILP	No	Genetic algorithm

calculated. To get accurate results, the on-street stops are considered because these stops are good candidates for charging due to the dwell time that can be utilized for charging. On the other hand, the timetable information on the selected service roads is collected. Table 12.3 demonstrates the arrangement of the collected time-table for the selected routes. In the disjoint scheduling approach, a scheduling model will be run for each route to find the optimal fleet size for that selected route. Joint scheduling on the other side requires merging all timetables and resorting them by compound sorting of departure and arrival. For example, if there are two trips to dif-ferent routes that have the same departure time but different arrivals, it is considered to sort them by arrival as well.

12.7 ELECTRIC BUS SCHEDULING MODEL

To understand how the electric bus scheduling model works, let's take the following example. Assuming that there is one bus line, the bus starts at 7:00 am and finishes at 10:00 am. The bus requires 15 minutes of charging at the terminal. Table 12.4 illustrates this scheduling example.

EB #1 is going to be ready for its next trip after 7:45 am due to charging time. Therefore, EB#1 cannot perform Trip #2, but it can do Trip #3. Therefore, another EB or EB #2 is required to perform Trip #2. Similarly, EB #1 can perform Trip #1,

TABLE 12.3
The OCCT Timetable Sample

Route No.	Departure	Arrival
1	7:40 AM	8:15 AM
1	8:10 AM	8:45 AM
1	8:40 AM	9:15 AM
2	7:00 AM	7:40 AM
2	7:15 AM	7:55 AM
2	7:30 AM	8:10 AM

TABLE 12.4
Illustrative Example of EB Scheduling

Route No.	Trip No.	Departure	Arrival	Charging time (minutes)
1	1	7:00 am	7:30 am	15
1	2	7:30 am	8:00 am	15
1	3	8:00 am	8:30 am	15
1	4	8:30 am	9:00 am	15
1	5	9:00 am	9:30 am	15
1	6	9:30 am	10:00 am	15

3, and 5, while EB #2 can do Trip #2, 4, and 6. Therefore, this transportation system requires two EBs. Let **I** be the set of all trips and **U** be the set of all feasible trips in the system. Let z_{ij}^l denote an EB finishing its trip i and arrive at the beginning of trip j on route $l \in \mathbf{L}$. Then z_{ij}^l can be presented as follows:

$$z_{ij}^l = \begin{cases} 1, \text{if trip } i \text{ is performed and the EB arrived to } 0_{ij}^s \\ 0, \text{Otherwise} \end{cases} \tag{12.1}$$

The objective of the scheduling model is to minimize the total number of EBs by satisfying the timetable. We refer the reader to reference [54] for the mathematical model.

$$\text{Min} \sum_{(0, j, 0_1) \in U} z_{0j}^{0_1} \tag{12.2}$$

Subject to:

$$\sum_{j, l: (i, j, l) \in U} z_{ij}^l = 1, i \in \mathbf{I} \tag{12.3}$$

$$\sum_{i, l: (i, j, l) \in U} z_{ij}^l = 1, j \in \mathbf{I} \tag{12.4}$$

$$z_{ij}^l \in \{0,1\}, (i, j, l) \in U \tag{12.5}$$

The objective function (12.2) is to minimize the total number of EBs in the system, which is the total number of EBs that depart from the depot. The constraint (12.3) is to make sure that every trip i is connected to only one successor trip j. The constraint (12.4) indicates that each trip j is joined with one predecessor trip i.

12.8 REACTIVE POWER OPTIMIZATION

It is crucial to differentiate between active and reactive power concepts. Active power is actual or real power that is used in the circuit, while reactive power (RP) bounces back and forth between load and source, which is theoretically useless. RP can be used to produce magnetic and electric field. RP is provided by generators, synchronous condensers, or electrostatic equipment, such as capacitors, and directly influences electric system voltage. RP is measured by volt-amps-reactive (VAR). Optimizing power loss requires advanced knowledge of the power distribution for at least 24 hours to minimize daily system power losses [87].

Several studies concentrate on minimizing power losses through limiting switching operations. This problem is common in electric vehicle (EV) charging systems due to frequent charging per day. The problem is classified as non-convexMINP, which is very hard to solve [88,89]. To reduce the computational time, the distribution network is divided into two parts, namely, feeders and substations. The switching sequence of capacitors in each feeder is first determined using a dynamic programming approach, and then the dispatch schedules of capacitors and transformers in a substation are obtained [90,91]. Another approach was observed in the literature through consideration of maximum allowable daily switching operation number (MADSON). The load curve was divided into several segments according to the distribution of losses at each hour, and then this was converted into an equivalent staircase curve based on an equal-area criterion, in which the numbers of steps satisfy a given MADSON constraint. By means of this mechanism, the multi-period optimization model is transformed into several single-period models [92]. A fuzzy-based RP optimization model is proposed and tested in a real distribution network [93]. To accelerate the required time for the solution, a round-off algorithm was used to transform the discrete variables into numerical ones [94]. Reference [87] developed a nonlinear optimization model to minimize daily energy losses based on the interior-point method and discretization penalties. The proposed algorithm yields competitive results compared to the Genetic algorithm and GAMS nonlinear solvers. Some of the constraints represent the limitation of the number of switching operations of transformer load tap changers (LTCs) and capacitors, which are modeled as discrete control variables [87]. The effectiveness of the proposed model was shown in practice, in the city of Guangzhou, China, since 2003 [87].

In the photovoltaic systems, power loss is a very common problem due to intermittent power from the solar system. Table 12.5 shows the recent optimization methods of reactive power control.

Voltage violation problem (VVP) is a common problem in the photovoltaic system due to the nature of generated power from these solar panels. Figure 12.5 shows the traditional and advanced methods used to address this problem.

On-load tap changer transformers (OLTC) are effective devices used to regulate voltage [99,100]. Step voltage regulators (SVR) are commonly used at the downstream

TABLE 12.5
Optimization Methods of Reactive Power Control

Refernce	Objective	Method/solver
[95]	To minimize power loss	GA
[96]	To optimize power loss, voltage violation, tap movement rate, and curtailed power of PV	Multi-objective Grey-Wolf-Lévy Optimizer
[97]	To control feeder's voltage and reduce customer demand	CPLEX 12.7
[98]	To control reactive power in power distribution systems Link: https://github.com/siemens/powergym	Reinforcement Learning Environment

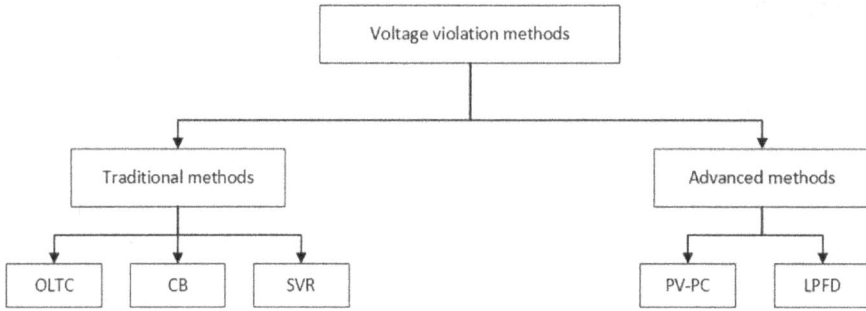

FIGURE 12.5 Voltage violation methods.

feeders. Due to redundant power fluctuation, the lifetime of these devices could significantly decline. The advanced voltage control methods include active PV power curtailment (PV-PC) during low demand and reactive power control of the PV inverter [91–101]. Another way is to add a leading power factor device (LPFD) [96].

12.9 CONCLUSION

Although there have been remarkable contributions to DWC planning, the literature review is limited to a single-depot DWC planning system. There is a need for more research to upgrade the single-depot system to a multiple-depot system, which practically represents a real-world bus network system. The benefit of potentially sharing dynamic charging with other EVs on the road and its impact on the adoption of EVs can be investigated as future work. For example, at Binghamton University, staff members and professors can drive and charge their electric vehicles from the same DWC facilities on the road. The project can be extended to metropolitan cities, such as New York City. Additionally, operational costs, such as driving and maintenance costs, are left for future studies. An alternative future study could investigate the emerging technologies of fast stationary charging and DWC to reduce the charging time required in the terminals. The multiple-depot scheduling problem itself is NP-hard; therefore, it is recommended in future studies to establish an appropriate heuristic method to handle the large-scale problems of this sort of transportation system, e.g., genetic algorithm or particle swarm optimization. Different fleet bus types, such as an EB fleet with different sizes or a mixed bus fleet (EB, hybrid, and diesel), can be investigated as another future work.

On the other hand, studies on investigating the impact of a power supply network in the EB systems considering DWC is rarely reported in the literature. There is an urgent need for further investigation in this field.

NOTES

1. MINP: Mixed-integer nonlinear programming.
2. PSO: Particle swarm optimization.
3. MIQCP: Mixed-integer quadratic constraint programming.
4. MISOCP: Mixed-integer second-order cone programming.

REFERENCES

[1] X. Tang, X. Lin, and F. He, "Robust scheduling strategies of electric buses under stochastic traffic conditions," *Transportation Research Part C: Emerging Technologies*, vol. 105, pp. 163–182, 2019.

[2] K. An, "Battery electric bus infrastructure planning under demand uncertainty," *Transportation Research Part C: Emerging Technologies*, vol. 111, pp. 572–587, 2020.

[3] J. H. Raimundo Atal, and Andrew Wilson, "Electrifying America's public transportation systems." Available: www.warren.senate.gov/imo/media/doc/DFP_Electrifying%20 Transport_2021.pdf, 2021.

[4] K. Murakami and H. Morita, "A column generation model for the electric and fuel-engined vehicle routing problem," in *2015 IEEE International Conference on Systems, Man, and Cybernetics*, pp. 1986–1991: IEEE, 2015.

[5] C. Iliopoulou and K. Kepaptsoglou, "Integrated transit route network design and infrastructure planning for on-line electric vehicles," *Transportation Research Part D: Transport and Environment*, vol. 77, pp. 178–197, 2019.

[6] M. Pternea, K. Kepaptsoglou, and M. G. Karlaftis, "Sustainable urban transit network design," *Transportation Research Part A: Policy and Practice*, vol. 77, pp. 276–291, 2015.

[7] S. Helber, J. Broihan, Y. J. Jang, P. Hecker, and T. Feuerle, "Location planning for dynamic wireless charging systems for electric airport passenger buses," *Energies*, vol. 11, no. 2, p. 258, 2018.

[8] Y. D. Ko and Y. J. Jang, "The optimal system design of the online electric vehicle utilizing wireless power transmission technology," *IEEE Transactions on Intelligent Transportation Systems*, vol. 14, no. 3, pp. 1255–1265, 2013.

[9] C. Posthoorn, "Vehicle scheduling of electric city buses: A column generation approach," 2016.

[10] Proterra. *Proterra fast chargers.* Available: www.proterra.com/press-release/proterra-introduces-new-high-power-interoperable-ev-charging-technology/, 2018.

[11] Z. Chen, Y. Yin, and Z. Song, "A cost-competitiveness analysis of charging infrastructure for electric bus operations," *Transportation Research Part C: Emerging Technologies*, vol. 93, pp. 351–366, 2018.

[12] J.-Q. Li, "Battery-electric transit bus developments and operations: A review," *International Journal of Sustainable Transportation*, vol. 10, no. 3, pp. 157–169, 2016.

[13] C. Zhang and P. Chen, "Economic benefit analysis of battery charging and swapping station for pure electric bus based on differential power purchase policy: A new power trading model," *Sustainable Cities and Society*, p. 102570, 2020.

[14] W. Jing, I. Kim, and K. An, "The uncapacitated battery swapping facility location problem with localized charging system serving electric bus fleet," *Transportation Research Procedia*, vol. 34, pp. 227–234, 2018.

[15] P. You, *et al.*, "Scheduling of EV battery swapping. Part I: Centralized solution," *IEEE Transactions on Control of Network Systems*, vol. 5, no. 4, pp. 1887–1897, 2017.

[16] V. Cirimele, M. Diana, F. Freschi, and M. Mitolo, "Inductive power transfer for automotive applications: State-of-the-art and future trends," *IEEE Transactions on Industry Applications*, vol. 54, no. 5, pp. 4069–4079, 2018.

[17] Y. J. Jang, "Survey of the operation and system study on wireless charging electric vehicle systems," *Transportation Research Part C: Emerging Technologies*, vol. 95, pp. 844–866, 2018.

[18] M. T. Sebastiani, R. Lüders, and K. V. O. Fonseca, "Evaluating electric bus operation for a real-world BRT public transportation using simulation optimization," *IEEE Transactions on Intelligent Transportation Systems*, vol. 17, no. 10, pp. 2777–2786, 2016.

[19] Y. He, Z. Song, and Z. Liu, "Fast-charging station deployment for battery electric bus systems considering electricity demand charges," *Sustainable Cities and Society*, vol. 48, p. 101530, 2019.

[20] Y. He, Z. Liu, and Z. Song, "Optimal charging scheduling and management for a fast-charging battery electric bus system," *Transportation Research Part E: Logistics and Transportation Review*, vol. 142, p. 102056, 2020.

[21] A. Kunith, R. Mendelevitch, and D. Goehlich, "Electrification of a city bus network: An optimization model for cost-effective placing of charging infrastructure and battery sizing of fast-charging electric bus systems," *International Journal of Sustainable Transportation*, vol. 11, no. 10, pp. 707–720, 2017.

[22] Y. Wang, Y. Huang, J. Xu, and N. Barclay, "Optimal recharging scheduling for urban electric buses: A case study in Davis," *Transportation Research Part E: Logistics and Transportation Review*, vol. 100, pp. 115–132, 2017.

[23] M. Xylia, S. Leduc, P. Patrizio, F. Kraxner, and S. Silveira, "Locating charging infrastructure for electric buses in Stockholm," *Transportation Research Part C: Emerging Technologies*, vol. 78, pp. 183–200, 2017.

[24] B.-R. Ke, C.-Y. Chung, and Y.-C. Chen, "Minimizing the costs of constructing an all plug-in electric bus transportation system: A case study in Penghu," *Applied energy*, vol. 177, pp. 649–660, 2016.

[25] F. Chen, N. Taylor, and N. Kringos, "Electrification of roads: Opportunities and challenges," *Applied Energy*, vol. 150, pp. 109–119, 2015.

[26] A. Gil and J. Taiber, "A literature review in dynamic wireless power transfer for electric vehicles: Technology and infrastructure integration challenges," in *Sustainable automotive technologies 2013*. Springer, pp. 289–298, 2014.

[27] S. Moon, B.-C. Kim, S.-Y. Cho, C.-H. Ahn, and G.-W. Moon, "Analysis and design of a wireless power transfer system with an intermediate coil for high efficiency," *IEEE Transactions on Industrial Electronics*, vol. 61, no. 11, pp. 5861–5870, 2014.

[28] F. van der Pijl, P. Bauer, and M. Castilla, "Control method for wireless inductive energy transfer systems with relatively large air gap," *IEEE Transactions on Industrial Electronics*, vol. 60, no. 1, pp. 382–390, 2013.

[29] T. Imura, T. Uchida, and Y. Hori, "Flexibility of contactless power transfer using magnetic resonance coupling to air gap and misalignment for EV," *World Electric Vehicle Journal*, vol. 3, no. 2, pp. 332–341, 2009.

[30] C. Zheng, H. Ma, J.-S. Lai, and L. Zhang, "Design considerations to reduce gap variation and misalignment effects for the inductive power transfer system," *IEEE Transactions on Power Electronics*, vol. 30, no. 11, pp. 6108–6119, 2015.

[31] W. Zhong and D. Xu, "Wireless EV charging system without air-gap and misalignment," in *2018 International Power Electronics Conference (IPEC-Niigata 2018-ECCE Asia)*, pp. 2569–2575: IEEE, 2018.

[32] K. A. Kalwar, M. Aamir, and S. Mekhilef, "Inductively coupled power transfer (ICPT) for electric vehicle charging: A review," *Renewable and Sustainable Energy Reviews*, vol. 47, pp. 462–475, 2015.

[33] D. Leskarac, C. Panchal, S. Stegen, and J. Lu, "PEV charging technologies and V2G on distributed systems and utility interfaces," *Vehicle-to-Grid: Linking Electric Vehicles to the Smart Grid*, vol. 79, pp. 157–209, 2015.

[34] G. A. Covic and J. T. Boys, "Modern trends in inductive power transfer for transportation applications," *IEEE Journal of Emerging and Selected Topics in Power Electronics*, vol. 1, no. 1, pp. 28–41, 2013.

[35] S. Y. R. Hui, W. Zhong, and C. K. Lee, "A critical review of recent progress in mid-range wireless power transfer," *IEEE Transactions on Power Electronics*, vol. 29, no. 9, pp. 4500–4511, 2014.

[36] S. Li and C. C. Mi, "Wireless power transfer for electric vehicle applications," *IEEE Journal of Emerging and Selected Topics in Power Electronics*, vol. 3, no. 1, pp. 4–17, 2015.

[37] T. Ching and Y. Wong, "Review of wireless charging technologies for electric vehicles," in *Power Electronics Systems and Applications (PESA), 2013 5th International Conference on*, pp. 1–4: IEEE, 2013.

[38] F. Musavi, M. Edington, and W. Eberle, "Wireless power transfer: A survey of EV battery charging technologies," in *Energy Conversion Congress and Exposition (ECCE), 2012 IEEE*, pp. 1804–1810: IEEE, 2012.

[39] C. Liu, A. P. Hu, B. Wang, and N.-K. C. Nair, "A capacitively coupled contactless matrix charging platform with soft switched transformer control," *IEEE Transactions on Industrial Electronics*, vol. 60, no. 1, pp. 249–260, 2013.

[40] J. Dai and D. C. Ludois, "Wireless electric vehicle charging via capacitive power transfer through a conformal bumper," in *Applied Power Electronics Conference and Exposition (APEC), 2015 IEEE*, pp. 3307–3313: IEEE, 2015.

[41] A. P. Hu, C. Liu, and H. L. Li, "A novel contactless battery charging system for soccer playing robot," in *Mechatronics and Machine Vision in Practice, 2008. M2VIP 2008: 15th International Conference on*, pp. 646–650: IEEE, 2008.

[42] C. Panchal, S. Stegen, and J. Lu, "Review of static and dynamic wireless electric vehicle charging system," *Engineering Science and Technology, an International Journal*, vol. 21, pp. 992–937, 2018.

[43] N. P. Suh and D. H. Cho, *The On-Line Electric Vehicle: Wireless Electric Ground Transportation Systems*. Springer, 2017.

[44] H. Xie, T. Ding, S. Liu, and Z. Bie, "Sequential power flow simulation of integrated dynamic wireless power transfer systems," in *2016 IEEE Power and Energy Society General Meeting (PESGM)*, pp. 1–5: IEEE, 2016.

[45] H. England, "Feasibility study: Powering electric vehicles on England's major roads," *Highways England Company*, 2015.

[46] Y. J. Jang, E. S. Suh, and J. W. Kim, "System architecture and mathematical models of electric transit bus system utilizing wireless power transfer technology," *IEEE Systems Journal*, vol. 10, no. 2, pp. 495–506, 2015.

[47] S. Jeong, Y. J. Jang, and D. Kum, "Economic analysis of the dynamic charging electric vehicle," *IEEE Transactions on Power Electronics*, vol. 30, no. 11, pp. 6368–6377, 2015.

[48] I. Hwang, Y. J. Jang, Y. D. Ko, and M. S. Lee, "System optimization for dynamic wireless charging electric vehicles operating in a multiple-route environment," *IEEE Transactions on Intelligent Transportation Systems*, vol. 19, no. 6, pp. 1709–1726, 2017.

[49] Y. J. Jang, S. Jeong, and Y. D. Ko, "System optimization of the on-line electric vehicle operating in a closed environment," *Computers & Industrial Engineering*, vol. 80, pp. 222–235, 2015.

[50] Y. D. Ko, Y. J. Jang, and M. S. Lee, "The optimal economic design of the wireless powered intelligent transportation system using genetic algorithm considering nonlinear cost function," *Computers & Industrial Engineering*, vol. 89, pp. 67–79, 2015.

[51] Y. Jang, S. Jeong, and M. Lee, "Initial energy logistics cost analysis for stationary, quasi-dynamic, and dynamic wireless charging public transportation systems," *Energies*, vol. 9, no. 7, p. 483, 2016.

[52] Y. Alwesabi, F. Avishan, İ. Yanıkoğlu, Z. Liu, and Y. Wang, "Robust strategic planning of dynamic wireless charging infrastructure for electric buses," *Applied Energy*, vol. 307, p. 118243, 2022.

[53] Y. Alwesabi, Z. Liu, S. Kwon, and Y. Wang, "A novel integration of scheduling and dynamic wireless charging planning models of battery electric buses," *Energy*, vol. 230, p. 120806, 2021.

[54] Y. Alwesabi, Y. Wang, R. Avalos, and Z. Liu, "Electric bus scheduling under single depot dynamic wireless charging infrastructure planning," *Energy*, pp. 118855, 2020.

[55] Y. Alwesabi, N. Mohammed, A. Alaa, and Y. Wang, "Energy demand estimation of electric buses considering dynamic wireless charging technology," *presented at the International Congress of Advanced Technology and Engineering (ICOTEN 2021)*, 2021.

[56] Z. Liu and Z. Song, "Robust planning of dynamic wireless charging infrastructure for battery electric buses," *Transportation Research Part C: Emerging Technologies*, vol. 83, pp. 77–103, 2017.

[57] N. Mouhrim, A. E. H. Alaoui, and J. Boukachour, "Optimal allocation of wireless power transfer system for electric vehicles in a multipath environment," in *2016 3rd International Conference on Logistics Operations Management (GOL)*, pp. 1–7: IEEE 2016.

[58] S. Mohrehkesh and T. Nadeem, "Toward a wireless charging for battery electric vehicles at traffic intersections," in *2011 14th International IEEE Conference on Intelligent Transportation Systems (ITSC)*, pp. 113–118: IEEE, 2011.

[59] A. Ashtari, E. Bibeau, S. Shahidinejad, and T. Molinski, "PEV charging profile prediction and analysis based on vehicle usage data," *IEEE Transactions on Smart Grid*, vol. 3, no. 1, pp. 341–350, 2011.

[60] A. L. Soyster, "Convex programming with set-inclusive constraints and applications to inexact linear programming," *Operations Research*, vol. 21, no. 5, pp. 1154–1157, 1973.

[61] N. Rezaei, A. Khazali, M. Mazidi, and A. Ahmadi, "Economic energy and reserve management of renewable-based microgrids in the presence of electric vehicle aggregators: A robust optimization approach," *Energy*, p. 117629, 2020.

[62] B. L. Gorissen, İ. Yanıkoğlu, and D. den Hertog, "A practical guide to robust optimization," *Omega*, vol. 53, pp. 124–137, 2015.

[63] D. Bertsimas, D. B. Brown, and C. Caramanis, "Theory and applications of robust optimization," *SIAM Review*, vol. 53, no. 3, pp. 464–501, 2011.

[64] A. Nemirovski, "Lectures on robust convex optimization," in *Lecture Notes: Georgia Institute of Technology*, 2012.

[65] A. Ben-Tal, L. El Ghaoui, and A. Nemirovski, *Robust Optimization*. Princeton University Press, 2009.

[66] A. Ben-Tal and A. Nemirovski, "Selected topics in robust convex optimization," *Mathematical Programming*, vol. 112, no. 1, pp. 125–158, 2008.

[67] F. Babonneau, J.-P. Vial, and R. Apparigliato, "Robust optimization for environmental and energy planning," in *Uncertainty and Environmental Decision Making*. Springer, 2009, pp. 79–126.

[68] D. Bertsimas, E. Litvinov, X. A. Sun, J. Zhao, and T. Zheng, "Adaptive robust optimization for the security constrained unit commitment problem," *IEEE Transactions on Power Systems*, vol. 28, no. 1, pp. 52–63, 2012.

[69] H. Sun, J. Yang, and C. Yang, "A robust optimization approach to multi-interval location-inventory and recharging planning for electric vehicles," *Omega*, vol. 86, pp. 59–75, 2019.

[70] S. Yan and C.-H. Tang, "Inter-city bus scheduling under variable market share and uncertain market demands," *Omega*, vol. 37, no. 1, pp. 178–192, 2009.

[71] Y. Yan, Q. Meng, S. Wang, and X. Guo, "Robust optimization model of schedule design for a fixed bus route," *Transportation Research Part C: Emerging Technologies*, vol. 25, pp. 113–121, 2012.

[72] X. J. Wang and D. J. Curry, "A robust approach to the share-of-choice product design problem," *Omega*, vol. 40, no. 6, pp. 818–826, 2012.

[73] A. Albadvi and H. Koosha, "A robust optimization approach to allocation of marketing budgets," *Management Decision*, 2011.

[74] A. Ben-Tal, B. Golany, A. Nemirovski, and J.-P. Vial, "Retailer-supplier flexible commitments contracts: A robust optimization approach," *Manufacturing & Service Operations Management*, vol. 7, no. 3, pp. 248–271, 2005.

[75] S. Lim, "A joint optimal pricing and order quantity model under parameter uncertainty and its practical implementation," *Omega*, vol. 41, no. 6, pp. 998–1007, 2013.

[76] A. Fredriksson, A. Forsgren, and B. Hårdemark, "Minimax optimization for handling range and setup uncertainties in proton therapy," *Medical Physics*, vol. 38, no. 3, pp. 1672–1684, 2011.

[77] D. Huisman, *Integrated and Dynamic Vehicle and Crew Scheduling* (no. 325). 2004.

[78] S. Bunte and N. Kliewer, "An overview on vehicle scheduling models," *Public Transport*, vol. 1, no. 4, pp. 299–317, 2009.

[79] H. Niu, X. Zhou, and X. Tian, "Coordinating assignment and routing decisions in transit vehicle schedules: A variable-splitting Lagrangian decomposition approach for solution symmetry breaking," *Transportation Research Part B: Methodological*, vol. 107, pp. 70–101, 2018.

[80] I. Steinzen, V. Gintner, L. Suhl, and N. Kliewer, "A time-space network approach for the integrated vehicle-and crew-scheduling problem with multiple depots," *Transportation Science*, vol. 44, no. 3, pp. 367–382, 2010.

[81] M. E. van Kooten Niekerk, J. van den Akker, and J. Hoogeveen, "Scheduling electric vehicles," *Public Transport*, vol. 9, no. 1–2, pp. 155–176, 2017.

[82] M. Janovec and M. Koháni, "Exact approach to the electric bus fleet scheduling," *Transportation Research Procedia*, vol. 40, pp. 1380–1387, 2019.

[83] Y. Lin, K. Zhang, Z.-J. M. Shen, B. Ye, and L. Miao, "Multistage large-scale charging station planning for electric buses considering transportation network and power grid," *Transportation Research Part C: Emerging Technologies*, vol. 107, pp. 423–443, 2019.

[84] L. Li, H. K. Lo, and F. Xiao, "Mixed bus fleet scheduling under range and refueling constraints," *Transportation Research Part C*, vol. 104, pp. 443–462, 2019.

[85] G.-J. Zhou, D.-F. Xie, X.-M. Zhao, and C. Lu, "Collaborative optimization of vehicle and charging scheduling for a bus fleet mixed with electric and traditional buses," *IEEE Access*, vol. 8, pp. 8056–8072, 2020.

[86] E. Yao, T. Liu, T. Lu, and Y. Yang, "Optimization of electric vehicle scheduling with multiple vehicle types in public transport," *Sustainable Cities and Society*, vol. 52, p. 101862, 2020.

[87] M. Liu, C. A. Canizares, and W. Huang, "Reactive power and voltage control in distribution systems with limited switching operations," *IEEE Transactions on Power Systems*, vol. 24, no. 2, pp. 889–899, 2009.

[88] F.-C. Lu and Y.-Y. Hsu, "Fuzzy dynamic programming approach to reactive power/ voltage control in a distribution substation," *IEEE Transactions on Power Systems*, vol. 12, no. 2, pp. 681–688, 1997.

[89] Y.-Y. Hsu and F.-C. Lu, "A combined artificial neural network-fuzzy dynamic programming approach to reactive power/voltage control in a distribution substation," *IEEE Transactions on Power Systems*, vol. 13, no. 4, pp. 1265–1271, 1998.

[90] R.-H. Liang and C.-K. Cheng, "Dispatch of main transformer ULTC and capacitors in a distribution system," *IEEE Transactions on Power Delivery*, vol. 16, no. 4, pp. 625–630, 2001.

[91] Y. Liu, P. Zhang, and X. Qiu, "Optimal volt/var control in distribution systems," *International Journal of Electrical Power & Energy Systems*, vol. 24, no. 4, pp. 271–276, 2002.

[92] Y. Deng, X. Ren, C. Zhao, and D. Zhao, "A heuristic and algorithmic combined approach for reactive power optimization with time-varying load demand in distribution systems," *IEEE Transactions on Power Systems*, vol. 17, no. 4, pp. 1068–1072, 2002.

[93] R.-H. Liang and Y.-S. Wang, "Fuzzy-based reactive power and voltage control in a distribution system," *IEEE Transactions on Power Delivery*, vol. 18, no. 2, pp. 610–618, 2003.

[94] M. Liu, S. Tso, and Y. Cheng, "An extended nonlinear primal-dual interior-point algorithm for reactive-power optimization of large-scale power systems with discrete control variables," *IEEE Transactions on Power Systems*, vol. 17, no. 4, pp. 982–991, 2002.

[95] X. Cheng, X. Chen, Y. Fan, Y. Liu, S. Li, and Q. Zhao, "Research on reactive power optimization control strategy of distribution network with photovoltaic generation," in *2019 Chinese Automation Congress (CAC)*, pp. 4026–4030: IEEE, 2019.

[96] K. Mahmoud, M. M. Hussein, M. Abdel-Nasser, and M. Lehtonen, "Optimal voltage control in distribution systems with intermittent PV using multiobjective Grey-Wolf-Lévy optimizer," *IEEE Systems Journal*, vol. 14, no. 1, pp. 760–770, 2019.

[97] R. R. Jha, A. Dubey, C.-C. Liu, and K. P. Schneider, "Bi-level volt-var optimization to coordinate smart inverters with voltage control devices," *IEEE Transactions on Power Systems*, vol. 34, no. 3, pp. 1801–1813, 2019.

[98] T.-H. Fan, X. Y. Lee, and Y. Wang, "Powergym: A reinforcement learning environment for volt-var control in power distribution systems," *arXiv preprint arXiv:2109.03970*, 2021.

[99] K. W. Kow, Y. W. Wong, R. K. Rajkumar, and R. K. Rajkumar, "A review on performance of artificial intelligence and conventional method in mitigating PV grid-tied related power quality events," *Renewable and Sustainable Energy Reviews*, vol. 56, pp. 334–346, 2016.

[100] T. Dragičević, X. Lu, J. C. Vasquez, and J. M. Guerrero, "DC microgrids. Part I: A review of control strategies and stabilization techniques," *IEEE Transactions on Power Electronics*, vol. 31, no. 7, pp. 4876–4891, 2015.

[101] N. Mahmud and A. Zahedi, "Review of control strategies for voltage regulation of the smart distribution network with high penetration of renewable distributed generation," *Renewable and Sustainable Energy Reviews*, vol. 64, pp. 582–595, 2016.

13 Telecommunications Connectivity for Electric Transportation Systems: Challenges, Technologies and Solutions

Alberto Sendin

CONTENTS

DOI: 10.1201/9781003293989-13

13.1 ELECTRIC TRANSPORTATION SYSTEMS IN THE CONTEXT OF SMART GRIDS

Electricity is today at the core of most human activities. Thus, electric power systems are instrumental in modern societies. The evolution of electric power systems is happening around the smart grid concept, and electric transportation Systems (ETSs) both benefit from it and, are affected by it as well.

The smart grid concept has extensively been covered in the literature. The smart grid is considered a container of all the modern and evolutionary changes in electric power systems as a consequence of the advent and adoption of new technologies, that progressively add new capabilities to the grid and help it to become a more efficient system [1]. Among these changes, some are associated with new power system classic elements (e.g., power electronics for electricity transportation and distribution, new electricity production assets, etc.); some others are associated with auxiliary but instrumental technologies that are required to efficiently integrate and operate these new elements of the grid (e.g., information and communication technologies [ICTs]); and others add novel requirements due to their appearance as a new need or service that the power system needs to deliver. ETSs are among these last ones.

ETSs are the materialization of the electrification of transport. ETSs do not just refer to road transport, but include other systems that have traditionally run on electricity, such as trains and trams, and may consider other less conventional road users, such as bicycles. If we take trains or trams, these systems are good examples of how early ETSs have needed a bespoke electricity power system with an associated infrastructure. However, these are not the ETSs that will impose new requirements on the grid. Most authors concur with the fact that electric vehicles (EVs; specifically understood as conventional cars, buses and trucks) will be the elements disrupting the traditional electric power grid [2].

EVs can be classified as plug-in hybrid EVs (PHEVs), battery EVs (BEVs) and hybrid EVs (HEVs). A PHEV is a type of EV with an internal combustion engine, an electric motor and a high-capacity rechargeable battery pack that can be charged by plugging in the car to an electric power grid. A BEV is a type of EV that uses rechargeable battery packs to store electrical energy and an electric motor for propulsion. Both PHEV and BEV, are the EVs considered here due to their effect and potential for the grid, and they will be referred to as PEVs (plug-in EVs) in this chapter. HEVs are not considered in this context, as they do not connect to the grid, due to their construction as the combination of a conventional combustion engine and an electric motor for propulsion.

PEVs are a disruption in traditional power systems for several reasons directly stressing the grid and utility procedures [3–4]:

- They will cause a major increase in electrification needs, specifically of the distribution transformers. Moderate-to-high penetration scenarios of PEVs may lead to increased voltage drops along distribution feeders that could cause low-voltage violations, specifically in areas far from distribution substations.

- They will affect grid power quality if harmonic distortion is not controlled, as a result of a large proliferation of inverter-based PEV charging facilities.
- They will increase the peak demand and network load of areas where they connect to the grid, as PEV charging periods tend to follow similar patterns across grid users.
- They will make demand more unpredictable, changing the classical load profile of customers when charging PEVs at home or at work. PEV mobility will just make this worse, due to their changes in location in a context where grid connection points have been static and have traditionally served static or seasonal needs.

On the positive side, PEVs have some features that can be beneficial for the grid. PEVs do not only reduce their carbon footprint but can also be managed as smart loads and as energy sources:

- The PEV as a load is inherently different from other deferrable loads, thus diversifying the flexible load fleet for demand response (DR)[1] programs.
- PEVs can potentially feed electricity into the grid, acting as distributed energy resources[2] (DERs). There could be incentives for PEV owners. However, PEVs need to be prepared for this functional requirement, and the industry is still far from it.
- Charging stations need to be equipped with controls that can provide system operators with, at least, voltage monitoring capabilities and, potentially, voltage response resources.

Therefore, the grid needs to be prepared to support PEVs, acknowledging that PEVs can be a source of opportunities. The answer to the PEV challenge will come within the smart grid, as explained in this chapter.

13.2 WHAT IS THE SMART GRID?

The smart grid concept was coined decades ago to comprehend the evolutionary changes in the power system, as a consequence of the advent and adoption of new technologies to progressively add new capabilities to the grid. The smart grid objectives are not static and help to stablish a common framework, more helpful than the commonly accepted definitions [5–14] themselves. The smart grid objectives intend to:

- Achieve a resilient electric power system.
- Facilitate grid infrastructure modernization.
- Pursue power quality assurance.
- Achieve an efficient power delivery system.
- Facilitate efficient customer consumption.
- Reduce the environmental impact of electricity production and delivery.
- Combine bulk power generation with DER.
- Integrate storage technologies, both bulk and grid edge.

- Automate operational processes.
- Increase the number of sensors and controls in the electricity system.
- Monitor and control critical and non-critical components of the power system.

Smart grid objectives have been fostered by the advances in electric grid technologies and the applicability of information and communication technologies (ICTs) to the power system. On the one hand, new grid technologies are being integrated in the different segments of the grid and have changed the traditional electricity delivery model (e.g., DERs have made consumers be an active part of the grid). On the other hand, continuous ICTs' innovations have also reached utilities in their way toward their digital transformation and the automation of their processes.

Thus, the smart grid integrates new power system–related technologies and ICT innovations to produce a better and smarter power system, with modernized and more powerful assets, enhanced with new applications and capable of delivering new services within a more efficient and environment-friendly operational framework.

Indeed, grid modernization (both new elements in the grid and existing grid assets refresh), the adoption of new grid-edge technologies [15], new technologies in energy storage and microgrid domains, large shares of renewable energy and new grid uses (e.g., EVs) have of recent emerged as smart grid principal components. Most of these changes happen in the distribution grid, the part of the grid that is closer to the customer in its edge. The distribution grid starts in the primary substations, and through the secondary substations with medium-voltage (MV) and low-voltage (LV) power lines (cables), reaches consumers connected to electricity meters. There is a wide consensus that the distribution grid is the segment of the electricity system where most of the challenges and evolution of the smart grid are.

Closer to the telecommunications domain, the utility industry has always been a special case [16]. While the incorporation of ICTs to the grid in its evolution toward a smarter grid can be interpreted in terms of "enablement" of the capabilities of existing and new grid components to seamlessly achieve their integration with the grid as an enhanced system, telecommunications in particular have evolved in utilities with a focus on two aspects. The first one, the need to provide connectivity to grid assets where no preexisting telecommunication networks, interest or access to the service existed (from remote bulk energy generation sites to rural areas, through assets in confined locations). The second is the mission-critical nature of many of the grid operational aspects (e.g., power line teleprotection services). Due to these circumstances, utilities have traditionally developed their own private networks to provide solutions for their special requirements with appropriate telecommunication services. This is an aspect not very commonly known in the industry, and often neglected.

Last but not least, there is no standardized or globally accepted definition of the smart grid, as there is also no universally accepted "standard" or implementation of "the grid". Grids all over the world have many similarities, but their structure, assets and operations differ. In different world regions, in different utilities, smart grids reflect different evolutions of the region's needs, affected by the grid starting point, regulatory framework and medium- and long-term objectives. Thus, smart grid implementations in utilities will be different one from the other, and even more in the segment of the grid where a wider variety of architectures and solutions exists, i.e., the distribution grid.

ETSs will be explicitly impacting the distribution grid segment and are to be added to the challenge of the smart grid in this domain. In this context, the smart grid connects with ETSs with the necessary grid hardening in the MV and LV segments, to reinforce its resiliency and remote operation capabilities, enhanced through the adoption of ICTs, specifically telecommunications connectivity, and autonomous and/or assisted automation intelligence.

13.3 TELECOMMUNICATION TECHNOLOGIES AND SYSTEMS IN SMART GRIDS AND ELECTRIC TRANSPORTATION SYSTEMS

Telecommunication technologies are used to build telecommunication networks or systems. Although the words "networks" and "systems" are used interchangeably ("system" is more prevalent in North America), they are used in telecommunications to refer to and define complete sets of assets, including telecommunications equipment, network cables, information technology components (servers, databases, etc.) and ancillary elements (batteries, antennas, etc.) [17–18]. Although telecommunication systems are, in essence, hybrid systems (they mix a variety of elements and technologies), it is common practice to classify both technologies and systems, based on a reference conceptual framework that delimits where each technology and/or system fits.

Telecommunication systems provide telecommunication services to their users. Telecommunication services can be defined as any communication capability offered to a user by a telecommunication network or system. There is no harmonized classification of telecommunication services, but there is a common simplification referring to their nature (voice, data or video—although modern telecommunication systems provide data service that can convey any type of final service), their data rate capacity (narrowband or broadband) and their timing limitations (latency and/or jitter). Any telecommunication service will eventually be defined with a QoS (quality of service), that together with the SLA (service level agreement) will affect the user QoE (quality of experience).

Telecommunication technologies are often used to define telecommunication systems. Even if in a telecommunication system more than one single telecommunications technology is used, each of these systems shows itself with a number of characteristics that root in the most prevalent technology used, the part of the telecommunication network where it is more often used, and the type of services it can deliver.

Different telecommunication technologies can be used in either private or public (commercial) domains. When they are part of a network (system) developed to provide telecommunication services to the general public on a commercial basis, they are known as public or commercial. On the contrary, when they are restricted to provide services to their owner and not for resale (in general), they are named as private networks (e.g., private companies such as utilities, state enterprises or government entities usually run private networks, at least for some of their more critical operations-related services). Both types of networks may have regulatory constraints, and public networks may be subject to obligations derived from the socioeconomic interest of the telecommunication service to be provided. This public vs. private nature of networks adds a new dimension to the hybrid nature of telecommunication systems, as private systems may make use of telecommunication services provided through public networks, and public networks may use services or assets provided by private networks.

13.3.1 TELECOMMUNICATIONS NETWORKS REFERENCE FRAMEWORK

There are several ideas that must be clarified to understand where the challenges are in telecommunication networks, to deliver a specific service. One of the initial difficulties when trying to present them is that they are not used homogeneously across the literature, both due to their origin and the interest to present them in a different way in systems that will try to influence and/or capture certain markets.

Telecommunication networks spread over territories connecting end users with the telecommunication services being delivered. Telecommunication networks are hierarchically organized, differentiate network areas and define system entities.

Traditional telecommunication networks were organized around the concept of backbone and access network areas. Backbone was used to refer to the long-distance point-to-point connectivity among network core elements; access network area referred to the last part of the network connecting end users.

In modern telecommunication systems, the term backbone has evolved into core [19–20], indicating the more network nature of this central part of the network (nowadays the name backbone refers to the "spine" connectivity of this core network element). While the name access has remained unaltered, the growth in size of networks and in the variety of services has naturally produced an "aggregation" zone between core and access, known as aggregation, typically at a metropolitan ("metro") level.

However, there is another trend that has fostered the emergence of a new set of terms. The influence of computer networks has coined network terminology around the concept of "area networks". First, the terms local area network (LAN) and wide area network (WAN) were used (LANs were interconnected with the WAN), referring to the more access- and backbone-related network segments, respectively. Eventually, and in parallel to the previous aggregation concept, metropolitan area networks (MANs) emerged to connect LANs in the same urban area, with WAN connecting MANs [21–23]. The evolution of this same idea gave way to all sort of xANs, such as PANs (personal area networks; closer to the customer), HANs (home area networks; inside customer homes), etc. To complicate things further, the letter "W" has been used to mean "wireless" (radio): WPAN, WLAN, etc. [24].

As mentioned before, different knowledge, systems or industry domains have also created their own terminology; specific to the smart grid and probably more important in a US context, the neighborhood area network (NAN) and field area network (FAN) concepts have their own meaning closely related to the telecommunication service being offered. NAN is often associated with smart metering, connecting utility premises with the utility WAN. FAN refers to the transmission and, more often, distribution grid interconnection, connecting substations and the different grid elements, mainly for distribution automation (DA) purposes.

To connect these definitions and make them meaningful for the discussions in this chapter, Figure 13.1 offers a perspective of this framework in the smart grid.

13.3.2 How Is Telecommunications Connectivity Built?

Frameworks are useful, as they allow us to make abstractions of the reality to make it evolve and improve. However, conceptual views need to be connected with the reality to be able to technically specify real networks and systems that are aware of the constraints imposed by the environment they will adapt to. Moreover, this link with the reality is necessary to understand the shortcomings of the real networks that will be built once the concepts produce network elements.

Figure 13.2 shows an example of a layered telecommunication network, with emphasis in its physical components and technologies. At the bottom of Figure 13.2, the physical transmission media allow the transmission of the telecommunication signals with telecommunication network elements, as the building blocks of telecommunication transport networks. In successive layers, different telecommunications technologies, each working in a different functional domain, implement the telecommunication system, providing different services at different levels and for different end users. The topology used by each layer (ring, full mesh, star, point-to-point, etc.) is part of the network design, and the selection depends on the results that need to be obtained (e.g., QoS).

Transport networks implement the point-to-point connectivity of network elements, to connect them over wireline (optical fiber, metallic cables, power lines, etc.) or wireless (radio) media. These transport technologies (in Figure 13.2, DWDM [dense wavelength division multiplexing] and SDH/SONET [synchronous digital hierarchy/synchronous optical network]) have evolved to increasingly allow the transmission of more efficient physical resources use and higher-rate telecommunication signals (bits per second [bps]), with improved quality (lower bit error rates [BER]) and enhanced route-protection mechanisms.

Switching and routing technologies have improved the aggregation and automatic route selection algorithms so that telecommunication services, varied in their nature

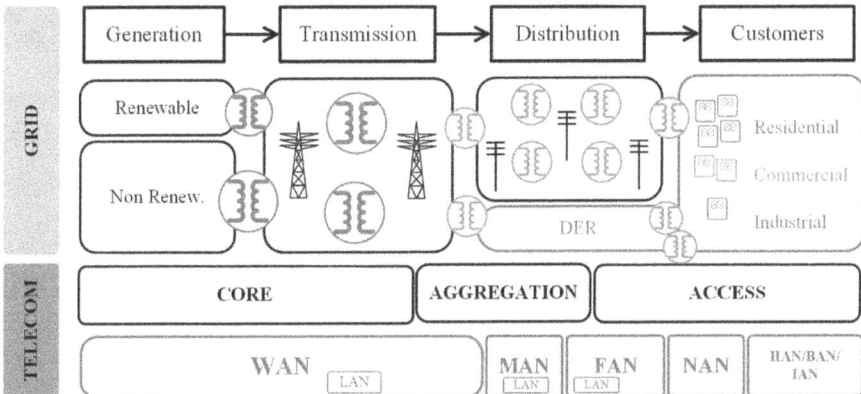

FIGURE 13.1 Smart grids and telecommunications conceptual architectures.

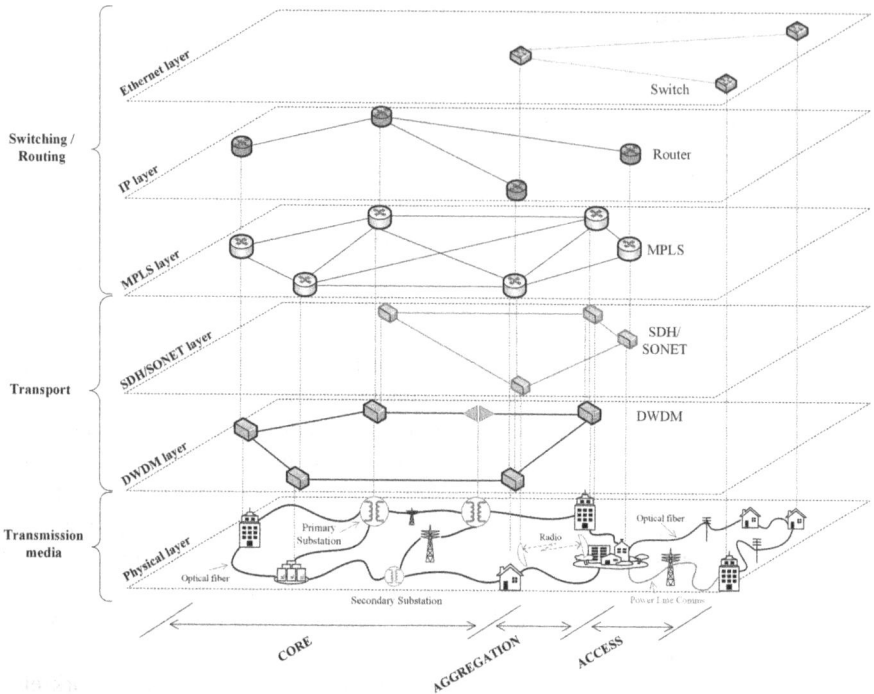

FIGURE 13.2 Telecommunications layered architecture.

(requiring different data rates, latency and QoS) and, carried over a unique telecommunication network, can reach their destination in an automatic way, with the information available within the data included in the information being carried.

In this network representation, it is evident that different physical media are combined and that the information can circulate through different paths to reach its destination. To achieve this, a network design is needed, together with the use of the appropriate technologies that may allow this to happen in real time.

Indeed, the QoS that a telecommunication system can produce does not only derive from the definition of any telecommunication system but from the real network design that is engineered and deployed on the field, the ancillary elements (e.g., battery backup), together with the proper operation and maintenance of the solution. Moreover, in some cases, it will also necessarily take into consideration the regulatory aspects of some telecommunication networks.

13.3.3 THE CORE VS. THE ACCESS CHALLENGE IN TELECOMMUNICATIONS

The telecommunications connectivity challenge usually depends on two major factors connected with the infrastructure supporting telecommunication services: the preexisting availability of public networks, and/or the availability of suitable infrastructure to easily deploy telecommunication networks when required. Often, these two considerations are interconnected; in the case of core networks, the first consideration strongly depends on the regulated mandates to offer telecommunication broadband services in non-commercially attractive areas.

Core networks–related infrastructure is normally associated with high bandwidth connectivity among distant network areas where end users are mostly connected. Very high bandwidth (typically from hundreds of Gbps to Tbps) is provided over long distances (tens of kilometers) with optical fiber–based technologies. Although wireless technologies can provide high-bandwidth solutions, their capacity to provide it reliably over equivalent distances is limited. Thus, while DWDM can provide multiple Tbps over 80 km with two network elements connected to a pair of optical fibers in an optical fiber cable with hundreds of them, radio point-to-point solutions can offer a maximum of a few Gbps in much shorter distances with state-of-the-art solutions. Consequently, long-distance communication can be favored if optical fiber cable–supporting infrastructures can be found. Luckily, there are infrastructures that can offer this support; traditional telephone poles, railway infrastructure, gas ducts infrastructure, motorways' sideways and power line towers are used to host these cables. The extent to which these are used for this purpose today, depends on the telecommunication needs of both commercial telecommunication companies, and on the infrastructure companies owning those assets. For the former, the need might be commercial when the location to get connected with this telecommunication network has a natural commercial interest, or when there is a regulatory mandate to provide this connectivity; for the latter, operational needs requiring highly reliable telecommunication services are usually the driver.

Access networks are typically connected with managing a high density of users concentrated in an area. Many of the locations where these users are connected may not be easily reachable by telecommunication means, and each user needs to get connected independently of where they need the network access. With wireline technologies, telecommunications reachability depends on the ability to lay a cable; apart from the cost of this individual user connection, the physical or property-related difficulties make this aspect instrumental. With wireless technologies, reachability depends on the penetration of the radio signal; depending on the frequency used, line of sight (LOS) may be needed, as physical obstacles prevent a good signal propagation or even basic connectivity.

13.3.4 WIRELINE TECHNOLOGIES–BASED SYSTEMS

Optical fiber–based technologies and power line communication (PLC) are the most future-proof access wireline fixed-network solutions for the smart grid.

Optical fiber is present in all the areas of a telecommunications network, as it is instrumental for core connectivity in long distance. Referring to the smart grid, there is a high synergy between the electricity service, based on power lines (cables) spread wherever the electricity service needs to reach, and carried on top of poles, or buried and, most commonly, ducted when needed, and the ability of this infrastructure to host optical fiber cables. However, the role of optical fiber in systems that take its name to be defined (e.g., passive optical network [PON] systems), is more focused on the access segment of the network, with PONs being used to carry high-bandwidth telecommunication services to end users.

PLC is the natural option for any access technology reaching devices connected to the power grid. Utilities combine optical fiber and PLC solutions as a way to supplement a lower cost to every end user. PLC-based systems use a transmission media (the grid) that is harsh in terms of telecommunication signal propagation, as no significant and consistent barriers have been placed to avoid wanted or unwanted noise signals with origin in electric appliances.

There are other wireline-based technologies that play a role in some implementations of the smart grid. Digital subscriber line (xDSL; copper cables) and hybrid fiber-coaxial (HFC) systems are mostly present in commercial telecommunication services and have been used in the last mile access to end users. However, these technologies are not prevalent in the smart grid domain. Furthermore, xDSL services are being deprecated by systems based on optical fiber, and HFC systems are losing traction in the commercial space in favor of all optical networks.

13.3.4.1 Passive Optical Network Technologies–Based Systems

A PON is a point-to-multipoint topology optical fiber network that minimizes the use of electronic devices in the last mile access domain, to deliver bidirectional high data rate flows from a single network connection point to multiple user end points [25].

PONs mainly consist of optical fiber and splitting/combining components and are different from active optical networks, as electrical power is only required at the transmitter and the receiver. Sometimes, the term optical distribution network (ODN) is also used, although to refer not only to PONs, but to extensions of it or even to essentially active optical networks; it extends the traditional LAN optical networking in the access domain. From a market perspective, the acronym FTTx (Fiber to the x—"x" refers to the access to any type of premise in the network edge; H for Home or FTTH, B for Building, C for Curb or Cabinet, N for Node, O for Office, P for Premises, etc.) is also used and, although it is not only comprehensive of PONs, passive networks are its most common implementation [26].

PONs were first developed in the 1980s, first using optical fiber rings and eventually gave way to optical tree architectures [27]. PON network architectures define an optical line terminal (OLT) at the access network edge, connecting end users' optical network terminals (ONTs; they are also known as ONUs [optical network units]).

PON standards are classified in two families, ITU and the IEEE families (see Table 13.1), promoted by FSAN (full-service access network) [28] and Ethernet Alliance [29], respectively. The different standards have mostly been deployed by telecommunication carriers in their public networks.

PON networks' characteristics can be summarized as follows:

- Ethernet (IEEE 802.3) with 10BASE-T, 100BASE-TX and 1000BASE-T interfaces, and Gbps speeds with reduced latencies can be delivered to OLTs.
- Ranges can be between 20 and 40 km (active optical networks may reach up to 100 km).
- The passive infrastructure facilitates technology upgrades and service enhancements, as higher throughputs can be provided upgrading the electronic parts of the system.

TABLE 13.1
Passive Optical Network (PON) Standards.

Standard	GPON	XG-PON	XGS-PON	NG-PON2	EPON	10G-EPON
Approval date	2003	2010	2016	2015	2004	2009
Family	ITU-T	ITU-T	ITU-T	ITU-T	IEEE	IEEE
	G.984	G.987	G.9807.1	G.989	802.3ah	802.3av
Standards	G.984.1/	G.987/	G.9807.1/	G.989/	802.3ah-2004	802.3av-2009
	G.984.2/	G.987.1/	G.9807.2	G.989.1/		
	G.984.3/	G.987.2/		G.989.2/		
	G.984.4/	G.987.3/		G.989.3		
	G.984.5/	G.987.4				
	G.984.6/					
	G.984.7					
Downstream (Gbps)	2.50	10	10	4x10	1.25	10
Upstream (Gbps)	1.25	2.5	10	4x10	1.25	10
Split ratio[3]	1:64	1:64	1:128	1:128	1:64	1:64
	(128)	(256)	(256)	(256)		1:128

- The powerless infrastructure lowers rollout and maintenance costs, that is implicitly high, as it is common with optical fiber networks. However:
  - Maintenance cannot be performed without disrupting the performance of network users, as they share a common optical fiber infrastructure.
  - Likewise, optical fiber failures will affect users down the feeders.

13.3.4.2 PLC Technologies–Based Systems

PLC is a general term to refer to a unique set of technologies that use the power system cables for telecommunication purposes. PLC, in essence, is as general a term as radio, and since its first experimentations at the end of the nineteenth century, and its application at the beginning of the twentieth century [30], it has found many technology expressions in the different frequency ranges of the power line spectrum, as well as in the different HV (high voltage), MV and LV segments of the grid.

PLC has demonstrated to be a reliable transmission mean in high-power (50 W) long-distance (tens of kilometers with frequencies from 50 to 300 kHz) over overhead HV power lines, a high data rate (tens of Mbps) reliable aggregation technology in urban and suburban MV connectivity (as far as 2 or 3 km in frequencies above 2 MHz and below 30 MHz), an effective smart metering connectivity technology (hundreds of kbps) in LV grids connecting hundreds of smart electricity meters working in a frequency range above 30 kHz and below 500 kHz, and a very high data rate (hundreds of Mbps in the frequency range below the 50 MHz in in-home environments). Some very special ultra-narrowband but very long-distance applications (hundreds of km), complete the multipurpose use of this PLC technology in its different forms.

PLC uses power lines to superimpose a telecommunication signal in parallel to the high-power 50 or 60 Hz electricity signal. From a pure telecommunications perspective, the system operates as an FDM transmission of a 50/60 Hz signal, and the rest of the PLC spectrum uses. PLC is not a core network technology, but it is mostly considered as an access technology for specific segments of the grid.

PLC can only be deployed using the grid architecture and thus follows its structure to define a telecommunications system. Unfortunately, there is no reference architecture in PLC, and the different combinations of the different systems that make use of it, and with other non-PLC technologies, depend on the utility and its grid structure and composition; it often varies across world regions. Thus, the varied combinations of grid segments with narrowband and broadband PLC ranges, configure different PLC network topologies. However, there is a common terminology referring to the PLC systems elements; head-end and repeaters are typically used to refer to the elements at the top of the PLC transmission hierarchy (and connecting to the WAN core network) and the signal extension devices, respectively.

There are several aspects that need to be considered when analyzing different PLC technologies and their applications. Focusing on the application of these technologies for modern smart grids, these are:

- Broadband vs. narrowband. Early PLC networks were designed to use low-frequency carriers (some tens of kHz). The need to arrange compatibility between radio spectrum and the spectrum in the power lines (signal leakage), the amplitude modulated (AM) radio broadcasting band (530 and 1,700 kHz) naturally created two natural and different frequency bands for PLC: narrowband (up to 530 kHz) and broadband (from 1.8 MHz onward).
- Power line voltage level. Grid safety needs to be considered for PLC signal coupling. These aspects (both the size of couplers and the possibility of connecting them) are constraints when finding where PLC devices can be connected.
- PLC standardization. It was common until the twenty-first century not to find PLC standard systems. Thus, literature references often lack consistency when offering reliable sources of technical and field-derived information. A very significant number of grid cables are just accessible to utilities, and the fact that the primary purpose of utilities is not to develop telecommunication systems, has hampered the evolution of PLC technologies.
- Separation of HV, MV and LV grids. PLC signal spectrum transmission has some natural barriers in the grid. While the 50/60 Hz signal smoothly propagates from generation to customers, higher frequency signals get stopped by transformers isolating different voltage sections. This "constraint" is in fact useful from a telecommunications perspective, as the frequency ranges can be reused in the different grid domains while the interference is controlled.

The most common high data rate technologies–narrowband and broadband PLC technologies—with on the field smart grid representation, are shown in Table 13.2 and 13.3, respectively [31].

TABLE 13.2
Narrowband PLC Technologies.

Technology	PRIME	G3-PLC
Standard	ITU-T G.9904	ITU-T G.9903
Band	Cenelec A	Cenelec A
	FCC	FCC
PHY max data rate	128 kbps	45 kbps
	1 Mbps	300 kbps

TABLE 13.3
Broadband PLC Technologies.

Technology/ Standard	OPERA project	IEEE 1901–2010	ITU-T G.hn	HomePlug AV	IEEE 1901.1	IEEE 1901–2020
Band	2–7 MHz 8–18 MHz	2–60 MHz	2–100MHz	2–30 MHz	< 12 MHz	< 100 MHz
Modulation	OFDM	FFT OFDM Wavelet OFDM	FFT OFDM	OFDM	FFT OFDM Wavelet OFDM	FFT OFDM Wavelet OFDM
PHY max data rate	Ten Mbps	500 Mbps	2 Gbps	200 Mbps	Tens Mbps	500 Mbps

13.3.5 Wireless-Based Technologies

Wireless technologies are an instrumental technology to allow quick and effective telecommunication service availability over a wide coverage area. The ability to reach end users depends on the location of the base stations (the elements transmitting and receiving the radio signals, connected to the core network), the frequency of the system and the signal captured by the end user antenna.

Cellular concept-based wireless systems have been widely used in modern systems as the way to overcome the limitations of the spectrum scarcity, through frequency channel reuse. 3GPP (3rd Generation Partnership Project) is the project that unites seven telecommunications standard development organizations and the global wireless industry, to develop cellular systems. These systems have the ambition to provide universal access to a variety of telecommunication services, and their successive generations (2G, 3G, 4G, 5G and eventually 6G, etc.) provide voice and data services, trying to adapt to different general public and industry environments, through technological changes and a non-stop appetite to absorb newer frequency bands. Typically, large, country-wide commercial networks are deployed using 3GPP technologies, and recently they have also been used for operational support in professional activities across different sectors and industries, specifically when spectrum is reserved for such kind of purposes (e.g., Industry 4.0 initiative).

On the other side, and taking inspiration from the Internet of Things (IoT) concept, a set of wireless technologies have appeared, both taking advantage of the opportunity to apply low-cost state-of-the-art narrowband wireless technologies to a plethora of applications that will benefit from a non-wireline ubiquitous connectivity, and boosted by the IoT concept intention of connecting anything with everything. Three dimensions define this set of technologies: range, achievable data rate and type of spectrum used (licensed or license-exempt). Depending on these dimensions different systems can be mentioned: short-range low–data rate technologies (e.g., Bluetooth low energy (BLE), Zigbee, Z-Wave, and Thread [32], for mass-market consumer IoT applications (e.g., healthcare/fitness wearables, smart home, appliances) and industrial applications (e.g., factory automation, critical infrastructure monitoring, asset tracking), high–data rate but short-to-medium coverage range technologies (e.g., Wi-Fi, based on the IEEE 802.11 standards, and the extensions focused on long-range, low-power IoT applications, such as IEEE 802.11ah Wi-Fi HaLow]), wide area coverage unlicensed spectrum-based LPWAN (low-power WAN) technologies (e.g., Sigfox and LoRaWAN, as a very low-cost alternative technology choice to cellular networks), and finally adaptations of cellular technologies to address applications requiring low-latency, high-reliability, and high data rate communications requirements (built upon mMTC [massive machine type communications] services and features, such as EC-GSM-IoT [EC as extended coverage], LTE-M and NB-IoT, known collectively as Cellular IoT [33]).

13.3.5.1 3GPP Systems

The first cellular networks were introduced in the early 1980s, and since then they have followed a ~10-year cycle of technology innovations that have made 1G systems evolve as far as 5G systems today (see Table 13.4), that will be eventually overcome by 6G.

Thus, while 1G systems fundamentally provided narrowband, circuit-switched voice services, 5G systems provide advanced services that are grouped in three domains to cover different general public and business needs. Enhanced mobile broadband (eMBB) services, which are the logical continuation and improvement of mobile broadband services that started to be shyly delivered by 2G networks and in a consistent way from 3G on; mMTC services, enabling the connectivity of low-complexity, low–energy consumption and low-cost devices intended to get to hard-to-reach locations); and ultra-reliable low latency communications (URLLC) services, able to deliver latencies in the millisecond scale or below, together with the necessary reliability mechanisms.

These cellular systems are referred to as 3GPP systems. 3GPP specifications are the complete description of the different generations (from 2G) of mobile communication systems, including radio access, core network and service capabilities, and take the form of technical specifications (normative documents) and technical reports (supporting documentation). Technical specifications are constantly evolving and are delivered in the form of sequential packages called releases every 1.5 or 2 years. 3GPP activity is complemented with the ITU development activity

TABLE 13.4
3GPP Standards Evolution.

3GPP Generation	1G	2G	3G	4G	5G
Technical standards	Multiple incompatible systems (e.g.: AMPS/ NMT/TACS/ etc.)	Several standards: GSM/D- AMPS/ cdmaOne/ PHS	Two main standards: UMTS/HSPA (3GPP) CDMA2000/ EV-DO (3GPP2)	Global standard: LTE (3GPP)	Global standard: NR and LTE Evolution (3GPP)
Main services	Voice	Voice, SMS and low–data rate services	Voice and data services	Mobile broadband (MBB)	Enhanced MBB (eMBB) ultra-reliable low-latency communications (URLLC) massive machine type communications (mMTC)
ITU-R program			IMT-2000	IMT-Advanced	IMT-2020

through IMT (International Mobile Telecommunications) standards (specifically IMT-2000 -3G-, IMT-Advanced -4G-, IMT-2020 -5G- and eventually IMT-2030 -6G-) and the identification of suitable frequency bands for these systems across world regions.

The high-level architecture of a cellular network has evolved (Figure 13.3), and it is named differently in the successive generations. However, it keeps a similar structure with a core network (CN) as the central element to which the radio access network (RAN) connects, to deliver services to user equipment (UE), i.e., any mobile terminal (e.g., smartphone, modem, IoT device) connecting to the mobile network via the radio interface of the appropriate generation version. UEs include a subscriber identity module (SIM), or SIM card, to identify user data (e.g., identities, security keys). The CN connects to external networks and performs most of the smart control functions of the network. The RAN is the flagship of cellular systems, its most visible part, being mainly composed of elements (base transceiver stations [or just base stations], NodeBs, evolved NodeBs [eNodeBs or eNBs], next generation eNodeBs or gNodeBs; the name changes across generations) acting as radio interfaces for UE network access.

Currently, 2G to 5G mobile cellular networks coexist within the mobile network operators' (MNOs) networks. Most of these networks are public in nature, and the spectrum is assigned by governments in auctions or beauty contests. Due to the need to adjust to the different countries' operation license conditions and, burdened by the high costs of running several networks in parallel, most MNOs are in the process of shutting down 3G networks (2G are favored by regulatory institutions due to the high number of users of this technology, whose transition cost would heavily impact the confidence of users in these kinds of public services). 4G and 5G (to a lesser extent) are the networks that are being actively deployed today.

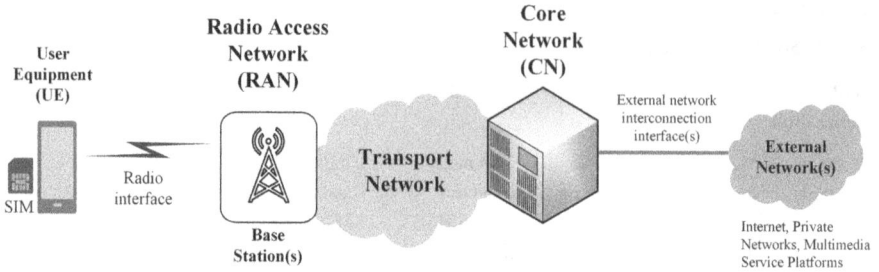

FIGURE 13.3 Simplified architecture of a 4G/5G 3GPP system.

With the recent interest of the private sector in high data rate wireless networks, there has been a dual phenomenon of equipment vendors reaching out to them offering 4G network infrastructure to be deployed and run with exclusive spectrum and the 3GPP focusing on industry expansion (i.e., "verticals"), offering new services that, e.g., include similar functionalities to a LAN within a 5G network [34–36], or the capability to use a public CN to connect private RANs. These are referred to as non-public networks (NPN) in the 3GPP specifications.

13.3.5.2 Wireless IoT Systems

There are three main families of wireless IoT systems relevant for smart grids:

- IEEE 802.15.4–based technologies, Zigbee and Wi-SUN being the most extended technologies.
- Unlicensed spectrum-based LPWAN, LoRaWAN and Sigfox being the more popular options.
- Cellular IoT, with LTE-M and NB-IoT as the most typically present in commercial proposals.

The characteristics of these three families are different and do not share too many commonalities. Table 13.5 summarizes some of their comparable aspects.

Zigbee and Wi-SUN (wireless smart ubiquitous network) are the most prominent examples of IEEE 802.15.4–based technologies, within the realm of the so-called low-rate wireless personal area networks (LR-WPANs) [37]. The IEEE 802.15.4 standard specifies PHY (with a wide variety of configurations, including more than 20 PHY layer options in different frequency bands) and MAC layers for LR-WPANs. Both Zigbee and Wi-SUN extend IEEE 802.15.4 specifying the networking and application layers sitting above the PHY and MAC layers.

Zigbee specifications are developed and promoted by the former Zigbee Alliance, now the Connectivity Standards Alliance (CSA). Wi-SUN is specified and promoted by the Wi-SUN Alliance. Both technologies define profiles to adapt to different markets or services. Thus, in the context of the smart grid, Zigbee offers the core Zigbee (Zigbee 3.0), Zigbee Smart Energy and Jupiter Mesh; Wi-SUN includes a HAN and a FAN profile.

TABLE 13.5
Main Wireless IoT Technologies in smart grids.

Aspect	Zigbee	Wi-SUN FAN	Sigfox	LoRa/LoRaWAN	NB-IoT	LTE-M
Baseline standards	IEEE 802.15.4	IEEE 802.15.4, IETF, ANSI	Proprietary	Proprietary	3GPP specifications	
Channel bandwidth (kHz)	600–5,000	100–600	125	50–125	200	1,400
Spectrum	Unlicensed	Unlicensed	Unlicensed	Unlicensed	Licensed. Some support for unlicensed.	
Frequency bands	Sub 1 GHz and 2.4 GHz	Sub 1 GHz	Sub 1 GHz	Sub 1 GHz	LTE operating bands	
Data rate (kbps)	250	50–300	0.6	0.3–50	200	1,000
Latency (multiples of milliseconds)	10x	10x	1,000x	1,000x	100x	10x

In terms of architecture, both systems work under the PAN concept, with one of the devices generally acting as the PAN coordinator for the rest of the nodes. Each system reserves different names for functions within the PAN, when they are different (e.g., Wi-SUN has a border router, router nodes and leaf nodes).

LoRaWAN and Sigfox[4] are the most market-spread unlicensed spectrum-based LPWAN systems. Using sub-GHz bands, these systems achieve a very good power budget balance to allow narrowband communications to spread over wide territories to devices needing low battery consumption. Both LoRaWAN and Sigfox use proprietary technology that has been well received by the market when no other alternative existed.

Sigfox was developed by a French firm of the same name and is commercially exploited by the Sigfox company itself (sometimes in partnership with other network operators). LoRaWAN makes use of a radio technology known as LoRa (long range) by the chip company Semtech Corporation. Contrary to Sigfox, public or private LoRaWAN networks can be deployed. The LoRa Alliance gathers the industry interest around the LoRa ecosystem.

Both LoRaWAN and Sigfox networks follow a star-of-stars topology. In LoRaWAN, end devices exchange messages with gateways, that relay them to and from a central network server. Sigfox has a core named Sigfox Support Systems and a layer of network equipment acting as base stations to connect to Sigfox devices.

LTE-M and NB-IoT (long-term evolution for machine-type communications, and narrowband Internet of Things, respectively) are extensions of 3GPP cellular technologies intending to provide mMTC services. LTE-M is a backward compatible extension of LTE that started to be developed when GSM/GPRS were still the main cellular technology used for machine-to-machine (M2M) communications in the WAN wireless domain. LTE-M focused on an LTE-based solution evolution, with low device cost and complexity [38]. NB-IoT, as a contrast, is a new and different radio-access technology that can be deployed on existing LTE networks but is not backward compatible. Its development started later, and its design target was to find a cellular solution to compete with LPWAN technologies, specifically LoRaWAN and Sigfox. LTE-M and NB-IoT address different applications: while LTE-M, besides voice communications, allows low data rate of a few Mbps and latencies of tens of milliseconds with handover (i.e., moving from one base station to any adjacent one), NB-IoT just offers a data rate of a few kbps maximum and message delivery times of seconds, with no handover.

Both technologies make use of the baseline LTE network architecture, with some new entities in the RAN (to allow additional radio connections) and in the core (in the evolved packet core [EPC] with the service capability exposure function [SCEF] and the MTC interworking function [MTC-IWF]). Large device numbers and reduced complexity are fundamentally allowed with these new entities, together with several options to deploy these new technologies in existing or new networks.

13.3.6 How Is Telecommunications Connectivity Solved in Transportation Systems?

The use of ICTs is pervasive in our society, and no aspect of it is today understood without their direct or indirect support. Transportation systems also benefit from the use of a wide range of technologies in the ICT domain and within the smart grid framework. This realm is usually known as intelligent transportation (transport) systems (ITSs). The experience gained in this area is helpful to solve ETSs' telecommunications for smart grid challenges. This is the case in some of the overlapping areas, as it will be shown when discussing vehicle-to-grid (V2G) interactions.

An ITS can be defined in different ways:

- As "an operational system of various technologies that, when combined and managed, improve the operating capabilities of the overall system" [39].
- As "a combination of technologies such as computers, telecommunications, positioning, and automation to improve the safety, management, efficiency, usability and environmental sustainability of terrestrial transportation systems" [40].
- As "the use of Information Technology, sensors and communications in surface transport applications—especially roads" [41].

The research [41] on the application of control and information technologies to surface transportation began in the 1950s, mainly in the USA, addressing the automatic control of the automobile and electronic route guidance. Navigation and mapping technologies, dynamic message signs, traffic management centers and global positioning systems were among the initial research vectors.

During the 1980s [39], technology became economically accessible, and based on this, wider availability of smarter solutions for improved traffic management emerged. This decade coined the term intelligent vehicle highway system (IVHS;

the IVHS concept can be considered the predecessor of the current ITS) to describe a group of technologies that included information processing, communications in a broad context, control and electronics connecting vehicles to infrastructure to improve the safety and efficiency of transport systems. At the end of this decade, both the USA and Europe understood the interest of formal programs and legislation to stimulate the progress in this ITS area. The US Congress legislated support for a program in IVHS. The European Commission announced its DRIVE (Dedicated Road Infrastructure for Vehicle Safety in Europe) research program, at a time when the first generation of mobile radio systems were being adopted and the second generation of fully digital cellular systems (e.g., GSM) was being developed.

An important sign of a concept realization and adoption is always when standards emerge. As a relevant milestone in 1996, the US Department of Transportation (DOT) established the ITS Standards Program. Among the relevant standards developing associations, the following can be named: American Association of State Highway and Transportation Officials (AASHTO), American National Standards Institute (ANSI), American Society for Testing & Materials (ASTM), Institute of Electrical and Electronics Engineers (IEEE), Institute of Transportation Engineers (ITE), National Electrical Manufacturers Association (NEMA), Society of Automotive Engineers (SAE), European Committee for Standardization (CEN [Comité Européen de Normalisation]), European Telecommunications Standards Institute (ETSI), International Organization for Standardization (ISO) and International Telecommunication Union (ITU). Each organization takes a different role, and they generally cooperate to avoid overlapping areas.

Standardization in ITSs [42] covers the following domains:

- Standardization of ITS services architectures.
- Radio communication systems.
- Formats and structure of message systems.
- Security and privacy in the technologies and systems.
- Interfaces and reference points.
- Database and data file structures.

A classification of the different groups of standards can be made as follows:

- Vehicle-to-driver communication.
- Vehicle-to-vehicle communication.
- Vehicle-to-roadside communication.
- Vehicle-to-infrastructure communication.
- Goods and vehicle information.
- Traveler information systems.
- Transport control systems.
- Public transport aspect, including emergency systems.

Due to the evident interest in communications in ITS as the enabler of the interaction among the different elements, a set of V2X (vehicle-to-everything) acronyms try to cover all communication interactions (even beyond the ones listed previously) from the vehicle to the rest of the transportation system elements [43–44]:

- Vehicle to vehicle (V2V). It enables the exchange of data in real time among vehicles to give them a 360-degree representation of their surroundings.
- Vehicle to pedestrian (V2P). It enables the protection of different vulnerable road user groups (e.g., pedestrians and bicyclists) from vehicle actions.
- Vehicle to device (V2D). It enables the exchange of information among any smart devices in the context of the vehicle.
- Vehicle to infrastructure (V2I). It is used for the bidirectional exchange of information between the vehicle and the road infrastructure.
- Vehicle to grid (V2G).[5] It provides bidirectional data exchange between EVs and the smart grid to support electrification. Balancing loads, bill reductions, grid support, etc. are among the possibilities.
- Vehicle to network (V2N). It allows the communication of the vehicle with management systems.
- Vehicle to cloud (V2C). It allows data exchange with the cloud to allow added value applications.

"Communications" are understood here in a broad context, meaning the infrastructure part (the part linked with telecommunications networks covered in previous sections of this chapter) and the protocol part (the intelligence that makes and uses the infrastructure part).

Referring to the infrastructure part, each telecommunication technology can play a different role depending on the target connectivity needed. Focusing on the use cases behind each group of standards, there are some that may naturally be connected to a fixed telecommunications infrastructure, while most of them find the most effective telecommunications support in radio communications. Thus, it is not surprising to see the attention paid to this part by both standardization bodies closely connected with radio and the industry groups that focus their attention on these systems. The systems that cover this radio communication need develop their solutions in the following contexts [41] [45]:

- Short-range communications. Specifically, dedicated short-range communication (DSRC) systems for vehicles and road infrastructure short ranges (less than 100 m). There are two types of DSRC, active and passive, where passive means that the vehicle radio does not have an oscillator to generate carrier frequencies and takes it from the roadside radio. The global standards for DSRC are in ITU-R M.1453–2 for 5.8 GHz band. WAVE (wireless access in the vehicular environment) is used in the USA and refers to IEEE 1609 standard series in 5.9 GHz band (the physical layer based in IEEE 802.11p).
- Wide area communications:
  - Broadcast communications. When, like in commercial radio broadcasting, a multiplicity of users does need to be reached unidirectionally,

broadcasting is used. Traditional frequency modulated (FM) broadcast is used to share traffic and travel information. Narrowband channels are used as part of the RDS (radio data system).

- Cellular radio. From 4G onwards, V2X communications have been developed (e.g., LTE V2X and eV2X; both further developed in 5G). Cellular mobile communication networks consist of user equipment (UE), base stations e-Node B, and roadside equipment (RSE). A UE may provide V2V, V2P, V2I, and V2N. Both traditional Uu interface and PC5 interface are used.

13.4 THE CHALLENGE FOR THE SMART GRID TO SUPPORT ELECTRIC TRANSPORTATION SYSTEMS

PEV charging is one of the most important interactions between ETSs and the electricity grid. In this context, PEV charging emerges as the single most important challenge for the grid. According to recent research, there are two main infrastructure enablers of PEV roll-out, the public charging network and the upgrades to the electricity distribution grid segment to cope with peak charging demand (analysts claim that electricity demand will increase between 15% and 25% due to the adoption of EVs in the full-electrification scenario by 2050).

However, the pace at which this will happen is unclear, and is a "chicken and egg" situation: PEV uptake is likely to grow once rapid and ultra-fast chargers become more widespread [46]. This happens because, from a consumer's perspective (private and fleet), charging infrastructure is one of the three main concerns, together with cost and range. And from a wider social perspective, as decarbonization ambitions continue to be tightened, the grid infrastructure build-out may need to happen faster.

Grid expansion needs are driven by the increase in electricity demand, and by its effect in terms of thermal and voltage aspects. Thermal aspects have a connection with the power flow in cables, and the capacity of distribution transformers. Voltage aspects depend on the increase in demand that will make the voltage down a power line decrease, thus dropping voltages below statutory limits. These effects need to be handled differently depending on the charging need (slow or fast): slow charging is handled at the secondary substation level; fast charging needs MV grid upgrades. However, although there are industry estimates that state that the reinforcement needs in the distribution grid may double the investment today, some others consider it a longer-term problem, as grid expansion is to become unavoidable when the PEV penetration reaches 30% [47]. Moreover, the use of ICT-enabled smart charging initiatives will contribute to reduce the problem and even make it become an opportunity.

All grid resources to solve the PEV charging challenge are within the scope of the smart grid [3], that includes the V2G interactions mentioned in ITSs. Traditionally, V2G interactions have been considered out of the core of the smart grid; however, as it has happened in the context of smart meters, the elements in the grid edge become increasingly meaningful for the grid and, similar to what happened with smart metering systems (considered the initial step toward the smart grid, and now considered an integral part of it), the V2G are an integral part of the smart grid concept.

13.4.1 Electric Vehicle Charging

Charging stations are referred to in the literature as EVSE, or electric vehicle supply equipment. Inside the EVSE, the SECC refers to its internal supply equipment communication controller, that interfaces with the EVCC, or electric vehicle communication controller.

PEV charging can be classified, depending on its nature, as private (e.g., if the charging station is at home) or public (if it is similar to the service provided by a gas station). PEV charging can also be classified considering location, which implicitly involves a certain charging regime:

- Single-family home (private parking). Multiple hours per day of charging.
- Multi-family home (private or shared parking). Multiple hours per day of charging.
- Workplace (shared parking). 2–10 hours of charging.
- Destination (public parking). Less than 4 hours of charging.
- On the go (public parking). Less than 1 hour of charging.
- Fleet depot (private parking). Charging dependent on fleet management policies.

There are different sets of charging-related standards for Europe and the USA to allow PEV charging from an electricity service perspective. IEC 61851 series, IEC 62196 (accessories, i.e, sockets, vehicle couplers, etc.; IEC 62196–2 for AC, and IEC 62196–3 for DC, for conductive charging systems as specified in IEC 61851–1) and SAE J1772 are the respective standards for power, voltage and current, including other different aspects [48]. The terminology they use is different (e.g., Europe refers to charging "modes" and the USA to "levels"), but there is in general a close relationship between solutions in SAE and IEC contexts for PEVs.

The USA refers to "levels" [49] and in different releases, their standards have enlarged the portfolio from Level 1 (regular household 120 Vac outlets to deliver maximum power below 1.92 kW, with the AC/DC converter inside the car) and Level 2 (same as Level 1 but for 240 Vac outlets to deliver maximum power below 19.2 kW) to direct DC solutions to avoid the converter in the PEV.

Europe refers to "modes". Some of them are not allowed in certain countries, and in some others, more restrictive areas power limits apply (e.g., country related):

- Mode 1. This method is intended for the connection of a PEV to a standard socket-outlet of an AC supply network, utilizing a cable and plug, both of which are not fitted with any supplementary pilot or auxiliary contacts. Rated values for current and voltage shall not exceed 16 A and 250 Vac (single-phase) or 16 A and 480 Vac (three-phase).
- Mode 2. This method is used for the connection of a PEV to a standard socket-outlet of an AC supply network utilizing an AC EVSE with a cable

and plug, with a control pilot function and system for personal protection against electric shock placed between the standard plug and the PEV. Rated values for current and voltage shall not exceed double the Mode 1 limits.

- Mode 3. This is the method used for the connection of a PEV to an AC EVSE permanently connected to an AC supply network, with a control pilot function that extends from the AC EVSE to the PEV.
- Mode 4. This is the method used for the connection of a PEV to an AC or DC supply network utilizing a DC EVSE, with a control pilot function (a control pilot circuit is the circuit designed for the transmission of signals or communication between the PEV and the PEV supply equipment) that extends from the DC PEV supply equipment to the PEV. It can be either permanently connected or connected by a cable and plug to the supply network.

Wireless charging using electromagnetic waves to charge batteries without physical cables is also being developed fast (it uses either visible or hidden charging pads connected to a fixed grid infrastructure, and a plate attached to the vehicle to charge the battery).

Charging will make use of the available EVSE and PEV standards but needs to be integrated with the wider smart grid. Thus, charging can be classified as unconstrained or as "smart". The unconstrained charging approach is the simplest form of PEV charging, allowing vehicles to get charged without any limitations. This non-restricted charging can be afforded today by the grid, due to the low penetration of PEVs. However, this approach cannot scale when penetration increases, even less depending on how aggressive the PEV adoption curve is.

The "smart" charging approach can also be referred to as "constrained" charging. It is limited by the charging strategies arranged between the electricity provider and the vehicle. The idea of this smart charging would be to compromise on the different interests of the grid and the vehicles' users, to maximize the economic efficiency of vehicle charging while providing a charging user experience within the already commonly accepted standards.

There are some prevalent strategies being implemented, specifically affecting the private charging domain:

- Financial. Time-of-use (TOU) pricing, critical peak pricing and real-time pricing [1] can be used to incentivize users to adapt their PEV charging behavior.
- Direct. Delayed charging. DR programs enabled by smart charging stations.
- Information-based. Helping users with information and signals to help them make informed decisions on the cost and impact of their PEV charge.

Smart charging strategies require communication between the PEV, vehicle owner, charging stations and the distribution system operator (DSO). Pricing, PEV state of charge, power control-related signals, etc. need real-time, reliable and appropriate throughput and latency telecommunications connectivity between each point of the grid where the PEVs may be connecting. With the proliferation of PEVs, this is potentially every grid supply point. Thus, telecommunication technologies, systems or infrastructures that have served their purpose in the traditional smart grid will not be able to offer the needed range of telecommunications services and will progressively give way to others that can cope.

Some authors argue that "the largest impact smart charging will have on the electric grid is associated with the communications requirements" [3]. This statement shows the case of a DSO requiring to centrally monitor and control all PEV users and charging stations to manage its smart grid and the users' behaviors in it.

13.4.2 V2G TECHNOLOGIES

In the more sophisticated and powerful charging options, communications from the PEV to the EVSE is mandated. This is, e.g., the case of IEC 61851–1 Mode 4 method. In this charging mode, digital communication needs be established between the PEV and the DC PEV charging station that validates the DC energy transfer. Interestingly, this digital communication is optional for Modes 1–3 in the same standard (the AC ones).

The digital communication mandated can be found in IEC 61851–24, "Digital communication between a DC PEV charging station and an electric vehicle for control of DC charging" and allows the PEV to control the EVSE. There are two ways to implement the communication between the DC PEV charging station and the PEV: via basic communication and via high-level communications.

- Basic communication is used to manage the key steps in the charging control process, such as start of charging and normal/emergency shutdown, with signal exchange via the control pilot lines in a DC PEV charging system.
- High-level communication is a "digital communication" that will also be equipped in the DC PEV charging station, in addition to basic communication, in order to exchange the control parameters for DC charging between the DC PEV charging station and the PEV (the so-called high level communication, or the digital communication as described in the ISO 15118 series). The following digital communication alternatives can be used (one or the other):
  - Control area network (CAN) over dedicated digital communication circuit according to ISO 11898–1.
  - Power line communications (PLC) over control pilot circuit, specifically Homeplug Green PHY v.1.1. (see SAE J2931/4 [broadband PLC communication for plug-in electric vehicles] and ISO 15118–3).

The ISO 15118 series, named "Road Vehicles—Vehicle to grid communication interface" is the international standard that describes the digital communication protocol for a PEV and the EVSE communications (the EVCC and the SECC modules, respectively). ISO 15118 covers a comprehensive set of charging-related use cases spanning across wired (AC and DC) and wireless charging applications in different types of PEVs (normal vehicles, buses, etc.). These use cases are applicable both for

EVSE to charge the PEV battery or for the PEV battery to supply energy through the EVSE to the home, to loads or to the grid. It is definitely the V2G standard and holds the intention to offer an interoperable PEV–EVSE interface to all e-mobility actors beyond the SECC, including utilities under the DSO concept. Thus, e.g., it includes utility-specific requirements, such as "power limiting for grid control or local energy control". This capability will be used to minimize overloading situations and local perturbations that may happen in LV and MV grids if a large number of PEVs are charging at the same time. With this functionality, the energy transfer schedule and maximum power profile can be controlled.

ISO 15118 provides a general overview and a common understanding of aspects influencing identification, association, charge or discharge control and optimization, payment, load leveling, cybersecurity and privacy. The series is comprehensive of:

- ISO 15118–1:2019. General information and use case definition.
- ISO 15118–2:2014. Network and application protocol requirements.
- ISO 15118–3:2015. Physical and data link layer requirements.
- ISO 15118–4:2018. Network and application protocol conformance test.
- ISO 15118–5:2018. Physical and data link layer conformance tests.
- ISO 15118–8:2020. Physical layer and data link layer requirements for wireless communication.
- ISO/DIS 15118–9. Physical and data link layer conformance test for wireless communication (under development).

13.5 THE ELECTRIC TRANSPORTATION SYSTEM-READY SMART GRID

Smart grid–related telecommunication services are different depending on the aspect of the smart grid they need to care about. In a traditional smart grid concept, remote control (i.e., telecontrol) and remote metering (smart metering) are the ones that stand out.

Remote control services are related to DA in the wide context, meaning both the control of all grid substations (primary and secondary) from the energy control centers with their SCADAs, and the automation of the operations of the distribution grid reaching out to the different internal substation components (substation automation [SA]) and the rest of the grid devices outside the substations (field devices such as circuit switches, circuit breakers, reclosers, etc.).

Remote metering means access to smart meters and their control with operations such as tariff changes and internal switches operations (opening and closing energy supply). This domain is applicable to commercial and industrial environments, but more importantly to domestic environments, in such a way that the resources needed for this (AMI [advanced metering infrastructure]) reach out to every single corner of the distribution grid edge.

There are other services that are also delivered in a classical smart grid, and Table 13.6 summarizes the most important ones. Interestingly, this table gathers key telecommunication service real-field delivery characteristics, showing the business objectives realization of real-field deployed networks and systems.

TABLE 13.6
Common Traditional Smart Grid Services Reference Performance.

Service	Bandwidth (kbps)	Latency	Availability (%)	Power supply backup
Advanced metering	10–100	2–15 sec	99–99.99	Not necessary
Distribution automation	9.6–100	100 ms–2 sec	99–99.999	24–72 hours
Demand response	14–100	500 ms–several minutes	99–99.99	Not necessary
Distributed generation	9.6–56	20 ms–15 sec	99–99.99	1 hour
Plug-in electric vehicle	9.6–56	2 sec–5 min	99–99.99	Not necessary

Thus, although the literature usually tends to show telecommunication services characteristics from the perspective of either the user initial theoretical or desired requirements, and the theoretical technological capabilities of the different systems, real-field solutions offer results that are acceptable for smart grid services as a compromise among different factors, including technology, cost and time-to-market solutions.

Table 13.6 includes telecommunications' defining aspects that are instrumental to select the right telecommunication solution. On the one side, bandwidth and latency characteristics determine the technology or group of technologies that can be used for each service. On the other, availability depends greatly on the network design, and SLAs when commercial services are used. Power supply aspects are associated with availability aspects, but mostly affect the cost of the ancillary elements deployed to complete a telecommunications solution.

Interestingly, Table 13.6 refers to the PEV. This reference needs to be understood in the context of a very low PEV penetration, and where they are not predominantly charged in the LV grid. Moreover, if we observe the comparison with the rest of the services, the similarity with AMI can be appreciated. This is clearly not the case for the smart grid needed for ETSs, where PEV penetration in the grid is pervasive.

13.5.1 Availability, Reliability and Resilience

Two different concepts mix when referring to resiliency in telecommunications: reliability and resilience. Before we refer to them, we need first to focus our attention on the availability concept, prevalent in the measurement of telecommunications service conditions.

Availability is a formal concept defined in ITU-T E.800. It is the ability "of an item to be in the state to perform a required function at a given instant of time or at any instant of time within a given time interval, assuming that the external resources, if required, are provided". A network needs to be "available" to perform, i.e., to deliver its expected services. These service characteristics can then be observed and measured, to control performance (throughput, latency, etc.).

ITU-T and ITU-R recommendations define availability objectives. However, there is not a commonly accepted reference for smart grid–related telecommunication services, although it is broadly assumed that the strictest smart grid services need 99.999% availability (315.36 seconds of unavailability during the year), and that 99% is the minimum to be accepted for a managed service [50].

In order to have networks and services "available", reliability is to be guaranteed. Reliability is also defined in ITU-T E.800. It is "the probability that an item can perform a required function under stated conditions for a given time interval". Reliability is often associated with resiliency, but it is different. Resiliency is "the ability of an entity, system, network or service to resist the effects of a disruptive event, or to recover from the effects of a disruptive event to a normal or near-normal state" [51]. While reliability has to do with network design trying to achieve always-on services, resilience focuses on recovering systems, networks or services when they fail.

Thus, keeping telecommunications services available involves carrier-grade (carrier-class) network elements (equipment with redundancy in its vital operational pieces), reliable design of network links (availability assurance), protection of connectivity paths through independent redundant routes, licensed spectrum usage (legally protected not to be interfered), battery backup, etc. These aspects can be properly controlled in private networks; they can just be controlled through SLAs in appropriately designed commercial services.

13.5.2 THE EVOLUTION TOWARD THE ETS-READY SMART GRID

If a network is to be designed to cope with a variety of telecommunication services with different characteristics, it will be the most stringent or demanding of them, the one that will force both technologies and system design and network dimensioning. If the network or its services can be expected to evolve (e.g., have new needs, if they are still in their infancy) or expand (be offered more abundantly).

Thus, in a classical smart grid, DA services are the most demanding, and the network technology and design must be prepared to offer them. This is leading to some solutions in the core part of the network that are complemented with their extension toward the edge (i.e., the access network). However, as the edge of DA today reaches in the most favorable case as far as the secondary substations, any "remote control" action into the distribution grid that needs to penetrate into the LV grid where the smart meters and/or the PEV chargers are located, needs an evolution of the access segment with technologies that can offer DA-like services, but more pervasively.

PEV charging, in a context where PEVs' penetration reaches mass market, will need to be controlled in the same fashion as the rest of the grid. It is not only that both the MV and the LV grid will expand, and consequently the need to control them, but also that the added value services that will make D2G services practical, need the same service level as those pertaining to the DA domain.

Thus, AMI will need to evolve in such a way that will adopt characteristics, in terms of its telecommunications aspects, evolutionary from services offered by smart grids for DA remote control and automation services. These will be higher bandwidth, lower latency and highly reliable telecommunication services.

Narrowband telecommunication services will not have a role in the ETS-ready smart grid. Narrowband technologies come associated with high latency and low reliability services that cannot provide bidirectional reliable telecommunications connectivity. Furthermore, these technologies cannot offer wideband telecommunication services and are not capable of progressing the broadband telecommunication technologies used to connect PEV charging points to the grid infrastructure, into the utilities' energy control centers.

Commercially provided telecommunication services will play a role if they are able to offer reliable and resilient services. This is not connected to the technologies to be used but to the network designs and license conditions allowing telecommunication companies to develop new solutions, and not just consider smart grid services as those comparable to any other "vertical" industry. The synergies that can be found between utilities and telecommunication companies are evident when one thinks in terms of the foundations of both businesses: the need for infrastructure to support optical fiber–based solutions and the need to have a reliable energy service where network infrastructure is deployed may find common needs and opportunities [52].

Power line communications are a great opportunity for utilities to expand their services in the LV grid. While optical fiber solutions offer a future-proof solution due to their ability to cope with the incremental bandwidth needs, PLC allows these to happen with a very reduced investment need and in an integrated way with the existing grid infrastructure. While part of the grid will be expanded with new MV sections for the new connections that will help new optical fiber cables to be laid in a cost-effective manner, the areas where PEV will connect without this grid enlargement (e.g., the LV grid) will not have this possibility. If we consider that many of these LV grid connection points are not easily accessible through wireless means, and that grid connectivity will be always present, PLC becomes a by-default solution.

Eventually, when PEV become widely spread, the AMI concept will vanish. In terms of the likely evolution, the technologies already deployed in the field will in some cases be applied closer to the edge, while the telecommunications connectivity conquers the grid edge with its expansion from the core. AMI as we know it today will ultimately be abandoned when the natural amortization periods of current smart meters (12–15 years is a common figure) will allow their replacement with broadband-connected evolved smart meters, within high bandwidth, low latency and high reliability telecommunications networks to eventually offer functions inherent to PEV chargers today.

13.6 SUMMARY

Electricity provided through power systems is instrumental in modern societies. ETSs are an example of this, adopting electricity as the foundation of evolved transportation systems.

Electric power systems are being enhanced with the smart grid concept. Specifically, the changes in the distribution grid are one of the most important aspects of the evolution of power systems into the smart prid. These changes are not only a consequence of the interest of utilities to improve the operation of their grids or their intention to perform it more efficiently in technical and economic terms, but a need of the widespread integration of DERs, including PEVs. The challenge of enhancing the remote monitoring capabilities, the observability and controllability of the grid, is even greater when the injection or demand of energy is difficult to predict both in the time and geographic dimensions, as it is the case of PEVs.

Electric transportation has always been on the horizon of the challenges and opportunities of utilities. Even if the opportunity of enlarging utility grids and prospective business of utilities is attracting the attention of stakeholders (investors among them), the technical challenges are not still fully solved. Not only challenges in purely EV-related technology, but more importantly on the expansion of the grid and operation, at an incrementally quicker pace.

Remote connectivity to every component of the grid is a must in the context of the smart grid. Most utilities are on a journey toward making their grids smarter. Many of them started their evolution with the opportunity that smart metering has offered to monitor the more external part of their grid edge (smart meters). Some of them have developed systems capable of operating those smart meters, and a more reduced number of utilities have taken the opportunity of smart metering programs to increase the penetration of their remote-control systems further down their primary substations into an important percentage of their secondary substations. These utilities have expanded the reach of their telecommunication connectivity and today operate hybrid and complex telecommunication networks, with different degrees of availability, performance and resilience, depending on their local regulations, their smart grid vision and the needs of the different grid-related services. But the challenge of high bandwidth, low latency and high reliability telecommunication services at the customer edge is still to be realized.

Telecommunication networks and services operated by utilities are hybrid in their use of technologies and systems and are also hybrid when it comes to their mix of private telecommunication networks and public commercial services. Different wireline or wireless-based telecommunications adapt to the smart grid and ETSs challenges, both in their technical features and in terms of their cost perspective. Their integration in telecommunications network designs is instrumental to provide the appropriate telecommunication connectivity services that the ETSs alone, and the smart grids as a result of their need to integrate the quick and massive access to electricity services of ETSs, need.

NOTES

1. Demand response (DR) refers to all the short-term activities designed to influence customer use of electricity to produce the desired changes in the grid load shape (pattern and magnitude). It is a term often considered interchangeable with demand side management (DSM), the latter having a wider scope [1].

2. DER (sometimes shortened in literature also as DR—not in this chapter—or distributed resources) is any source of electric power that is not directly connected to a bulk power system, and includes both generators and energy storage technologies capable of exporting active power to an electric power system.

3. Values in brackets show the split ratio supported by the protocols. They do not imply feasibility of the system.

4. At the moment of writing this chapter, Sigfox is facing some business challenges.

5. V1G term is not used in this chapter. However, it is commonly used to refer to PEV being charged from the grid. V2G implies that the PEV may actually provide electricity to the grid.

REFERENCES

[1] Sendin, A., Sanchez-Fornie, M.A., Berganza, I., et al. (2016). Telecommunication Networks for the Smart Grid. London, UK: Artech House.

[2] World Economic Forum (2018). Electric Vehicles for Smarter Cities: The Future of Energy and Mobility (January 2018) [Online]. www3.weforum.org/docs/WEF_2018_%20Electric_For_Smarter_Cities.pdf (accessed 13 February 2022).

[3] Borlase, S. (ed.) (2017). Smart Grids: Advanced Technologies and Solutions, 2nd ed. Boca Raton: Taylor & Francis Group, CRC Press.

[4] Hatziargyriou, N., Poor, H.V., Carpanini, L., and Sanchez-Fornie, M.A. (2016). Smarter Energy: From Smart Metering to the Smart Grid. London: The Institution of Engineering and Technology.

[5] Madrigal, M., Uluski, R., and Mensan Gaba, K. (2017). Practical Guidance for Defining a Smart Grid Modernization Strategy: The Case of Distribution, Revised ed. Washington, DC: The World Bank.

[6] E.DSO (2020). European technology platform for smartGrids. About Us. www.edsoforsmartgrids.eu/policy/eu-steering-initiatives/smart-grids-european-technology-platform/ (13 February 2022).

[7] US Department of Energy (2020). About Us. www.energy.gov/about-us (13 February 2022).

[8] International Energy Agency (IEA) (2020). About. www.iea.org/about (13 February 2022).

[9] International Energy Agency—IEA (2011). Technology Roadmap: Smart Grids. Paris, France: OECD Publishing.

[10] World Economic Forum (2020). Our Mission. www.weforum.org/about/world-economic-forum/ (13 February 2022).

[11] Franz, O., Wissner, M., Büllingen, F., et al. (2006). Potenziale der Informations-und Kommunikations-Technologien zur Optimierung der Energieversorgung und des Energieverbrauchs (eEnergy). Bad Honnef: WIK.

[12] Park, M.-G., Cho, S.-B., Chung, K.-H., et al. (2014). Electricity market design for the incorporation of various demand-side resources in the Jeju smart grid test-bed. J. Electr. Eng. Technol. 9 (6): 1851–1863. https://doi.org/10.5370/JEET.2014.9.6.1851.

[13] Goel, S., Bush, S.F., and Bakken, D. (eds.) (2013). IEEE Vision for Smart Grid Communications: 2030 and Beyond. IEEE.

[14] GRID 2030 (2003). A National Vision for Electricity's Second 100 Years (July) [Online]. www.energy.gov/oe/downloads/grid-2030-national-vision-electricity-s-second-100-years (accessed 13 February 2022).

[15] World Economic Forum (2017). The Future of Electricity. New Technologies Transforming the Grid Edge (March) [Online]. www3.weforum.org/docs/WEF_Future_of_Electricity_2017. pdf (accessed 13 February 2022).

[16] Sendin, A., Matanza, J., and Ferrús, R. (2021). Smart Grid Telecommunications: Fundamentals and Technologies in the 5G Era. Chichester, UK: Wiley-IEEE Press.

[17] ITU-T Q.9: Vocabulary of Switching and Signalling Terms. www.itu.int/rec/T-REC-Q.9/en (accessed 13 February 2022).

[18] ITU-T Q.72: Stage 2 Description for Packet Mode Services. www.itu.int/rec/T-REC-Q.72/en (accessed 13 February 2022).

[19] Telecom Network Planning for Evolving Network Architectures. Reference Manual Part 1. Draft version 5.1. (30 January 2008) [Online]. www.itu.int/ITU-D/tech/NGN/Manual/Version5/NPM_V05_January2008_PART1.pdf (accessed 13 February 2022).

[20] Telecom Network Planning for Evolving Network Architectures. Reference Manual Part 2. Draft version 5.1 (30 January 2008) [Online]. www.itu.int/ITU-D/tech/NGN/Manual/Version5/NPM_V05_January2008_PART2.pdf (accessed 13 February 2022).

[21] (2000). IEEE 100: The Authoritative Dictionary of IEEE Standards Terms. Standards Information Network, 7th ed. New York: IEEE Press.

[22] Bidgoli, H. (ed.) (2004). The Internet Encyclopedia. Hoboken, NJ: Wiley.

[23] IEEE LMSC (2012). Overview and Guide to the IEEE 802 LMSC [Online]. www.ieee802.org/IEEE-802-LMSC-OverviewGuide-02SEPT%202012.pdf (accessed 13 February 2022).

[24] Hanes, D., Salgueiro, G., Grossetete, P., et al. (2017). IoT Fundamentals: Networking Technologies, Protocols, and Use Cases for the Internet of Things. Indianapolis, IN: Cisco Press.

[25] VIAVI Solutions Inc. www.viavisolutions.com/en-uk/passive-optical-network-pon (accessed 13 February 2022).

[26] Farmer, J., Lane, B., Bourg, K., and Wang, W. (2016). FTTx Networks: Technology Implementation and Operation. Cambridge, MA: Morgan Kaufmann.

[27] Hood, D. and Trojer, E. (2012). Gigabit-Capable Passive Optical Networks. Hoboken, NJ: Wiley Online Library.

[28] Full Service Access Network (FSAN). www.fsan.org (accessed 13 February 2022).

[29] Ethernet Alliance. ethernetalliance.org (accessed 13 February 2022).

[30] Sendin, A., Peña, I., and Angueira, P. (2014). Strategies for power line communications smart metering network deployment. Energies 7 (4): 2377–2420.

[31] López, G., et al. (2019). "The role of power line communications in the smart grid revisited: Applications, challenges, and research initiatives," IEEE Access 7: 117346–117368. doi: 10.1109/ACCESS.2019.2928391.

[32] Chew, D. (2019). The Wireless Internet of Things: A Guide to the Lower Layers. Hoboken, NJ: Wiley-IEEE Standards Association.

[33] Liberg, O., Sundberg, M., Wang, Y.-P. E., et al. (2018). Cellular Internet of Things Technologies, Standards, and Performance. London, UK: Academic Press, Elsevier Ltd.

[34] Chandramouli, D. (2020). 5G for Industry 4.0 (May 2020) [Online]. www.3gpp.org/news-events/2122-tsn_v_lan (accessed 13 February 2022).

[35] Prakash, R. (Qualcomm) (2020). Transforming Enterprise and Industry with 5G Private Networks (October).

[36] 3GPP TR 23.734 v16.2.0 (2019). Study on Enhancement of 5G System (5GS) for Vertical and Local Area Network (LAN) Services (Release 16) (June).

[37] IEEE Std 802.15.4–2020 (2020). IEEE Standard for Low-Rate Wireless Networks (Revision of IEEE Std 802.15.4–2015), pp. 1–800 (23 July). doi: https://doi.org/10.1109/IEEESTD.2020.9144691.

[38] 3GPP TR 36.888 v12.0.0 (2013). Study on Provision of Low-Cost Machine-Type Communications (MTC) User Equipments (UEs) Based on LTE (Release 12) (June).

[39] Ashley Auer, A., Feese, S., Lockwood, S., et al. (2021). History of the Intelligent Transportation System, May 2016; Updated November 2021, U.S. Department of

Transportation, Report No. FHWA-JPO-16–329. [Online]. www.its.dot.gov/history/pdf/HistoryofITS_book.pdf (accessed 13 February 2022).

[40] ITU-R M.1890 (2011) Intelligent Transport Systems: Guidelines and Objectives (April). www.itu.int/dms_pubrec/itu-r/rec/m/R-REC-M.1890-0-201104-S!!PDF-E.pdf (accessed 13 February 2022).

[41] (2015). Intelligent Transportation Systems Report for Mobile. GSMA Connected Living Programme. [Online]. www.gsma.com/iot/wp-content/uploads/2015/06/ITS-report.pdf (accessed 13 February 2022).

[42] Evensen, K. and Schmitting, P. (2014). D3.5b: Standardisation Handbook, iMobility Support (December). [Online]. www.normes-donnees-tc.org/wp-content/uploads/2015/04/D3.5b-Standardisation-handbook.pdf (accessed 13 February 2022).

[43] (2021). Global V2X in Automotive Market, By Communication Type (V2C, V2G, V2P, V2I, V2V, V2D), By Connectivity Type (DSRC Connectivity and Cellular Connectivity), By Offering Type, By Technology Type, By Propulsion Type, By Region, Competition Forecast & Opportunities, 2026. Report ID: 5105827 (August). [Online]. www.researchandmarkets.com/reports/5105827/global-v2x-in-automotive-market-by-communication?utm_source=BW&utm_medium=PressRelease&utm_code=xmh7bx&utm_campaign=1603399+-+Global+V2X+in+Automotive+(V2C%2c+V2G%2c+V2P%2c+V2I%2c+V2V%2c+V2D)+Markets%2c+Competition+Forecast+%26+Opportunities%2c+2026&utm_exec=chdo54prd (accessed 13 February 2022).

[44] Engineering, IT, RGBSI. [Blog] 7 Types of Vehicle Connectivity. [Online]. https://blog.rgbsi.com/7-types-of-vehicle-connectivity (accessed 13 February 2022).

[45] (2021). Handbook on Land Mobile: Volume 4, 2021 Edition: Intelligent Transport Systems. [Online]. www.itu.int/hub/publication/R-HDB-49-2021/ (accessed 13 February 2022).

[46] (2019). Hurry Up and . . . Wait: The Opportunities around Electric Vehicle Charge Points in the UK. Deloitte, 2019. [Online]. www2.deloitte.com/content/dam/Deloitte/uk/Documents/energy-resources/deloitte-uk-Electric-Vehicles-uk.pdf (accessed 13 February 2022).

[47] Friedl, G., et al. (2018). BLACKOUT E-MOBILITÄT SETZT NETZBETREIBER UNTER DRUCK. [Online]. www.oliverwyman.de/content/dam/oliver-wyman/v2-de/publications/2018/Jan/2018_OliverWyman_E-MobilityBlackout.pdf (accessed 13 February 2022).

[48] Bopp, K., Bennett, J., Lee, N. (2020). Electric vehicle supply equipment: An overview of technical standards to support lao PDR electric vehicle market development, national renewable energy laboratory (NREL). Meeting (September). [Online]. www.nrel.gov/docs/fy21osti/78085.pdf (accessed 13 February 2022).

[49] Hauke Engel, H., Hensley, R., Knupfer, S., et al. (2018). Charging ahead: Electric-Vehicle Infrastructure Demand (August). [Online]. www.mckinsey.com/industries/automotive-and-assembly/our-insights/charging-ahead-electric-vehicle-infrastructure-demand (accessed 13 February 2022).

[50] Gellings, C.W., Samotyj, M., and Howe, B. (2004). The future's smart delivery system. IEEE Power and Energy Magazine. 2 (5).

[51] (2011). Enabling and Managing End-to-End Resilience. Heraklion: Greece [Online]. www.enisa.europa.eu/publications/end-to-end-resilience/@@download/fullReport (accessed 12 February 2022).

[52] Artech House, Alberto Sendin. (2020). [Blog] The Chicken and the Egg. Telecommunications and Electricity in Smart Grids (February). [Online]. https://blog.artechhouse.com/2020/02/04/the-chicken-and-the-egg-telecommunications-and-electricity-in-smart-grids-with-alberto-sendin/ (accessed: 12 February 2022).

14 Smart Control of EV Charging in Power Distribution Grids

Ali Moradi Amani, Samaneh Sadat Sajjadi,
Najmeh Bazmohamadi, and Mahdi Jalili

CONTENTS

14.1 INTRODUCTION

Recent advances in energy generation and storage technologies, such as photovoltaic (PV) solar panels and battery energy storage systems (BESS), have made them cost effective to be installed in residential premises [1]. In this context, the concept of distributed energy resources (DERs) has been emerged whereby small-scale generation units locally supply consumers in the distribution grid [2]. Higher penetration of these technologies in the distribution grid is expected considering the recent commitment by worldwide countries to net-zero carbon emissions by 2050 [3]. This will change the power distribution grid from a 'network of consumers' to a 'network of prosumers' in which residential or commercial entities can export electricity to the grid during some times of the day, e.g., at noon when solar generation is in maximum, while they will be still consumers at night.

Electrification is one of the main strategies toward net-zero emissions. Consequently, the emergence of e-mobility, either in the form of electric public transport or electric vehicles (EVs), expected to be quicker than what distribution network operators (DNOs) were initially expecting [4, 5]. In fact, the high EV penetration will cause a significant increase in demand on the power distribution grid for charging. A UK study has predicted a doubling of electricity demand when EV market share reaches 40%–70%.[1] This brings several concerns for the DNO regarding whether the distribution grid asset, such as transformers, are ready for such a high increase.

In addition, the operation of EVs in the grid as mobile batteries, with rather uncertain and unpredictable charging/discharging process, makes the problem more

DOI: 10.1201/9781003293989-14

complicated. With fossil fuel–based cars, we get used to access of petrol stations at any desired time and get our car full in a few minutes. With this mindset, people may tend to connect their cars to home chargers whenever they want and expect their car to get fully charged quickly. If such behavior happens during peak demand times, a significant demand increase will happen, which may not be tolerable for the distribution grid. In other words, reliability of the distribution grid will be affected since the asset, such as power transformers, are not ready for this demand.

In such a complicated environment, local management of PV, BESS and other energy management systems may help to mitigate undesired effects of EV charging. The concept has been called non-wire alternatives in some literature, see e.g. [6], and aims to efficiently use flexibilities in the grid to accommodate more DERs and reduce planning and upgrade costs. In the case of EV charging, there is often flexibility in the charging timing [7]. It means that the EV is plugged in to the charger for a time slot that is much longer than what is required to get fully charged. Therefore, a smart charging algorithm can assign different charging slots to different EVs and manage the demand. A Dutch analysis shows that 59% of EV charging demand can be delayed for more than 8 hours [8], which means there is a considerable load shifting possibility in the EV charging demand.

14.2 MITIGATING EV CHARGING DEMAND

Emergence of new generation and storage technologies, such as PV solar panels and battery energy storage systems (BESS), has changed the distribution grid environment. Traditional residential and commercial electricity consumers can generate electricity during the daytime and even participate in local markets. Figure 14.1 shows how these technologies change the power consumption profile of households, based on the result of a study in Australia. The net active power consumption data of households is considered based on the data recorded by smart meters of households in Victoria, Australia, from July 1, 2018, to March 30, 2020.

Graphs show that consumption profiles of the households are closer to each other when they do not have PV or BESS technologies. This makes grid management and planning much easier for electricity distributors. However, customers with these technologies may behave very differently. Some of them export significant amounts of power to the grid during the daytime. In the grids with high PV penetration, simultaneous energy export of households may cause overvoltage problems in the grid [9]. Also, consumers' behaviors are different during early evening peak times. Therefore, customer engagement approaches for load shifting and valley filling, such as demand response, is interesting for electricity distributors. Electric vehicles can make this environment more complicated considering the almost unpredictability of the charging time and the customer's mindset to charge their vehicle at any time and at any level they want, in the same way as they have for their combustion engine–based cars. In the worst case, all EV owners in a region may connect their EVs to the grid at the evening peak time when they come back home from work. This means that advanced orchestration algorithms should be developed to let people charge their cars at home while the impact on the distribution grids are minimized.

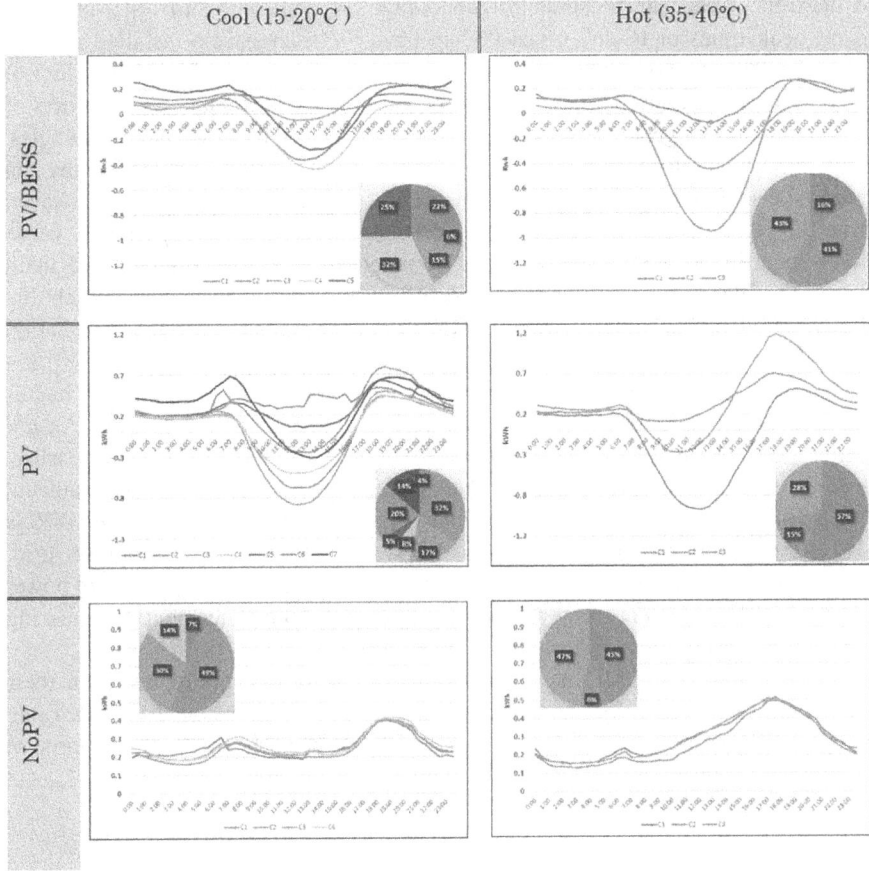

FIGURE 14.1 Consumption profile for Australian households with PV solar panels, PV and BESS, and none of these technologies during cold or hot weekdays.

There are two different approaches in dealing with EV charging demand in distribution grids. The first one is the *smart charging and scheduling*, which controls the charging start time, duration and rate. The second one is to use *BESS to support the grid* during EV charging. Indeed, the extra PV generation during the daytime is stored in the local BESS of the household and is used for EV charging when required.

14.2.1 EV SMART CHARGING AND SCHEDULING

To mitigate the impact of EV charging on the grid, several research activities have been reported. The first group is extracting the load curve for unmanaged charging, which has typically been achieved using the Monte Carlo technique and can be used by DNOs for load-prediction and planning purposes [10]. The second group focuses on demand-side management, using tariffs. Different electricity tariff policies, such as the time of use, can be applied to encourage EV owners to shift their charging to off-peak times or to day times if they have PV solar panels installed in their premises. This group of technics should carefully address the intrinsic conflict of interests between household owners and DNOs in using DERs and EVs. Owners of these technologies generally seek to maximize their benefits from the technologies. They may expect, for instance, to export electricity to the grid as much as they can during PV solar generation, or to have their EVs fully charged all the time. However, from the DNO perspective, maximum electricity export from premises may cause overvoltage in the grid; thus, it should be managed or simply limited. The main objective of demand-side management technics is to tune the tariff policy such that the owners' benefits coincide with the DNO's requirements to keep the distribution grid reliable.

The third group of DER and EV management techniques includes smart scheduling and charging algorithms with several technical and economic benefits. An EV charging process may work based on the first-come-first-served (FCFS) or earliest deadline first (EDF) prioritizing mechanisms. Based on the selected mechanism, scheduling algorithms assign optimal charging/discharging times to either BESSs [11] or EVs [12] for a customer-beneficial EV charging over a reliable power grid. These algorithms can be implemented using peak-shaving and valley-filling of power consumption in different frameworks, such as day-ahead management. Minimizing the energy consumption costs can be another objective.

Smart charging algorithms have one additional design parameter than their scheduling peers. As well as the start time and duration of charging, smart charging algorithms can tune the rate of charge of EVs. It has been shown that this degree of freedom can reduce the number of EV chargers in a charging station [13].

14.2.2 BESS Scheduling and Control

Increasing self-sufficiency of households using rooftop PVs and storage systems is one of the main strategies in facilitating the integration of these technologies in power distribution grids. In this context, intelligent scheduling and control algorithms for BESSs, as well as optimal use of EV charging flexibility, have attracted much attention [10]. Although the algorithms inside these systems optimally use flexibility of household consumers, such as that for EV charging, they need to consider grid stability constraints and electricity costs for a sustainable and beneficial operation. Research studies related to home energy management systems (HEMSs) mainly focus on the benefits of individual households, while those on central algorithms keep grid stability in their mind. This point is where a conflict of interest might happen between household owners and grid operators.

A number of research studies have been reported on BESS scheduling and control considering a single or multiple objectives. Control of BESS is an efficient method to reduce the impact of EV charging on the voltage profile of the grid [14]. Considering the uncertainty in EV charging time and duration, techniques based on forecasting and prediction are of high interest [15]. In [16–18], a hierarchical iterative distributed MPC strategy has been designed such that the state of charge (SoC) of a network of residential-scale batteries is optimally estimated. It has then been used to minimize variability in the accumulated power exchange with the main grid.

14.3 SYSTEM CONFIGURATION

Smart charging algorithms have been implemented in two schemes: centralized and distributed (sometimes called decentralized). In the centralized architecture (Figure 14.2), an EV aggregator is assumed to receive data, such as SoC of the battery, from all EV charging stations. The smart algorithm calculates the optimal charging time and rate by solving an optimization or control algorithm and sends these commands back to the local EV chargers for action. The aggregator has access to data from all EV charging stations, which makes implementation of advanced optimization and control algorithms easier. However, a communication infrastructure is required for this scheme, which adds to the cost. In addition, communications over this network make the system vulnerable to attacks. Aggregators may also need to seek the owners' consent for control and scheduling actions since EV chargers are normally behind the meter.

In the distributed scheme, each premises is equipped with a local intelligent system, and the scheduling/control algorithm is partly run in each of them (see Figure 14.1). This reduces the amount of communications, thus making the system more robust than the centralized case against attacks. The local algorithms can be integrated to HEMSs, which are developed to balance the local supply and demand requirements of each household and maximize the exploitation of renewable energy resources. In this case, each owner can create their own charging schedule and only some settings, for example, limitations on the charging rate, may be forced by the

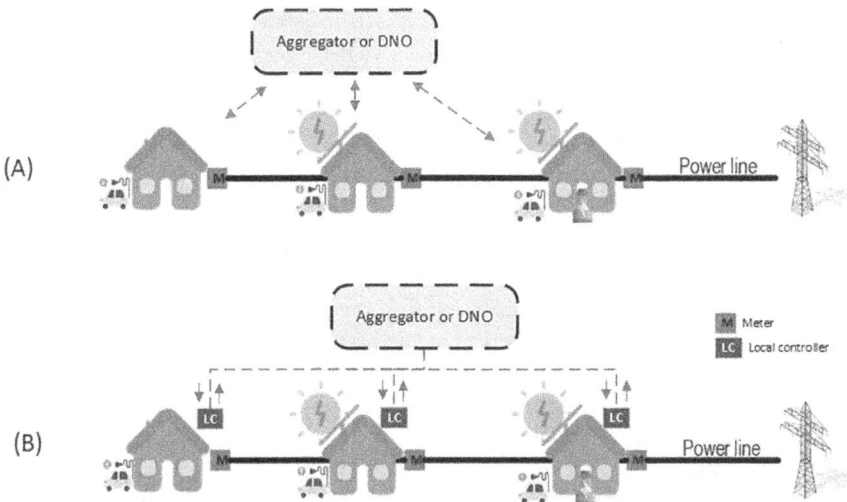

FIGURE 14.2 (A) Centralized vs. (B) Decentralized smart charging configuration.

aggregator or the DNO. Security in keeping privacy of the owners' data is also an important advantage of the distributed scheme. An extreme case of the distributed smart charging algorithms can be implemented when there is no communication with the aggregator or DNO. In other words, owners can flexibly choose any schedule and control setting in a specific range. Despite the benefit of reducing the cost of the communication infrastructure, it is difficult, if not impossible, to guarantee the optimal solution using only local actions.

14.4 PROBLEM FORMULATION AND CONTROLLER DESIGN

A residential power grid including households with different DER technologies, such as PV solar panels, BESSs and EVs, is considered. Households are classified into three groups according to their DERs. Group I includes those with both rooftop PV panels and BESSs, while Groups II and III refer to households with only PVs and conventional houses with neither PV nor BESSs, respectively. It is assumed that EV users are randomly distributed among these groups. According to Figure 14.2, $H(k) = \sum_{i=1}^{I} h_i(k)$ is the aggregated power exchange of the residential area with the main grid in which $I \in N$ shows the total number of households and $h_i(k)$ is the net power consumption of the ith household as

$$h_i(k) = g_i(k) + u_i(k). \tag{14.1}$$

In Eq. (14.1), $g_i(k)$ refers to the power consumption of the ith household minus the power this household exports to the grid. The signal $u_i(k)$ is the BESS rate of charge and can be either positive or negative; positive values show the power that the BESS absorbs for charging, while negative values represent discharging power. Clearly, $u_i(k) = 0$ for Groups II and III of the households, i.e., those without BESS. By efficient control of $u_i(k)$, charging/discharging of BESSs and EVs can be coordinated to reduce the charging load on the grid, i.e., EVs get charged using the energy stored in BESSs. They can also be tuned with the PV power generation profile and consumer load patterns to achieve maximum local consumption and become closer to a microgrid.

Local residential power grids are normally connected to an upstream distribution power system. From a reliability perspective, network operators need to always have a rather precise estimate of power consumption. They also prefer to limit the consumption value such that the stability of the grid is satisfied and power quality is acceptable. From a BESS and EV control and scheduling perspective, this means that the control objective in the residential grid is to make sure that the power exchange with the main distribution grid follows a desired profile, let's say $\hat{\Phi}(j)$. For example, when the total power consumption of the households goes higher than $\hat{\Phi}(j)$, the control algorithm should discharge BESS (and possibly EVs if they are V2G enabled) to reduce the total demand from the upstream point of view. Therefore, we define the following objective function to minimize the violation of the overall network average net load $\bar{\varphi}(k)$ from $\hat{\Phi}(j)$ over the prediction horizon N.

$$min_{\tilde{u}(.)} \sum_{j=k}^{k+N-1} \left(\overline{\varphi}(k) - \hat{\Phi}(j) \right)^2, \qquad (14.2)$$

$$\hat{\Phi}(k) = \frac{1}{I} \sum_{i=1}^{I} h_i(k), \qquad (14.3)$$

$$\overline{\varphi}(k) = \frac{1}{I \times N} \sum_{i=1}^{I} \sum_{j=k}^{k+N-1} g_i(j). \qquad (14.4)$$

To achieve this goal, different operation management strategies can be followed. Here, we assume the configuration of Figure 14.2 (B) and adopt an MPC strategy for the aggregator. The controller is responsible for controlling battery processes such that the desired net profile $\hat{\Phi}(j)$ is followed, even if unexpected load variation, such as significant unmanaged EV charging, happens.

MPC is a model-based control strategy that makes explicit use of the process model to predict the system's output trajectory in response to the future control actions [14]. A typical household with rooftop PV and BESS systems can be modeled as [15],

$$\begin{cases} f_i(k+1) = f_i(k) + T u_i(k), \\ h_i(k) = g_i(k) + u_i(k), \end{cases} \qquad (18.5)$$

where f_i is the SOC of the BSS in kilowatt-hours (kWh) with initial condition $f_i(0) = f_i^0$, and i is the household index. The control signal u_i is the battery charge/discharge rate in kilowatt (kW), and T is the period between two consecutive commands, e.g., $T = 0.5$ for sending commands every half an hour. At every sampling time t, a sequence of control actions $u_i(t+k \mid t)$, k $\{0, K, N\ 1\}$ where N is the control (prediction) horizon is determined by optimizing the operation criterion in Eq. (14.2) under the following constraints.

$$\begin{cases} 0 \le f_i(k) \le C_i, \\ L_i \le u_i(k) \le U_i \end{cases} \qquad (14.6)$$

Here, C_i and U_i are non-negative real constants and demonstrate maximum capacity and rate of charge of the BESS, respectively, and L_i is a non-positive minimum BESS charging rate in each household $i \in \{1, 2, \dots, I\}$. It is commonly recommended to maintain battery energy above 20% of its energy capacity to avoid deep discharge and thereby premature degradation of the battery.

14.5 SIMULATION RESULTS

In this section, MPC has been applied to a residential grid considering real power consumption data related to about 60 households in Victoria, Australia, recorded from July 1, 2018, to March 30, 2020 [16]. The data is collected from smart meters that provide the aggregated power consumption of prosumers/consumers every 30 minutes. The simulations have been done for a residential network, including 60 households, for one week when the temperature was between 20°C and 40°C. Twenty households are from Group I, i.e., with PV and BESS, 20 are from Group II, PV only, and the rest are conventional households with no BESS or PV.

First, we consider the network in normal operation, i.e., without the EV charging load. The objective is to study the performance of the BESS control in the presence of real-world consumption profile data and compare it with the unmanaged BESS charging. Then, in a more complicated scenario, it is supposed that 10 of the consumers have EVs and connect them simultaneously for charging. Two charging times at 11:30 AM and 6:00 PM are simulated, while the charging period is one hour with a charging rate of 5 kW. We consider the root mean square errors (RMSE) for one week as $E = \sqrt{\sum_{j=0.5}^{7\times48} e(j)^2 / (7\times48)}$, where e (j) is the jth instant tracking error. Figures 14.3 and 14.4 show the average load profiles and instant tracking error signals for the network in normal operation and in the presence of EV charging load, respectively. It is illustrated that the MPC strategy provides the less RMSE in tracking the distributors' desired load profile compared to having no control over battery SoC. As it is depicted in Figure 14.4, a big jump in the consumption profile happens at 11:00 AM and 6:00 PM due to EV charging demand. This jump is unexpected for both uncontrolled and MPC-based BESS. Figure 14.4 shows that the MPC strategy performs better in mitigating the unexpected demand.

Figure 14.5 shows the average state of charge and average rate of charge of BESS of Group I households when 10 of them are EV users. As it depicts, when EVs are connected to the network to charge, the battery charge rate becomes negative, which means that the BESSs are discharged to compensate the EV charging load and support the network by reducing the load.

It is worth noting that all BESS systems in our simulations are supposed to be under control, which means the controller's degree of freedom (DOF) is 100%. Now, as the final simulation, we study the impact of DOF on the performance of the MPC. To this end, two networks, 60% and 100% PV penetration, are considered and are called Type I and Type II, respectively, in the rest of this section. We assume that only households with PV are eligible to install BESS. For each of these networks, we increase the DOF and assess the RMSE. For example, 60% PV penetration in a residential network with 60 households means that 36 houses have PV solar panel. Now, 100% DOF means that all 36 houses have also BESS, while 50% DOF means that only 18 of 36 have BESS on their properties. Table 14.1 presents the impact of

TABLE 14.1
Impact of DOF on the Tracking Error in Networks of Type I and II

DOF	10%	50%	70%	80%	90%	100%
E in Type I	1.0172	0.1316	0.1119	0.1943	0.3137	0.1368
E in Type II	2.0928	0.3288	0.1728	0.1998	0.2487	0.2263

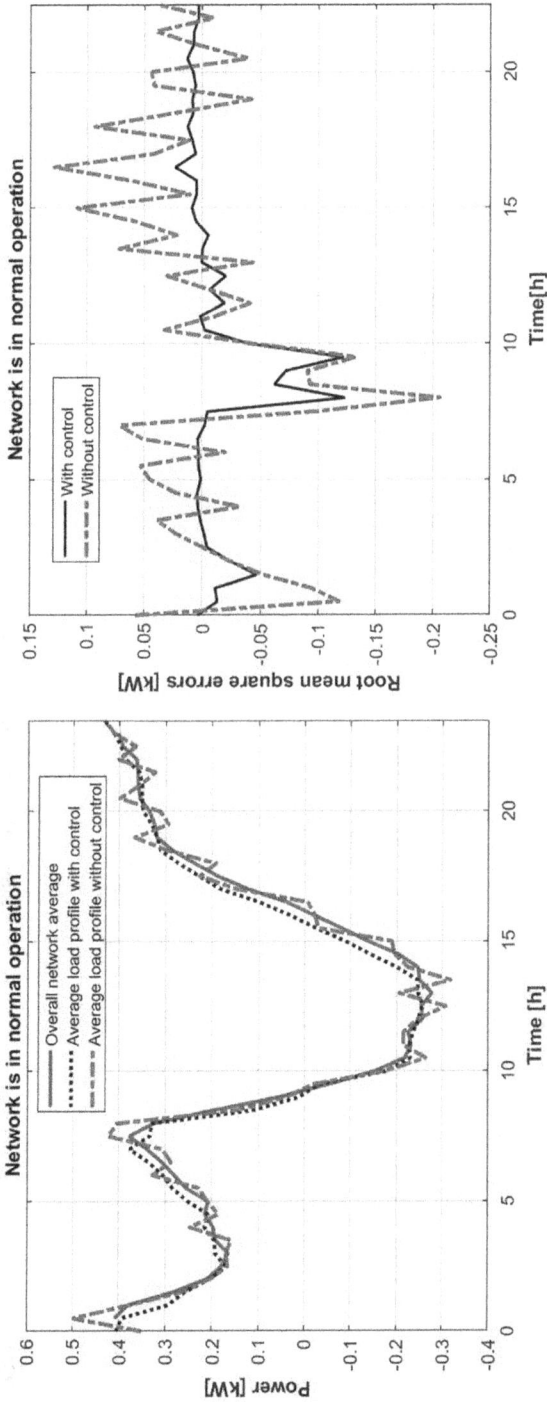

FIGURE 14.3 Average net power profile for 60 residential houses and the tracking error signal for the network in normal operation.

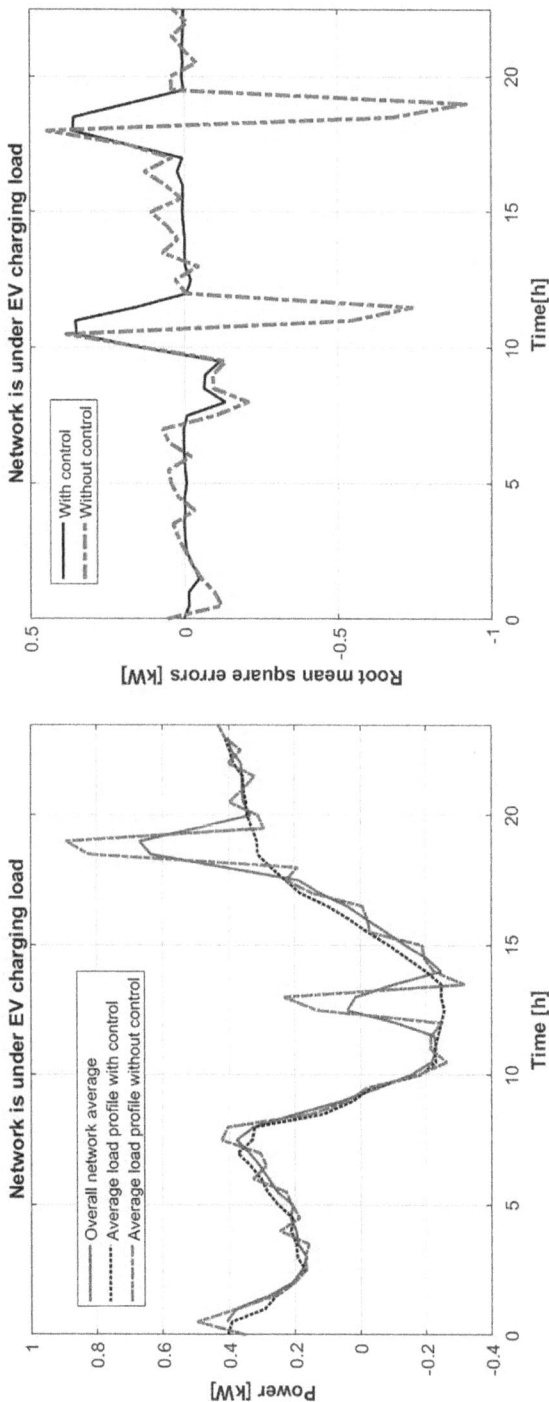

FIGURE 14.4 Average net power profile for 60 residential houses and the tracking error signal for the network under EV charging load.

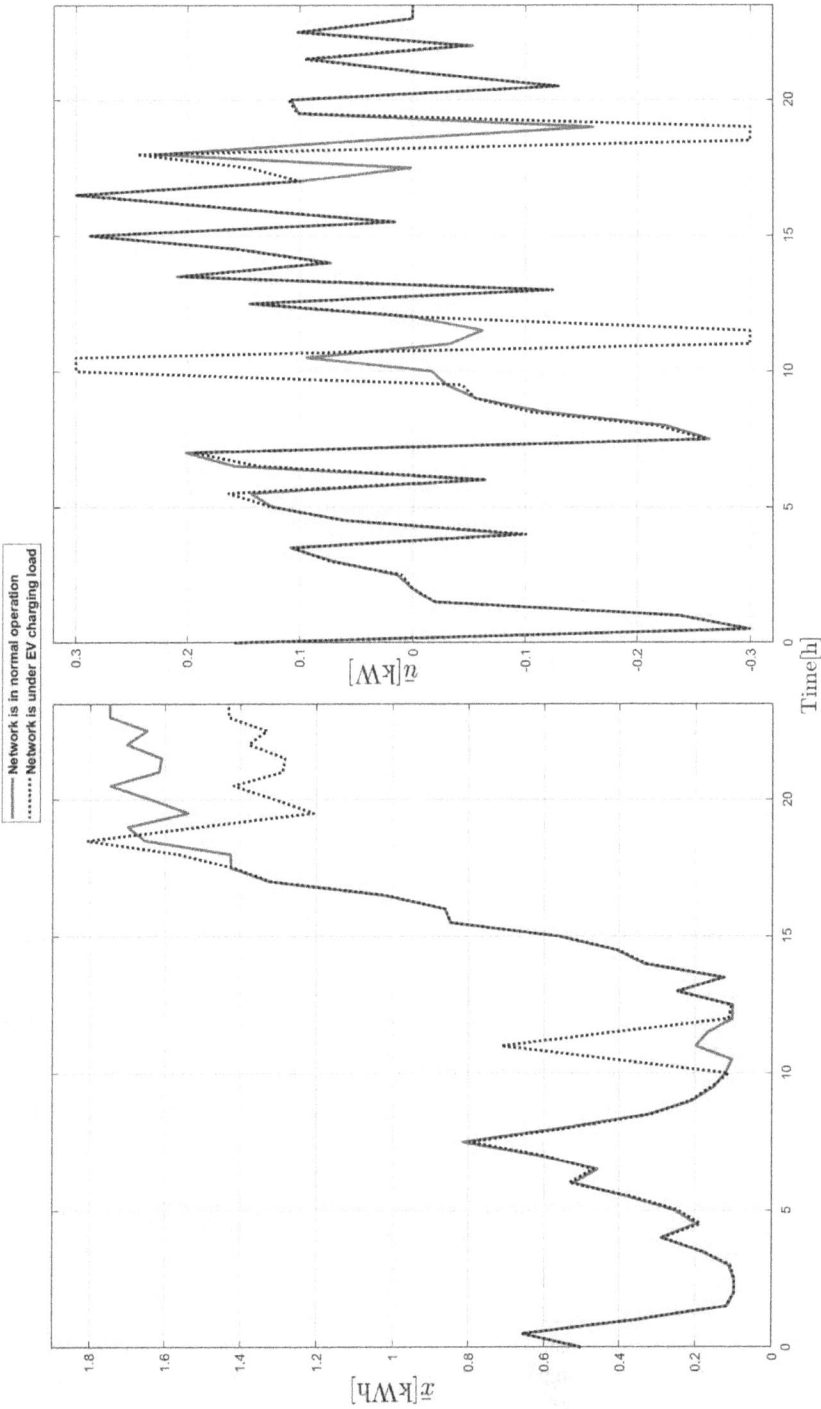

FIGURE 14.5 Average state of charge and average rate of charge of batteries installed in 20 households in which 10 of them are EV users.

DOF on the tracking error in networks without EV charging demand. Interestingly, it shows that the minimum tracking error happens when DOF is around 70%. In other words, increasing the percent of BESS to more than 70% of the PV generation does not result in a better control performance. In other words, increasing the DOF does not necessarily improve the tracking error in a normal operation without EVs.

NOTE

1. *Electricity Statement of Opportunities, A Report for the National Electricity Market.* Available at: https://aemo.com.au/en/energy-systems/electricity/national-electricity-market-nem/nem-forecasting-and-planning/forecasting-and-reliability/nem-electricity-statement-of-opportunities-esoo.

REFERENCES

[1] B. Bai, S. Xiong, B. Song, and M. Xiaoming, "Economic analysis of distributed solar photovoltaics with reused electric vehicle batteries as energy storage systems in China," *Renewable and Sustainable Energy Reviews*, vol. 109, pp. 213–229, 2019.
[2] Z. A. Arfeen, *et al.*, "Energy storage usages: Engineering reactions, economic-technological values for electric vehicles: A technological outlook," *International Transactions on Electrical Energy Systems*, vol. 30, no. 9, 2020.
[3] J. Liu, N. Zhang, C. Kang, D. Kirschen, and Q. Xia, "Cloud energy storage for residential and small commercial consumers: A business case study," *Applied Energy*, vol. 188, pp. 226–236, 2017.
[4] N. Liu, X. Hu, L. Ma, and X. Yu, "Vulnerability assessment for coupled network consisting of power grid and EV traffic network," *IEEE Transactions on Smart Grid*, 2021.
[5] A. M. Amani and M. Jalili, "Power grids as complex networks: Resilience and reliability analysis," *IEEE Access*, 2021.
[6] K. Mahani and F. Farzan, "Unwiring the country: The United States' alternatives today," *IEEE Power Energy Magazine*, vol. 20, no. 2, pp. 14–22, 2022.
[7] F. G. Venegas, M. Petit, and Y. Perez, "Active integration of electric vehicles into distribution grids: Barriers and frameworks for flexibility services," *Renewable Sustainable Energy Reviews*, vol. 145, p. 111060, 2021.
[8] M. K. Gerritsma, T. A. AlSkaif, H. A. Fidder, and W. G. van Sark, "Flexibility of electric vehicle demand: Analysis of measured charging data and simulation for the future," *World Electric Vehicle Journal*, vol. 10, no. 1, p. 14, 2019.
[9] S. Hashemi and J. Østergaard, "Methods and strategies for overvoltage prevention in low voltage distribution systems with PV," *IET Renewable Power Generation*, vol. 11, no. 2, pp. 205–214, 2017.
[10] D. Betancur *et al.*, "Methodology to evaluate the impact of electric vehicles on electrical networks using Monte Carlo," *Energies*, vol. 14, no. 5, p. 1300, 2021.
[11] K. N. Qureshi, A. Alhudhaif, and G. Jeon, "Electric-vehicle energy management and charging scheduling system in sustainable cities and society," *Sustainable Cities Society*, vol. 71, p. 102990, 2021.
[12] K. G. Firouzjah, "Profit-based electric vehicle charging scheduling: Comparison with different strategies and impact assessment on distribution networks," *International Journal of Electrical Power Energy Systems*, vol. 138, p. 107977, 2022.
[13] A. Khaksari, G. Tsaousoglou, P. Makris, K. Steriotis, N. Efthymiopoulos, and E. Varvarigos, "Sizing of electric vehicle charging stations with smart charging

capabilities and quality of service requirements," *Sustainable Cities and Society*, vol. 70, 2021.

[14] K. Wang, W. Wang, L. Wang, and L. Li, "An improved SOC control strategy for electric vehicle hybrid energy storage systems," *Energies*, vol. 13, no. 20, p. 5297, 2020.

[15] T. Morstyn, B. Hredzak, and V. G. Agelidis, "Control strategies for microgrids with distributed energy storage systems: An overview," *IEEE Transactions on Smart Grid*, vol. 9, no. 4, pp. 3652–3666, 2016.

[16] P. Braun, L. Grüne, C. M. Kellett, S. R. Weller, and K. Worthmann, "A distributed optimization algorithm for the predictive control of smart grids," *IEEE Transactions on Automatic Control*, vol. 61, no. 12, pp. 3898–3911, 2016.

[17] K. Worthmann, C. M. Kellett, P. Braun, L. Grune, and S. R. Weller, "Distributed and decentralized control of residential energy systems incorporating battery storage," *IEEE Transactions on Smart Grid*, vol. 6, pp. 1914–1923, 2015.

[18] N. Al Khafaf, A. Asgharian Rezaei, A. M. Amani, M. Jalili, B. McGrath, L. Meegahapola, and A. Vahidnia, "Impact of battery storage on residential energy consumption: An Australian case study based on smart meter data", *Renewable Energy*, vol. 182, pp. 390–400, 2022.

15 Centralized Supervision and Coordination of Load/Frequency Control Problems in Networked Multi-Area Power Systems

Alessandro Casavola, Giuseppe Franzè, and Francesco Tedesco

DEDICATION

Ugait luptatumsan

ex eum ad tat lortie tat acilit velis nis autatum nulput am at volorerit luptation vel delit ip exeros

CONTENTS

DOI: 10.1201/9781003293989-15

15.1 INTRODUCTION

Actual multi-area power systems are large-scale and extremely complex networked interconnected systems. They are composed by generation units, tie-lines and loads organized in a multi-layered hierarchical manner, which exhibits a variety of dynamic behaviors at different time and space scales. A corresponding hierarchical control structure, with many local and some wide-area controllers, is required to effectively handle all adverse phenomena, such as imbalance, fluctuation, disturbances, etc.

Typical local control objectives include [1]: power system protection, power production stabilization, local voltage control and power flow control. These control actions provide basic small-signals compensation and are computed on the basis of local information only. They form the so-called *primary control layer*. On the contrary, wide-area control actions are superimposed to the former and are usually referred to as the *secondary control layer* in power systems jargon. Amongst them, the load/frequency and the regional voltage control problems are the most considered in the literature. As a common feature, they require a regional communication infrastructure to be effective because their actions depend on data spread out in the power grid at faraway locations.

15.1.1 LOAD/FREQUENCY CONTROL

Load/frequency control (LFC) has been one of the earlier wide-area control applications in the power industry. Its basic objective is the global matching between power generation and demand, which must be maintained as close as possible regardless of load fluctuations. If an imbalance results, large frequency deviations may occur, with serious impacts on system operations. Power imbalances are caused by unusual load profiles, fault/failure events on generating units and/or breakdowns of the tie-lines connecting the utilities amongst the areas. All the aforementioned phenomena are hard to anticipate and, if not promptly detected and corrected, may cause massive power shortages or excesses that result in unacceptable frequency deviations exceeding maximum and minimum limits [2]. This is especially true when a massive integration of plug-in electrical vehicles (PEVs) to the grid have to be considered. In fact, their power demand profiles cannot be anticipated, unlike the residence power demand profiles [3], and can produce huge, unexpected power demand changes. This, if not managed, may lead to protection relay activation with the corresponding disconnection of many components of the power grid and the breakup of the entire system into relatively small, disjointed islands.

Typical LFC actions are devoted to keeping the frequency deviations within acceptable and secure limits during these events. However, it is fair commenting that the LFC scope is limited to the stabilization of small imbalances around the nominal equilibrium. In fact, these controllers are not usually capable, for example, of taking the system from a pre-fault equilibrium and steering it into another stable

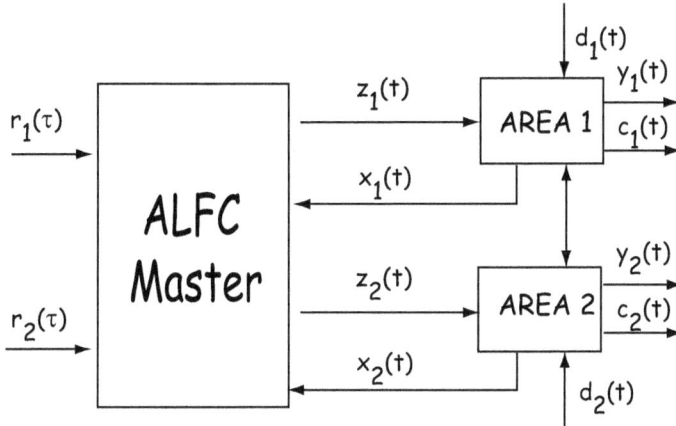

FIGURE 15.1 A distributed tertiary LFC control structure for a two-area power system.

post-fault equilibrium, outside the stability region of the former [4]. Current LFC control architectures become ineffective during extreme events (e.g., huge power demand or tie-lines breakdown) and, as a consequence, undesirable transient phenomena could compromise the system's integrity if the action of the tripping devices would not take place.

Details on LFC can be found, e.g., in [2–5]. In essence, the LFC requirements are satisfied by the so-called ALFC (automatic load frequency control) stage, which is implemented in the primary and secondary control layers. The primary loop provides basic compensation for power production but is not able to compensate for frequency deviations. To this end is devoted to the secondary LFC stage, which collects, via dedicated data links, information on frequencies and power flows at relevant points of the grid and generates, usually in a centralized fashion (in a central station), appropriate set-point adjustments for each generation unit of the multi-area power system.

15.1.2 A TERTIARY LFC SUPERVISING STRATEGY

In this chapter we summarize our past research activity on the study of tertiary LFC supervisory solutions based on constrained control ideas aimed at developing wide-area LFC control schemes for supervising remotely located generation units, connected by distribution lines to remotely distributed loads, under coordination requirements that enforce pointwise-in-time constraints on the evolutions of relevant system variables. In particular, the proposed LFC scheme is able to directly enforce constraints on the maximum allowable frequency and power deviations from nominal values and has some reconfiguration capability under system faults. Moreover, it will be shown in Chapter 16 that this approach is also effective in the presence of time-varying communication time delay and possible data loss. The content of this chapter is mainly based on [7].

We focus on the centralized scenario depicted in Figure 15.1, where the ALFC represents a supervisor in charge of taking decisions based on data coming from the areas via data networks, namely, suitable state vectors $x_i(t)$, where each area in Figure 15.1 represents a micro-grid to be supervised. Primary and secondary standard control structures are still present and are not modified by the addition of this proposed tertiary control level.

In Figure 15.1, the signals r_i, d_i, x_i, y_i and c_i represent, respectively, nominal set-points, loads, states, performance-related and coordination-related outputs for the i-th area. In such a context, the supervisory task can be expressed as the requirement of satisfying some tracking performance, viz. $y_i(t) \approx r_i(t)$, whereas the coordination task consists of enforcing pointwise-in-time constraints of the form $c_i(t) \in \mathscr{C}_i$ on each area and/or $f(c_1(t), c_2(t), \ldots, c_N(t)) \in \mathscr{C}$ on the overall evolution of the power grid. These constraints can be used to impose all safety, operative and coordination requirements that ensure feasible evolutions to the power grids and prevent the actions of the tripping devices.

For each area, the ALFC action consists of a vector $\hat{z}_i := \begin{bmatrix} \hat{g}_i' & \hat{\theta}_i' \end{bmatrix}'$, where $\hat{g}_i$ represents the best approximation of r_i compatible with the previous constraints. The modification of r_i in g_i takes place if the equilibrium corresponding to r_i is not any longer sustainable given the actually changed conditions. On the contrary, if g_i is applied instead of r_i, it doesn't produce constraints violation and allows the overall network to reach a new feasible and sustainable equilibrium. The vector $\hat{\theta}_i$ is an additional optional command. If used, it represents an offset sequence to be added at the input terminal of the generation units in order to enlarge the set of feasible evolutions for that area and for the overall power grid. Finally, we denote with $z_i(t)$ the vector of commands computed and applied at time t.

The previous ALFC scheme makes use of an LTI-TD dynamic model of each area for computing the predictions of future evolutions of the power grid. Moreover, unlike most LFC schemes reported in the literature, here the models have been defined in terms of absolute variables. The motivation of such a choice is related to the possibility of taking advantage of set-point adjustments w.r.t. their nominal values in order to enlarge the set of feasible transients. This is of paramount importance in the presence of faults and/or unexpected large load changes for avoiding the intervention of the tripping devices.

It is expected that the previously described ALFC strategy, in response to critical events, be capable to reconfigure the set-points and the additional offsets for the local primary/secondary control laws in such a way as to maintain the overall system evolutions always coordinated within the prescribed safety and operative constraints.

15.1.3 A Predictive Control Approach

The core of the previous ALFC supervisory strategy is based on predictive control ideas used to synthesize *reference governor* (RG) [8–13] and *parameter governor* (PG) units [14] in more traditional contexts. The idea here is to combine in a single unit both strategies referred to hereafter as *reference-offset governor* (ROG). The basic ROG control structure is depicted in Figure 15.2. There, the ROG unit is an add-on memoryless control device whose action is computed on the basis of the actual reference $r(t)$, current state $x(t)$ and prescribed constraints $c(t)$.

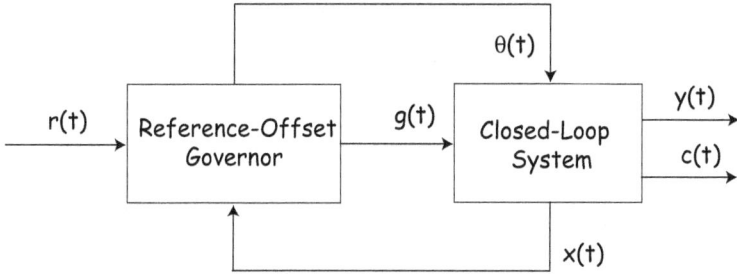

FIGURE 15.2 The reference-offset governor (ROG) © John Wiley & Sons, Ltd.

The aim of the ROG device is to modify, whenever necessary, the reference and the offset to the nominal control law in order to enforce pointwise-in-time constraints. In particular, the constraints are fulfilled via the join actions provided by the parameter $\theta(t)$, which represents an offset on the nominal control input, and the reconfigured reference $g(t)$. The ROG parameters are online computed via a receding horizon minimization of a quadratic selection index: the design methodology follows the same lines of RG [8–13] and PG frameworks [14].

Moreover, in order to take care of the time-varying communication latency from the control center to each area, an extension of this strategy will be presented in Chapter 16, there referred to as *distributed reference offset governor* (DROG), by resorting to the master/slaves approach of [8].

The chapter is organized as follows. In Section 15.2, a two-area power system model is detailed and the LFC problem is formulated. In Section 15.33, the ROG units are discussed and their relevant properties summarized. Section 15.4 reports simulation results of a detailed investigation undertaken both on a two- and four-area power system. Finally, some conclusions end the chapter.

15.2 POWER SYSTEM MODEL AND PROBLEM FORMULATION

In this section, a detailed description of a power network and its traditional LFC scheme is presented. For the sake of simplicity, we consider the two-area power system depicted in Figure 15.3, whose dynamics is described in (15.1).

$$\begin{cases} \dot{x}(t) = Ax(t) + B_1 f_{ref}(t) + B_2 \theta(t) + FP_D(t), \\ y(t) = Cx(t) \end{cases} \tag{15.1}$$

where $f_{ref} = [f_1 re_f f_2 re_f]^T$, $\theta = [\theta_1\ \theta_2]^T$, $y = [y_1\ y_2]^T = [f_1 f_2]^T$, $P_D = [P_D1\ P_D2]^T$, $x = [f_1\ P_T1\ P_v1\ P_c1\ P_{tie} f_2\ P_T2\ P_v2\ P_c2]^T$.

The vectors f_{ref} and θ represent, respectively, the frequency set-points and the additional offset vectors to be added on the nominal control inputs. All other variables are standard, and their descriptions can be found in [15].

The task of the primary/secondary control layers is to maintain the power system within acceptable operating limits by achieving the load balance $P_T \approx P_D$ and maintaining the area frequencies f_i at their nominal values f_{iref}. When the power balance

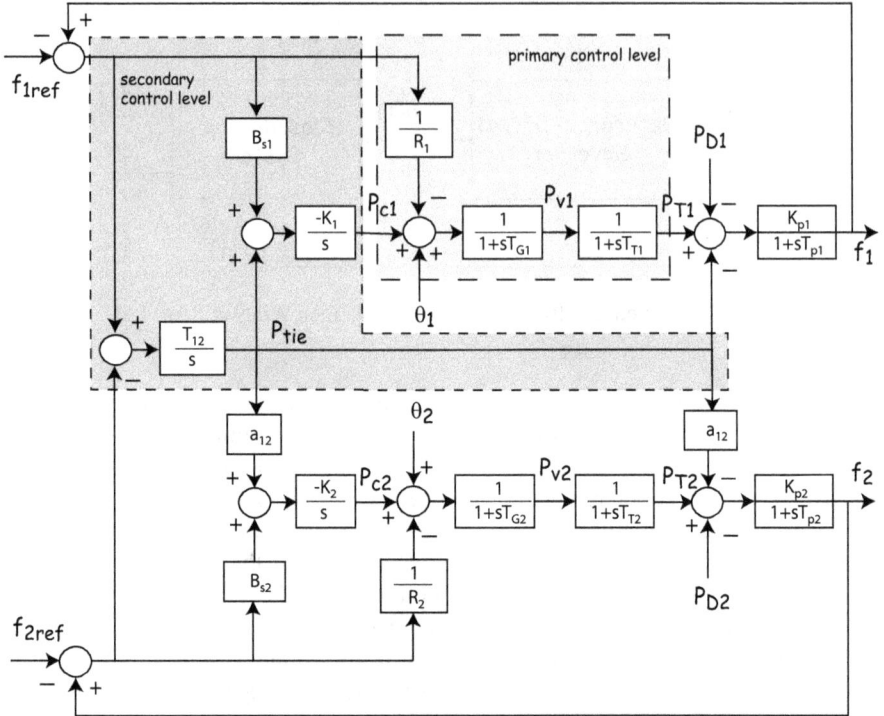

FIGURE 15.3   Block diagram of a two-area power system © John Wiley & Sons, Ltd.

is no longer preserved, frequency deviations arise, which lead to a lower quality of the delivered electrical energy and also to damaging vibrations in the turbines (at frequencies lower than ≈ 57–58 Hz).

The LFC requirements are traditionally satisfied by the ALFC stage (primary and secondary loops) included in the scheme (15.1), which typically take the form of PI local controllers. The purpose of the primary loop is to achieve fast adjustments on the produced power P_T in response to frequency deviations. To this end, the speed governor measures the frequency f and produces a command that causes a new position of the control valve P_V and, in turn, of P_T. In multi-area systems, as that depicted in Figure 15.3, the secondary ALFC loop is based on an error signal, referred to as *area control error* (ACE), which in addition to frequency deviations also contains information on the exchanged power between the areas via the tie-line [16].

The aforementioned ALFC control loops perform in a satisfactory way under small power imbalances arising during normal operating conditions. On the contrary, when the power system is in an emergency state due to a sudden generator loss or to a tie-line disconnection, the ALFC stage is ineffective because turbines have a slow response. In practice, for handling such situations, the power system is equipped with the so-called *emergency control*, whose task is to recover the normal operating state. However, such efforts could fail, and the system could end up in a total blackout.

Therefore, in order to avoid the such a scenario, it is important to define a set of operative and safety constraints that characterize more precisely the *normal conditions* one would like to impose. Then, a tertiary supervisory ALFC strategy will be designed in such a way as to maintain all evolutions of the power network, originating from load changes and/or faults/failures or other adverse conditions as much as possible within the constraints by acting on the nominal set-points and on the offset variables. Moreover, at the supervisory ALFC scheme the determination of a new equilibrium compatible with the changed conditions at a minimum distance from the pre-fault equilibrium is demanded.

Hereafter, the following constraints will be considered

$$f_{min,i} \leq f_i(t) \leq f_{max,i}, i = 1,2, \tag{15.2}$$

$$|P_{tie}(t)| \leq \beta_{max} \tag{15.3}$$

$$\gamma_{min,i} \leq P_{T_i}(t) \leq \gamma_{max,i}, i = 1,2. \tag{15.4}$$

Specifically, constraint (15.2) prescribes bounds on the maximum frequency deviations for the *i-th* area, (15.3) imposes bounds on the power flow exchanged between the areas via the tie-line, while (15.4) limits the maximum generable power in each area. The latter springs from the existing limitations on each generation unit, which can be identified as a saturation on the power production.

In the sequel, the purpose is to show that the action of the proposed tertiary ALFC stage allows one to better face with critical phenomena, avoiding the drawbacks related to the traditional ALFC stage. To this end, the next section is devoted to discussing a predictive control architecture of interest here.

15.3 REFERENCE-OFFSET GOVERNOR (ROG) DESIGN

A ROG control scheme, with plant, primal controller (equipped with an integral action) and ROG device, is depicted in Figure 15.4. Consider the following linear, time-invariant system of the plant regulated by the primal controller.

$$\begin{cases} x(t+1) = \Phi x(t) + G_g g(t) + G_\theta \theta(t) + G_d d(t) \\ \qquad\qquad y(t) = H_y x(t) \\ c(t) = H_c x(t) + L_g g(t) + L_\theta \theta(t) + L_d d(t) \end{cases} \tag{15.5}$$

where $x(t) \in \mathscr{R}^n$ is the state vector (which includes the controller states); $g(t) \in R^n$ is the manipulable reference, which, if no constraints were present, would essentially coincide with the reference $r(t) \in \mathscr{R}^m$; $\theta(t) \in \mathscr{R}^m$ is an adjustable offset on the nominal control input, which we assume be selected from a given convex and compact set Θ, with $0_m \in int\ \Theta$; $d(t) \in \mathscr{R}^{n_d}$ is an exogenous bounded disturbance satisfying $d(t) \in \mathscr{D}, \forall t \in \mathscr{X}_+$ with $\mathscr{D}$ a specified convex and compact set such that $0_{n_d} \in \mathscr{D}$; $y(t) \in \mathscr{R}^m$ is the output, viz. a performance related signal; and $c(t) \in \mathscr{R}^{n_c}$ is the constrained vector.

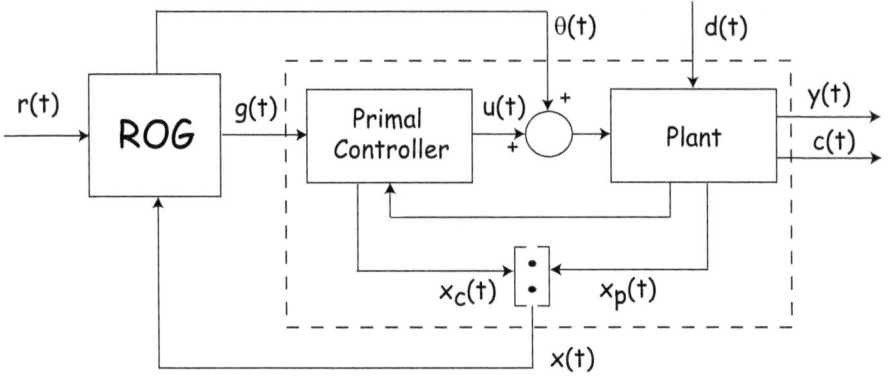

FIGURE 15.4   The ROG control scheme © John Wiley & Sons, Ltd.

$$c(t) \in \mathscr{C}, \forall t \in \mathscr{Z}_+ \qquad (15.6)$$

where $\mathscr{C} \subset \mathscr{R}^n$ is a prescribed convex and compact set. It is assumed that

$$(A1) \begin{cases} 1) & \Phi \text{ is a stable matrix} \\ 2) & \text{System (15.5)is offset-free w.r.t. } g(t) \text{i.e.} \\ & H_y (I_n - \Phi)^{-1} G_g = I_m \end{cases}$$

In the sequel, let us consider the following state description of (15.5).

$$\begin{cases} x(t+1) & = & \Phi x(t) + Gz(t) + G_d d(t) \\ y(t) & = & H_y x(t) \\ c(t) & = & H_c x(t) + Lz(t) + L_d d(t) \end{cases} \qquad (15.7)$$

where

$$z(t) = \begin{bmatrix} g(t) \\ \theta(t) \end{bmatrix} \in \mathscr{R}^{2m}$$

is the ROG output and the following matrices are defined:

$$G = \begin{bmatrix} G_g & G_\theta \end{bmatrix}, L = \begin{bmatrix} L_g & L_\theta \end{bmatrix}.$$

The ROG design problem consists of generating, at each time instant t, the command $z(t)$ as a static function of the current state $x(t)$ and reference $r(t)$

$$z(t) := \underline{z}(x(t), r(t)) \tag{15.8}$$

in such a way that, under suitable conditions, the constraints (15.6) are fulfilled for all possible disturbance sequences $d(t) \in \mathcal{D}$ and possibly $y(t) \approx r(t)$ is achieved.

Moreover, it is required that: 1) $g(t) \to \hat{r}$ whenever $r(t) \to r$, with $\hat{r}$ the best feasible approximation of r and $\theta(t) \to 0_m$; and 2) the ROG have a finite settling time, viz. $g(t) = \hat{r}$ for a possibly large but finite t whenever the reference stays constant after a finite time.

By linearity, one is allowed to separate the effects of the initial conditions and inputs from those of disturbances, e.g., $x(t) = \bar{x}(t) + \tilde{x}(t)$, where $\bar{x}(t)$ is the disturbance-free component and $\tilde{x}(t)$ depends only on the disturbances. Then, we adopt the following notations

$$
\begin{aligned}
\bar{x}_z &:= \left(I_n - \Phi\right)^{-1} Gz \\
\bar{y}_z &:= H_y \left(I_n - \Phi\right)^{-1} Gz \\
\bar{c}_z &:= H_c \left(I_n - \Phi\right)^{-1} Gz + Lz
\end{aligned}
\tag{15.9}
$$

for the disturbance-free equilibrium solutions of (15.5) to a constant command $z(t) = z$. Consider next the following set recursion

$$
\begin{aligned}
\mathscr{C}_0 &:= \mathscr{C} \sim L_d \mathscr{D} \\
\mathscr{C}_k &: \quad \mathscr{C}_{k-1} \sim H_c \Phi^{k-1} G_d \mathscr{D} \\
\mathscr{C}_\infty &:= \bigcap_{k=0}^{\infty} \mathscr{C}_k
\end{aligned}
\tag{15.10}
$$

where $\mathcal{A} \sim E$ is defined as $\{a : a + e \in \mathcal{A}, \forall e \in \mathcal{E}$. It can be shown that the sets $\mathscr{C}_k$ are nonconservative restrictions of $\mathscr{C}$, such that $\bar{c}(t) \in \mathscr{C}_\infty, \forall t \in \mathcal{Z}_+$, implies that $c(t) \in \mathscr{C}, \forall t \in \mathcal{Z}_+$. Thus, one can consider only disturbance-free evolutions of the system and adopt a "worst-case" approach. For reasons that have been clarified in [10], it is convenient to introduce the following sets for a given $\delta > 0$

$$\mathscr{C}^\delta := \mathscr{C}_\infty \sim \mathscr{B}_\delta \tag{15.11}$$

$$\mathscr{W}_\delta := \left\{ z \in \mathcal{R}^{2m} : \bar{c}_z \in \mathscr{C}^\delta \right\} \tag{15.12}$$

where $\mathscr{B}_\delta$ is a ball of radius δ centered at the origin. Let us assume that there exists a possibly vanishing $\delta > 0$, such that $\mathscr{W}_\delta$ is non-empty. In particular, $\mathscr{W}_\delta$ is the set of all commands whose corresponding steady-state solution satisfies the constraints with margin δ. From the foregoing definitions and assumptions, it follows that $\mathscr{W}_\delta$ is closed and convex.

The main idea here is to choose at each time step a constant virtual vector $z(\cdot) \equiv z$, with $z \in \mathscr{W}_\delta$, such that the corresponding virtual evolution fulfills the constraints over a semi-definite horizon; moreover, it is required that the offset on the control law and the distance between the command reference $g(t)$ and the constant reference $r(t)$ are minimal. The ROG commands are applied, a new state is measured and the procedure is repeated. In this context we define the set $\mathscr{V}(x(t))$ as

$$\mathscr{V}(x(t)) = \left\{ z \in \mathscr{W}_\delta : \bar{c}(k, x(t), z) \in \mathscr{C}_k, \forall k \in \mathscr{Z}_+ \right\} \tag{15.13}$$

where

$$\bar{c}(k, x(t), z) = H_c \left(\Phi^k x(t) + \sum_{i=0}^{k-1} \Phi^{k-i-1} Gz \right) + Lz \tag{15.14}$$

is to be understood as the disturbance-free virtual evolution at time k of the constrained vector from the initial condition $x(t)$ at time zero under the constant command $z(\cdot) \equiv z$. Consequently, $\mathscr{V}(x(t)) \subset \mathscr{W}_\delta$, and if non-empty, it represents the set of all constant virtual sequences in $\mathscr{W}_\delta$ whose evolutions starting from $x(t)$ satisfy the constraints also during transients. It can also be shown that such a set is finitely determined, viz. there exists a positive integer k_0, such that (15.13) is identically characterizable by restricting $k \in \{0, \ldots, k_0\}$, with k_0 computable off-line, as described in [11].

The ROG output is based on the minimization of a cost function subject to the constraints imposed by (15.13). The cost function has the following form:

$$J(z, r) = \| g - r \|_{\psi_g}^2 + \| \theta \|_{\psi_\theta}^2 \tag{15.15}$$

where $\psi_g = \psi_g^T > 0_m$, $\psi_\theta = \psi_\theta^T > 0_m$ and $\| v \|_\psi := v^T \Psi v$. Thus, at each time $t \in \mathscr{Z}_+$, the ROG output is chosen according to the solution of the following constrained optimization problem:

$$z(t) := \arg \min_{z \in \mathscr{V}(x(t))} J(z, r(t)). \tag{15.16}$$

The following properties hold true for the previously described ROG.

THEOREM 15.1

Consider system (15.7) along with the ROG selection rule (15.16). Let assumptions (A1) be fulfilled and $\mathscr{V}(x(0))$ be non-empty. Then:

1. The minimizer in (15.16) uniquely exists at each $t \in \mathbb{Z}_+$ and can be obtained by solving a convex constrained optimization problem, viz. $\mathscr{V}(x(0))$ non-empty

implies $\mathcal{V}(x(t))$ non-empty along the trajectories generated by the ROG command (15.16).

2. The set $\mathcal{V}(x(t)), \forall x(t) \in \mathbf{R}^n$, is finitely determined, viz. there exists an integer k_0 such that if $\bar{c}(k,x(t),z) \in \mathcal{C}_k, k \in \{0, 1, \ldots k_0\}$, then $\bar{c}(k,x(t),z) \in \mathcal{C}_k \ \forall k \in \mathbb{Z}_+$. Such a *constraint horizon* k_0 can be determined off-line.

3. The constraints are fulfilled for all $t \in \mathbb{Z}_+$.

4. The overall system is asymptotically stable; in particular, whenever $r(t) \equiv r$, $\lim_{t\to\infty} \theta(t) = 0_m$ and $g(t)$ converges either to r or to its best steady-state admissible approximation $\hat{r}$, with

$$\hat{z} := \begin{bmatrix} \hat{r} \\ 0_m \end{bmatrix} := \arg\min_{z \in \mathcal{W}_\delta} J(z,r) \tag{15.17}$$

Consequently, by the offset-free condition (A1.2), $\lim_{t\to+\infty} \bar{y}(t) = \hat{r}$, where $\bar{y}$ is the disturbance-free component of y. ∎

Proof—A proof was given in [6] and is reported here for completeness. The proof of points 1–3 can be determined by following the same arguments used in the standard RG approach [8–13]. Point 4 first addresses the asymptotic stability of the overall system. To this end, let us assume the reference is kept constant, $r(t) \equiv r, \ \forall t \geq t^*$. Consider the following candidate Lyapunov function:

$$V(x(t)) := \min_{z \in \mathcal{V}(x(t))} J(z,r) = J(z(t),r) \tag{15.18}$$

At time $t + 1$, $z(t)$ is an admissible solution for (15.16). In fact, $z(t) \in \mathcal{V}(x(t+1))$, but $z(t)$ need not be the minimizer for $\min_{z \in \mathcal{V}(x(t+1))} J(z,r)$. Consequently, it results that

$$V(x(t+1)) = \min_{z \in \mathcal{V}(x(t+1))} J(z,r) \leq J(z(t),r) = V(x(t)), \forall t \geq t^*. \tag{15.19}$$

Hence, $V(x(t))$, being non-negative monotonically non-increasing, this implies that

$$\lim_{t\to\infty} [V(x(t)) - V(x(t+1))] = 0,$$

and $V(x(t))$ has a finite limit.

$$V(x(\infty)) = \lim_{t\to\infty} \left[\| g(t) - r \|_{\Psi_g}^2 + \| \theta(t) \|_{\Psi_\theta}^2 \right] \geq 0. \tag{15.20}$$

Note that $\lim_{t\to\infty} \| \theta(t) \|_{\Psi_\theta}^2 = 0$. In fact, a proof by contradiction can be given by following the same lines of [14]. To this end, suppose that $\lim_{t\to\infty} \| \theta(t) \|_{\Psi_\theta}^2 = \theta_{lim} > 0$, with θ_{lim}

being sufficiently large. The idea is to compare the optimal control offset value at time t, $\theta *(t)$, with an alternative solution

$$\hat{\theta}(t) := \theta^*(t) \frac{\theta_{lim}}{\theta_{lim} + \varepsilon}$$

and to show that if $\varepsilon > 0$ is sufficiently small, $\hat{\theta}(t)$ is a feasible choice at time t with a lower value result for (15.15). In fact, let us consider the difference of the cost corresponding to $\hat{z}(t) = [g(t)\hat{\theta}(t)]^T$ and $z^*(t) = \left[g(t)\theta^*(t) \right]^T$:

$$J\left(\hat{z}(t),r\right) - J\left(z^*(t),r\right) = \left(\hat{\theta} - \theta^*\right)^T \Psi_\theta \left(\hat{\theta} + \theta^*\right) = -\varepsilon \frac{\left\| \theta^*_\Psi \right\|}{\left(\theta_{lim} + \varepsilon\right)^2} \left(2\theta_{lim} + \varepsilon\right).$$

Note that, since $0_m \in int\ \Theta$, $\hat{\theta} \in \Theta$ is a feasible choice at time t. Therefore, if $\theta_{lim} > 0$ and $\varepsilon > 0$ are sufficiently small, one has that

$$J\left(\hat{z}(t),r\right) < J\left(z^*(t),r\right),$$

which contradicts the assumption that $\mathscr{I} *(t)$ is the optimal offset at time t. Consequently,

$$\lim_{t \to \infty} \theta(t) = 0_m \tag{15.21}$$

and

$$V(x(\infty)) = \lim_{t \to \infty} \| g(t) - r \| \psi_g. \tag{15.22}$$

From (15.22), one can state that $g(t)$ converges to r, if $V(x(\infty)) = 0$, or to its best steady-state admissible approximation $\hat{r}$, with $V(x(\infty)) > 0$. Moreover, by (15.21) and the offset-free condition (A1.2), it results that

$$\lim_{t \to \infty} \bar{y}(t) = \lim_{t \to \infty} g(t)$$

and this concludes the proof.

15.4 SIMULATIONS

In this section, the aim is at analyzing the behavior of the supervised networked power system and verifying the capability of the proposed tertiary ALFC scheme of reconfiguring the frequency set-points and control offsets in the case of large load deviations from nominal conditions and/or because of faults.

We consider first the two-area power system described in Section 15.2 with the standard primary and secondary ALFC stages acting as primal controller in the ROG framework. Then, the system (15.1) along with the constraints (15.2)–(15.4) can be rewritten as

$$
\begin{cases}
x(t+1) &=& \Phi x(t) + Gz(t) + G_d P_D(t) \\
y(t) &=& H_y x(t) \\
c(t) &=& H_c x(t) + Lz(t) + L_d P_D(t)
\end{cases}
\tag{15.23}
$$

once discretized with period $T_c = 0.1\ s$. In (15.23), $z(t) = \left[f_{ref}^T(t)\theta^T(t) \right]$, $\Phi = \bar{A}$, $G = \left[\bar{B}_1 \quad \bar{B}_2 \right]$, $G_d = \bar{F}$ and $H_y = \bar{C}$, with $\left(\bar{A}, \bar{B}_1, \bar{B}_2, \bar{F}, \bar{C} \right)$ a discrete-time realization of (15.1). In particular, the constrained vector

$$
c(t) = \left[f_1(t) f_2(t) P_{tie}(t) P_{T_1}(t) P_{T_2}(t) \right]^T
$$

is instrumental to describe the constraints (15.2)–(15.4) and is characterized by the following matrices

$$
H_c = \begin{bmatrix} H_y \\ e_5^T \\ e_2^T \\ e_7^T \end{bmatrix}, L_d = 0_{5\times2}, L = 0_{5\times4},
$$

where e_i is the canonical base of $\mathscr{R}^9$.

In all simulations, we have assumed that the nominal frequency of each area is $f_{ire\ f}(t) = 60\ Hz$, $i = 1, 2, \forall t$, [2]. The power demand has the following form: $P_{D_i}(t) = \bar{P}_{D_i} + \tilde{P}_{D_i}(t), i = 1, 2$, where the same nominal load, $\bar{P}_{D_i} = 3MW$, for each area has been considered. On the contrary, $\tilde{P}_{D_i}(t)$ allows one to consider different load variations on each area. It is worth pointing out that each area of the power system in Figure 15.3 has the capability to autonomously balance its own nominal load.

In all simulations, we have considered that the power demands could vary to a certain extent with respect to their nominal values. To exactly characterize the set of admissible load variations $\tilde{P}_D$ (in MW) of interest, a convex and compact region of the following form:

$$
\mathscr{D}_{\mathscr{P}_D} := \left\{ \tilde{P}_D \in \mathscr{R}^2 : U\tilde{P}_D \leq h \right\},
\tag{15.24}
$$

with $U = \begin{bmatrix} I_2 \\ -I_2 \end{bmatrix}$ and h to be specified later, will be used. It is worth pointing out that the bound h represents the maximum disturbances that the ROG is asked to handle.

The following initial state condition $x(0) = [60\ 3\ 0\ 0\ 0\ 60\ 3\ 0\ 0\ 0]^T$ and ROGs' parameters $\delta = 10^{-6}$, $\Psi_g = I_2$ and $\Psi_\theta = I_2$ were chosen. The constraint horizon $k_0 = 300$ was computed via the procedure given in [11].

Three distinct scenarios will be considered in order to analyze the proposed tertiary ALFC scheme, which represents the three most frequent phenomena that have been proven to cause power shortages and possibly system disintegration:

1 Load changes (step changes);
2 Transmission line failure; and
3 Power generation failure.

All simulations have been carried out with Matlab and Simulink.

15.4.1 CENTRALIZED ROG SOLUTIONS

In this first set of simulations, we have considered that the power demand could vary no more than 33% with respect to its nominal value. Consequently, the region of admissible load variations (15.24) is given by

$$h = \begin{bmatrix} 1 \\ 1 \\ 1 \\ 1 \end{bmatrix} [MW]. \tag{15.25}$$

Moreover, the following set of constraints (expressed in Hz and MW)

$$58.5 \le f_1(t) \le 61.5, \quad 58.5 \le f_2(t) \le 61.5,$$
$$|P_{\text{tie}}(t)| \le 0.48, \tag{15.26}$$
$$2.2 \le P_{T_1}(t) \le 3.8, \qquad 2.2 \le P_{T_2}(t) \le 3.8$$

are of interest.

The initial solution of the ROG synthesis problem was determined in $g_i(0) = 60$ Hz, $i = 1, 2$ and $\theta_i(0) = 0$, $i = 1, 2$, which correspond to the nominal equilibrium.

15.4.1.1 Load Changes

The aim of this experiment is to show how the overall power network behaves under the proposed tertiary ALFC scheme in the presence of critical load changes in both areas. The profiles of the considered load demands are depicted in Figure 15.5.

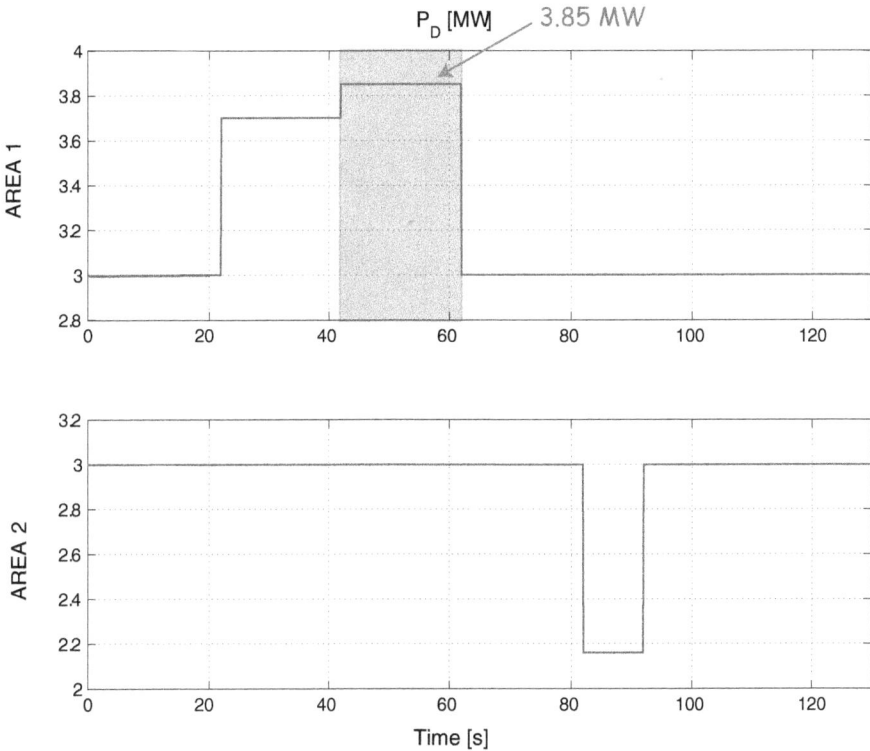

FIGURE 15.5 Load demands.

We have considered first a load variation in Area 1 followed by a load reduction in Area 2. The two variations are not simultaneous for ease of presentation. However, both of them are characterized by a step change aimed at modeling plug-in/plug-out operations of electrical vehicles. It is worth noting that the load demand in Area 1 (starting from $t = 42$ s until $t = 62$ s) is higher than the maximum generable power for that area, whereas in Area 2 (from $t = 82$ s to $t = 92$ s) it is lower (see constraints (15.26)) than the minimum. In principle, such a situation could prevent matching between production and demand. In order to make this more clear, let us consider first the behavior of the overall system when the only standard primary/secondary ALFC scheme is used as a coordination device.

In Figures 15.6 and 15.7, the evolutions of all constrained variables are reported. As it clearly results, a constraints violation follows any occurrence of abrupt load changes. As expected, in order to maintain the power balance, the power generation in each area also tracks the load demand if out of the prescribed limits. It is worth commenting that such limits spring from turbine saturation, i.e., the practical limits on the generation capability of a turbine, and their violation could lead to dangerous evolutions of the power system.

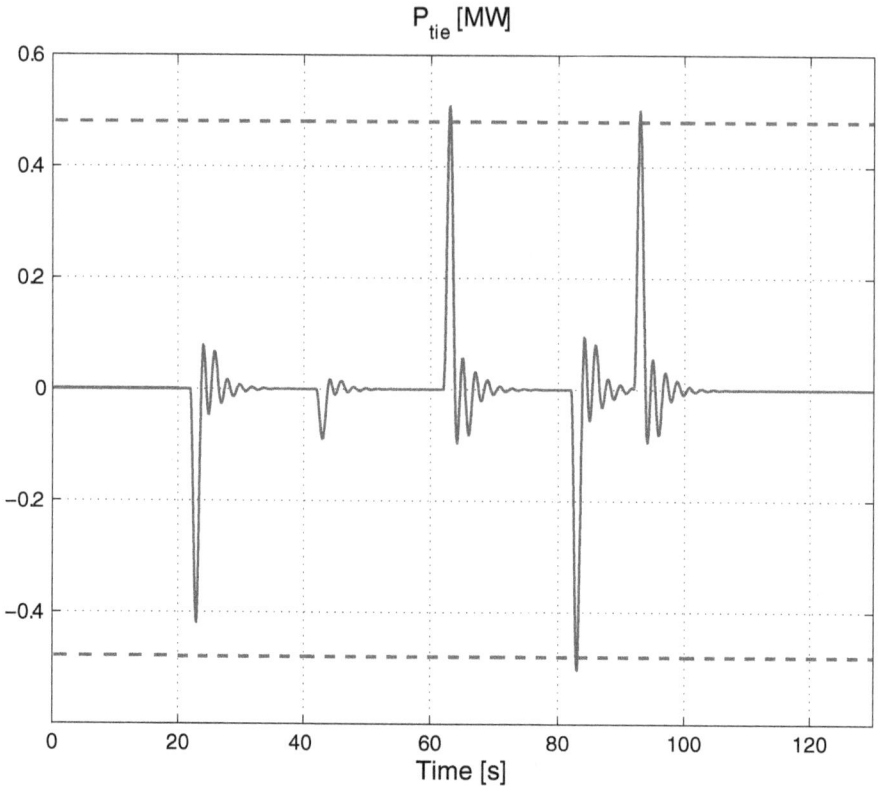

FIGURE 15.6 Tie-line power flow: without ROG. The dashed lines represent the boundaries of the prescribed constraint.

This anomalous behavior is avoided when the ROG unit is added (see Figures 15.8 and 15.9). First of all, observe that all variables are always constrained inside their bounds. Regarding the generation side, the ROG device imposes the power balance without any constraints violation. This is obtained by acting on the tie-line power transfer. In particular, consider the case of a 3.85 MW load demand in Area 1 (from $t = 42$ s to $t = 62$ s). In this time interval, at the steady state, the power generation reaches a value of 3.73 MW and the remaining power required to balance the load is obtained from the generator of Area 2 (Figure 15.8), which exactly increases its production of 0.12 MW, and such a fraction is transferred to Area 1 via the tie-line (Figure 15.9). It is to be pointed out that this new equilibrium has been determined automatically by the ROG unit, and it is the best feasible approximation of the nominal equilibrium given the changed load conditions. Similar remarks can be made for the load reduction in Area 2.

Figure 15.10 shows the ROG actions, viz. how the set-points and offsets have been reconfigured. It is important to note that the control offsets are modified from their nominal values ($\theta_i(0) = 0$, $i = 1, 2$) in order to reduce the frequency deviations.

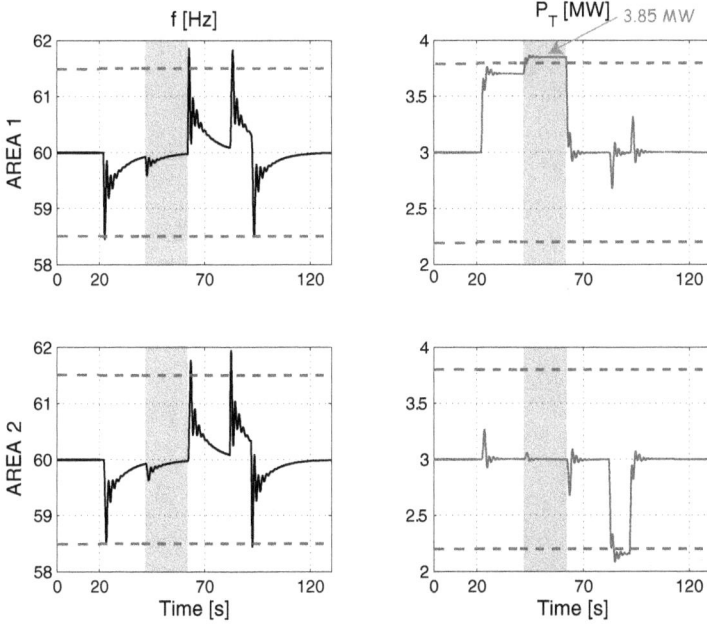

FIGURE 15.7 Frequency (left side) and generated power (right side): without ROG. The dashed lines represent the boundaries of the prescribed constraint.

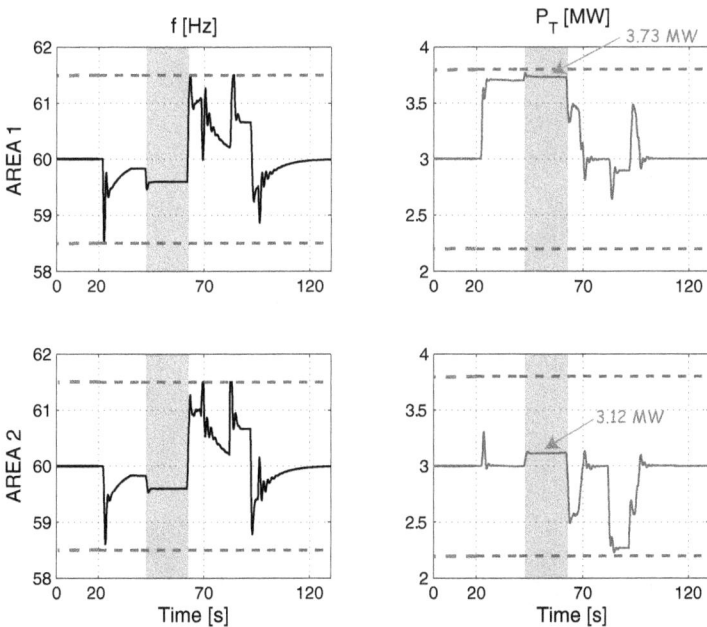

FIGURE 15.8 Frequency (left side) and generated power (right side): with ROG. The dashed lines represent the boundaries of the prescribed constraint.

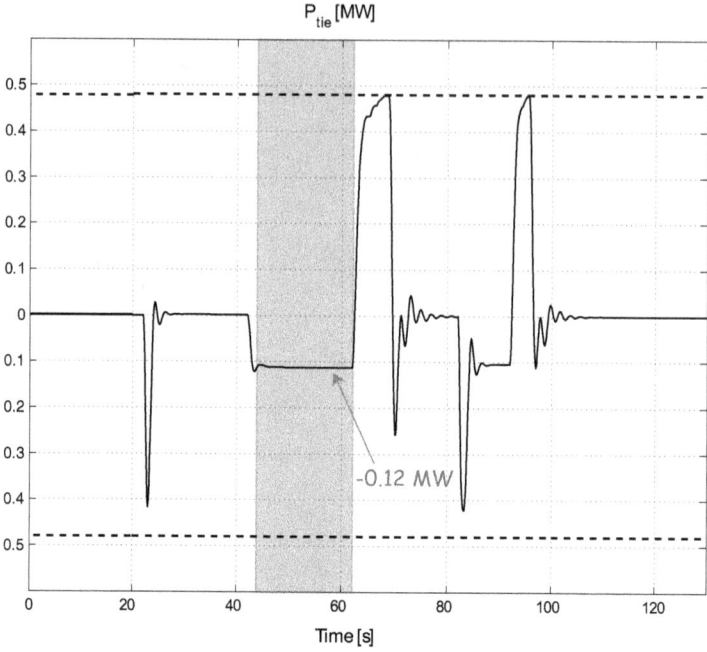

FIGURE 15.9 Tie-line power flow: with ROG. The dashed lines represent the boundaries of the prescribed constraint.

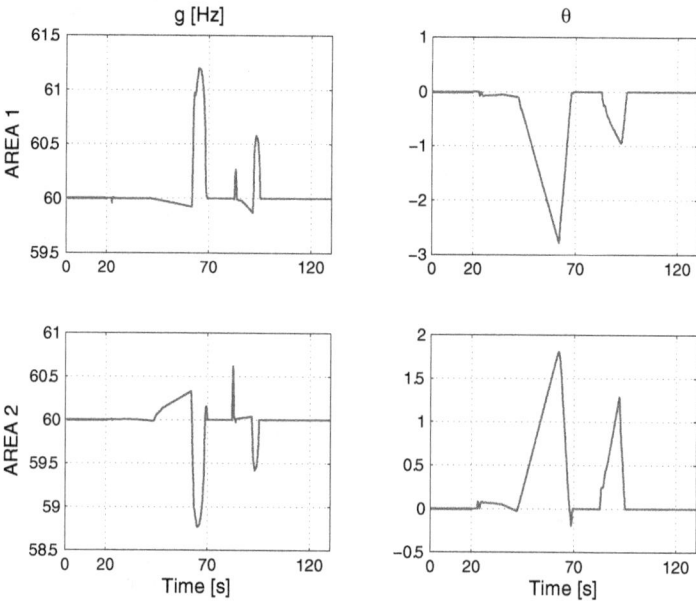

FIGURE 15.10 Outputs of the ROG unit. Reconfigured frequency set-points (left side) and control offset (right side).

15.4.1.2 Transmission Line Failure

In this example, we want to investigate how a failure on the transmission line (e.g., breakdown of the power flow) can affect whole-system evolution. Notice that in this case, the existence of a solution of the ROG synthesis problem cannot be ensured because the experiment is out of the sufficient conditions of Theorem 1. However, it has a solution in many cases, depending on the load profiles and given constraints, and is considered here to highlight the fault-tolerant properties of the proposed tertiary ALFC scheme and its ability at reconfiguring the set-points also in case of faults.

The hypothesis here is that the two areas cannot exchange power due to a failure in the tie-line. When such an event occurs, the two areas are losing the possibility of exchanging power. Such a situation implies that the load scenario of the previous experiment (see Figure 15.5) is not manageable (the ROG problem has not a solution), because the power flow between the two areas is interrupted and the power balance cannot be achieved. Therefore, we consider the new load demand of Figure 15.11.

The c-evolutions, with and without the ROG unit are respectively depicted in Figures 15.12 and 15.13. First, we would point out that the generator units are always

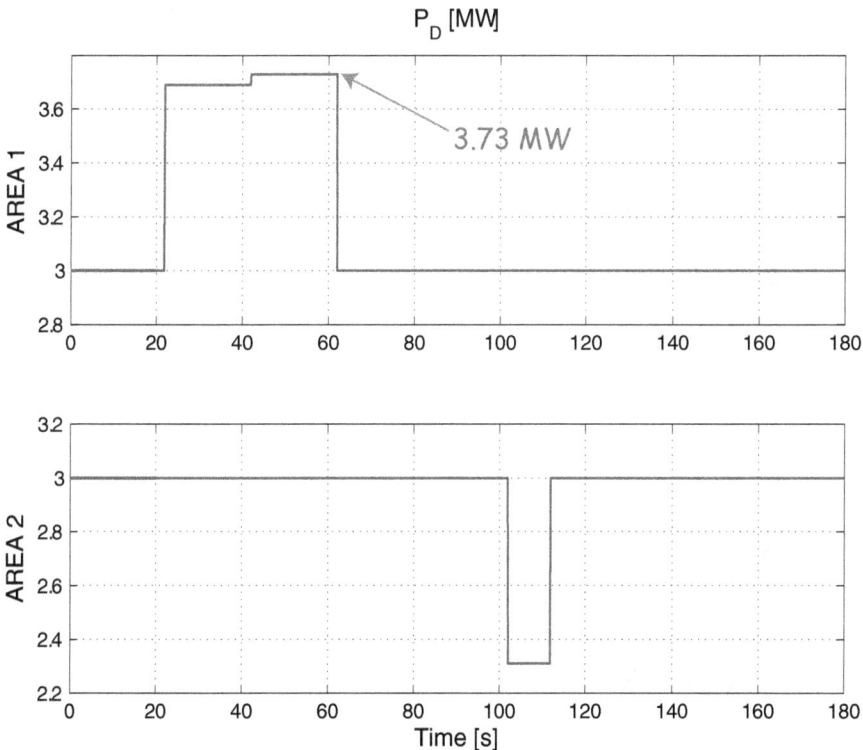

FIGURE 15.11 Load demands.

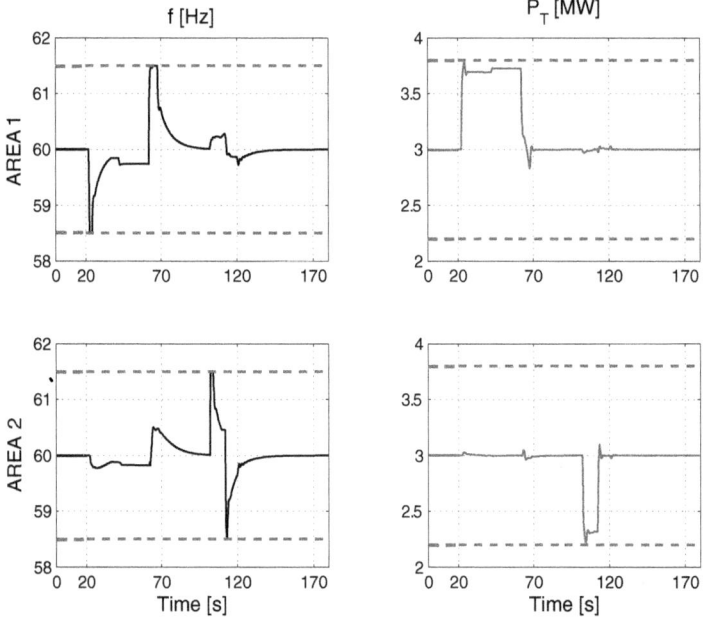

FIGURE 15.12 Frequency (left side) and generated power (right side): with ROG. The dashed lines represent the boundaries of the prescribed constraint.

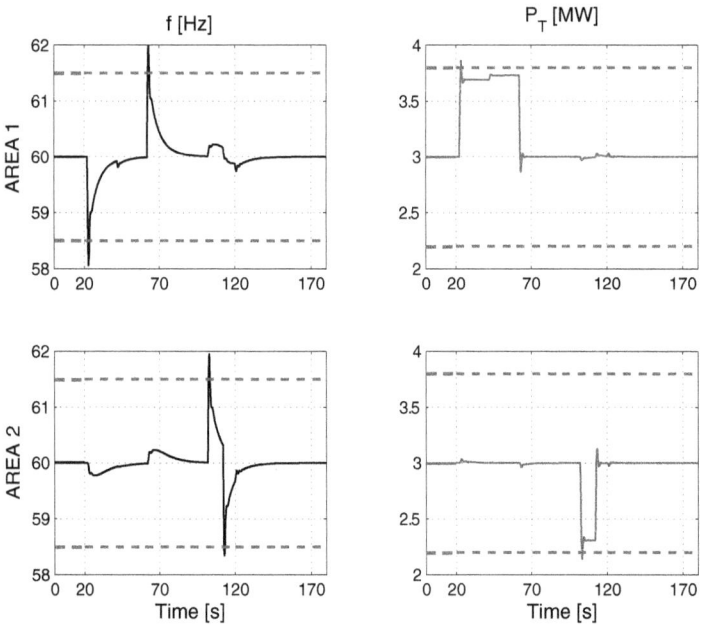

FIGURE 15.13 Frequency (left side) and generated power (right side): without ROG. The dashed lines represent the boundaries of the prescribed constraint.

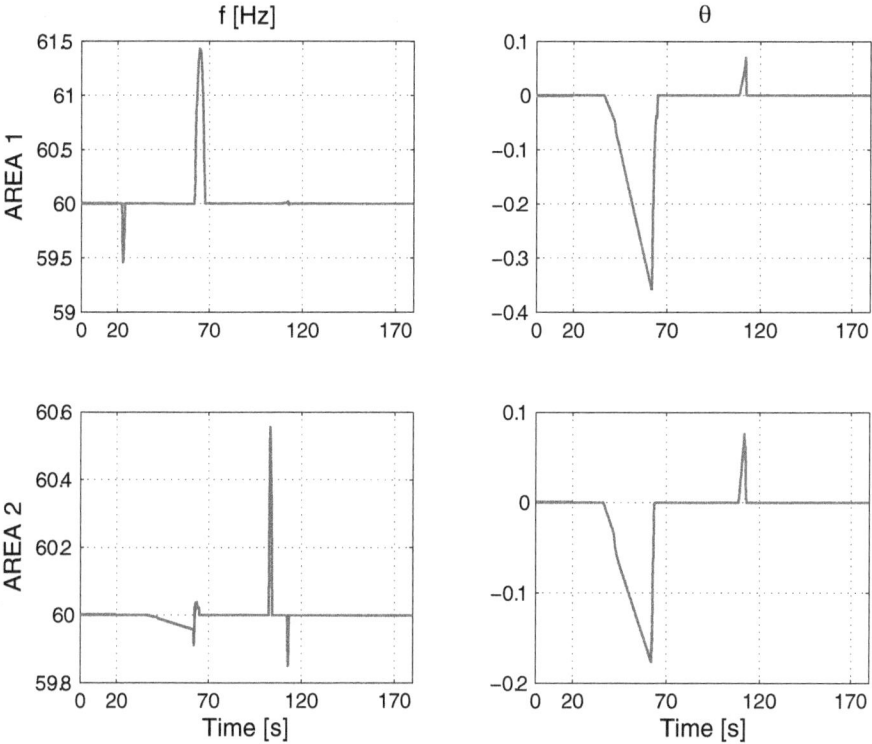

FIGURE 15.14 Outputs of the ROG unit. Reconfigured frequency set-points (left side) and control offset (right side).

able to satisfy the local load demand, as imposed in (15.26). Therefore, at the steady state, the prescribed constraints are not violated even when only the traditional primary/secondary ALFC scheme is used. Nevertheless, during the transients, loss of interconnection causes high peaks both in frequencies and power productions, and constraints violations occur (see Figure 15.13). On the contrary, the use of ROG allows one to handle this unpredictable event by suitably modifying the frequency set-points and adjusting the offset parameters (Figure 15.14).

15.4.1.3 Power Generation Failure

This final experiment is instrumental to show the ability of the ROG solution to reallocate the power generation amongst the areas in case a generator fails or is out of order. Specifically, here it is assumed that the generator of Area 1 exhibits a 90% reduction in its power production from time instants $t = 23$ s to $t = 43$ s due to a fault. This implies that it is able to generate only 10% of power w.r.t. the no-faulty condition. Moreover, a load addition (step change) of 0.5 MW occurs in Area 1 from $t = 23$ s until $t = 43$ s (see Figure 15.15). Note that we have considered that the load disturbance and the power generation failure occur in the same time interval.

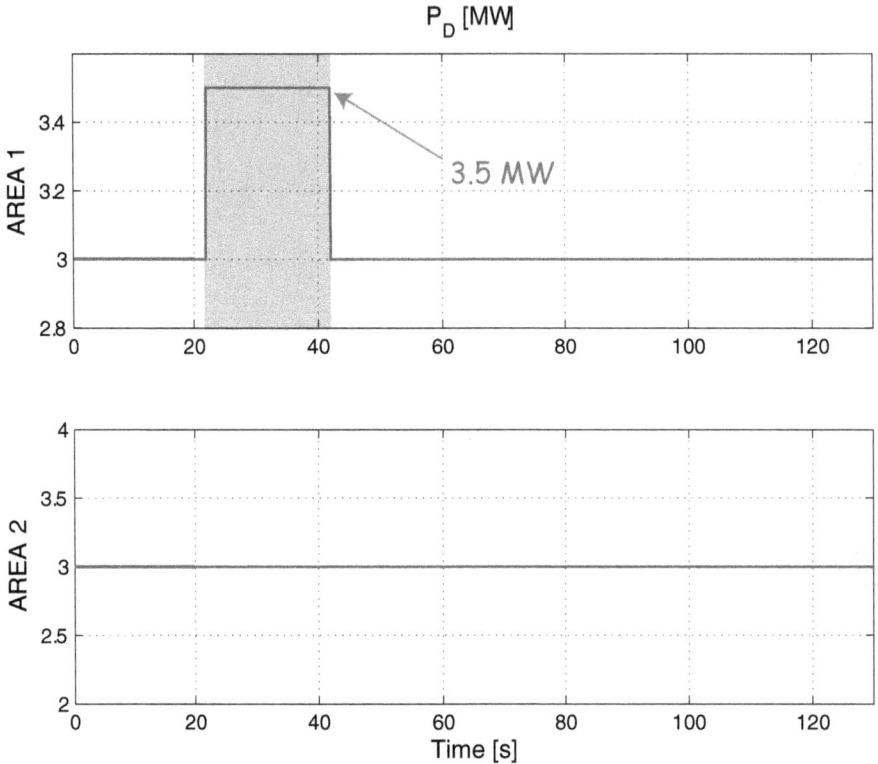

FIGURE 15.15 Load demands.

In this case, the generator of Area 1 is no longer capable of satisfying the local load demand. Therefore, a new equilibrium has to be determined. Again, it turns out that the standard primary/secondary ALFC action alone is not capable of enforcing the constraints during transients. In fact, starting from $t = 23$ s the power balance of the overall network would not have been ensured under the prescribed constraints without a ROG unit (see Figures 15.16 and 15.17).

On the contrary, the ROG unit is capable of modifying the nominal frequency set-points and control offsets (see Figure 15.18) and ensuring a new feasible power balance. In fact, in response to this situation, the generator of Area 2 increases its power production w.r.t its local demand and the surplus is transferred to Area 1 via the tie-line (Figure 15.19) so as to compensate for the faulty condition (Figure 15.20). It is worth commenting that the ROG unit is not informed of the fault occurrence, therefore such an approach is able to manage, under some circumstances, unpredictable events.

Remark 1—It is worth pointing out that in the previous critical cases of abrupt load changes, transmission lines and generation failures are abnormal situations that are out of the normal operating hypotheses of the ROG scheme. Then, the supervision algorithm could not be able to reconfigure the grid and found a feasible

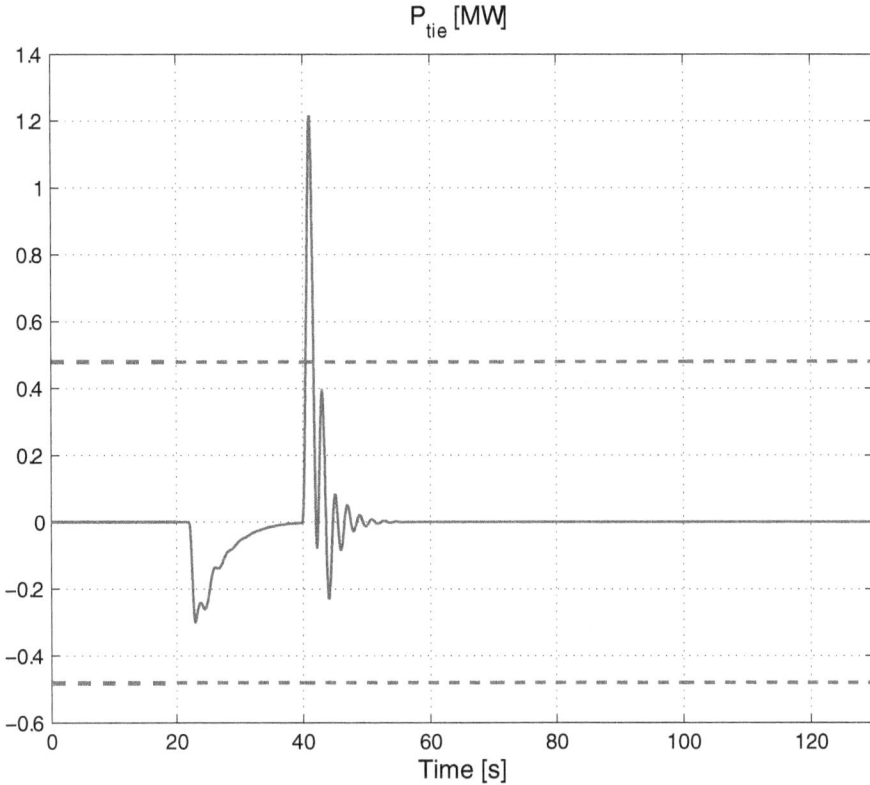

P_{tie} [MW]

FIGURE 15.16 Tie-line power flow: without ROG. The dashed lines represent the boundaries of the prescribed constraint.

post-fault equilibrium if the perturbations are too big. Thus, the analysis needs to be performed case by case. However, in many reasonable cases, the ROG scheme is able to feasibly reconfigure the grid, and the batteries of PEVs could help in providing the extra power required for the power balance. They could be discharged in critical situations to alleviate the amount of power transferred via the tie-lines to satisfy the load demand P_D. In order to investigate how fast the strategy is able to reconfigure the frequency and generated power set-points and how long the corresponding feasible equilibria can be maintained, one could look, e.g., at the Figures 15.8 and 15.9 related to the abrupt load change case. A similar analysis can be undertaken for the other cases.

In those figures, one can observe that the post-fault feasible equilibria are reached in a few seconds from the start of the gray area. Notice also that at the equilibrium there is a perfect balance between the generated and consumed powers. Then such an equilibrium can be maintained as long as required if no other adverse situations occur.

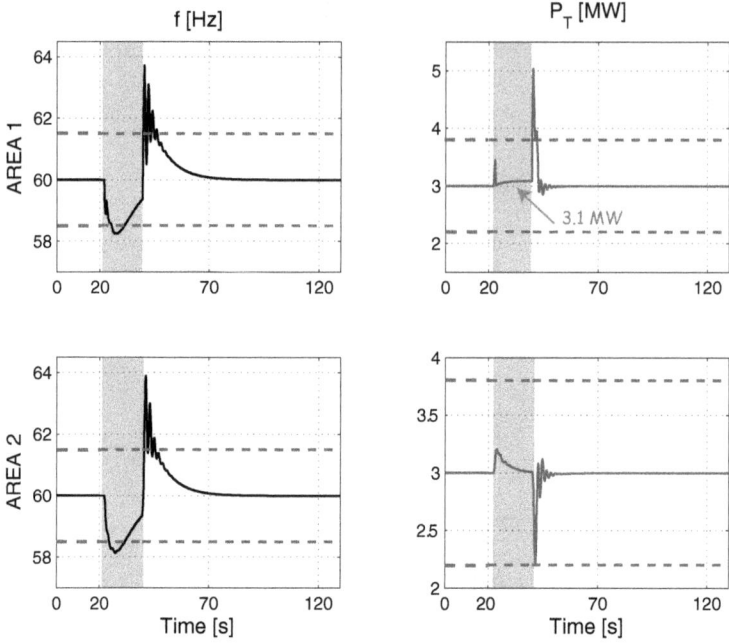

FIGURE 15.17   Frequency (left side) and generated power (right side): without ROG. The dashed lines represent the boundaries of the prescribed constraint.

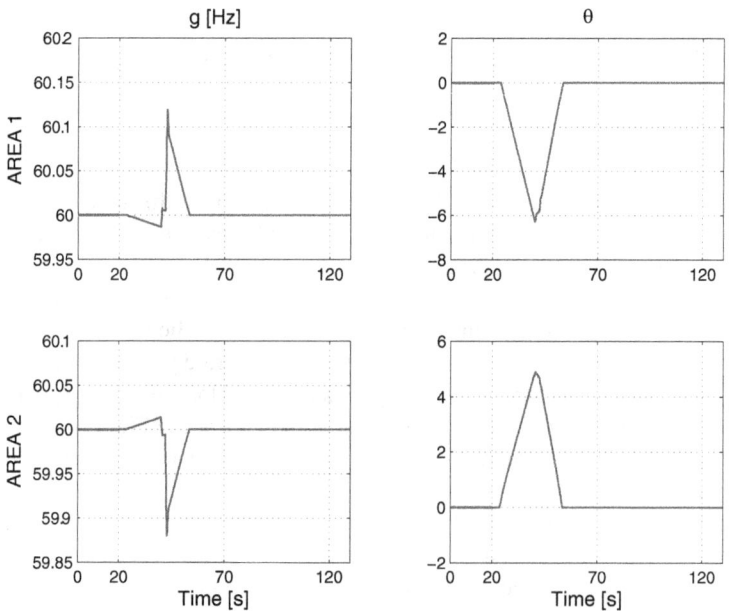

FIGURE 15.18   Outputs of the ROG unit. Reconfigured frequency set-points (left side) and control offset (right side).

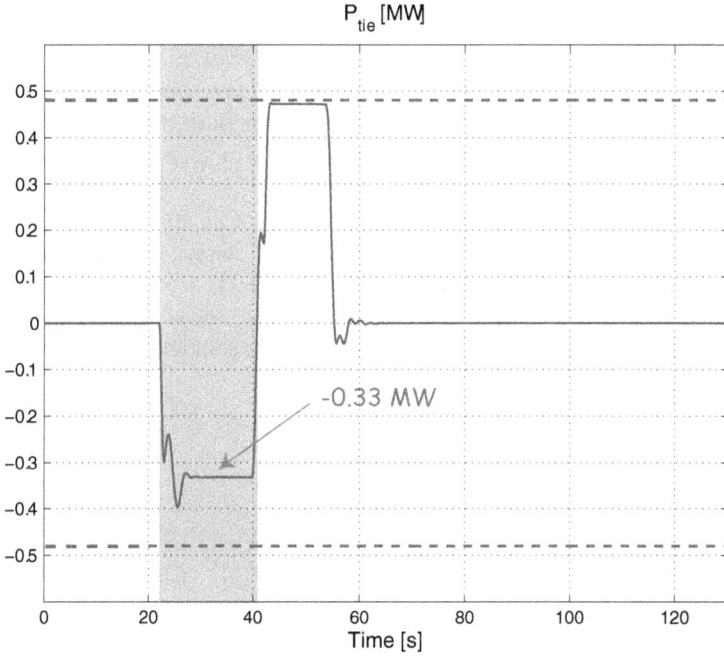

FIGURE 15.19 Tie-line power flow: with ROG. The dashed lines represent the boundaries of the prescribed constraint.

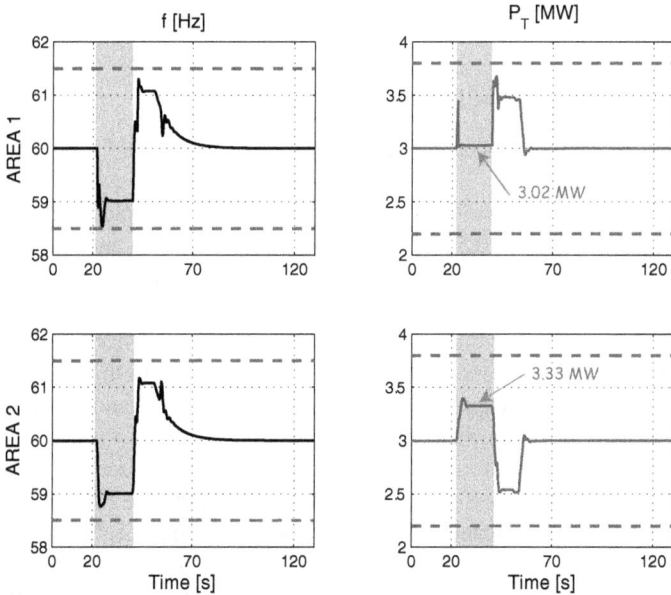

FIGURE 15.20 Frequency (left side) and generated power (right side): with ROG. The dashed lines represent the boundaries of the prescribed constraint.

15.5 ROG VERSUS RG

As pointed out in the introduction, the ROG can be viewed as a generalization of the standard RG scheme [8–13] in that an additional degree of freedom (an offset on the control law) is taken into consideration. The main merit of the ROG w.r.t RG is to enlarge the set of feasible system evolutions and, in turn, the set of critical scenarios that can be handled.

In the case of the power system in Figure 15.3, with frequency references f_{iref} $(t) = 60\ Hz$, $i = 1$, 2, $\forall t$ and constraints (15.26), the standard RG problem has not solutions for the given disturbances (in the set (15.24)–(15.25)), viz. $\mathcal{V}(x(0))$ is empty. Consequently, in order to make the RG problem solvable, the set of allowable disturbances must be reduced. It can be showed that solutions exist when they are restricted to

$$
h = \begin{bmatrix} 0.7 \\ 0.7 \\ 0.7 \\ 0.7 \\ 0.7 \end{bmatrix} [MW]. \tag{15.27}
$$

This implies that the addition of an offset parameter effectively enlarges the feasibility set and provides less conservative solutions when dealing with constraints and disturbances. Furthermore, even when both the schemes admit feasible solutions, the ROG approach ensures a larger viable set for the evolutions of the networked power system, which translates into better transient performances. This will be shown via two examples.

15.5.1 Power Generation Failure

We consider the following scenario: a power generation failure during a load change (step change).

Load changes: We assume for Area 1 the load variations depicted in Figure 15.21, where a step increment of 0.5 *MW* on power demand occurs at time $t = 10\ s$ and finally, after $t = 40\ s$, the load demand settles down at its nominal value.

Power generation failure: Area 1 exhibits a 20% reduction in its power production from $t = 10\ s$ to $t = 40\ s$ due to a fault. Thus, it is able to generate only 80% of its power w.r.t. the no-faulty condition for the same command. In this case, the generator of Area 1 is no longer capable to satisfy the local load demand. Therefore, a new equilibrium has to be determined.

Figures 15.22–15.24 describe the dynamic behaviors of frequencies, produced and exchanged powers when either the ROG or RG units are used as supervisory units. First of all, note that the safety constraints (15.26) are fulfilled at each time instant in both cases. Nonetheless, some differences arise between the schemes. In fact, the system evolutions under the RG action (dashed line) highly fluctuate in response to the failure occurrence, while the ROG action (continuous line) results in smother responses that could prevent component malfunctions.

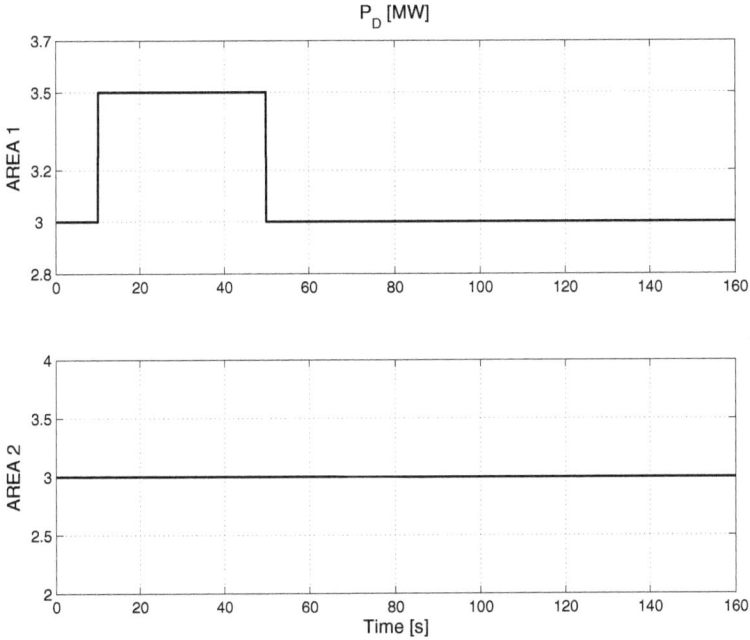

FIGURE 15.21   Load demands.

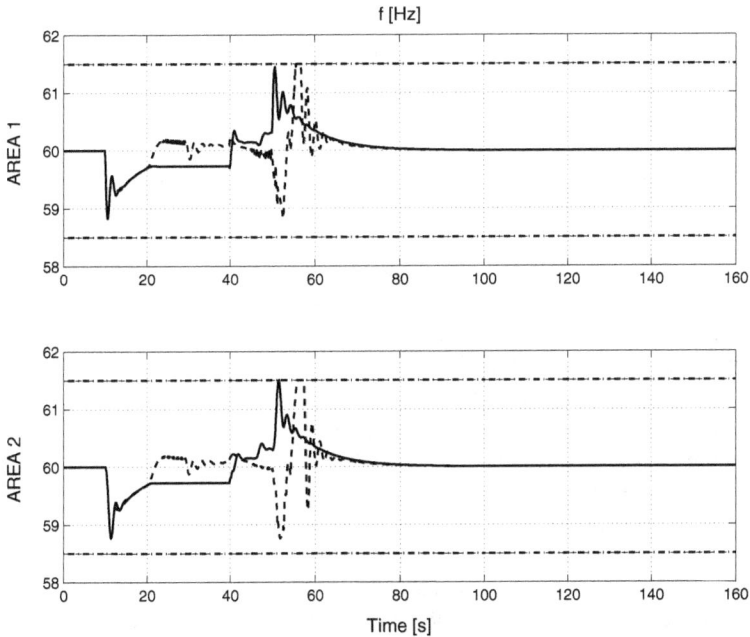

FIGURE 15.22   Frequency: with ROG (continuous line) and RG (dashed line) units. The dash-dot lines represent the prescribed constraints.

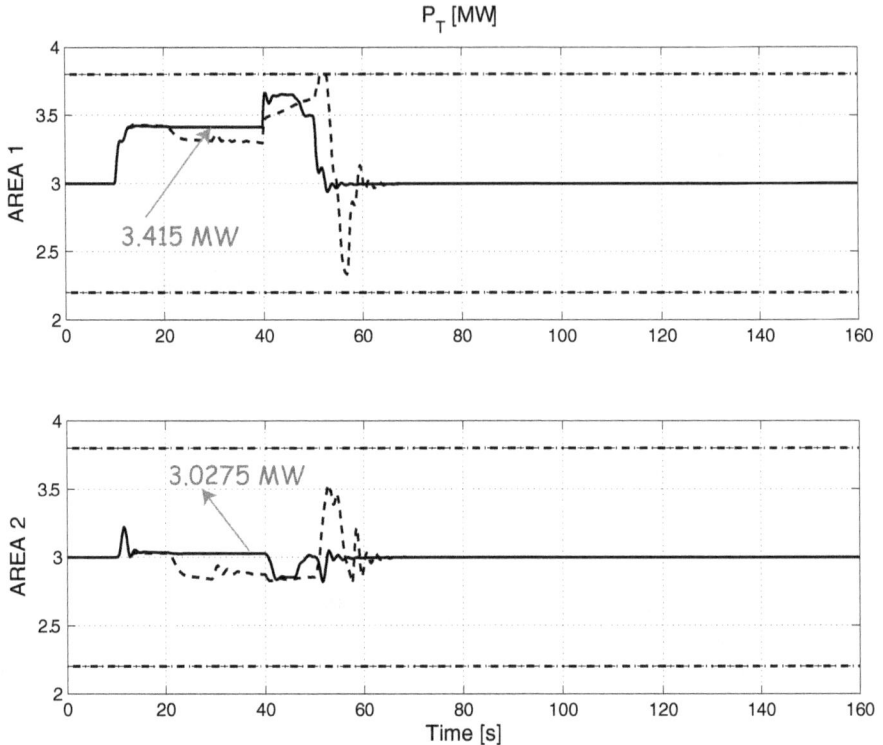

FIGURE 15.23 Generated power: with ROG (continuous line) and RG (dashed line) units. The dash-dot lines represent the prescribed constraints.

Importantly enough, the ROG unit is capable to steer the system into a new equilibrium compatible with the faulty condition, as is evident in Figures 15.22–15.24 (continuous lines). To clarify the latter, let us consider the time interval [20, 40] s, where frequencies, produced and exchanged powers reach the new steady-state setpoints. In particular, Area 1 produces 3.415 MW and receives via the tie-line 0.0275 MW produced by Area 2. Therefore, on the whole, it is supplied by 3.4425 MW, while the local demand is 3.5 MW. This means that, due to generation reduction in Area 1, it is not possible to match the load request. Anyway, the ROG automatically reaches the best feasible equilibrium compatible with the failure occurrence. Note also that the operating frequency settles to 59.73 Hz both for Area 1 and Area 2.

On the contrary, the RG is able only to ensure constraints fulfillment without any reconfiguration capabilities. This confirms that the ROG approach rises to less conservative results than the RG one: in fact, at each instant t, the set $\mathcal{V}(x(t))$, computed under the ROG framework, contains a larger number of feasible commands. For the sake of completeness, the ROG and RG commands are reported in Figures 15.25 and 15.26, respectively.

It is worth pointing out that the bound (15.27) represents the maximum on the manageable disturbances for the RG scheme. To highlight the latter, let us

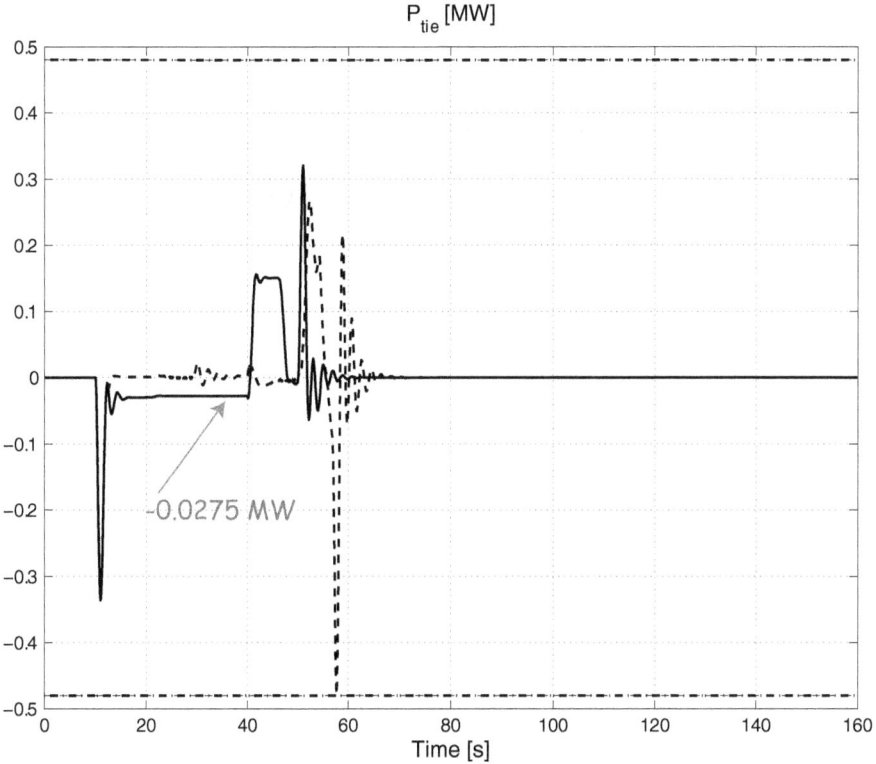

FIGURE 15.24 Tie-line power flow: with ROG (continuous line) and RG (dashed line) units. The dash-dot lines represent the prescribed constraints.

suppose that power demands could vary up to about 33% with respect to $\bar{P}_{D_i}, i = 1, 2$. Consequently, the set of admissible load disturbances (15.27) is characterized by the following matrices:

$$U = \begin{bmatrix} I_2 \\ -I_2 \end{bmatrix} \text{ and } h = \begin{bmatrix} 1111 \end{bmatrix}^T. \tag{15.28}$$

In this case, the RG scheme, while of ROG, has no solutions, viz. $\mathcal{V}(x(0))$ is empty.

In order to analyze the ROG responses in this case, we consider the following scenario: the power generation failure described in the previous simulation plus the load change (step change) depicted in Figure 15.27.

Load changes: In Area 1, we suppose first an increment of 0.69 MW on power demand occurring at time $t = 10$ s, then a further increment of 0.73 MW at time $t = 50$ s and finally, after $t = 70$ s, the load demand settles down at 3 MW. The system evolutions are reported in Figures 15.28–15.30, and the prescribed constraints are always fulfilled. Consider first the time window [10, 40] s, which includes the failure

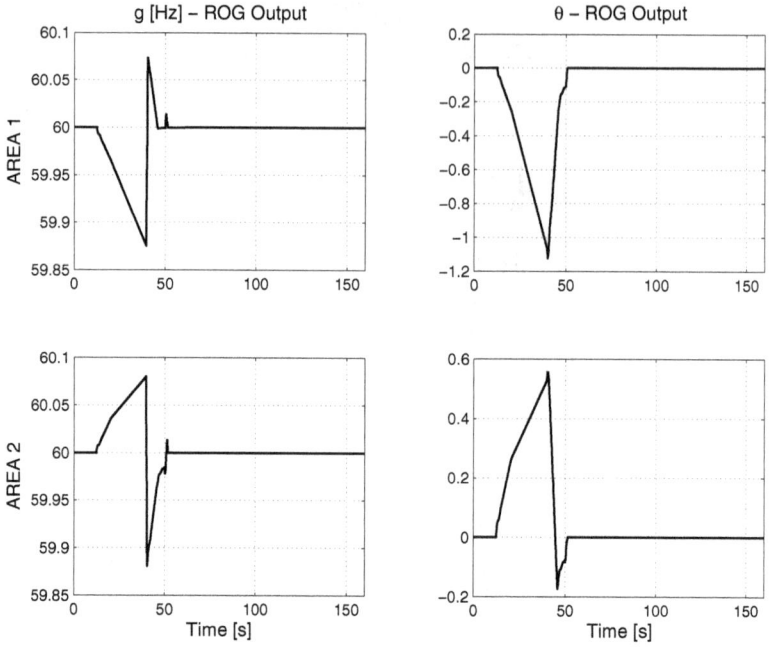

FIGURE 15.25   ROG actions: reconfigured frequency set-point (left side) and offset (right side).

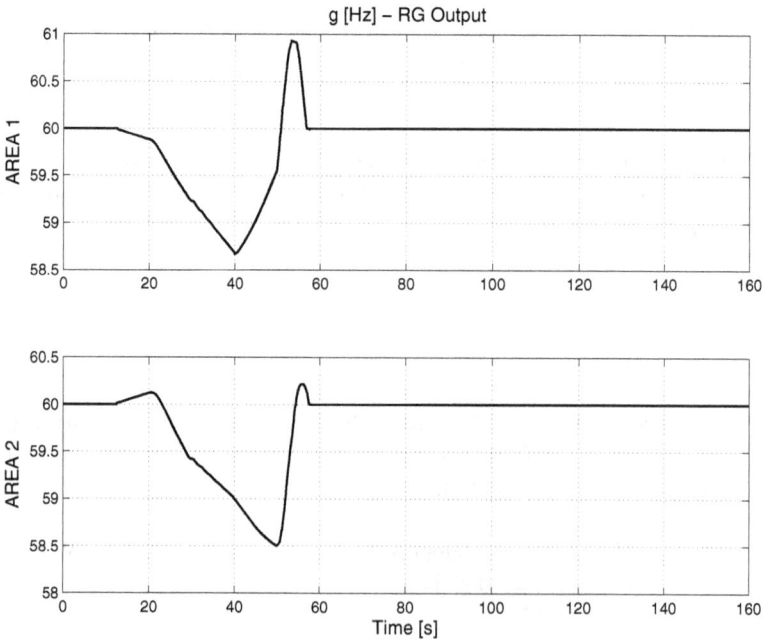

FIGURE 15.26   RG action: reconfigured frequency set-point.

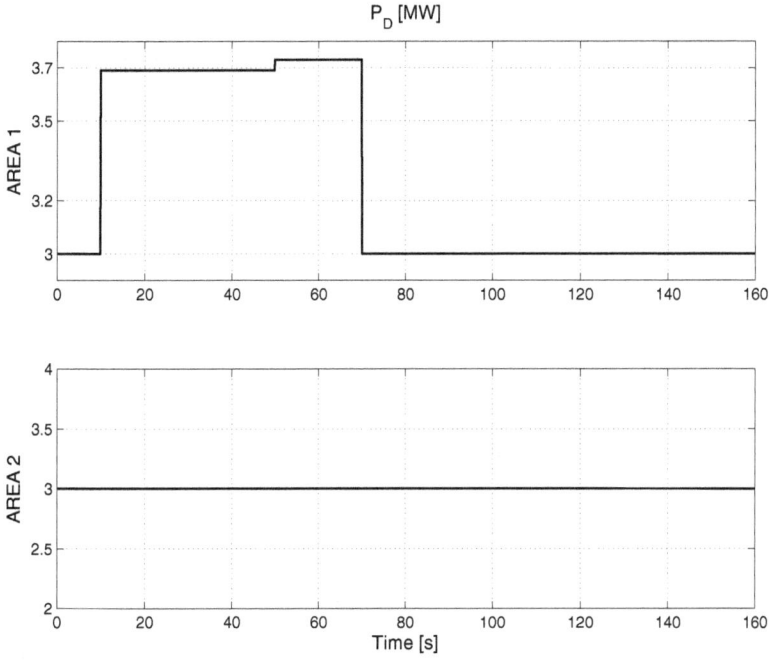

FIGURE 15.27 Load demands.

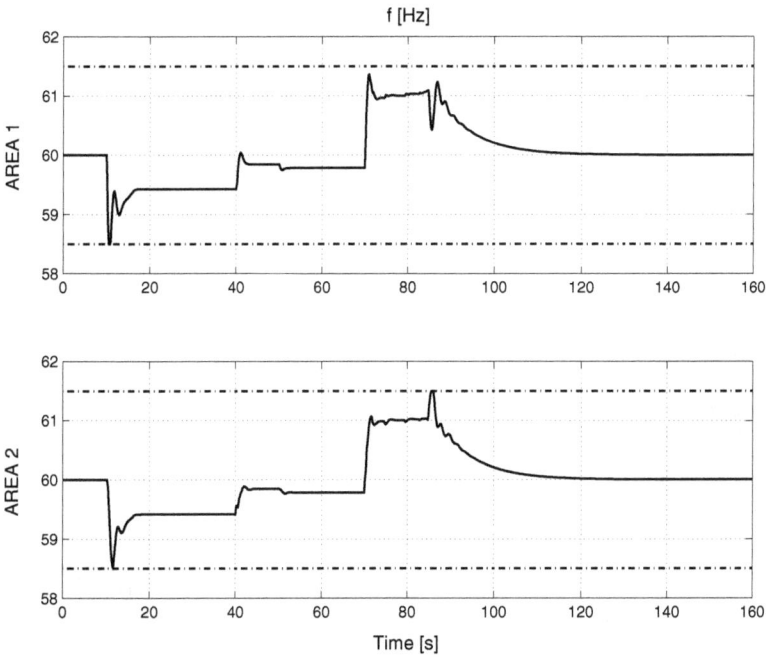

FIGURE 15.28 Frequency (continuous line) and prescribed constraints (dash-dot lines).

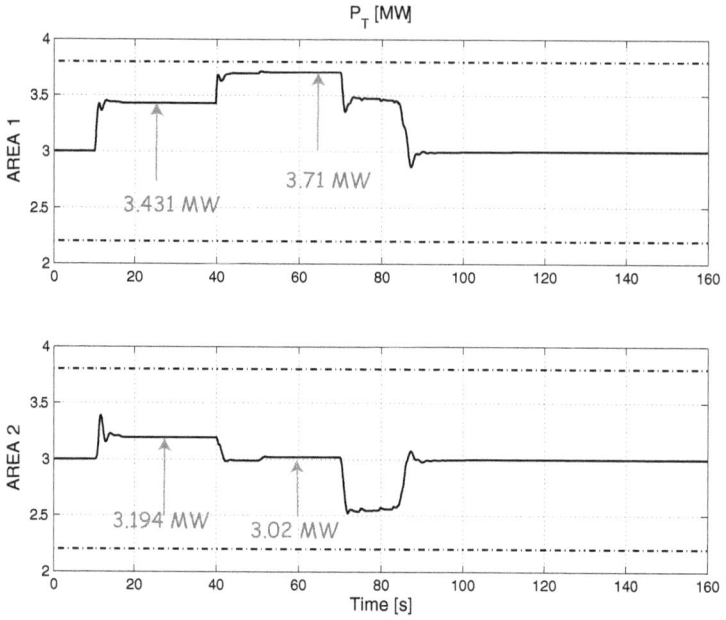

FIGURE 15.29 Generated power (continuous line) and prescribed constraints (dash-dot lines).

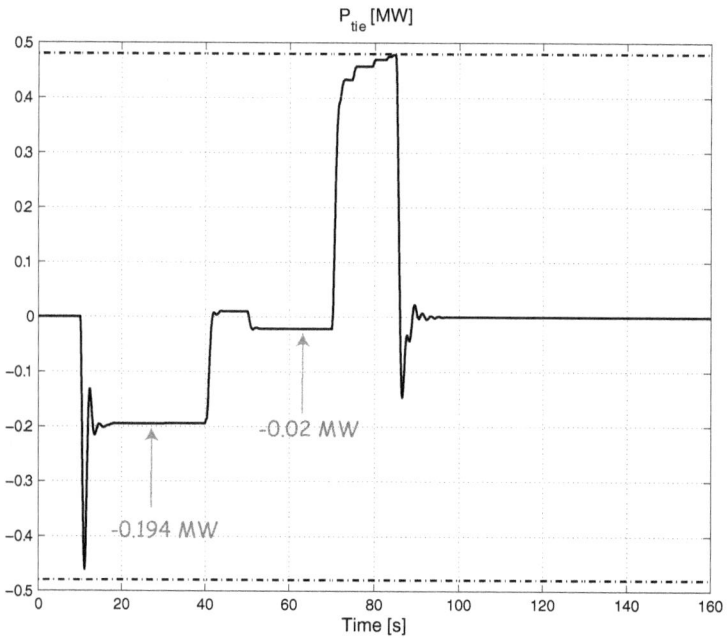

FIGURE 15.30 Tie-line power flow (continuous line) and prescribed constraints (dash-dot lines).

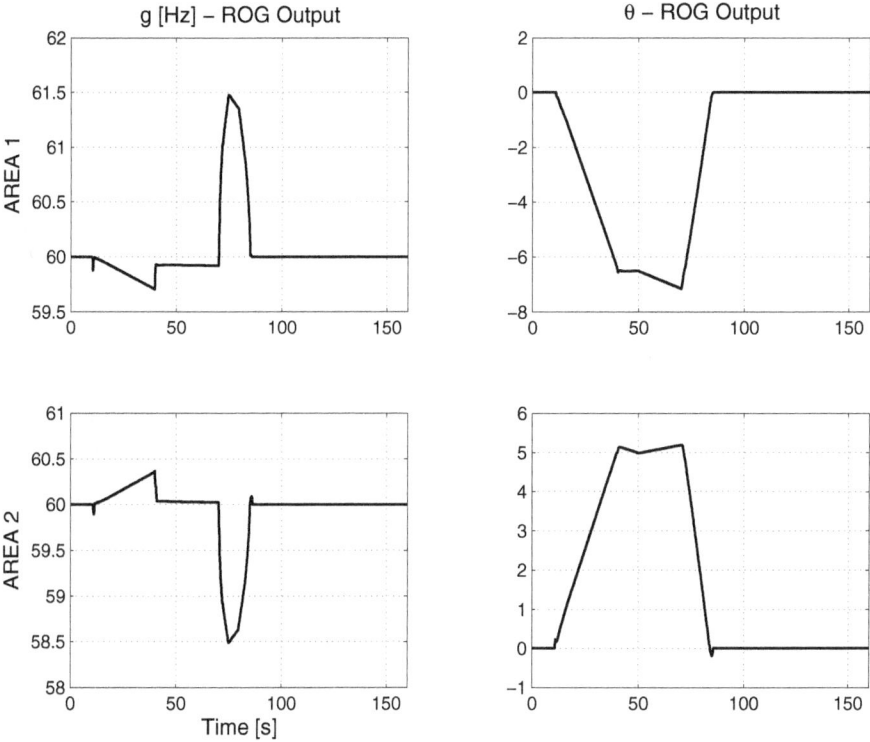

FIGURE 15.31 ROG actions: reconfigured frequency set-point (left side) and offset (right side).

occurrence. As it has been already remarked, due to the generator malfunction in Area 1, it is not possible to match the total power request (3.69 MW), and the ROG automatically defines a new feasible equilibrium (3.625 MW). In particular, Area 1 produces 3.431 MW, and 0.194 MW is transferred via the transmission lines from Area 2. On the contrary, during the nominal generator operation, the produced and requested powers can always be balanced.

Consider now the interval [50, 70] s, when the Area 1 local demand is 3.73 MW. At steady state, Area 1 power generation reaches a value of 3.71 MW, and the remaining power required to balance the load is obtained from the generator of Area 2 (see Figure 15.29), which exactly increases its production of 0.02 MW and transfers this fraction via the tie-line (Figure 15.30). For the sake of completeness, the ROG actions are finally depicted in Figure 15.31.

15.5.2 TRANSMISSION LINE FAILURE

Let us consider the set of constraints in (15.26) and the admissible disturbances set defined by (15.24) and (15.27). The load demand is depicted in Figure 15.32, and we have assumed that a transmission line breakdown occurs during all the simulations.

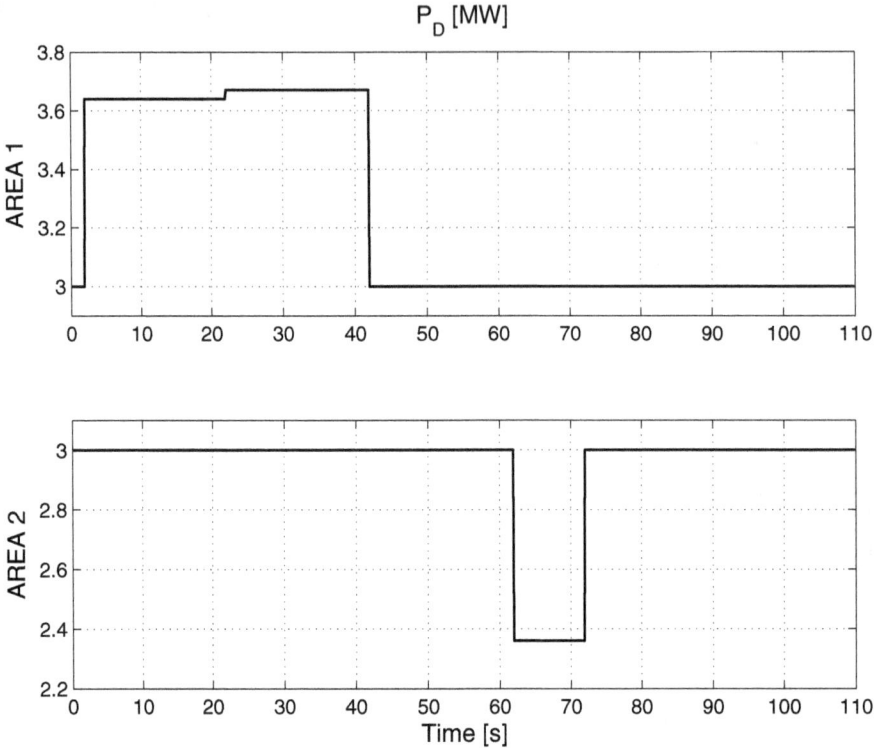

FIGURE 15.32 Load demands.

Figures. 15.33 and 15.34 depict the dynamic behavior of frequency and generated power in each area. It is worth highlighting that the use of a ROG unit allows the overall system to promptly respond to any load change, whereas the RG approach requires a larger settling time to cope with new power demand. Moreover, undesirable fluctuations on the frequency and produced power are avoided (see Figures 15.33 and 15.34, in particular during the time interval [40, 50] s), which is crucial in order to prevent damages to the power system equipment.

15.5.3 A FOUR-AREA POWER SYSTEM EXPERIMENT

In this simulation, an interconnected four-area power system has been used to confirm the effectiveness of the proposed approach in a more realistic power system scenario. A simplified representation for an interconnected system in a general form, including both ring [17] and bus [18-20] interconnections, is shown in Figure 15.35. In particular, Area 1 is depicted in detail in Figure 15.36 with its interconnections toward the other areas.

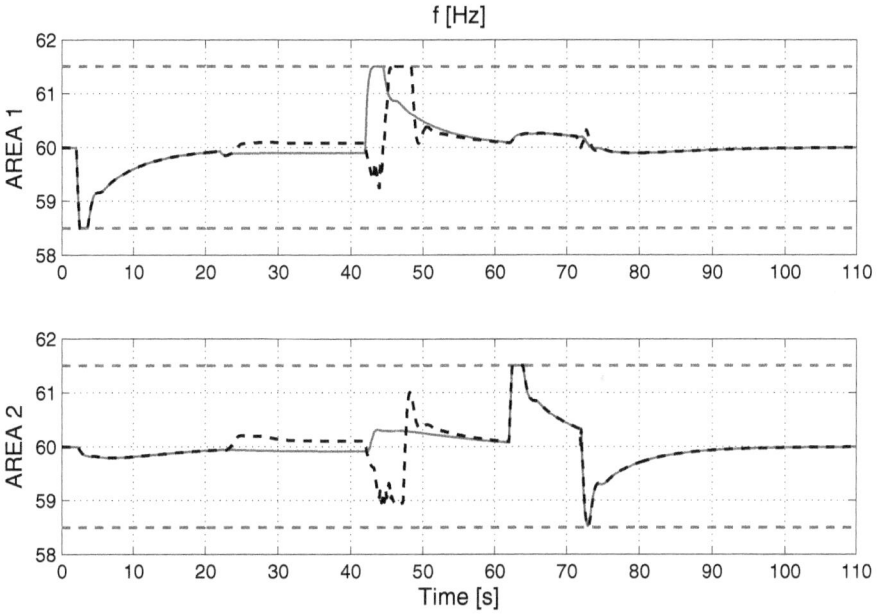

FIGURE 15.33 Frequency constraints: ROG (continuous line) and RG (dashed line). The horizontal dashed lines represent the boundaries of the prescribed constraints.

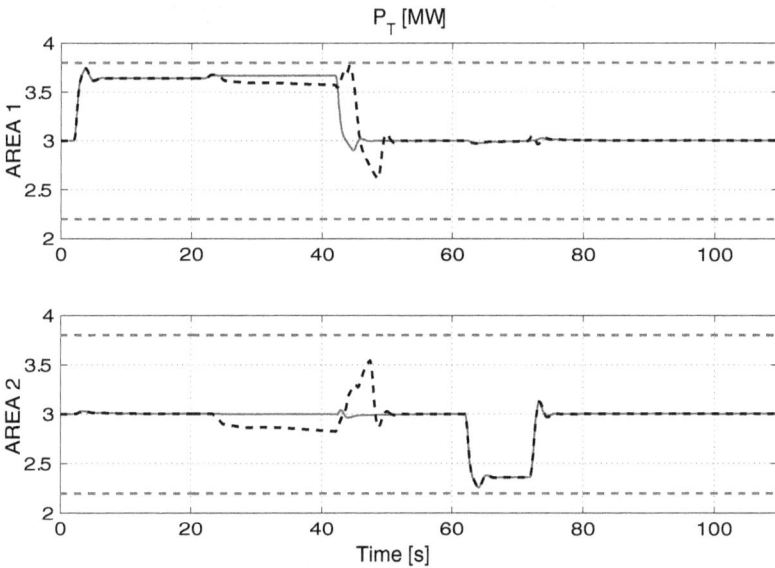

FIGURE 15.34 Generated power constraints: ROG (continuous line) and RG (dashed line). The horizontal dashed lines represent the boundaries of the prescribed constraint.

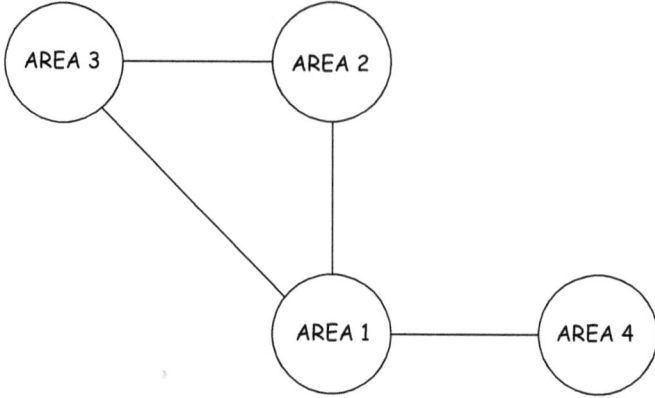

FIGURE 15.35 A four-area power system.

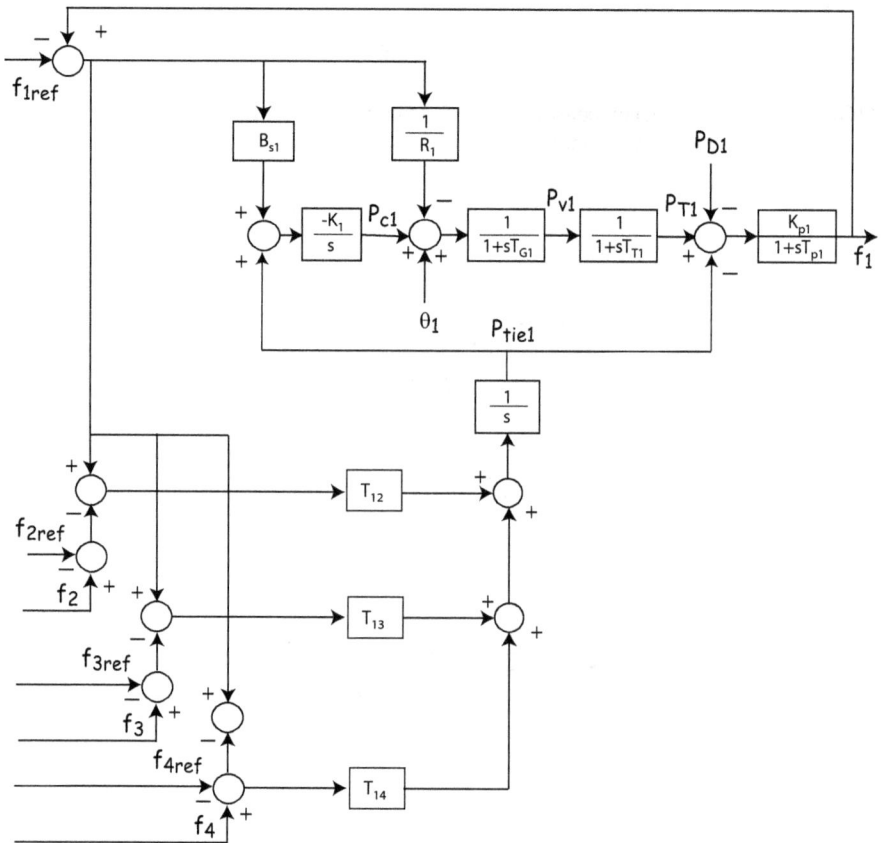

FIGURE 15.36 Block diagram of Area 1.

The state-space model for this system has been described in [15]. For the sake of comprehension, we report the expression for the power exchanged via the tie-lines with the $i - th$ area P_{tiei}

$$P_{tiei} = \sum_{j \neq i} \frac{T_{ij}}{s} \left(\left(f_i - f_{iref} \right) - \left(f_j - f_{jref} \right) \right) = \sum_{j \neq i} P_{tie,ij} \qquad (15.29)$$

Then, P_{tiei} represents the total power exchanged by the i-th area via the tie-lines, whereas $P_{tie,Ij}$ is the specific contribution between the i-th and j-th areas. It is important to note that the previous expression is a linear combination, under the weights T_{ij}, of the frequency difference between the areas. Clearly, the following power balance is always satisfied:

$$\sum_i P_{tiei} = 0.$$

In the simulations, we have assumed that the nominal operating frequency for each area is $f_{iref}(t) = 60$ Hz, $i = 1, \ldots, 4 \; \forall t$. The power demand has the following form: $P_{D_i}(t) = \bar{P}_{D_i} + \tilde{P}_{D_i}(t), i = 1,\ldots,4$ where the same nominal load $\bar{P}_{D_i} = 3MW$, for each area has been considered, while $\tilde{P}_{D_i}(t)$ identifies possible load variations. It is worth pointing out that the power system scheme of Figure 15.35 is based on the assumption that each area has the capability to autonomously balance its own nominal load. In the simulations, we have supposed that power demands could vary up to 33% with respect to $\bar{P}_{D_i}, i = 1,\ldots,4$. Consequently, the following convex and compact region

$$\mathcal{D}_{\tilde{P}_D} := \left\{ \tilde{P}_D \in \mathcal{R}^4 : U\tilde{P}_D \leq h \right\}, \qquad (15.30)$$

where $U = \begin{bmatrix} I_4 \\ -I_4 \end{bmatrix}$ and $h = [11111111]^T$ characterizes the set of admissible load disturbances (in MW). Observe that such a bound represents the maximum on the manageable disturbances when the ROG approach is used. The following set of constraints (expressed in Hz and MW)

$$58.5 \leq f_i(t) < 61.5,$$
$$\left| P_{tiei}(t) \right| \leq 0.7, \qquad (15.31)$$
$$2.2 \leq P_{T_i}(t) \leq 3.8, \quad i = 1,\ldots 4.$$

to be fulfilled pointwise-in-time are of interest. Finally, the initial state condition

$$x(0) = [603000\;603000\;603000\;603000]^T$$

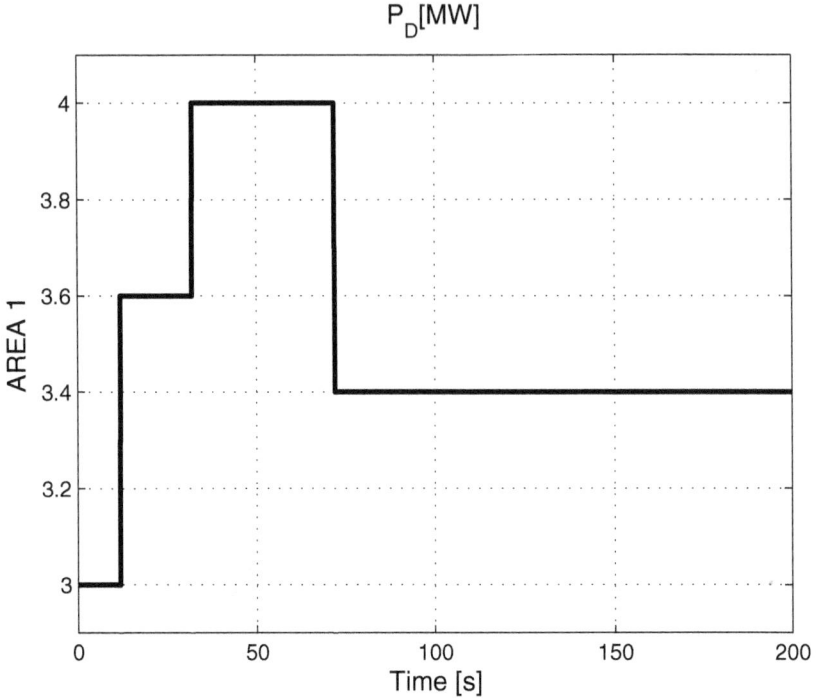

FIGURE 15.37 Load demand of Area 1.

has been chosen, and the initial admissible ROG command vector has been determined as $g_i(0) = 60$, $i = 1,\ldots, 4$ and $\theta_i(0) = 0$, $i = 1,\ldots, 4$.

We assume for Area 1 the load variation w.r.t. the nominal condition depicted in Figure 15.37. In particular, first a step increment of 0.6 MW on power demand occurs at 12 s, then a further increment of 0.4 MW is simulated at 32 s and finally after 40 s, the load demand settles down at 3.2 MW.

15.5.3.1 Transmission Line Failure
First, we focus hereafter on a transmission line failure, which seems to be the most relevant case in order to show the benefits coming out from the interconnection among many areas.

We impose breakdowns on the transmission lines between Area 1 and Areas 2 and 4 during all the simulation time windows (see Figure 15.38).

In order to clarify the power system response, we restrict our attention to the highest load request (4 MW), which overcomes the maximum power production of Area 1 (see constraints (15.31)). It is worth pointing out that such a critical situation cannot be managed in the two-area power system of Section 15.5.1.2 because the physical redundancy is not sufficient.

To underline the relevance of the inter-area links, we simulate the four-area power system in both faulty and no-faulty conditions. The corresponding system evolutions are reported in Figures 15.39–15.41.

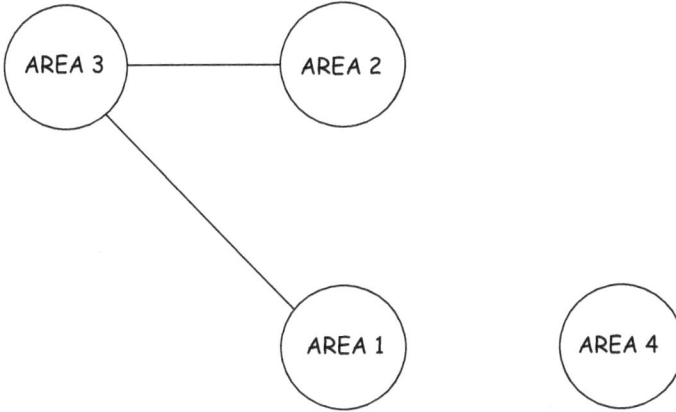

FIGURE 15.38 The four-area power scheme under transmission line failures.

In the no-faulty case (dashed line), Area 1 (Figure 15.39) produces at steady state (around 65 s) 3.515 MW, whereas the power demand is 4 MW. The remaining power to reach the balance is generated by the other three areas and transferred via the tie-lines. In fact, P_{tie1} is equal to 0.485 MW, which is the sum of the contributions of each area $P_{tie12} + P_{tie13} + P_{tie14} = 0.173 + 0.1625 + 0.149$. Note that $P_{tie23} = -P_{tie32} \approx 0$ because the local demands of Area 2 and Area 3 do not change. Therefore, almost all their extra produced power is transferred to Area 1.

Consider now the faulty case (continuous line) in Figure 15.38. It is evident that Area 4 is disconnected from the others; consequently, it is not influenced by the remaining network, and its tie-line is always zero (Figure 15.40).

In the faulty network topology, Area 1 is directly connected only to Area 3. Again, Area 1 is not capable to produce the necessary power to balance its local demand and the power fraction required to accomplish the request is furnished via the tie-line P_{tie1}. In particular $P_{T}1 = 3.55$ MW and $P_{tie1} = 0.45$ MW, while $P_{tie2} = 0.215$ MW and $P_{tie3} = 0.235$ MW. Recalling that $P_{tie3} = P_{tie31} + P_{tie32}$ and observing that $P_{tie2} = P_{tie23} = -P_{tie32}$, and $P_{tie1} = P_{tie13} = -P_{tie31}$, one can conclude that the overall power balance is satisfied. In fact, as is evident from Figure 15.39, both Areas 2 and 3 produce more power than their local needs, and the surplus contributes to match the demand of Area 1 via the path *Area* 2 → *Area* 3 → *Area* 1. This coincides with a new power routing, which can be influenced by the choice of the weighting matrices Ψ_g and Ψ_θ in the cost function (15.15).

For the sake of completeness, the frequency evolutions are also shown in Figure 15.41, where it clearly shows that constraints are never violated. However, some coupling amongst areas is still present, as it results from unexpected frequency adjustments in Area 4 (see Figure 15.41) not interested during the faulty condition to any power exchange.

15.5.3.2 Power Generation Failure

Here, the simulations are instrumental to show the capability of the ROG to reallocate the power generation amongst the areas in the case of out-of-order or failed generators. Moreover, we want to mark the ability of the ROG to automatically reach the best feasible solution compatible with critical unpredictable phenomena by resorting

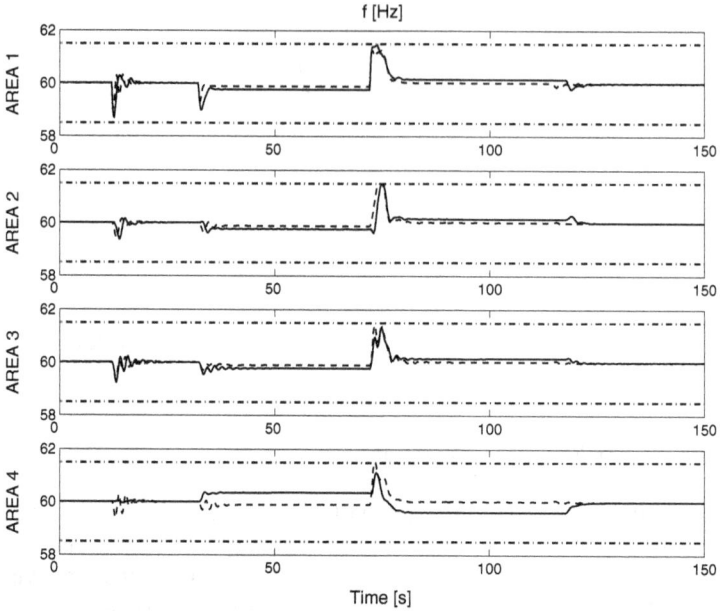

FIGURE 15.39 Generated powers: faulty (continuous line) and no-faulty (dashed line) cases. The dash-dotted lines represent the boundaries of the prescribed constraints.

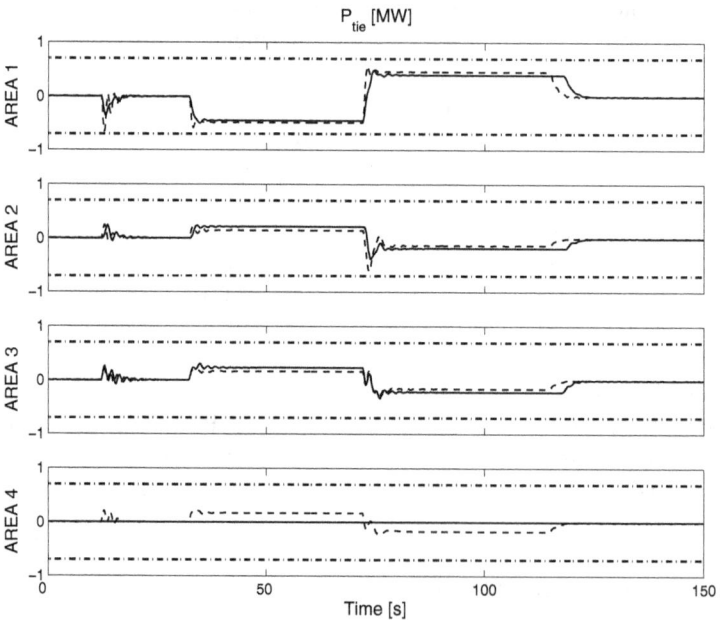

FIGURE 15.40 Tie-lines: faulty (continuous line) and no-faulty (dashed line) cases. The dash-dotted lines represent the boundaries of the prescribed constraints.

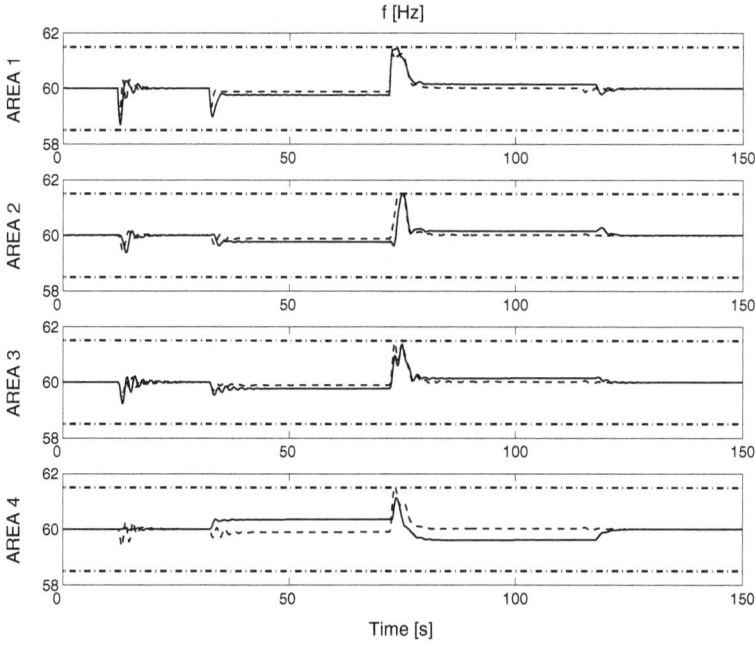

FIGURE 15.41 Frequencies: faulty (continuous line) and no-faulty (dashed line) cases. The dash-dotted lines represent the boundaries of the prescribed constraints.

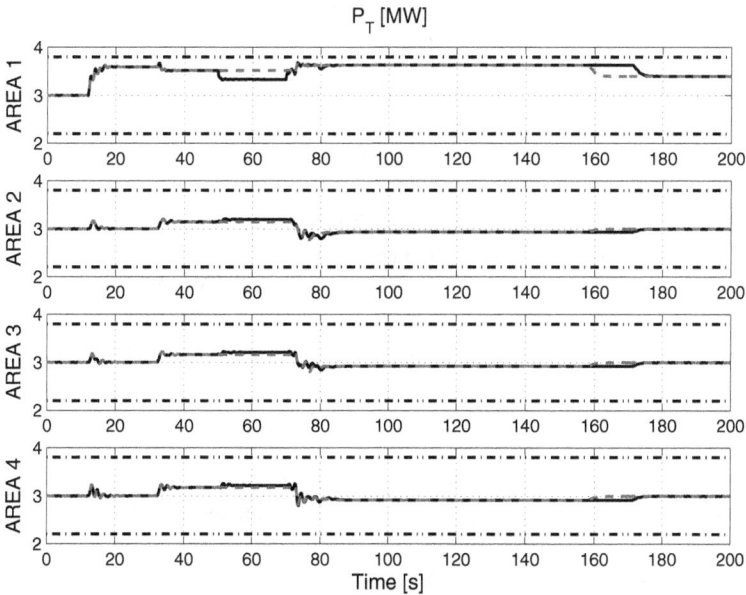

FIGURE 15.42 Generated powers with ROG: faulty (continuous line) and no-faulty (dashed line) cases. The dash-dotted lines represent the boundaries of the prescribed constraints.

to redundancies both in power generation units and in inter-area links. To this end, we consider a load change (step change) scenario during two operating conditions:

Normal operations;
Power generation failure occurrence.

Load changes: We assume for Area 1 the load variation depicted in Figure 15.37. *Power generation failure occurrence*: Area 1 exhibits a 15% reduction in its power production from $t = 50$ s to $t = 70$ s due to a fault. Thus, it is able to generate only 85% of its power w.r.t. the no-faulty condition for the same command. In this case, the generator of Area 1 is no longer capable to satisfy the local load demand. Therefore, a new equilibrium has to be determined.

The four-area power system has been simulated in both faulty and no-faulty conditions, and the corresponding system evolutions are reported in Figures 15.42–15.45.

In order to clarify the power system response, we restrict our attention to the highest load request (4 MW), which overcomes the maximum power production of Area 1 (see constraints (15.31)). It is worth pointing out that the no-faulty case (dashed line) can be managed thanks to the power generation redundancy. In fact, the growing load demand of Area 1 can be supplied by the generators of Areas 2 and Area 3 via network links. Note that Area 1 (Figure 15.42) produces at steady state (around 60 s) 3.517 MW, whereas the power demand is 4 MW. The remaining power to reach the balance is generated by the other three areas and transferred via the tie-lines as $P_{tie1} = P_{tie12} + P_{tie13} + P_{tie14} = 0.147 + 0.163 + 0.173 = 0.483$ MW. Note that $P_{tie23} = -P_{tie32} \approx 0$ because the local demands of Area 2 and Area 3 do not change. Therefore, almost all their extra power productions are transferred to Area 1. Consider now the faulty case (continuous line) when the power generation in Area 1 suffers a 15% reduction. In this case, the ROG unit is capable of steering the system into a new equilibrium compatible with the faulty condition, Figures 15.42–15.45. To clarify the latter, let us consider the time window [50, 70] s where, at the steady state, Area 1 produces 3.33 MW and receives via the tie-line $P_{tie1} = 0.63$ MW. In particular, all the other three areas produce more power than their local needs, i.e., $P_{T_2} = 3.20MW$, $P_{T_3} = 3.21MW$, $P_{T_4} = 3.22MW$ (see Figure 15.43), and the surpluses contribute to match the demand of Area 1 via the inter-area links, $P_{tie2} = 0.20$ MW, $P_{tie3} = 0.21$ MW and $P_{tie4} = 0.22$ MW.

Therefore, on the whole, Area 1 is supplied by 3.96 MW, while the local demand is 4 MW. This means that, due to generation reduction in Area 1, it is not possible to match the load request. Anyway, the ROG automatically reaches the best feasible equilibrium compatible with the power failure. For the sake of completeness, the frequency evolutions are also shown in Figure 15.45, where it clearly shows that constraints are never violated. It is worth pointing out that the ROG is not informed of the power generation failure, and it steers the system to new compatible equilibria in response to load changes and/or failures based on its intrinsic reconfiguration capability by modifying the set-points and control offsets (Figures 15.46 and 15.47). To stress the importance of the proposed supervisory control strategy, in Figures 15.48–15.50, the evolutions of all constrained variables without the actions of the ROG unit in the case of tie-lines failures are reported. First, as is expected, in order to maintain the power balance, the power generation in each area tracks the load

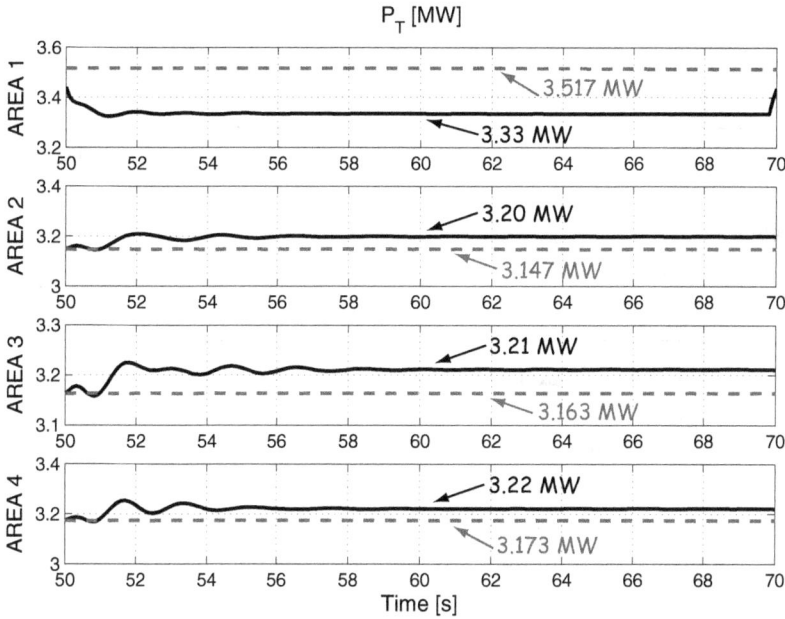

FIGURE 15.43 Generated powers with ROG: zoom on the time interval [50, 70] *s*. Faulty (continuous line) and no-faulty (dashed line) cases.

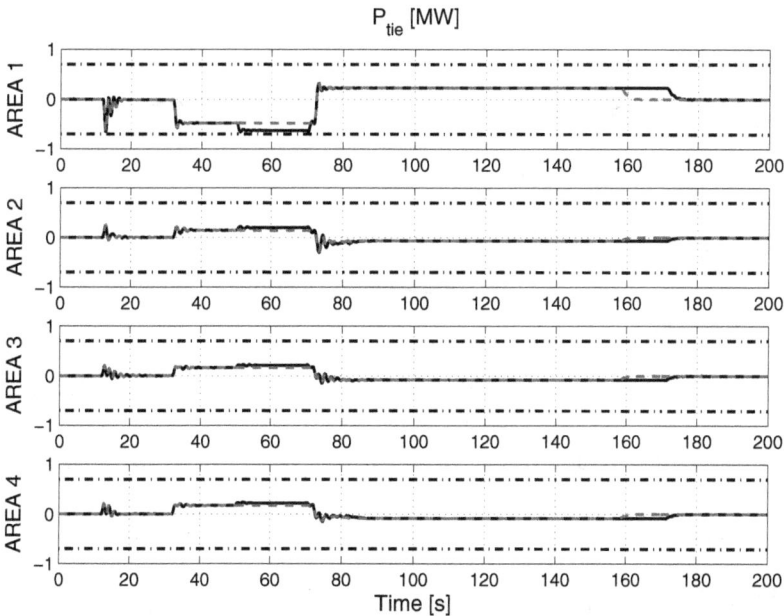

FIGURE 15.44 Tie-lines with ROG: faulty (continuous line) and no-faulty (dashed line) cases. The dash-dotted lines represent the boundaries of the prescribed constraints.

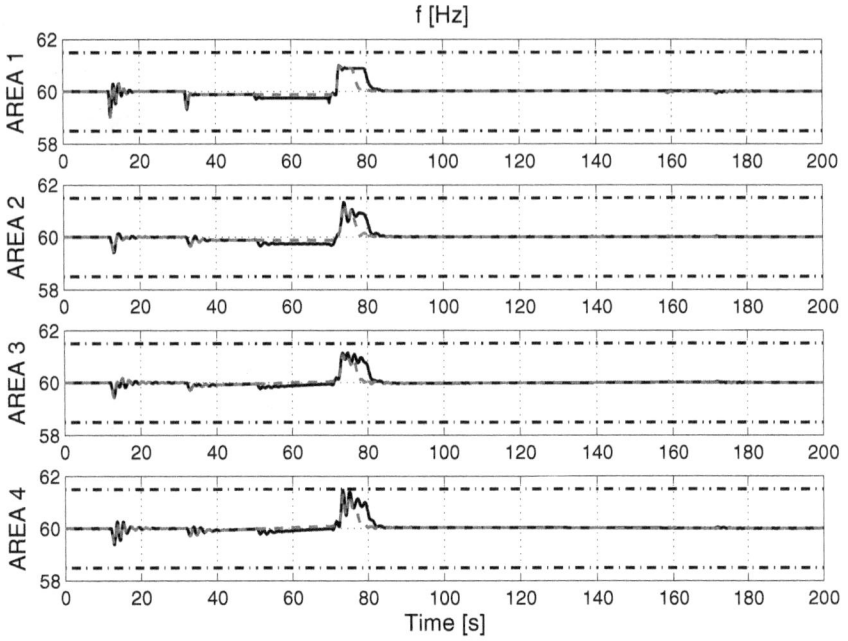

FIGURE 15.45   Frequencies with ROG: faulty (continuous line) and no-faulty (dashed line) cases. The dash-dotted lines represent the boundaries of the prescribed constraints.

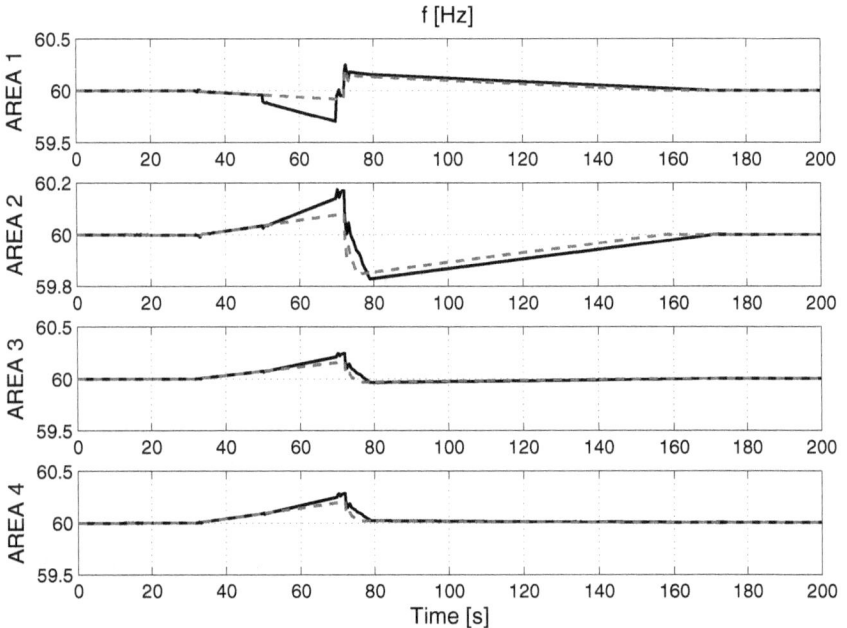

FIGURE 15.46   ROG frequency commands: faulty (continuous line) and no-faulty (dashed line) cases.

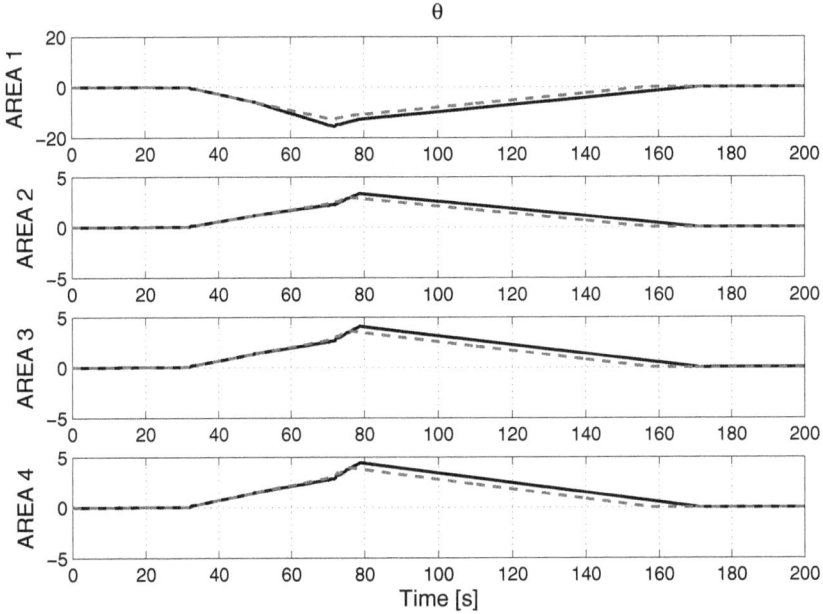

FIGURE 15.47 ROG offset commands: faulty (continuous line) and no-faulty (dashed line) cases.

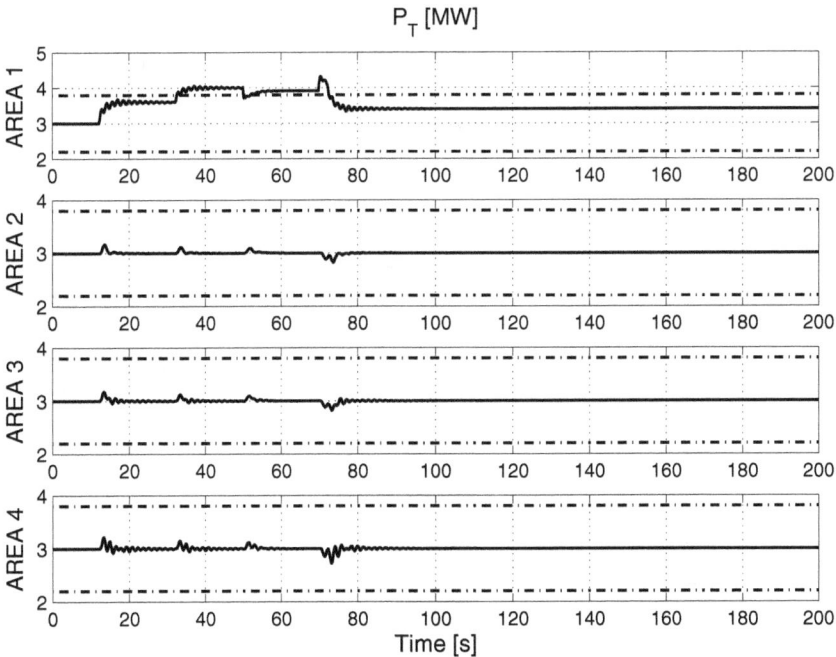

FIGURE 15.48 Generated powers without ROG: faulty case. The dash-dotted lines represent the boundaries of the prescribed constraints.

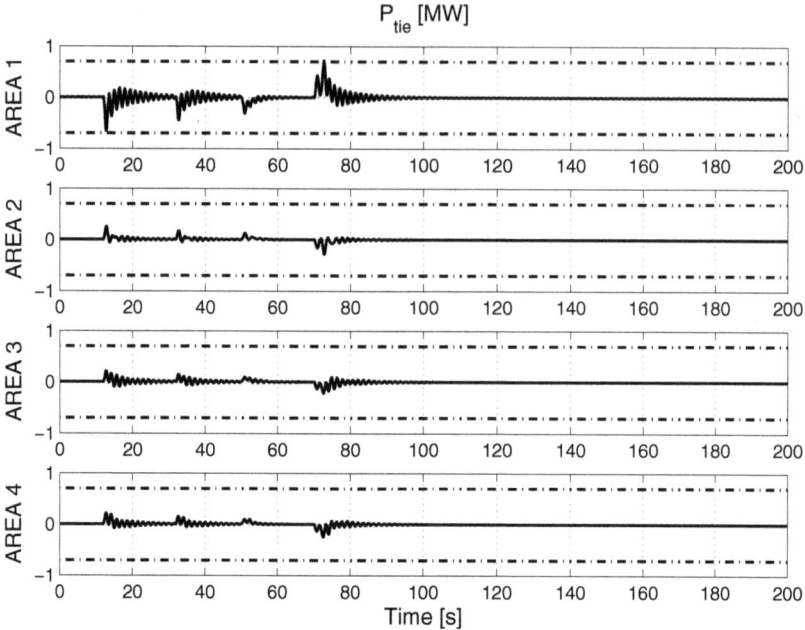

FIGURE 15.49 Tie-lines without ROG: faulty (dashed line) case. The dash-dotted lines represent the boundaries of the prescribed constraints.

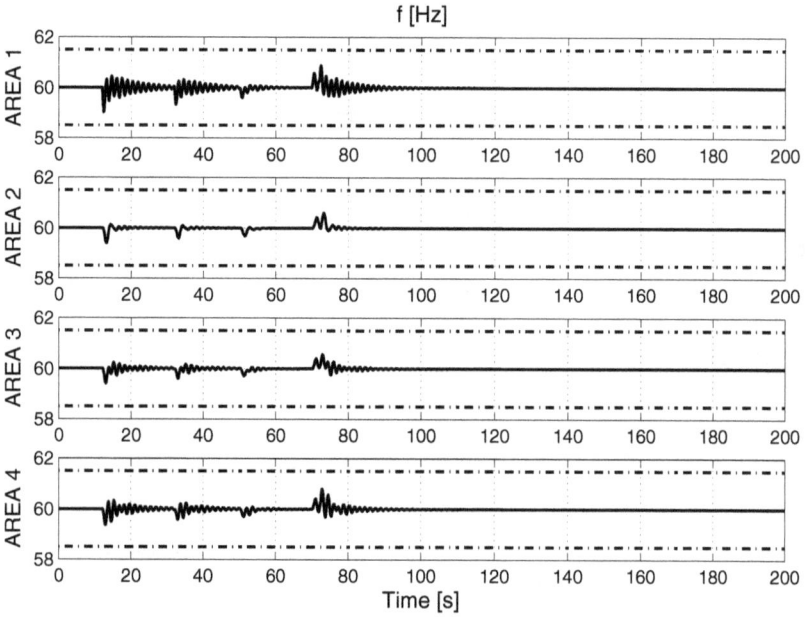

FIGURE 15.50 Frequencies without ROG: faulty case. The dash-dotted lines represent the boundaries of the prescribed constraints.

demand, also if out of the prescribed limits as it occurs in Area 1 (Figure 15.48). Note that such limits spring from turbine saturation, and their violation could lead to dangerous evolutions of the power system. Then, the abrupt load changes affect frequency variations, which result in undesirable transient phenomena marked out by excessive oscillations (Figure 15.50).

15.6 CONCLUSIONS

In this chapter, a centralized supervisory strategy for a wide-area LFC control problem has been presented. Details on its effectiveness in coordinating the power generation units and maintaining relevant variables of the grid within prescribed safety constraints have been reported. Special efforts have been devoted to investigating how this class of strategies behaves under critical events, such as faults, failure or abrupt changes in the load demand.

Relevant examples have been presented where hard-to-predict adverse phenomena have been considered as benchmark problems for the proposed strategies. It has been shown that the standard primary/secondary ALFC control schemes are ineffective, whereas the proposed tertiary level has been proven to be effective in handling such failure events.

In particular, simulation results have shown that the proposed approaches ensure viable evolutions to the overall networked system with respect to safety and operative constraints, despite remarkable changes in the load demand.

APPENDIX

TWO-AREA SYSTEM

T_T: turbine time constant, $T_{T_1} = T_{T_2} = 0.3s$;
T_G: governor time constant, $T_{p_1} = T_{p_2} = 20s$;
T_p: power system time constant, $T_p1 = T_p2 = 20\ s$;
R: regulation parameter, $R_1 = R_2 = 2.4$ Hz/MW;
K_p: power system gain, $K_p1 = K_p2 = 120$ Hz/MW;
T_{12}: synchronizing bias parameter, $T_{12} = 0.545$ MW;
B_S: frequency bias parameter, $B_{S_1} = B_{S_2} = 0.425$ MW / Hz;
K: integration gain, $K_1 = K_2 = 1$;
a_{12}: the ratio between the base values of the two areas, $a_{12} = -1$.

FOUR-AREA SYSTEM

$T_{p1} = 20\ s\ K_{p1} = 120,\ T_{T1} = 0.3,\ T_{G1} = 0.08,\ R_1 = 2.4$;
$T_{p2} = 25\ s\ K_{p2} = 112.5,\ T_{T2} = 0.33,\ T_{G2} = 0.072,\ R_2 = 2.7$;
$T_{p3} = 20\ s\ K_{p3} = 125,\ T_{T3} = 0.35,\ T_{G3} = 0.07,\ R_3 = 2.5$
$T_{p4} = 15\ s\ K_{p4} = 115,\ T_{T4} = 0.375,\ T_{G4} = 0.085,\ R_4 = 2$
$T_{12} = T_{13} = T_{14} = T_{21} = T_{23} = T_{31} = T_{32} = T_{41} = 0.545$
$T_{24} = T_{34} = T_{42} = T_{43} = 0$;
$K_1 = K_2 = K_3 = K_4 = 0.6$;
$B_{S_1} = B_{S_2} = B_{S_3} = B_{S_4} = 0.425$ Hz/MW.

REFERENCES

[1] K. Tomsovic, D.E. Bakken, V. Venkatasubramanian and A. Bose, "Designing the next generation of real-time control, communication and computations for large power systems", *Proceedings of the IEEE*, Vol. 93, N. 5, pp. 965–979, 2005.

[2] G. Andersson, "Modelling and analysis of electrical power systems", EEH-Power Systems Laboratory, ETH Zürich, 2004.

[3] Y. Shi, H.D. Tuan, A.V. Savkin, T.Q. Duong and H.V. Poor, "Model predictive control for smart grids with multiple electric-vehicle charging stations", *IEEE Transactions on Smart Grids*, Vol. 10, N. 2, pp. 2127–2136, 2019.

[4] M.D. Ilic, H. Allen, W. Chapman, C.A. King, J.H. Lang and E. Litvinov, "Preventing future blackouts by means of enhanced electric power systems control: From complexity to order", *It Proceedings of the IEEE*, Vol. 93, N. 11, pp. 1920–1941, 2005.

[5] E. Nobile, A. Bose and K. Tomsovic, "Feasibility of a bilateral market for load following", *IEEE Transactions on Power Systems*, Vol. 16, N. 4, pp. 782–787, 2001.

[6] G. Franz and F. Tedesco, "Constrained load/frequency control problems in networked multi-area power systems", *Journal of the Franklin Institute*, Vol. 348, N. 5, p. 832–852, 2011.

[7] F. Tedesco and A. Casavola, "Fault-tolerant distributed load/frequency supervisory strategies for networked multi-area microgrids", *International Journal of Nonlinear and Robust Control*, Vol. 24, N. 8–9, pp. 1380–1402, 2014.

[8] A. Casavola, M. Papini and G. Franze', "Constrained supervision of dynamic systems in spatial networks", *IEEE Transaction on Automatic Control*, Vol. 51, N. 3, pp. 421–437, 2006.

[9] E.G. Gilbert, I.V. Kolmanovsky and K. Tin Tan, "Discrete-time reference governors and the nonlinear control of systems with state and control constraints", *International Journal on Robust and Nonlinear Control*, Vol. 5, N. 5, pp. 487–504, 1995.

[10] A. Bemporad, A. Casavola and E. Mosca, "Nonlinear control of constrained linear systems via predictive reference management", *IEEE Transactions on Automatic Control*, Vol. 42, N. 3, pp. 340–349, 1997.

[11] E.G. Gilbert and K. Tin Tan, "Linear systems with state and control constraints: The theory and applications of maximal output admissible sets", *IEEE Transactions on Automatic Control*, Vol. 36, N. 9, pp. 1008–1020, 1991.

[12] E. Garone, F. Tedesco and A. Casavola, "Sensorless supervision of linear dynamical systems: The feed-forward command governor approach", *Automatica*, Vol. 47, N. 7, pp. 1294–1303, 2011.

[13] A. Casavola, E. Mosca and D. Angeli, "Robust command governors for constrained linear systems", *IEEE Transactions on Automatic Control*, Vol. 45, N. 11, pp. 2071–2077, 2000.

[14] I.V. Kolmanovsky and J. Sun, "Parameter governors for discrete-time nonlinear systems with pointwise-in-time state and control constraints", *Automatica*, Vol. 42, N. 5, pp. 841–848, 2006.

[15] T.C. Yang, Z.T. Ding and H. Yu, "Decentralized power system load frequency control beyond the limit of diagonal dominance", *Electrical Power and Energy Systems*, Vol. 24, N. 3, pp. 173–184, 2002.

[16] O.I. Elgerd, "Control of electric power systems", *IEEE Control Systems Magazine*, Vol. 1, N. 2, pp. 4–16, 1981.

[17] N. Cohn, *Control of generation and power flow on interconnected systems*, New York: Wiley, 1959.

[18] T. Hiyana, "Design of decentralized load frequency regulators for interconnected power systems", *IEE Proceedings C*, Vol. 129, N. 1, pp. 17–23, 1982.

[19] K.Y. Lim, Y. Wang and R. Zhou, "Decentralized robust load-frequency control in coordination with frequency-controllable HVDC links", *Electrical Power and Energy Systems*, Vol. 19(7), 1997, pp. 423–431.

[20] G.C. Walsh and H. Ye, "Scheduling of networked control systems", *Em IEEE Control Systems Magazine*, Vol. 21, pp. 57–65, 2001.

16 Distributed Supervision and Coordination of Load/Frequency Control Problems in Networked Multi-Area Power Systems

Alessandro Casavola, Giuseppe Franzè, and Francesco Tedesco

DEDICATION

CONTENTS

16.1 INTRODUCTION

In this chapter, the class of constrained coordination and supervision problems described in Chapter 15 is generalized to master/slave architectures that can better face communication latency and packet drops due to the network. Here, the network latency is taken into account and is abstractly modeled as a time-varying time delay, which is allowed to become unbounded in the case of data loss. When the network's protocol is equipped with a time-stamping mechanism, the Linear Time-Invariant

DOI: 10.1201/9781003293989-16

425

(LTI) discrete-time paradigm can be used for system description. Amongst many, this approach has the merit of allowing the analysis and the design of networked control systems without being hampered by network and protocol details. In fact, it has a low impact on the existing control methodologies, which can be used without modifications apart from the presence of the time delay, expressed now as a multiple of the sampling time. See [1] and references therein for further discussions and theoretical justifications.

In networked control applications, time delay usually affects both directions, viz. the collection of data from the field and also the transmission of commands to the individual units, but usually the latter direction is the most critical and needs special care for avoiding loss of performance or even stability. The control problem is even more complex because the statistical properties of the time delay can be time varying. In fact, the time delay is typical constant only on dedicated physical links, such as radio or optical links, when no possibility of congestion or data-loss exists. Over packet networks, it is typically random and stationary when the traffic is sparse and the bandwidth high. Under congestions or denial-of-service attacks, the time delay abruptly increases and it cannot any longer be assumed stationary.

The necessity to face the problem of delayed communication has received increasing importance from the researcher community and is now well understood that it can compromise the system integrity by driving the power system toward instability or other unacceptable behaviors when authoritative wide-area control actions are implemented. See, e.g., [2–5] for studies that have analyzed the presence of communication time delay in power system networks and limitedly to normal operating scenarios. Hereafter, we present a quite general approach for wide-area load/frequency control (LFC) under delayed communication yet having large possibilities of customization with respect to the available *a priori* information on the time delay statistics.

The presence of a network requires a more general distributed supervision architecture with respect to that discussed in Chapter 15 (see Figure 16.1). In particular, we focus on the master/slave scenario depicted in Figure 16.1, where the ALFC master represents a supervisor in charge of taking decisions based on delayed information

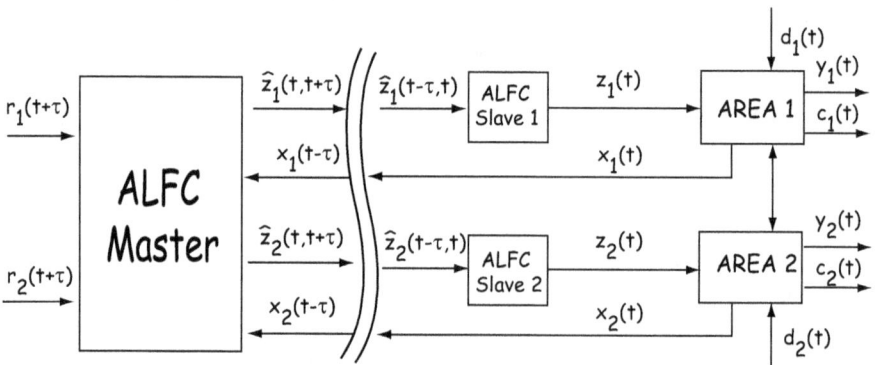

FIGURE 16.1   A distributed tertiary LFC control structure for a two-area power system ©
Elsevier, Ltd.

coming from the areas via data networks, namely, suitable state vectors $x_i(t - \tau)$, with τ being the time-delay representing the network latency. At remote sides, the ALFC slave units are those parts of the strategy that require only local, though updated, information. Moreover, each area in Figure 16.1 represents a micro-grid to be supervised. Then, they are implemented distributively in the proximity of each generating unit. Primary and secondary standard control structures are still present and are not modified by the addition of this proposed tertiary control level.

In Figure 16.1, the signals r_i, d_i, x_i, y_i and c_i represent, respectively, nominal setpoints, loads, states, performance-related and coordination-related outputs for the i-th area. In such a context, the supervisory task can be expressed as the requirement of satisfying some tracking performance, viz. $y_i(t) \approx r_i(t)$, whereas the coordination task consists of enforcing pointwise-in-time constraints of the form $c_i(t) \in C_i$ on each area and/or $f(c_1(t), c_2(t), \ldots, c_N(t)) \in C$ on the overall evolution of the power grid. These constraints can be used to impose all safety, operative and coordination requirements that ensure feasible evolutions to the power grid and prevent the actions of the tripping devices.

For each area, the ALFC master action consists of a vector $\hat{z}_i := \begin{bmatrix} \hat{g}_i' & \hat{\theta}_i' \end{bmatrix}'$, where $\hat{g}_i$ represents the best approximation of r_i compatible with the previous constraints. The modification of r_i in g_i takes place if the equilibrium corresponding to r_i is not any longer sustainable given the actually changed conditions. On the contrary, if g_i is applied instead of r_i, it doesn't produce a constraints violation and allows the overall network to reach a new feasible and sustainable equilibrium. The vector $\hat{\theta}_i$ is an additional optional command. If used, it represents an offset sequence to be added at the input terminal of the generation unit in order to enlarge the set of feasible evolutions for that area and for the overall power grid. It is worth commenting that, because of time delay, the computed vector $\hat{z}_i(t, t+\tau)$ is doubled indexed. Such a notation indicates that they are computed at time t for being applied exactly at time $t + \tau$ or nevermore. It will be shown that their exact application at time $t + \tau$ is mandatory to ensure constraints fulfillment and one cannot depart from it. Finally, because of possible data loss, if the vector $\hat{z}_i(t-\tau, t)$ is not available at time t at the remote side, the ALFC slave logic is typically instructed to replace it with the last applied vector $z_i(t-1)$. Then, we denote with $z_i(t)$ the vector of commands really applied at time t and allow it to be different from $\hat{z}_i(t-\tau, t)$ in general.

The previous ALFC scheme makes use of an LTI-TD dynamic model of each area for computing the predictions of future evolutions of the power grid. Moreover, unlike most LFC schemes reported in the literature, here the models have been defined in terms of absolute variables. The motivation of such a choice is related to the possibility of taking advantage of set-point adjustments w.r.t. their nominal values in order to enlarge the set of feasible transients. This is of paramount importance in the presence of faults and/or unexpected large load changes for avoiding the intervention of the tripping devices.

It is expected that the previously described ALFC strategy, in response to critical events, be capable to reconfigure the set-points and the additional offsets for the local primary/secondary control laws in such a way as to maintain the overall system evolutions always coordinated, within the prescribed safety and operative constraints.

The core of the previous ALFC supervisory strategy is based on the ROG strategy described in Chapter 15. Here, in order to take care of the time-varying communication latency from the control center to each area, an extension of this strategy is presented, hereafter referred to as *distributed reference offset governor* (DROG), by resorting to the master/slave approach of [6].

The chapter is organized as follows. In Section 16.2, the DROG design problem is discussed and its relevant properties summarized. Section 16.3 reports simulation results of a detailed investigation on a two-area power system in the presence of network latency. Finally, some conclusions end the chapter. The content of this chapter is mainly based on [7, 8].

16.2 DISTRIBUTED REFERENCE-OFFSET GOVERNOR (DROG) DEVICE

For ease of presentation, the DROG approach is presented for a single-slave system. The extension to the general multi-slave case is straightforward and follows similar ideas of [6]. We focus on the master/slave scenario depicted in Figure 16.2. The typical system structure for the slave side consists of a primal compensated plant (equipped with an integral action), described by the following state-space representation:

$$
\begin{cases}
x(t+1) &= \Phi x(t) + G_g g(t) + G_\theta \theta(t) + G_d d(t) \\
y(t) &= H_y x(t) \\
c(t) &= H_c x(t) + L_g g(t) + L_\theta \theta(t) + L_d d(t)
\end{cases}
\tag{16.1}
$$

In particular, $x(t) \in \mathbf{R}^n$ is an extended state that possibly collects plant and compensator states; $g(t) \in \mathcal{R}^m$ is the manipulated reference that the ROG slave logic really applies to the plant at time t. If no ROG units were present, it would essentially coincide with the reference $r(t) \in \mathcal{R}^m$; $\theta(t) \in \mathcal{R}^m$ is an adjustable offset applied to the nominal control action, which we assume to be selected from a given convex and compact set Θ, with $0_m \in \text{int}\,\Theta$; $d(t) \in \mathcal{R}^{n_d}$ is an exogenous bounded disturbance satisfying $d(t) \in \mathcal{D}, \forall t \in \mathcal{Z}_+$ with $\mathcal{D}$ a specified convex and compact set, such that $0_{n_d} \in \mathcal{D}$; $y(t) \in \mathcal{R}^m$ is the output, viz. a performance-related signal; and, finally,

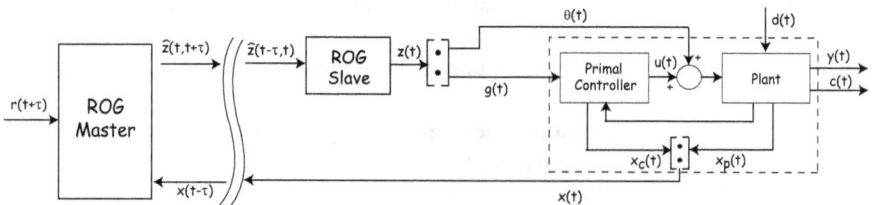

FIGURE 16.2 The distributed reference offset governor (DROG) © Elsevier.

$c(t) \in \mathbf{R}^{n_c}$ is the prescribed constraint output vector, viz. $c(t) \in \mathscr{C}$, $\forall t \in \mathbb{Z}_+$, with $\mathscr{C}$ a specified convex and compact set. It is assumed that

$$(A2) \begin{cases} 1. & \text{System (16.1) is asymptotically stable;} \\ 2. & \text{System (16.1) is offset free w.r.t. } g(t), \text{viz.} \\ & H_y(I-\Phi)^{-1}G_g = I_m \end{cases} \qquad (16.2)$$

In the sequel, let us consider the following state description of (16.1):

$$\begin{cases} x(t+1) &= \Phi x(t) + Gz(t) + G_d d(t) \\ y(t) &= H_y x(t) \\ c(t) &= H_c x(t) + Lz(t) + L_d d(t) \end{cases} \qquad (16.3)$$

where $G = [G_g \ G_\theta]$, $L = [\ L_g \ L_\theta]$, and $z(t) = \left[g^T(t)\theta^T(t) \right]^T \in \mathscr{R}^{2m}$ is the command that the ROG slave logic really applies to the plant at time t. At the master side, the system representation (16.3) is used as a model for computing the ROG master actions $\hat{z}(t,t+\tau) = \left[\hat{g}^T(t,t+\tau)\hat{\theta}^T(t,t+\tau) \right]^T \in \mathbf{R}^{2m}$ (see Figure 16.2), which has to be read as: commands generated at time t for being applied at time $t + \tau$. Typically it would be $z(t) = \hat{z}(t - \tau, t)$ under normal conditions. However, if $\hat{z}(t - \tau, t)$ would not be available at the remote side at time t for congestions or data loss, the ROG slave logic is instructed to apply the previously applied command, viz. $z(t) = z(t - 1)$.

As before, it is required that: 1) $\hat{g}(t,t+\tau) \to \hat{r}$ whenever $r(t + \tau) \to r$, with $\hat{r}$ the best admissible approximation of r and $\hat{\theta}(t,t+\tau) \to 0_m$; and 2) the DROG have a finite settling time, viz. $\hat{g}(t,t+\tau) \to \hat{r}$ and $\hat{\theta}(t,t+\tau) \to 0_m$ for a possibly large but finite t whenever the reference stays constant after a finite time.

In order to make our discussion more general, we allow hereafter the time delay to be possibly time varying. To this end, let $\tau_f(t)$ and $\tau_b(t)$ be, respectively, the forward and backward time delays at each time instant t (expressed in sampling units), viz. $\tau_f(t)$ is the delay from the master to the slave unit, whereas $\tau_b(t)$ is the delay in the opposite direction. We assume further that the following least upper-bounds $\bar{\tau}_f$ and $\bar{\tau}_b$

$$\tau_f(t) \le \bar{\tau}_f \text{ and } \tau_b(t) \le \bar{\tau}_b \qquad (16.4)$$

are either known due to the nature of communication channel or prescribed as operative limits so as to ensure some level of tracking performance. In the latter case, we can distinguish two different operative modes

Normal case: (16.4) holds true $\forall t \in \mathbb{Z}_+$ (16.5)

Abnormal case: (16.4) is occasionally violated.   (16.6)

Notice that an abnormal case occurs when one or both of the previous conditions are violated for at least a time instant. However, the forward time delay limits the performance much more severely than the backward one. This asymmetry is justified by the fact that occurrences of abnormal events, even for a single time instant, prevent the delivered commands from being applied at their scheduled instants. This occurrence is in fact fully equivalent to data loss in our scheme. On the contrary, the same occurrences in the backward direction are much more tolerable, consisting of an extra delay in the flow of data coming from the slaves and on the consequent predictions updating. The backward time delay $\bar{\tau}_b$, however, has to be considered for feasibility reasons deriving from the presence of the disturbances.

The idea here is that the DROG master logic device acts as if the time delay would not be present by modifying, whenever necessary, the reference $r\left(t+\bar{\tau}_f\right)$ into $\hat{g}(t,t+\bar{\tau}_f)$ and by adding an offset $\hat{\theta}(t,t+\bar{\tau}_f)$ on the nominal control action. In so doing, at each time t, an admissible command $\hat{z}(t,t+\bar{\tau}_f)$ is generated. The computation of the DROG master action relies on the future state and constrained vector predictions from $t+\bar{\tau}_f$ onward, computed under the assumption that a constant vector z is applied. To this end, let $t_b \leq t$ be the most recent sampling instant at which the master has received a piece of information from the slave and $x(t_b)$ its state at that time. Notice that in the normal case (16.5), the following inequality $t+\bar{\tau}_f -t_b \leq \bar{\tau}_b +\bar{\tau}_f$ holds true which represents the maximum round-trip delay on the network. Again, by linearity, one is allowed to separate the effects of the initial conditions and inputs from those of disturbances, e.g., $x(t) = \bar{x}(t)+\tilde{x}(t)$, where $\bar{x}(t)$ is the disturbance-free component and $\tilde{x}(t)$ depends only on disturbances. Then, the remote disturbance-free plant state at $t+\bar{\tau}_f$ can easily be computed as

$$\bar{x}\left(t+\bar{\tau}_f \mid t_b\right) \quad \Phi^{t+\bar{\tau}_f -t_b}x\left(t_b\right)+ \sum_{i=t_b}^{t+\bar{\tau}_f -1} \Phi^{t+\bar{\tau}_f -i-1}G\hat{z}\left(i-\bar{\tau}_f,i\right) \tag{16.7}$$

under the assumption that all commands $\hat{z}$ delivered from t_b to $t+\bar{\tau}_f -1$ have been received and timely applied.

Consequently, the disturbance-free state and c-vector future predictions emanating from $\bar{x}\left(t+\bar{\tau}_f \mid t_b\right)$ under a whatever constant command z are given by

$$\begin{aligned}
\bar{x}\left(k,\bar{x}\left(t+\bar{\tau}_f \mid t_b\right),z\right) &:= \Phi^k\bar{x}\left(t+\bar{\tau}_f \mid t_b\right)+ \sum_{i=0}^{k-1} \Phi^{k-i-1}Gz \\
\bar{c}\left(k,\bar{x}\left(t+\bar{\tau}_f \mid t_b\right),z\right) &:= H_c\bar{x}\left(k,\bar{x}\left(t+\bar{\tau}_f \mid t_b\right),z\right)+ Lz.
\end{aligned} \tag{16.8}$$

Finally, as in the previous section, consider the set recursion in (15.10) along with the sets (15.11) and (15.12). Then, if non-empty, the convex set

$$\mathscr{W}\left(\bar{x}\left(t+\bar{\tau}_f \mid t_b\right)\right) = \left\{z \in \mathscr{W}_\delta : \bar{c}\left(k,\bar{x}\left(t+\bar{\tau}_f \mid t_b\right),z\right) \subset \mathscr{C}_{k+\bar{\tau}_b+\bar{\tau}_f}, \forall k \in \mathscr{L}_+\right\} \tag{16.9}$$

collects all constant virtual commands in $\mathscr{W}_\delta$ whose corresponding c-evolutions starting at time $t+\bar{\tau}_f$ from the state $\bar{x}\left(t+\bar{\tau}_f \mid t_b\right)$ satisfy the constraints for all $k \in \mathscr{L}_+$.

Notice also that the set $\mathscr{C}_{k+\bar{\tau}_b+\bar{\tau}_f}$ is used in (16.9) in the place of $\mathscr{C}_k$ as (15.13). This is required for feasibility reasons in that we have used the disturbance-free state prediction $\bar{x}\left(t+\bar{\tau}_f \mid t_b\right)$ instead of $x\left(t+\bar{\tau}_f\right)$, that doesn't contain information on the disturbances, as a starting state in defining the virtual predictions (16.8) at each time t. However, because of linearity, we can equivalently take care of the effect of disturbances on the state evolution from t_b to $t+\bar{\tau}_f$ in (16.7) by restricting the feasibility conditions in (16.9) by an amount corresponding to the effect of the disturbances acting for a time equal to $\bar{\tau}_b+\bar{\tau}_f$, which is an upper-bound of $t+\bar{\tau}_f-t_b$ during normal phases.

Also in this case, it can also be shown that $\mathscr{W}\left(\bar{x}\left(t+\bar{\tau}_f \mid t_b\right)\right)$ is finitely determined. Then, the DROG master problem becomes that of determining, at each time instant t,

$$\hat{z}\left(t, t+\bar{\tau}_f\right)=\arg \min_{z \in \mathscr{W}\left(\bar{x}\left(t+\bar{\tau}_f \mid t_b\right)\right)}\left\|g-r\left(t+\bar{\tau}_f\right)\right\|_{\Psi_g}^2+\|\theta\|_{\Psi_\theta}^2 \qquad (16.10)$$

with $\Psi_g=\Psi_g^T>0_m$, $\Psi_\theta=\Psi_\theta^T>0_m$ and $\|v\|_\Psi:=v^T\Psi v$, while the slave part of the ROG logic is far more simpler and reduces to

$$z(t)=\begin{cases}\hat{z}\left(t-\bar{\tau}_f, t\right), & \text{if available}\\ z(t-1) & \text{otherwise}\end{cases} \qquad (16.11)$$

This strategy is justified by the fact that $\mathscr{W}\left(\hat{x}\left(\bar{\tau}_f \mid 0_b\right)\right)$ non-empty implies that $\mathscr{W}\left(\hat{x}\left(t+\bar{\tau}_f \mid 0_b\right)\right)$ is non-empty along system evolutions generated by the DROG strategy (16.10)–(16.11). The main properties of the DROG strategy under **normal conditions** are summarized next in Theorem 16.1.

THEOREM 16.1

Consider system (16.3) along with the DROG logic (16.10)–(16.11). Let assumptions (A2) be fulfilled and $\mathscr{W}\left(\bar{x}\left(t_0+\bar{\tau}_f \mid t_{0_b}\right)\right)$ be non-empty, where $t_0 b$ is the time instant at which the master receives the first piece of information from the slave, generated at time $t_0 b \le t_0$. Then:

1. All the properties enunciated in Theorem **15.1** are preserved. In particular, the constraints are fulfilled $\forall k \in \mathscr{Z}_+$, the tracking performance optimized and the overall asymptotical stability ensured.
2. The strategy achieves a complete time-delay compensation, in that all commands generated at time t at the master side will be applied to the slave with exactly $\bar{\tau}_f$ sampling steps of delay. $\blacksquare$

Proof—The proof is straightforward by noting that the DROG logic (16.10)–(16.11) essentially implements the ROG scheme with (15.8) replaced by

$$z\left(t+\overline{\tau}_f\right)=\hat{z}\left(t,t+\overline{\tau}_f\right):=\underline{z}\left(x\left(t+\overline{\tau}_f\right),r\left(t+\overline{\tau}_f\right)\right).$$

Therefore, all properties of Theorem **15.1** trivially follow. Moreover, because $z\left(t+\overline{\tau}_f\right)=\hat{z}\left(t,t+\overline{\tau}_f\right)$, $\forall t$, the stated time-delay compensation property also follows.

16.2.1 HANDLING OF ABNORMAL CASES

The main drawback of the previous strategy is that it cannot handle command-missing situations that are the result of forward abnormal events or data loss. In such cases, one or more slaves could not find the scheduled commands to apply, and the tracking operations had to be aborted.

A way to ameliorate the situation is that of equipping the strategy with a data re-synchronization mechanism, which allows the master ALFC logic to continue to effectively operate after such a re-synchronization phase. The basic idea for a single slave is depicted in Figure 16.3.

During this phase, the normal tracking mode is suspended. This new mode of operations, hereafter referred to as *resync* mode, is advocated by the slave ALFC logic every time it discovers that the command scheduled to be applied at the current time instant is missing. In particular, during the resync mode, the slave CG logic keeps applying the last applied command and sends a re-synchronization request to the master. Once received, the master stops sending new commands, updates the predictions accordingly to the resync accompanying data and starts again to generate new commands in a standard way. Essentially, the resync consists of updating the predictions by taking care that a constant command has been applied to the slave from a specified initial state and for a specified amount of time. Specifically, along with the resync request, the master receives information regarding the instant of data missing, say t_b, the initial state $x(t_b)$ and the applied command $z(t_b) = z(t_b - 1)$.

Then, the state prediction at time $t+\overline{\tau}_f$ can be resynchronized as follows:

$$\overline{x}\left(t+\overline{\tau}_f \mid t_b\right):=\Phi^{t+\overline{\tau}_f-t_b}x\left(t_b\right)+\sum_{k=0}^{t+\overline{\tau}_f-t_b-1}\Phi^k\overline{G}z\left(t_b\right) \tag{16.12}$$

Along with the first command sent after resync, the master also sends an acknowledgment that ends the resync mode and restores the normal tracking mode.

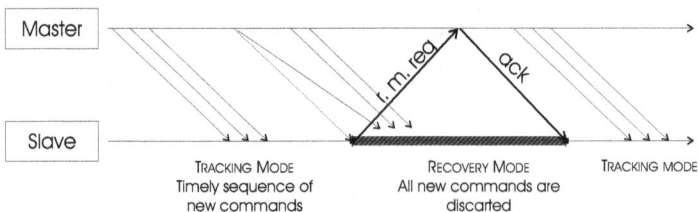

FIGURE 16.3 Resync mode.

The aforementioned, quite simple, resync mechanism requires attention in order to be implemented in a multi-area context. In fact, to be effective it is necessary that all slave logics enter in the resync mode at the same time instant, even if only one of them has experimented with the occurrence of a missing command. Details on how to implement such a distributed agreement protocol can be found in [8].

Also, backward abnormal events can prejudge the feasibility retention property of the method because it is not any longer true that $t + \overline{\tau}_f - t_b \leq \overline{\tau}_b + \overline{\tau}_f$. Then, the effect of the disturbances can be underestimated in the predictions. A simple way to avoid feasibility problems is that of modifying the definition of $\mathcal{W}(x)$ in (16.9) as follows:

$$\mathcal{W}\left(\overline{x}\left(t + \overline{\tau}_f \mid t_b\right)\right) = \left\{ z \in \mathcal{W}_\delta : \overline{c}\left(k, \overline{x}\left(t + \overline{\tau}_f \mid t_b\right), z\right) \subset \mathcal{C}_\infty, \forall k \in \mathcal{L}_+ \right\} \qquad (16.13)$$

This change is justified by the fact that $\mathcal{C}_\infty$ is a restriction that takes care of the effect of the disturbances acting for an infinite time.

16.3 SIMULATIONS

Here, the two-area power system described in Chapter 15 is used. As previously described, the set-point signals are sent from the control center via dedicated links, therefore a bounded latency is a valid assumption. We have modeled it as identical forward and backward random time-varying delays with upper-bounds $\overline{\tau}_f = \overline{\tau}_b = 2$ (sampling steps), as shown in Figure 16.4. The admissible load disturbance is of the form (15.24) with

$$h = \begin{bmatrix} 0.7 \\ 0.7 \\ 0.7 \\ 0.7 \end{bmatrix} [MW]. \qquad (16.14)$$

It is worth pointing out that the allowable disturbances are notably reduced w.r.t the delay-free case of Chapter 15. This is not surprising because the occurrence of induced time delay degrades the system performance (e.g., stability degree, constraints satisfaction). Consequently, if the allowable disturbance set is too big the problem may not have a solution, viz. the set of feasible initial states $\mathcal{V}\left(\hat{x}\left(\overline{\tau}_f \mid 0_b\right)\right)$ is empty. In particular, we have been forced to reduce the maximum bound h to the value given in (16.14). Moreover, the following set of constraints (expressed in Hz and MW) are of interest:

$$59.3 \leq f_1(t) \leq 60.7, \quad 59.3 \leq f_2(t) \leq 60.7,$$
$$|P_{tie}(t)| \leq 0.22, \qquad (16.15)$$
$$2.6 \leq P_{T_1}(t) \leq 3.4, \qquad 2.6 \leq P_{T_2}(t) \leq 3.4.$$

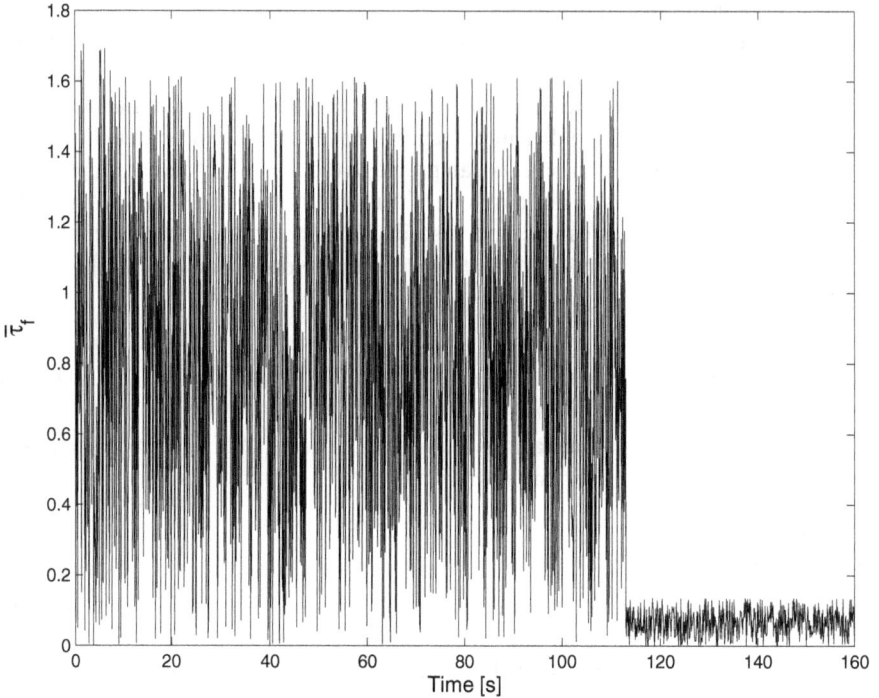

FIGURE 16.4   Communication latency [6].

The initial admissible DROG command vector has been determined as $w(0) = [60\ 60\ 0\ 0]^T$.

16.3.1 Load Changes

The load demand scenario here considered is depicted in Figure 16.5. We have supposed first a load addition in Area 1 (from $t = 2\ s$ to $t = 42\ s$) and then a load reduction in Area 2 (from $t = 62\ s$ to $t = 72\ s$). The load demand in Area 1 (starting from $t = 22\ s$ until $t = 42\ s$) is higher than the maximum generable power for Area 1's generator, whereas in Area 2 it is lower (see constraints (16.15)) than the minimum and such a situation could prevent the matching between production and demand.

In Figure 16.6, it is worth noting how the DROG modifies the frequency setpoints and control offsets from their nominal values, allowing feasible system evolutions in response to any disturbance prescribed in (15.24) and (16.14). In particular, let us consider the case of 3.49 MW load demand in Area 1 (see Figure 16.5); it is interesting to observe that at steady-state the generated power is 3.345 MW and the remaining power required to balance the load (0.145 MW) is produced by Area 2 (see Figure 16.7) and transferred via the tie-line, as shown in Figure 16.8. Similar comments can be made for the load reduction in Area 2. Note that Figure 16.9 has

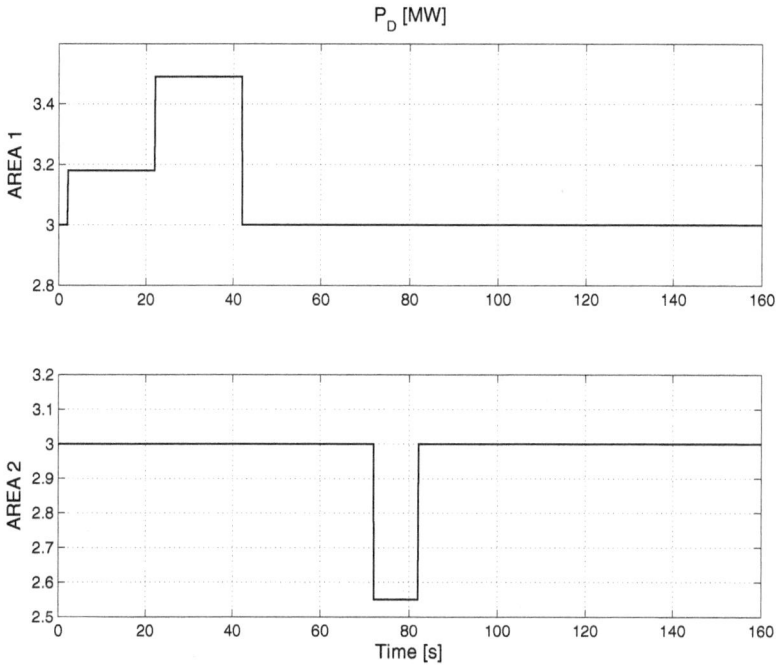

FIGURE 16.5 Load demands.

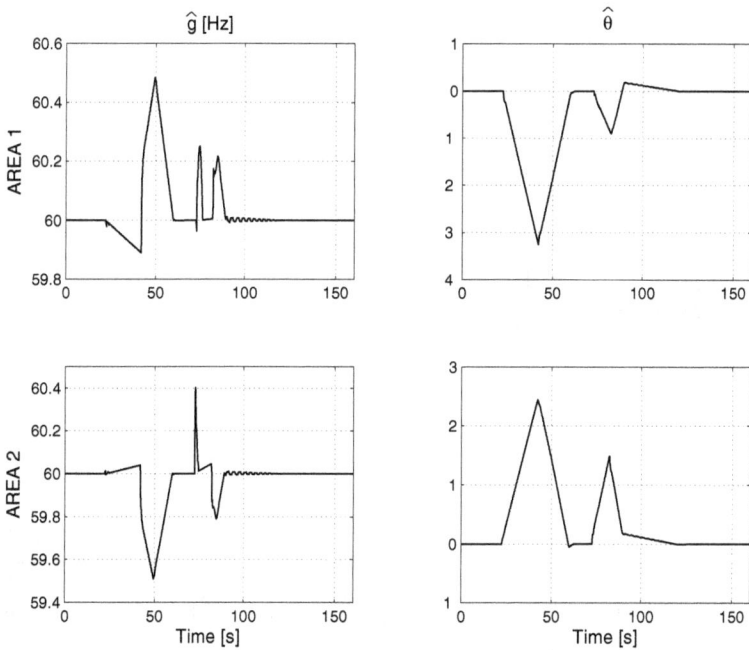

FIGURE 16.6 Frequency (left side) and offset (right side) commands.

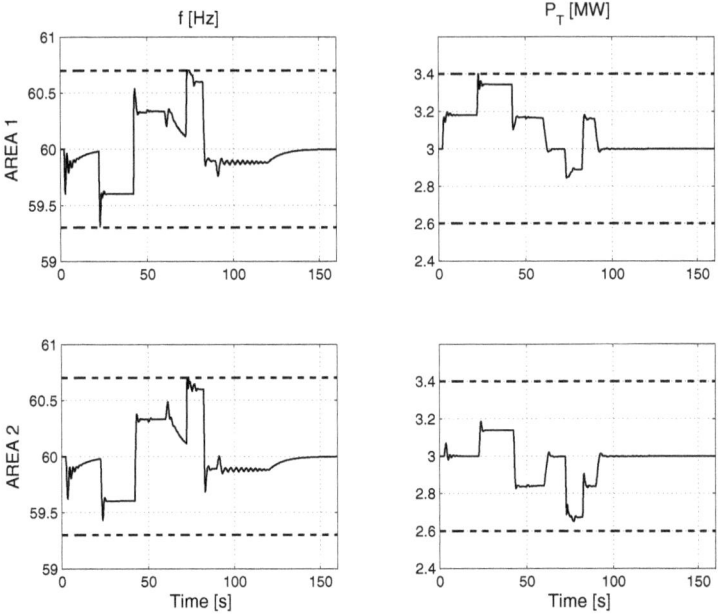

FIGURE 16.7   Frequency (left side) and generated power (right side): with DROG.

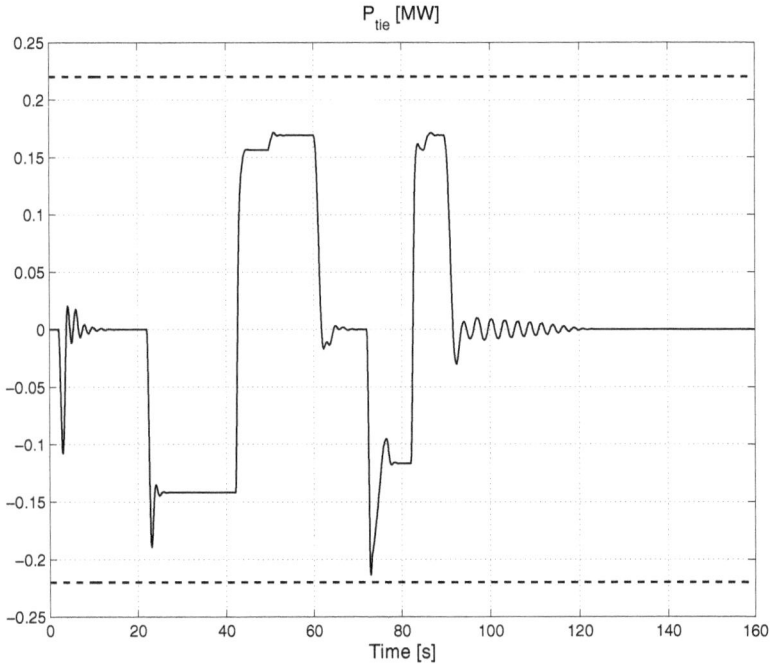

FIGURE 16.8   Tie-line power flow: with DROG.

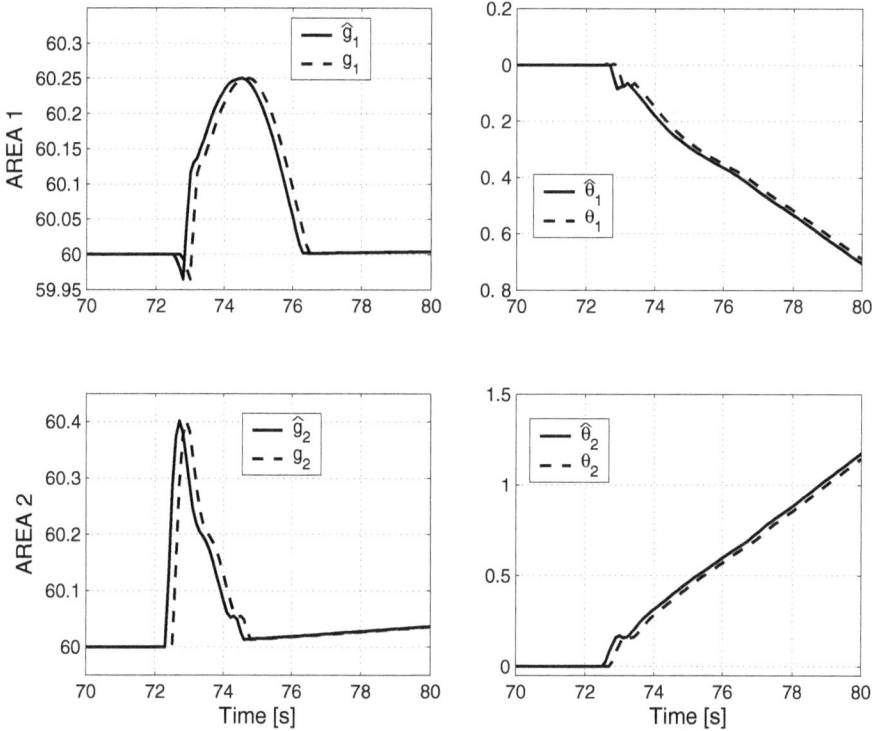

FIGURE 16.9 Computed $\hat{w}$ (continuous line) and applied w (dashed line) commands. Frequency (left side) and control offset (right side): zoom on the time interval $[70, 80]\ s$.

been added to show the relationships between the computed commands $\hat{w}$ and the applied commands w. Notice that the applied command w is just a translated version of $\hat{w}$, that is, $w(t) = \hat{w}\left(t - \overline{\tau}_f, t\right)$.

Finally, Figures 16.10 and 16.11 depict the response of the power system from the initial state $x(0)$ under the control of the ALFC stage only, that is, without any DROG action. Under the same scenario, violations of the prescribed frequency and power constraints come out.

16.3.2 TRANSMISSION LINE FAILURE

In this example, similarly to Section 15.4.1.2 the aim is to investigate how a failure on the transmission line can affect the system evolutions when the latency cannot be neglected. Figure 16.12 depicts the load demand here taken into consideration; it is important to remark that due to such a failure the power exchanged between the two areas is interrupted. Consequently, the load scenario of Figure 16.5 is not manageable.

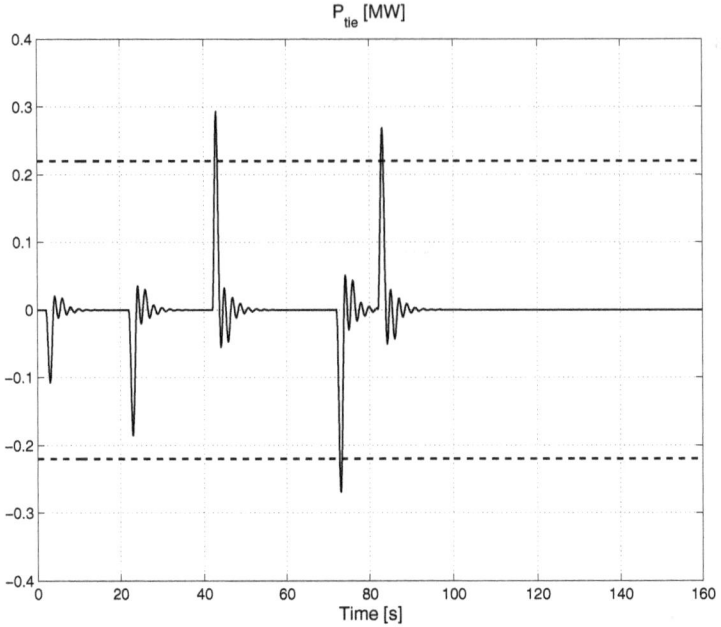

FIGURE 16.10 Tie-line power flow: without DROG.

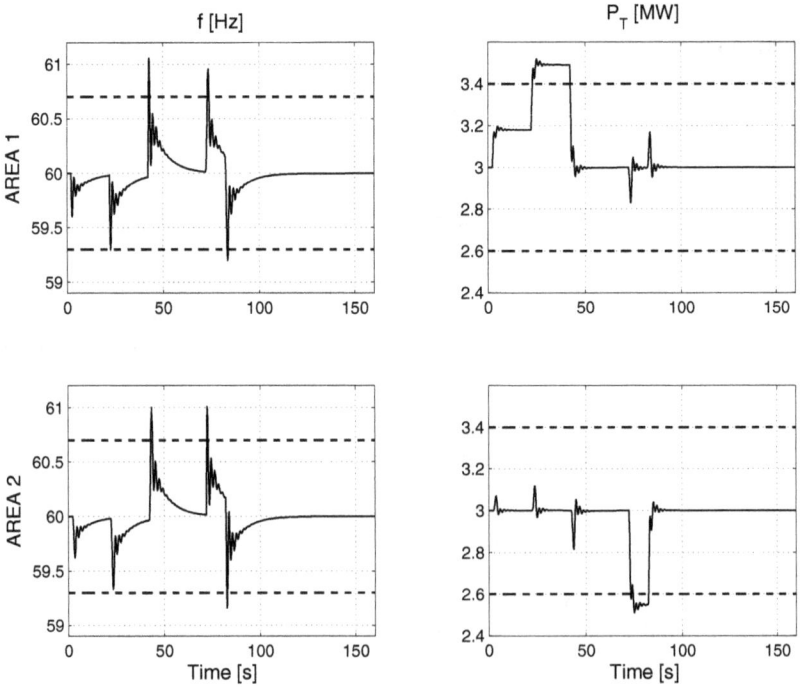

FIGURE 16.11 Frequency (left side) and generated power (right side): without DROG.

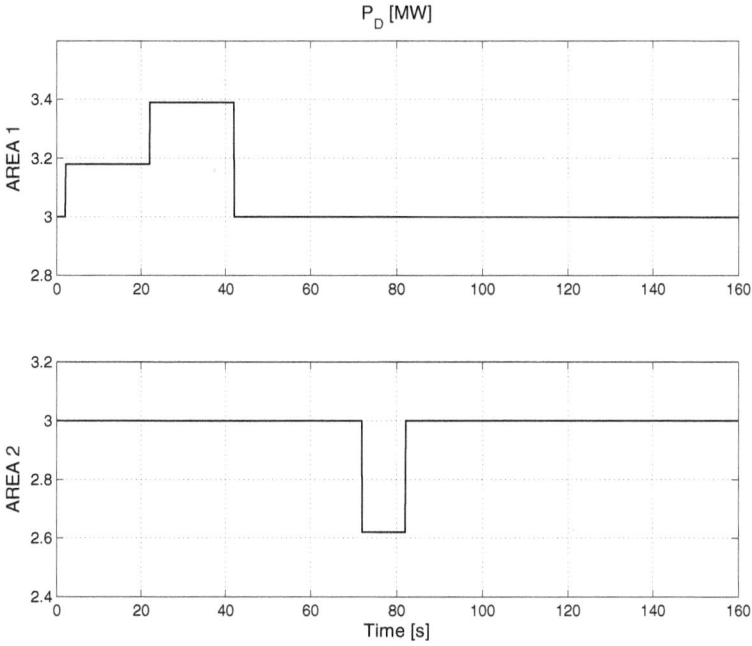

FIGURE 16.12   Load demands.

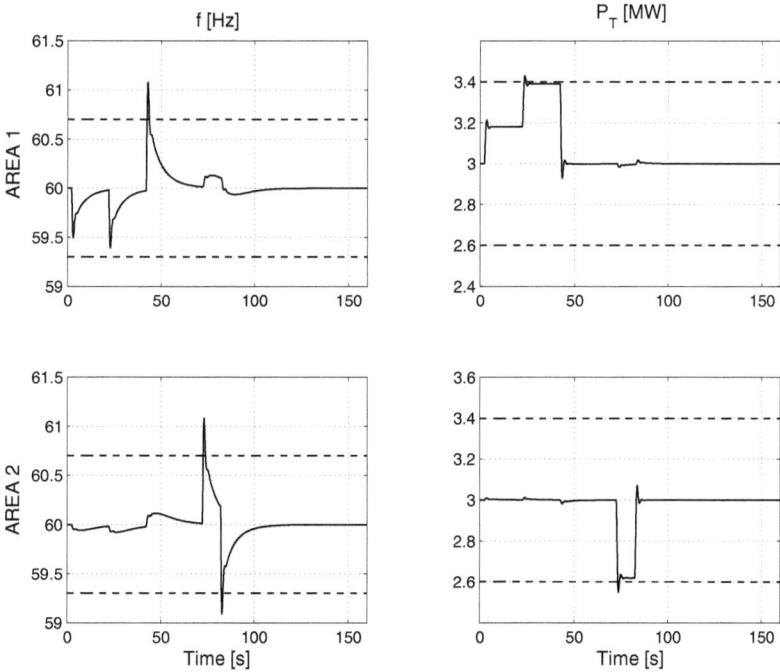

FIGURE 16.13   Frequency (left side) and generated power (right side): without DROG.

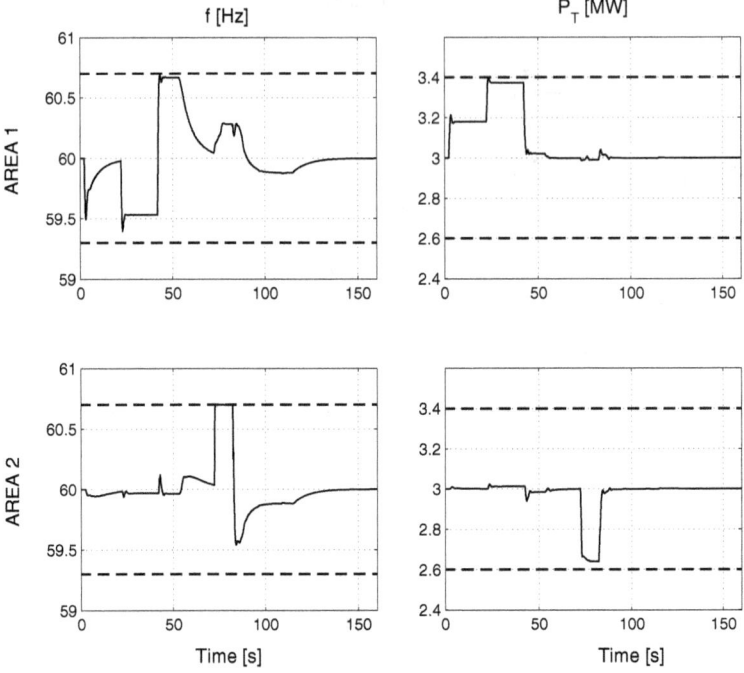

FIGURE 16.14 Frequency (left side) and generated power (right side): with DROG.

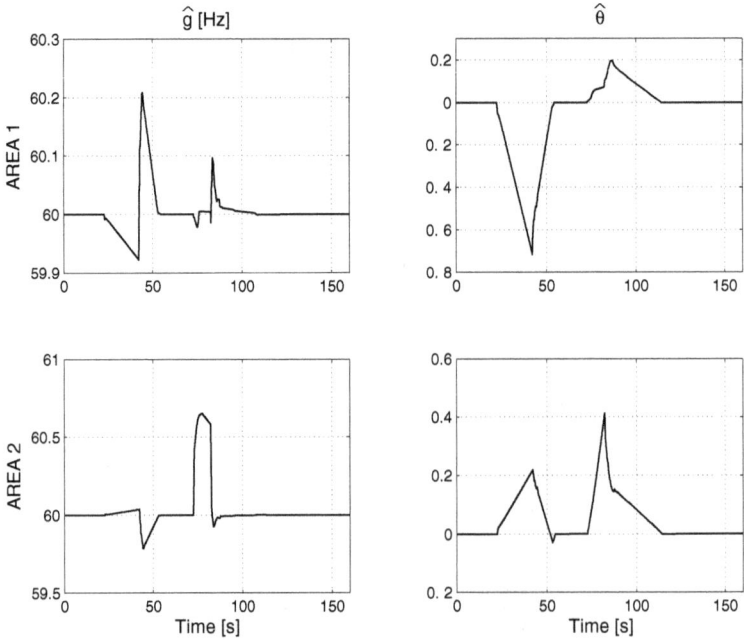

FIGURE 16.15 Frequency (left side) and offset (right side) commands.

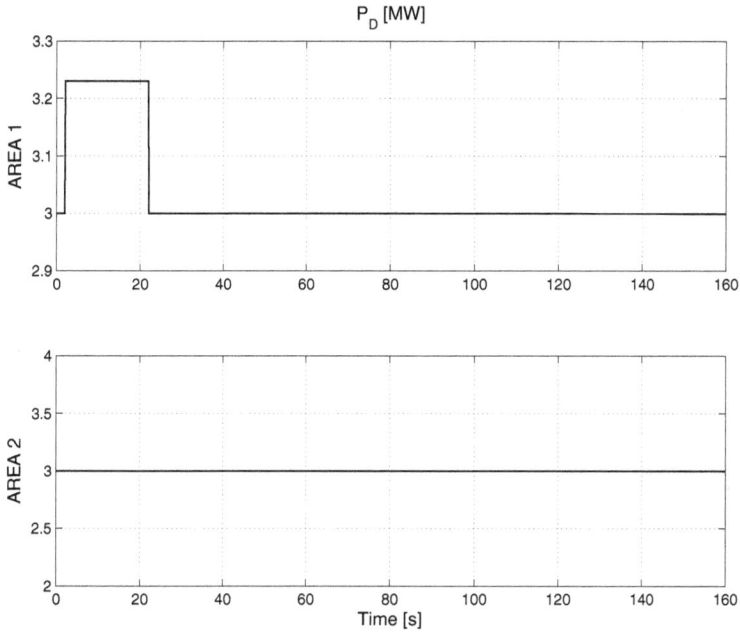

FIGURE 16.16 Load demands.

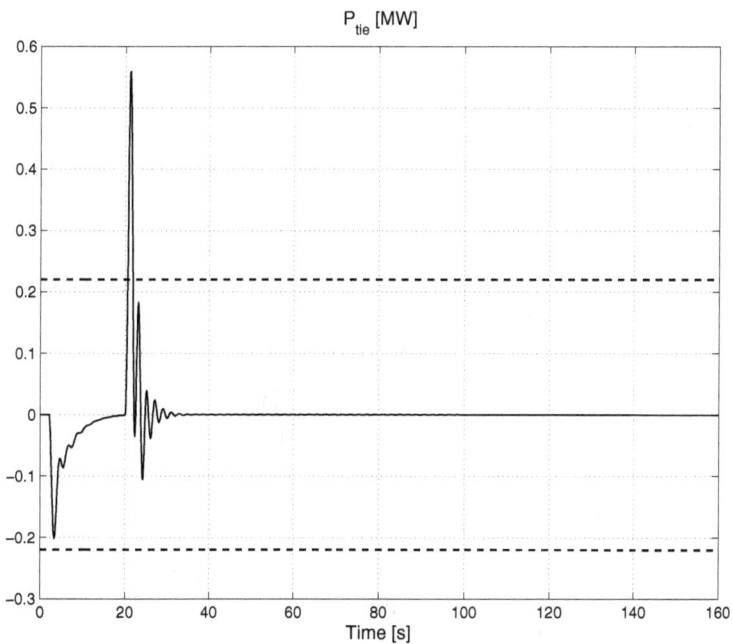

FIGURE 16.17 Tie-line power flow: without DROG.

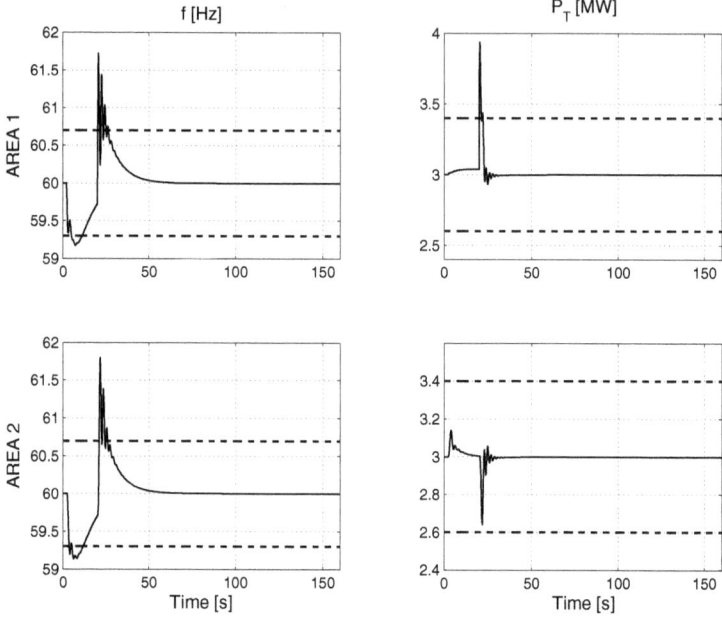

FIGURE 16.18 Frequency (left side) and generated power (right side): without DROG. The dashed lines represent the constraints' boundaries.

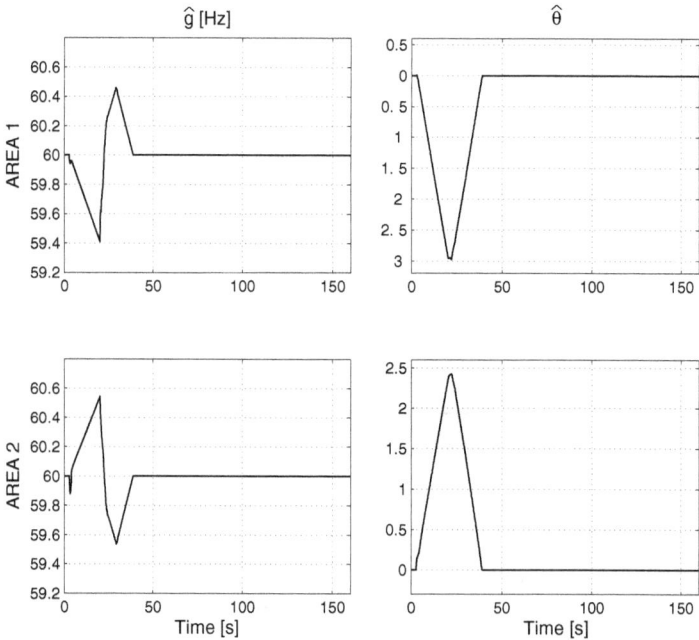

FIGURE 16.19 Frequency (left side) and offset (right side) commands.

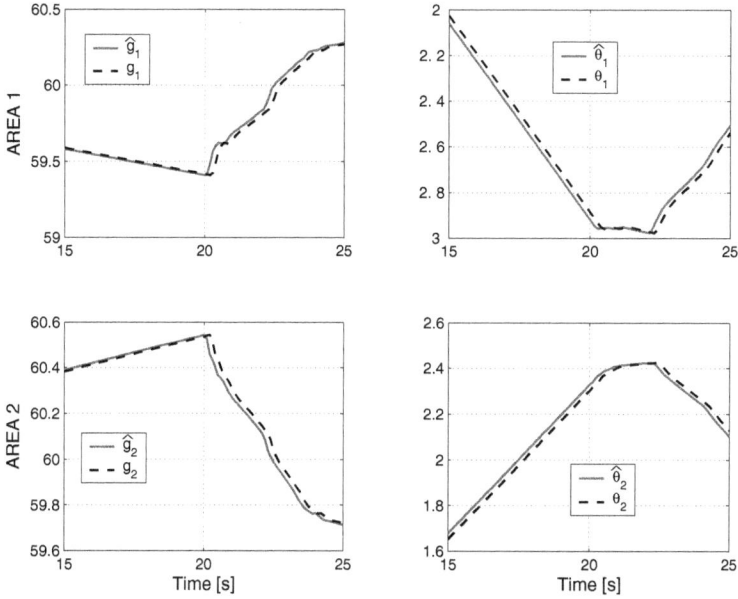

FIGURE 16.20 Computed $\hat{w}$ (continuous line) and applied w (dashed line) commands. Frequency (left side) and power (right side): zoom on the time interval $[15, 40]$ s.

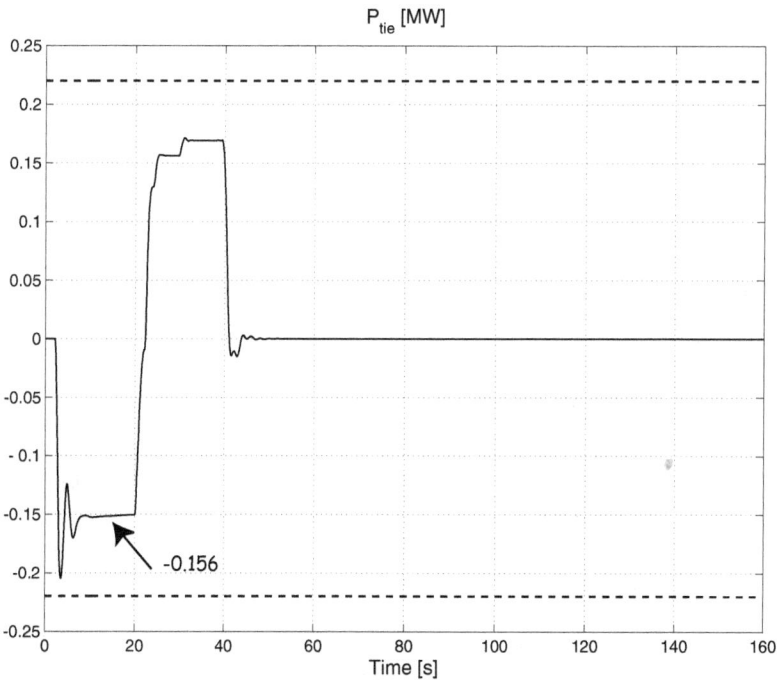

FIGURE 16.21 Tie-line power flow: with DROG.

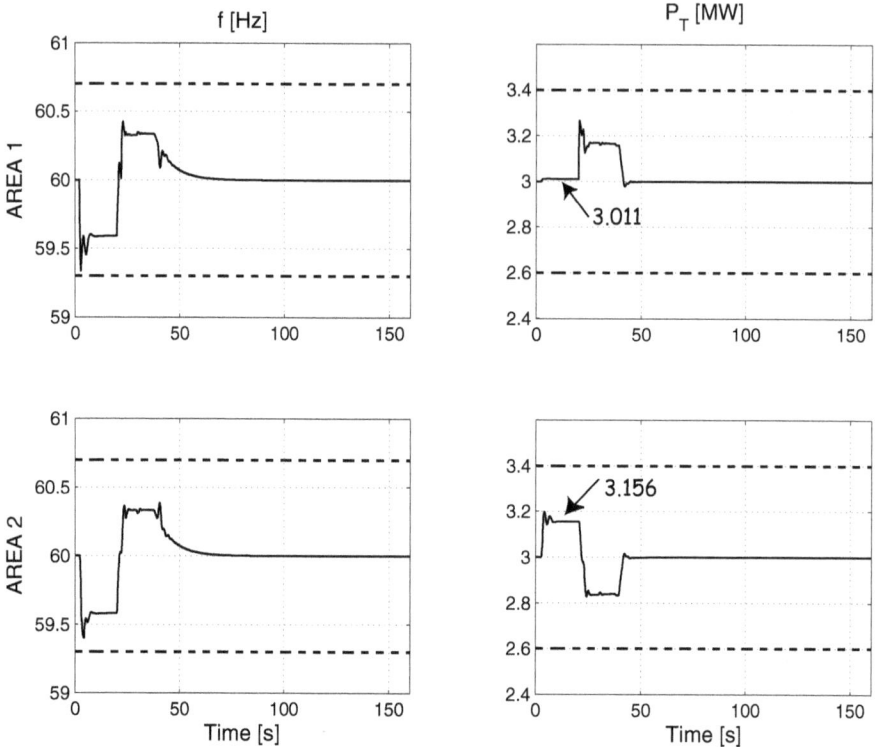

FIGURE 16.22 Frequency (left side) and generated power (right side): with DROG. The dashed lines represent the constraints' boundaries.

As it clearly results from Figure 16.13, during the transients, loss of interconnection causes high peaks both in frequencies and power production when no DROG units are used.

On the contrary, the DROG is able to ensure constraint fulfillments (Figure 16.14) by a suitable modification of the frequency set-points and control offsets (Figure 16.15).

16.3.3 POWER GENERATION FAILURE

The final experiment shows the effects on the two-area power system when the generator of Area 1 exhibits a 90% reduction in its gain due to a fault from $t = 3$ s to $t = 20$ s. This implies that it generates only the 10% of power w.r.t. the no-faulty condition. Moreover, a load addition (step change) of 0.5 MW occurs in Area 1 from $t = 2$ s until $t = 22$ s (see Figure 16.16). Note that we have considered that the load disturbance and power generation failure occur during the same time interval.

In such a simulation, the generator of Area 1 is no longer capable of satisfying the local load demand. Therefore, starting from $t = 3$ s the power balance of the overall network would not have been ensured under the prescribed constraints without a DROG unit (see Figures 16.17 and 16.18). In fact, the DROG unit is capable of modifying the nominal frequency set-points and control offsets (see Figure 16.19) and ensuring a new power balance. Consequently, the generator of Area 2 increases its power production, and a fraction of that is transferred to Area 1 via the tie-line (Figure 16.21) so as to compensate for the faulty condition (Figure 16.22) and balance the load. It is worth commenting that the DROG is not informed of the fault occurrence, and its behavior hinges on its intrinsic reconfiguration capability. As a conclusion, this DROG approach is able to manage, under some circumstances, unpredictable events, such as faults/failures.

16.4 CONCLUSIONS

In this chapter, a master/slave distributed supervisory strategy for wide-area LFC control problems has been presented, well suited for dealing with communication latency effects arising during the data transmission. A detailed methodological analysis has been reported, and the main properties of the strategy have been rigorously proven. Under mild conditions, the DROG scheme has been shown to be able to coordinate the evolutions of the slave micro-grids within prescribed operative safety constraints in the presence of bounded persistent disturbances and bounded time delays.

Simulation results have shown that the proposed master/slave distributed approach ensures viable evolutions to the overall networked system in the presence of large load variations and faulty situations also in the case of communication time delays and packet dropouts.

REFERENCES

[1] G.C. Walsh and H. Ye, "Scheduling of networked control systems", *IEEE Control Systems Magazine*, Vol. 21, N. 1, pp. 57–65, 2001.
[2] M.D. Ilic', H. Allen, W. Chapman, C.A. King, J.H. Lang and E. Litvinov, "Preventing future blackouts by means of enhanced electric power systems control: From complexity to order", *Proceedings of the IEEE*, Vol. 93, N. 11, pp. 1920–1941, 2005.
[3] S. Bhowmik, K. Tomsovic and A. Bose, "Communication models for third party load frequency control", *IEEE Transactions on Power Systems*, Vol. 19, N. 1, pp. 543–548, 2004.
[4] C.H. Hauser, D.E. Bakken and A. Bose, "A failure to communicate: Next generation communication requirements, technologies, and architecture for the electric power grid", *Power and Energy Magazine*, Vol. 3, N. 2, pp. 47–55, 2005.
[5] X. Yu and K. Tomsovic, "Application of Linear Matrix Inequalities for load frequency control with communication delays", *IEEE Transactions on Power Systems*, Vol. 19, N. 3, pp. 1508–1515, 2004.
[6] A. Casavola, M. Papini and G. Franzè, "Constrained supervision of dynamic systems in spatial networks", *IEEE Transaction on Automatic Control*, Vol. 51, N. 3, pp. 421–437, 2006.
[7] G. Franzè and F. Tedesco, "Constrained load/frequency control problems in networked multi-area power systems", *Journal of the Franklin Institute*, Vol. 348, N. 5, pp. 832–852, 2011.
[8] F. Tedesco and A. Casavola, "Fault-tolerant distributed load/frequency supervisory strategies for networked multi-area microgrids", *International Journal of Nonlinear and Robust Control*, Vol. 24, N. 8–9, pp. 1380–1402, 2014.

17 Integrated Power and Transportation Systems Targeted by False Data Injection Cyberattacks in a Smart Distribution Network

Ehsan Naderi and Arash Asrari

ABBREVIATIONS

ACSE	AC estate estimation
BOOP	Bi-objective optimization problem
BCS	Best compromise solution
DCU	Data control unit
DFR	Distribution feeder reconfiguration
DMO	Distribution-level market operator
DSO	Distribution system operator
EV	Electric vehicle
FDI	False data injection
GAMS	General algebraic modeling system
GWO	Grey wolf optimizer
HGWO-PSO	Hybrid grey wolf optimizer-particle swarm optimization
ICT	Information and communication technology
PMU	Phasor measurement unit
PSO	Particle swarm optimization
PV	Photovoltaic
RAS	Remedial action scheme
RTU	Remote terminal unit
SCADA	Supervisory control and data acquisition
WT	Wind turbine

CONTENTS

17.1 INTRODUCTION

17.1.1 BACKGROUND AND MOTIVATION

Depletion of fossil fuels and global warming awareness are among the most important reasons to upgrade both power and transportation networks in modern-day unified energy systems. Toward this end, the concepts of smart grids and smart cities are scrutinized more than ever considering that the assets of conventional transportation systems are being upgraded via different forms of electric vehicles (EVs), such as public E-buses, leading to mobile energy storage units in electricity markets [1–2]. However, the intersection of mobility and electricity introduces new challenges for both modern power and transportation systems [3]. One of the main challenges in the process of rapid improvement of smart grid technology and electric transportation systems is oriented toward implementing a high level of information and communication technology (ICT), which can result in several issues, such as higher vulnerability of the entire network to malicious cyberattacks [4]. Among different types of malicious cyber threats targeting smart grids, false data injection (FDI) attacks in both forms of false positive and false negative are easier to be launched since they do not normally follow a set of strict technical rules; hence, attackers can simply penetrate to the cyber layer of the smart grids and compromise the sensors' readings via injecting malicious data. Moreover, they can keep the injected false data undetectable, negatively affecting the recorded measurements and the process of calculating the state variables in state estimations [5]. To this end, in the domain of modern power grids integrated with smart transportation systems, it is necessary to 1) model complicated attack frameworks targeting such unified systems for analyzing the impacts of cyberattacks on the normal operation of power grids and 2) propose suitable remedial action schemes (RASs) against those cyberattacks in order to recover the grid to the normal situation after the presence of false data in the server computers is confirmed.

17.1.2 LITERATURE REVIEW

17.1.2.1 Related Works Regarding Cyberattacks Targeting Transportation Systems

In [6], the performance of transportation systems targeted by cyberattacks was investigated according to four different criteria, including efficiency, safety, emissions, and fuel consumption. Based on the results reported in [6], the impacts of cyberattacks on the traffic flow could be noticeable by targeting more EVs via severe attacks, consequently leading to higher risk of rear-end collision, more air pollutants, and inefficient traffic operations. In [7], a security model for an intelligent transportation system was presented against different network attacks targeting connected cars, autonomous vehicles, and smart roads, resulting in a catastrophic impact on public safety. A potential cyberattack targeting automated vehicles was proposed in [8], where the impacts of cyber threats on autonomous vehicles, cooperative automated vehicles, and intelligent transportation systems were scrutinized from different aspects. In [9], the effectiveness of two change point models (i.e., expectation maximization and cumulative summation algorithms) for real-time vehicle-to-infrastructure cyberattack detection in a connected vehicle was investigated against three different types of cyberattacks comprising denial of service, impersonation, and false information attacks. A monitoring/controlling algorithm was developed in [10] in order to measure and control the system states via the Internet of Things (IoT)–embedded smart sensors and actuators for a transportation system targeted by cyberattacks. Finally, a comprehensive survey on the security challenges and solutions for a typical intelligence transportation system was presented in [11], where effective recommendations to protect the entire transportation system were also provided.

17.1.2.2 Related Works Regarding FDI Attacks Targeting Smart Power Systems

In [12], a FDI attack model, oriented toward a nonlinear physical-constraint model, was introduced to launch undetectable cyberattacks targeting state estimation in two unbalanced distribution smart grids (i.e., IEEE 13-bus and WSCC 9-bus systems). According to the obtained results reported in [12], not only was the false-positive rate of the bad data detection algorithm 100%, but the physical consequence of the attack was severe as well. In [13], an enhanced algorithm based on a random metric approach was proposed to construct the vector of stealthy FDI cyberattack targeting smart grids in the face of limited information used by attackers. An attack strategy leading to overloaded branch chains and cascading failures was developed in [14], where a game-theoretical bilevel problem was designed for the purpose of decision-making interactions between attacker and security constraints associated with an economic dispatching system. In [15], an attack model was introduced for an FDI cyberattack resulting in subsequent failures (e.g., sequential power outages) in a typical power system, where the attacker launched an optimal FDI attack to purposely trigger the protection devices of a targeted branch, leading to tripping of multiple branches as a consequence of the attack. In the earlier steps of this research, we proposed two different attack frameworks in order to target a smart transmission system [16] and a smart microgrid [17–18]. In [16], a FDI model along with an appropriate RAS was introduced to 1) target the DC state estimation process of a smart transmission network via launching stealthy cyberattacks leading to tie-line congestions and power outages and 2) mitigate the impacts of the cyberattacks by taking advantage of thyristor-controlled series capacitors optimally allocated in the transmission system to provide the power system operator with flexibility to upgrade the reactance of the targeted tie-lines to alleviate the congestion. In [17], a hardware-in-the-loop testbed was developed for a lab-scale microgrid located at Southern Illinois University, Carbondale, IL, USA, to study the impacts of FDI cyberattacks on the normal operation of the renewable-based microgrid when the attacker unexpectedly opened and closed different circuit breakers to intentionally change the level of demand and generation in the system. Finally, a market-based RAS was proposed in [18] to experimentally validate the impacts of FDI cyberattacks on the developed lab-scale microgrid and then handle the attack by re-dispatching the generation units in a secondary market.

17.1.3 RESEARCH QUESTION, RESEARCH GAP, AND CONTRIBUTIONS OF THIS CHAPTER

Although extensive research works have recently been proposed by researchers on modeling of cyberattacks targeting smart grids, a composite framework is yet to be proposed to scrutinize the impacts of an organized FDI cyberattack targeting EV aggregators in a power grid integrated with smart transportation systems to result in congestions and potentially cascading power outages on the distribution side of the electricity network. That is to say, the following question has not yet been addressed in the existing literature: *"How can we model an organized FDI attack targeting smart power and transportation sectors of a smart distribution grid leading to congestions and potentially cascading outages?"* Therefore, the main goal of this study is to assist power system operators to handle cyber threats since it is essential for them to be in attackers' shoes and scrutinize various possible approaches of cyberattacks to be 1) familiar with the impacts of such attacks on different sectors of the system and 2) able to proactively propose appropriate remedial actions to cope with them. To that end, the motivation of this chapter is to present an attack framework to address the indicated research gap. Hence, the main contribution of this chapter is twofold, which is elaborated as follows:

- An organized attack framework to target EV aggregators in a distribution-level electricity market by launching a FDI cyberattack resulting in system congestions and potentially cascading power outages.
- A suitable RAS based on coordination of distribution-level market operator (DMO) and EV aggregators to mitigate the negative impacts of the successful attack via running a secondary distribution-level electricity market and taking advantage of the EVs as mobile distributed energy storage units for congestion management.

17.1.4 ORGANIZATION OF THE CHAPTER

The proposed framework in this chapter is elaborated in Section 17.2. Section 17.3 proposes the corresponding problem formulation, which presents the objective functions and the associated constraints and limitations. Section 17.4 introduces the case study and analyzes the obtained simulation results. Finally, the conclusions and the future steps of this research are outlined in Section 17.5.

17.2 PROPOSED FRAMEWORK

17.2.1 OVERVIEW OF THE PROPOSED FRAMEWORK

Distribution grids were initially designed as passive systems to distribute the electrical energy among end-use customers after receiving the bulk energy from transmission networks [19]. However, renewable energy resources, along with different types of energy storage devices, have introduced a new variant of distribution systems, called active grids [20]. In an active system, not only are the end-users consumers of power, but they can also generate and store electric energy in collaboration with different aggregators to 1) reduce their overall energy cost and 2) contribute to alleviate the peak time stress by selling their extra power to the market. A typical active distribition system is illustrated in Figure 17.1. It is necessary to note that aggregator refers to a retailer having contract with owners (e.g., EV owners, PV owners, and WT owners in this chapter) to participate in the electricity market on their behalf and sell their available electrical energy to increase their profits. In addition, the distribution system operator (DSO) centrally manages the economy and security of the entire power distribution grid.

Referring to Figure 17.1, the proposed FDI attack framework in this chapter is illustrated in Figure 17.2. From Figure 17.2 (a), one can infer that attackers try to target the smart grid encompassing an EV-based transportation system by lanching FDI attacks targeting 1) data control units (DCUs) of EV aggregators to manipulate the (to be submitted to the electricity market) power and price associated with EV owners who have contract with the aggregator and 2) server computers in the supervisory control and data acquisition (SCADA) system to compromise the demand data associated with different buses of the system by bypassing AC estate estimation (ACSE). The intention of this organized attack is to push the distribution system toward overloading of assets, congestion, and possible power outage after the electricity market is cleared by the DSO with false information. In other words, as the DSO clears the market with

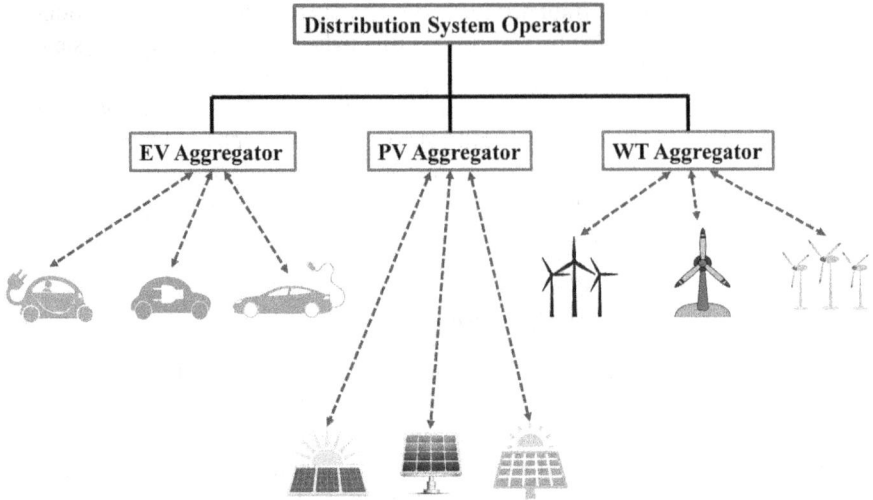

FIGURE 17.1   An active distribution system.

malicious information, the market outcome will be infeasible, resulting in distribution system experiences congestions in different branches as well as a shortage of power in different buses. However, according to Figure 17.2 (b), which is associated with the proposed RAS, the DMO in collaboration with EV aggregators try to mitigate the impacts of successful cyberattacks using a secondary market. In this regard, all of the EV aggregators will be requested to submit their updated packages in a new repository established to protect the updated information from attackers' access. In the next step, the DMO runs the secondary market and determines the new winning participants collaborating with DSO to cope with the congestions or power outages imposed by the FDI cyberattacks. Hence, based on the new obtained schedule, DSO can take advantage of the EVs as mobile energy storage units to handle the cyberattack without applying any further remedial actions, like distribution feeder reconfiguration (DFR). As an illustration, EV aggregator a can purchase the extra power of WT aggregator x depicted in Figure 17.2 (b). Likewise, EV aggregator b can purchase the required power from the grid to alleviate the congestion on the other side of the distibution system due to lack of enough local generation. As a result, the accomplished power exchange between EV aggregators and other market participants relives the congested branches without any need for expensive reaction procedures (e.g., DFR). More information about the impacts of cyberattacks targeting EV aggregators and the proposed RAS on the IEEE test system will be provided in Section 17.4.

17.2.2 Assumptions of the Attack Framework

According to [21], cyberattacks targeting different sectors of modern-day energy systems need to be investigated from different standpoints (e.g., from a computer science point of view, from a power system viewpoint, etc.) to effectively assist system operators coping with them; however, since the scope of this chapter is smart power and transportaion systems, it is necessary to adopt a set of typical assumptions, which are

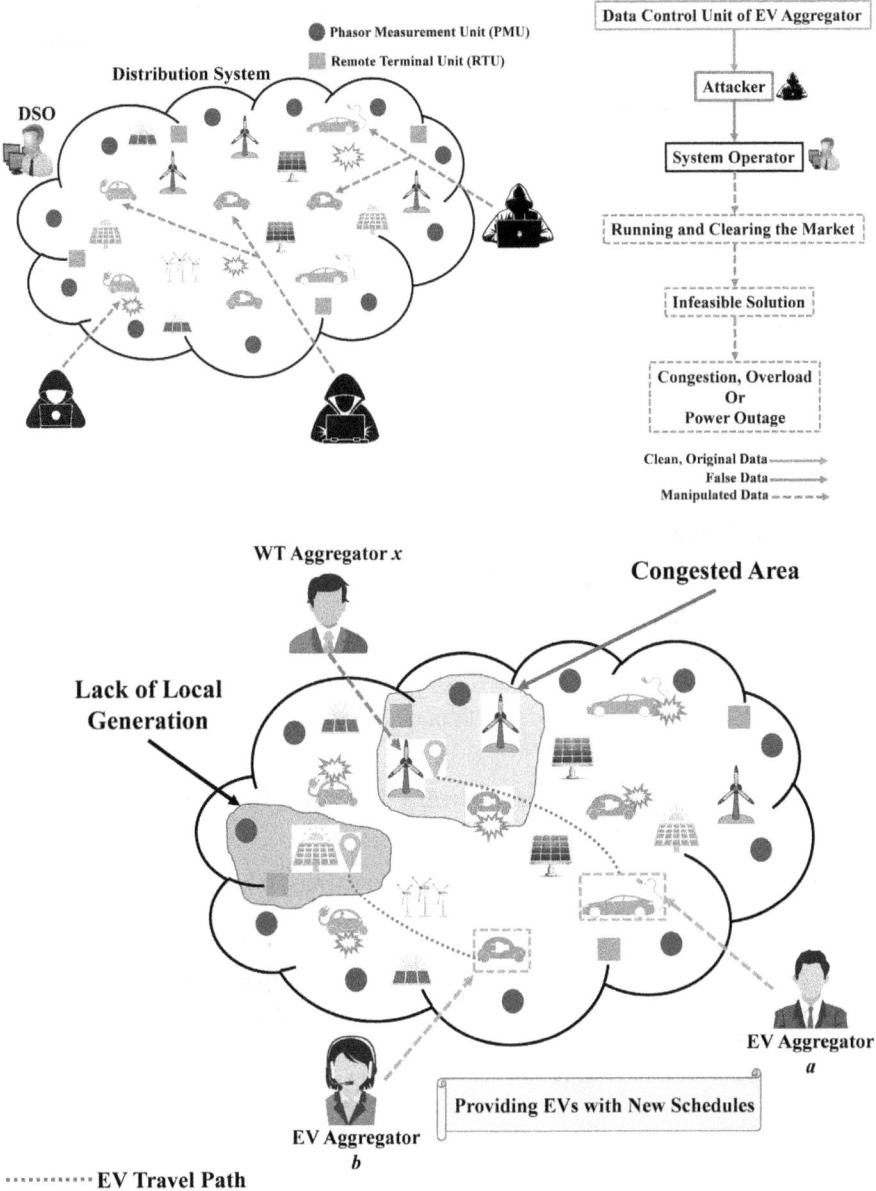

FIGURE 17.2 The proposed framework for (a) false data injection (FDI) cyberattack and (b) proposed RAS in this chapter.

called *realistic* assumptions [22–23], to develop the attack framework. The considered assumptions in this chapter are itemized hereunder.

- Attackers are aware of the detailed configuration of power networks, smart grids, and EV-based transportation systems. Hence, addressing this important question "*How can attackers obtain the confidential information about the operation of energy systems?"* is not in the scope of this paper. According to [24–25], bypassing the virtual private networks and infiltrating the firewalls are among the common methods to gain access to server computers.

- The magnitude of injected false data to target different assets (e.g., load centers, generation units, etc.) is limited to a certain range of the rated amounts to increase the stealthiness of the cyberattack [26].

17.3 PROPOSED PROBLEM FORMULATION

This section presents the objective functions associated with FDI attacks targeting 1) the EV aggregators to manipulate their offered price-bid package related to their clients and 2) the datasets of the smart distribution grid to compromise the demand data associated with a system's buses considering bypassing the ACSE. The following subsections elaborate on the objective function and the corresponding technical constraints for each FDI.

17.3.1 FDI CYBERATTACK TARGETING EV AGGREGATORS IN DAY-AHEAD ELECTRICITY MARKET

The optimization problem in this section is presented in (17.1), where the first, second, and third terms of Sigma sign, respectively, present the total cost of energy procurement from day-ahead market, battery degradation taken into account for selling energy to the market, and the benefit obtained via selling power at a fixed price.

$$\max \left\{ (1-\Psi)\left(\sum_{t=1}^{24} P_t \times \rho_t \times \Delta t + \Phi_t - FSP \times SOC_t^{DL} \right) \right\}, \qquad (17.1)$$

where ψ is the weighing factor; P_t and ρ_t are, respectively, the power purchased from day-ahead electricity market (kW) and the corresponding price ($/kWh) at tth time interval; Φ_t indicates the battery degradation at time slot t; FSP is the fixed selling price; and SOC_t^{DL} is the loss in state of charge (SOC) due to driving the EV.

Objective function (17.1) needs to be optimized from the attacker's standpoint by considering a set of constraints associated with energy balance (i.e., (17.2)) and technical limitations of EVs (i.e., (17.3)–(17.6)), which are provided as follows:

$$\sum_{t=1}^{24} P_t = \sum_{t=1}^{24} P_t^{EV-C} \qquad (17.2)$$

$$P_t^{EV-C} \times \Delta t + P_t^{EV-D} \times \Delta t \leq P_{\max}^{C-D}; \forall t \qquad (17.3)$$

$$SOC_t = SOC_{t-1} + P_t^{EV-C} \times \Delta t \times \Upsilon^{GB} - \frac{P_t^{EV-D} \times \Delta t}{\Upsilon^{BG}} - P_t^{Loss} \times \Delta t; \forall t \qquad (17.4)$$

$$SOC_{\min} \leq SOC_t \leq SOC_{\max} \qquad (17.5)$$

$$\sum_{t=1}^{24} SOC_t = SOC^{Final}, \qquad (17.6)$$

where P_t^{EV-C} and P_t^{EV-D}, respectively, denote the total power of charging and discharging a typical EV at tth time interval; P_{max}^{C-D} indicates the maximum power rate associated with charge and discharge processes; SOC_t and SOC_{t-1} are, respectively, the total state of charge associated with all EVs at tth and $(t-1)$th time intervals; Υ^{GB} and Υ^{BG} are, respectively, the grid-to-battery and battery-to-grid efficiencies; P_t^{Loss} is the total power loss due to EV's displacement from one place to another; SOC_{min} and SOC_{max} are the minimum and maximum boundaries of SOC; and SOC^{Final} is the final SOC at the end of day.

It is necessary to note that, in the normal operation of the electricity market, an EV aggregator can be considered as a price-maker agent who minimizes the cost of purchasing energy from the electricity market (by offering the power-price package on behalf of the clients) to increase the profits of his or her clients [27]. However, the attacker tries to maximize the EV aggregator's objective function (see (17.1)) to result in infeasible solutions after the market is cleared in the day-ahead horizon. Hence, after solving the optimization problem presented in (17.1), the attacker will come up with the false packages (i.e., power and price for the EV aggregators) to be substituted for the original data submitted by the EV aggregators for the next-day electricity market. The result of such data manipulations will be system congestions or power outages, which will be elaborated on in Section 17.4.3.

17.3.2 STEALTHY FDI CYBERATTACKS TARGETING LOAD CENTERS

To obtain the attack vector comprising the data manipulations, the attacker needs to minimize (17.7) and maximize (17.8) simultaneously for the set of buses having load centers and all of the branches, respectively.

$$\min\left\{\sqrt{\sum_{b\in\Theta}\left(\Delta P_b^{Demand}\right)^2}\right\} \tag{17.7}$$

$$\max\left\{\sum_l\left|PF_{BA}^l - PF_{AA}^l\right|\right\}, \tag{17.8}$$

where ΔP_b^{Demand} denotes (to be injected into the server computer of the smart grid) the level of manipulation of the active power associated with bth bus; Θ signifies the set of targeted load centers; and PF_{BA}^l and PF_{AA}^l are, respectively, the power flow associated with lth branch before and after the FDI cyberattack.

Although the objective functions presented in (17.7) and (17.8) need to be simultaneously optimized, it can be inferred that the presented objective functions are not in accord with each other. That is to say, optimizing one objective can deteriorate the other one; hence, a set of optimal solutions should be obtained and stored in a repository instead of only one unique optimal solution. After solving the problem, the attacker can opt for his or her desirable solution to be in line with the cyberattack goals. Moreover, the objective functions introduced in (17.7) and (17.8) have numerical values in different ranges; thus, their values need to be converted into the same range (i.e., [0–1]). To this end, Figure 17.3 demonstrates the trapezoidal membership function (i.e., (17.9)), which is utilized in this chapter to handle the normalization process [28].

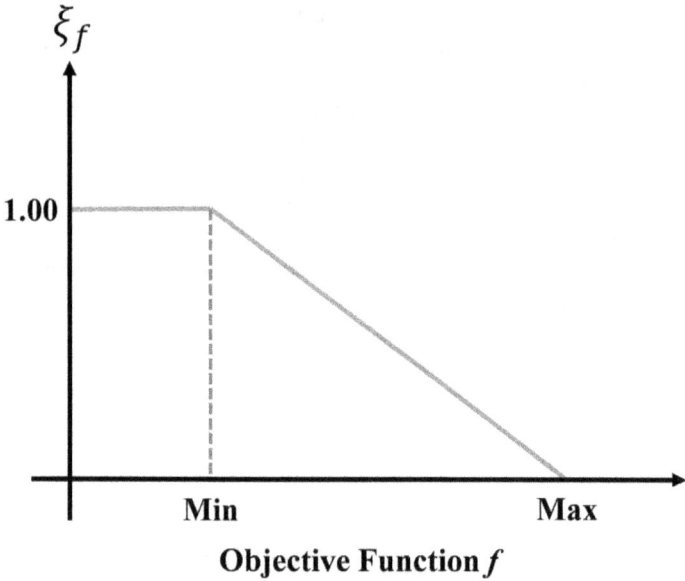

FIGURE 17.3 Trapezoidal membership function for objective functions (17.7) and (17.8) [28].

$$\xi_f = \begin{cases} 0 \\ \dfrac{OFV_f^{\max} - OFV_f}{OFV_f^{\max} - OFV_f^{\min}} \\ 1 \end{cases} \qquad \begin{aligned} & OFV_f \leq OFV_f^{\min} \\ & OFV_f^{\min} \leq OFV_f \leq OFV_f^{\max}; \forall f \\ & OFV_f \geq OFV_f^{\max}, \end{aligned} \qquad (17.9)$$

where ξ_f denotes the normalized value for fth objective function in the range of [0–1]; OFV_f, $OFV_f^{\min}$, and $OFV_f^{\min}$ are, respectively, the numerical value of fth objective function and its corresponding lower and upper limits.

To store the non-dominated optimal solutions (i.e., Pareto-front) after each iteration, a repository needs to be defined to save and maintain the elements of the Pareto-front based on (17.10), which is the decision-maker in the proposed method to approach the bi-objective optimization problem presented in (17.7) and (17.8) [29].

$$\Gamma_\xi(f,n) = \frac{\displaystyle\sum_{f=1}^{2} \omega_f \times \xi_f^n}{\displaystyle\sum_{n=1}^{S_{POF}} \sum_{f=1}^{2} \omega_f \times \xi_f^n}, \qquad (17.10)$$

where ω_f signifies the weight factor associated with fth objective function to control the importance of each objective function over the other; and S_{POF} is the size of the Pareto-optimal front (i.e., the number of non-dominated solutions stored in the repository).

The objective functions presented in (17.7) and (17.8) are concurrently optimized subject to bypassing the bad data detection (BDD) algorithm embedded in ACSE, which normally calculates the Euclidian norm of the measurement residual vector to identify the malicious data (see (17.11)). Hence, the attacker needs to take into account (17.12)–(17.14) while optimizing the objective functions (17.7) and (17.8) to make sure the obtained attack vector does not trigger the security alarms related to BDD in the ACSE process. It is noted that (17.14) is the necessary condition for the FDI cyberattacks to remain stealthy in ACSE [30]. In other words, the level of data manipulation (i.e., $\Delta \tilde{m}$), which leads to compromising the measurement function (i.e., $h(\tilde{x}+\Omega)$), should meet $\Delta m - h(\tilde{x}+\Omega) + h(\tilde{x}) = 0$ to bypass the BDD, as the blue terms are equal to zero.

$$\|m - h(\tilde{x})\| \geq \partial, \text{ where } m = h(x) + \varepsilon \tag{17.11}$$

$$\|m_{AA} - h(\tilde{x}_{AA})\| < \partial, \text{ where } m_{AA} = m + \Delta \tilde{m}, \text{ where } \tilde{x}_{AA} = \tilde{x} + \Omega \tag{17.12}$$

$$\|m + \Delta m - h(\tilde{x}+\Omega) + h(\tilde{x}) - h(\tilde{x})\| \leq \partial, \text{ where } m - h(\tilde{x}) \approx 0. \tag{17.13}$$

Therefore,

$$\Delta m - h(\tilde{x}+\Omega) + h(\tilde{x}) = 0, \tag{17.14}$$

where m is the set of measurements; $h(x)$ indicates a non-linear function of the measurements, which encompasses active and reactive power injections to the buses, active and reactive power flows into branches, and voltage variables (i.e., amplitude and phase angle); ∂ is the threshold of BDD algorithm embedded in ACSE; $\tilde{x}$ indicates the vector of estimated state variables; ε is the vector of measurement error; m_{AA} is the measurement tampered by injecting false data (i.e., $\Delta \tilde{m}$); and Ω denotes the vector of alteration on estimated values.

17.3.3 THE PROPOSED REMEDIAL ACTION SCHEME (RAS)

Once the FDI cyberattack targeting EV aggregators is confirmed or the presence of malicious data in the DCU of EV aggregators is detected, DMO will manage a secondary back-up electricity market, called distribution-level electricity market, to handle the cyberattack and mitigate its impacts. Therefore, EV aggregators will be asked to resubmit their updated packages in a new repository established to protect the updated information from attackers' access. The main objective function associated with the back-up market is presented in (17.15), where the first term of the Sigma sign indicates the energy cost purchased from the main power grid, the second term denotes the aggregated energy purchased from the EV aggregators, and the third and fourth terms, respectively, signify the aggregated energy purchased form PV aggregators and WT aggregators participated in the electricity market. One of the most important aims of the secondary market is to 1) enhance the share of distributed generation aggregators in serving the demand and 2) decrease the reliance of the distribution system on the main grid because in case of future cyber-attacks there will be fewer regions in the system suffering from noticeable shortages of power.

$$\min\left\{\sum_{t=1}^{24}\left(E_t^G \times \sigma_t^G + \sum_{i=1}^{N_{EVAG}} E_{i,t}^{EV} \times \sigma_{i,t}^{EV} + \sum_{j=1}^{N_{PVAG}} E_{j,t}^{PV} \times \sigma_{j,t}^{PV} + \sum_{k=1}^{N_{WTAG}} E_{k,t}^{WT} \times \sigma_{k,t}^{WT}\right)\right\},\quad (17.15)$$

where E_t^G and σ_t^G, respectively, denote the energy (kWh) and price ($/kWh) associated with the main grid at tth time interval; N_{EVAG}, N_{PVAG}, and N_{WTAG} are the total number of EV, PV, and WT aggregators, respectively; $E_{i,t}^{EV}$ and $\sigma_{i,t}^{EV}$ are, respectively, the aggregated energy and the corresponding price for ith EV aggregator at tth time interval; $E_{j,t}^{PV}$ and $\sigma_{j,t}^{PV}$, respectively, represent the aggregated energy and the associated price for jth PV aggregator at tth time interval; and $E_{k,t}^{WT}$ and $\sigma_{k,t}^{WT}$ are, respectively, the aggregated energy and the related price for kth WT aggregator at tth time interval.

The back-up market objective function (i.e., (17.15)) is minimized considering a set of technical constraints, which are elaborated in (17.16)–(14.22).

$$E_t^G + \sum_{i=1}^{N_{EVAG}} E_{i,t}^{EV} + \sum_{j=1}^{N_{PVAG}} E_{j,t}^{PV} + \sum_{k=1}^{N_{WTAG}} E_{k,t}^{WT} = \sum_{b=1}^{N_{Bus}} P_{b,t}^{Demand} + P_t^{Loss}; \forall t \quad (17.16)$$

$$P_{b',t} = \sum_{b'=1}^{N_{Bus}} V_{b',t} \times V_{b'',t} \times Y_{b'b''} \times \cos\left(\theta_{b'b''} - \theta_{b',t} + \theta_{b'',t}\right); \forall t,b'' \quad (17.17)$$

$$Q_{b',t} = \sum_{b'=1}^{N_{Bus}} V_{b',t} \times V_{b'',t} \times Y_{b'b''} \times \sin\left(\theta_{b'b''} - \theta_{b',t} + \theta_{b'',t}\right); \forall t,b'' \quad (17.18)$$

$$E_{i',\tau_0}^{EV} + \sum_{\tau=\tau_0}^{\tau_1} E_{i',\tau}^{EV} \geq 0.7 \times E_{i'}^{EV,\max}; \tau_0 < \tau_1 \quad (17.19)$$

$$V_{b,t}^{\min} \leq V_{b,t} \leq V_{b,t}^{\max}; \forall t \quad (17.20)$$

$$P_{l,t}^{\min} \leq P_{l,t} \leq P_{l,t}^{\max}; \forall t \quad (17.21)$$

$$I_{l,t} \leq I_{l,t}^{\max}; \forall t, \quad (17.22)$$

where $V_{b',t}$ and $\theta_{b',t}$ are, respectively, voltage amplitude and phase angle at buses b'; $V_{b'',t}$ and $\theta_{b'',t}$ are, respectively, voltage amplitude and phase angle at buses b''; $Y_{b'b''}$ and $\theta_{b'b''}$ are the amplitude and phase angle associated with the branch connecting buses b' and b'', respectively; $V_{b,t}^{\min}$ and $V_{b,t}^{\max}$ are, respectively, the minimum and maximum boundaries for bth bus at tth time interval; $P_{l,t}^{\min}$ and $P_{l,t}^{\max}$ are, respectively, the lower and upper limits of power flow for lth branch at tth time interval; and $I_{l,t}^{\max}$ denotes the maximum acceptable current without tripping the protective relays corresponding to lth branch at tth time interval.

17.4 CASE STUDY AND SIMULATION RESULTS

17.4.1 UTILIZED SOLVERS

To approach the developed optimization problems in Section 17.3, a combination of MATLAB® [31] and general algebraic modeling system (GAMS) [32] is implemented to take advantage of the strong points of each coding language. In this regard, the fast and efficient CPLEX solver [33] of GAMS has been used to solve the mixed-integer linear programming problems. However, to effectively handle the non-linear optimization problems (i.e., (17.7)–(17.22)), an efficient combination of grey wolf optimizer (GWO) and particle swarm optimization (PSO) algorithm in the form of a unified hybrid metaheuristic algorithm (i.e., HGWO-PSO) is utilized [20]. Furthermore, GDX-MRW functions [34] have been used to provide a flexible link between GAMS and MATLAB environments for exchanging data between two platforms. That is to say, the main script has been developed in MATLAB; however, GAMS-related subscripts are embedded in the body of different loops of the main MATLAB M-file to take advantage of commercial GAMS solvers (e.g., CPLEX) to solve the optimization problems.

17.4.1.1 GAMS Software

According to [32], GAMS is "one of the leading tool providers for the optimization, which is used by companies, universities, research institutions and governments in many different areas, including energy; GAMS is known as the first software system to combine the language of mathematical algebra with traditional concepts of computer programming in order to efficiently describe and solve complex optimization problems". More specifically, we utilized the CPLEX solver of GAMS to handle the parts of the developed framework that are inherently linear and mixed-integer linear problems (e.g., (17.1)–(17.6)).

17.4.1.2 The Proposed HGWO-PSO

According to [35], GWO, which mimics the social behavior of a group of grey wolves while hunting their prey, normally comprises four sets of solutions. The first three solutions (i.e., α, β, and δ) are the top solutions, whereas the remaining solutions are classified as η. Therefore, the inappropriate solutions (i.e., η) update their positions from the prey based on the location of the top three wolves (i.e., α, β, and δ). The hunting process can be mathematically written as (17.23)–(17.29).

$$\Lambda_\alpha = \left| C_1 \times W_\alpha - W \right| \tag{17.23}$$

$$\Lambda_\beta = \left| C_2 \times W_\beta - W \right| \tag{17.24}$$

$$\Lambda_\delta = \left| C_3 \times W_\delta - W \right| \tag{17.25}$$

$$W_1^T = W_\alpha - E_1 \times \Lambda_\alpha \tag{17.26}$$

$$W_2^T = W_\beta - E_2 \times \Lambda_\beta \tag{17.27}$$

$$W_3^T = W_\delta - E_3 \times \Lambda_\delta \tag{17.28}$$

$$W^{T+1} = Avg\left(W_1^T, W_2^T, W_3^T\right), \tag{17.29}$$

where Λ is the distance between the hunter (i.e., η) and the top solutions (i.e., α, β, and δ) at Tth iteration; W and W^{T+1} are, respectively, the position of the wolf at Tth and $T+1$th iterations; and C and E are coefficients to adjust the position of wolf with respect to the prey. Interested readers are directed to [35] for more details about GWO.

Based on [36], PSO, which mimics the social behavior of a category of birds or fishes, has a straightforward mechanism to converge to the optimal solution. In the PSO algorithm, each member (e.g., bird or fish) of the swarm updates its position and velocity via (17.30) and (17.31), where χ_m^T and χ_m^{T+1} are, respectively, the position of mth member at Tth and $T+1$th iterations; v_m^T and v_m^{T+1} are the velocity of mth member at Tth and $T+1$th iterations, respectively; κ, $\underline{\kappa}$, and $\overline{\kappa}$ are, respectively, the inertia factor and its minimum and maximum limits; μ_1 and μ_2 are personal and social learning factors in the range of [0–2]; Π_m^T and Π^T are, respectively, the personal best position for mth member and the global best position among all members; and $\overline{T}$ is the maximum number of iterations.

$$\chi_m^{T+1} = \chi_m^T + v_m^{T+1} \tag{17.30}$$

$$v_m^{T+1} = \kappa \times v_m^T + \mu_1 \times rand(1 \times 1) \times \left(\Pi_m^T - \chi_m^T\right) + \mu_2 \times rand(1 \times 1) \times \left(\Pi^T - c_m^T\right), \tag{17.31}$$

$$\kappa = \overline{\kappa} - T \times \frac{\overline{\kappa} - \underline{\kappa}}{\overline{T}}.$$

In the PSO algorithm, the best personal and global solutions (i.e., Π_m^T and Π^T) are improved toward global optima in iteration basis. Interested readers are directed to [37] for detailed information about PSO algorithm.

According to [20] and [38], the proposed HGWO-PSO algorithm, which is capable of expeditiously solving a verity of small-scale and large-scale optimization problems, has two unique characteristics with respect to the original GWO and original PSO: 1) it is fast enough for large-scale, complicated optimization problems, and 2) it can simply scape from the local optima by its parallel structure. To obtain a better perspective, Figure 17.4 demonstrates the parallel hybridization process of GWO and PSO in this chapter.

It is necessary to note that, in the HGWO-PSO algorithm, the size of the population and maximum number of iterations are, respectively, set to 300 and 100; the inertia factor (i.e., κ) is limited to a range of [0.4–0.9]; and both learning factors (i.e., μ_1 and μ_2) are set to 1.49618.

17.4.2 CASE STUDY

The effectiveness of the proposed attack framework introduced in Section 17.3 is scrutinized on the IEEE 33-bus distribution system, which has been upgraded by adding different distributed generation retailers (e.g., small-scale wind turbines

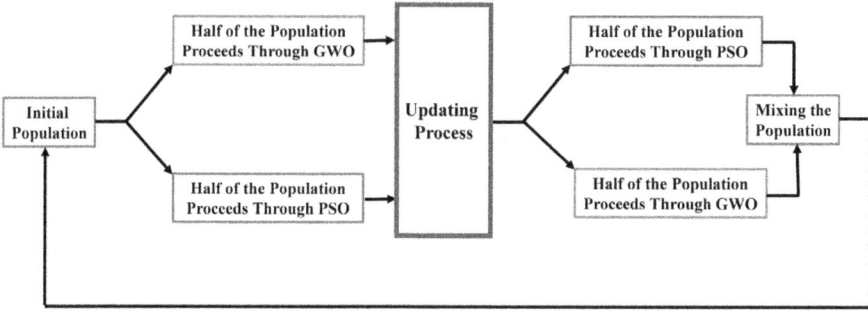

FIGURE 17.4 Flowchart of the hybridization process of GWO and PSO in the proposed HGWO-PSO algorithm [38].

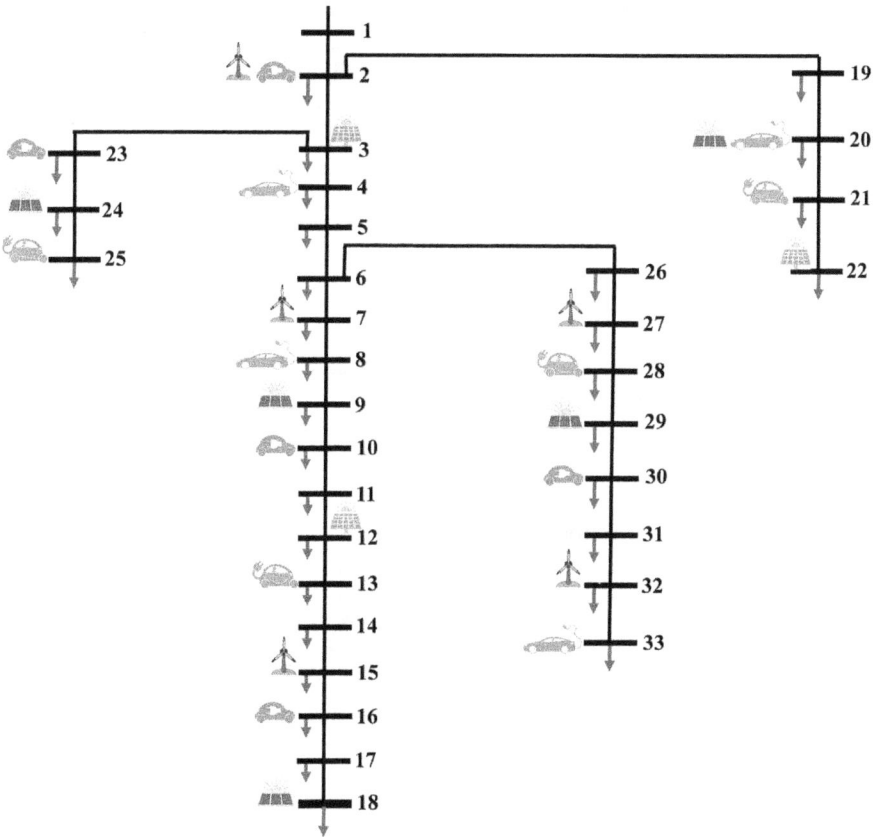

FIGURE 17.5 IEEE 33-bus distribution system modified by EV, PV, and WT units.

(WTs) and photovoltaic (PV) modules) and also EV aggregators. To obtain a better perspective, Figure 17.5 demonstrates the case study in this chapter. It is noted that the PV modules embedded in Figure 17.5 reflect the characteristics of Kyocera models [39] with a rated power of 135 W and 12.9% cell efficiency. The required land for each PV module is equal to 5,000 m² throughout the distribution system. Furthermore, the corresponding inverters are opted for Xantrex [40], with a maximum power of 1.5 kW and 92% efficiency. The rated power and blade diameter of WTs manufactured by Bonus [41] are, respectively, 300 kW and 33.4 m. To calculate the output power associated with PV and WT units, the solar irradiance and wind speed information are extracted from [42–43]. The essential information associated with EV aggregators is provided in Table 17.1. The remaining parts of the system's data for the IEEE 33-bus distribution grid (i.e., impedance of the branches, etc.) are extracted from [44]. It is necessary to note that the EV icon displayed in Figure 17.5 indicates the EV aggregator, who might have contract with EV owners up to 80 clients (see Table 17.1). For example, EV aggregator #10 denotes the agent connected to bus #10, who has 52 clients (i.e., EV owners) and is able to provide up to 450 kW (i.e., the aggregated capacity of EVs' batteries). The mentioned row has been bold faced in Table 17.1 to facilitate the reading process. In addition, all of the EVs utilized in the case study follow constraint (17.19), meaning that the EV owners leave home at 6:00 AM and get home at 6:00 PM, considering that the remaining part of electrical energy in the EV's battery at 6:00 PM is at least 70% of the rated capacity of the battery. Interested readers are directed to [45], the earlier step of this research, for more information about the EV patterns.

17.4.3 SIMULATION RESULTS, DISCUSSION, AND ANALYSIS

By solving the optimization problem formulated in (17.1)–(17.6), the attacker obtains a set of false packages (i.e., power and price) to be injected into the DCU of the targeted EV aggregators. In this situation, since the submitted information of the EV aggregators' demand is not valid, the entire market will be cleared with infeasible

TABLE 17.1
Description of the EVs in the Case Study (i.e., Figure 17.5)

EV Aggregator ID	Number of Associated EV Owners	Aggregated Power (kW)
2	35	400
4	10	215
8	44	435
10	**52**	**450**
13	40	430
16	67	465
20	25	335
21	55	457
23	70	470
24	35	400
28	48	445
30	80	505
33	15	300

results, leading to congestion, power shortages, and other technical issues (e.g., over-loading of components). To have a better perspective, Figure 17.6 demonstrates the submitted packages (i.e., malicious information) by EV aggregators after the FDI cyberattack. By comparing Figure 17.6 and Table 17.1, it can be concluded that the attacker has been successful since the submitted packages (i.e., power and price) are intentionally manipulated. As an illustration, the actual aggregated power for EV aggregator #2 is 400 kW (see Table 17.1); however, the malicious value obtained by the attacker after solving (17.1)–(17.6) is equal to 344 kW (see the red arrow displayed in Figure 17.6). In addition, the malicious aggregated power for EV aggregator #13 is increased from 430 kW to 445 kW (see the black arrow in Figure 17.6). Moreover, the original information of EV aggregators #2 and #24 remain the same after the cyberattack (see the purple arrows depicted in Figure 17.6). Hence, the submitted data for EV aggregators #2 and #24 and the rest of participants (i.e., PV and WT aggregators) are not manipulated.

To increase the chance of severe congestion or possible power outages, in the next step, the attacker targets the load centers of the distribution system by launching another FDI cyberattack. Toward this end, by solving the bi-objective optimization problem (BOOP) presented in (17.7)–(17.14), the attacker tries to weaken the DSO to handle the cyberattack. The importance of solving the problem as a BOOP is that the attacker needs to find a trade-off between the level of manipulations (i.e., ΔP_b^{Demand}) and the rate of intentional congestion. In other words, the attacker needs to increase the rate of congestion (i.e., (17.8)) by reducing the level of manipulations and the number of compromised load centers (i.e., (17.7)) to be manipulated by injecting false data. The Pareto-optimal front associated with the BOOP solved by the attacker is illustrated in Figure 17.7. It can be gathered from this figure that the attacker has a

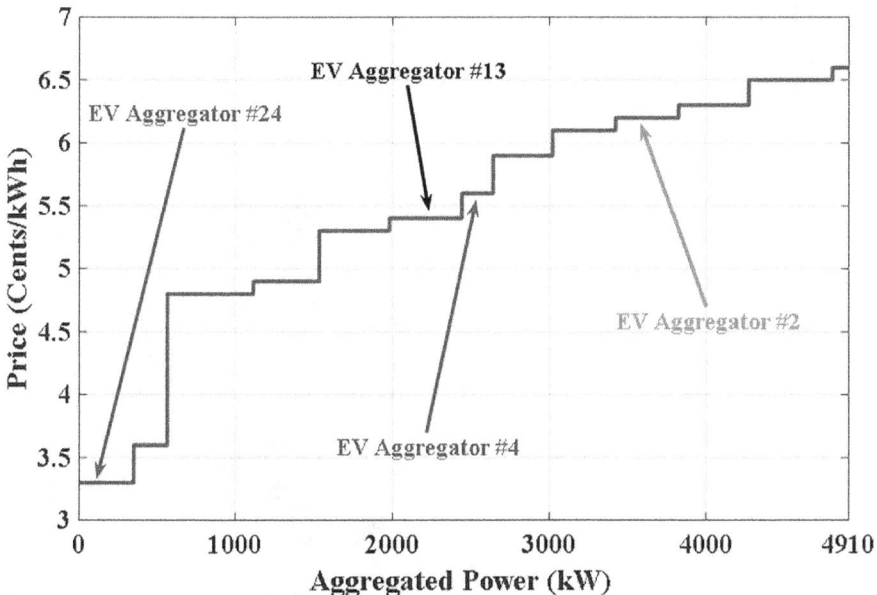

FIGURE 17.6   The submitted packages of EV aggregators after the FDI attack.

set of optimal solutions instead of only one unique optimal solution. Hence, he or she can opt for the best compromise solution (BCS) based on his or her desire. To obtain a better perspective about the BCS, the red pentagon displayed in Figure 17.7 is associated with $\omega_1 = \omega_2 = 0.5$ (see (17.10)) meaning that both objective functions (17.7) and (17.8) have the same importance.

According to Figure 17.7, it can also be inferred that the Pareto-optimal front is well distributed, providing the attacker with a range of data manipulations and imposed congestions or possible power outages. The FDI vector associated with the red pentagon demonstrated in Figure 17.7 is presented in Figure 17.8, showing a unique pattern of data manipulations to keep the cyberattack stealthy by bypassing the BDD embedded in ACES (i.e., (17.11)–(17.14)).

Although only a group of EV aggregators and a group of load centers are targeted by the attacker, the results of market clearing will be noticeably affected. In this regard, Figure 17.9 compares the market winners before and after the FDI cyberattack targeting EV aggregators. According to this figure, one can infer that if

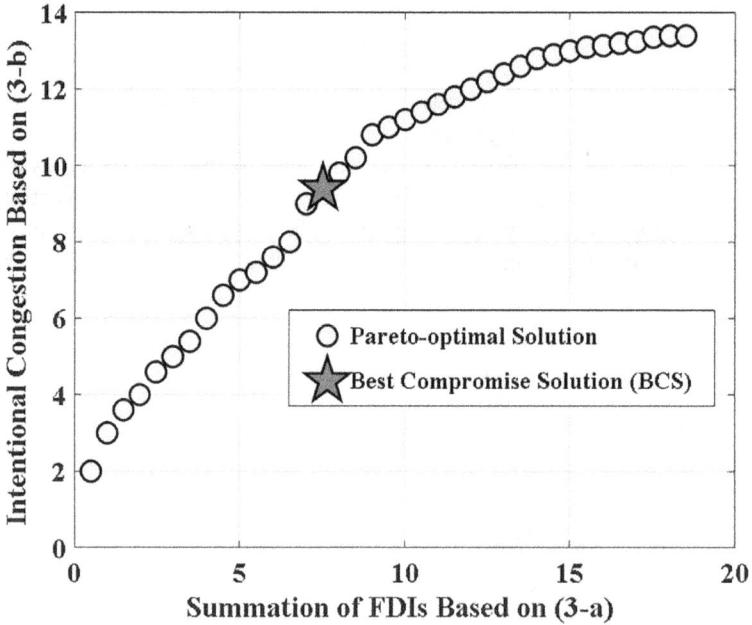

FIGURE 17.7 The two-dimensional Pareto-optimal front from the attacker's standpoint.

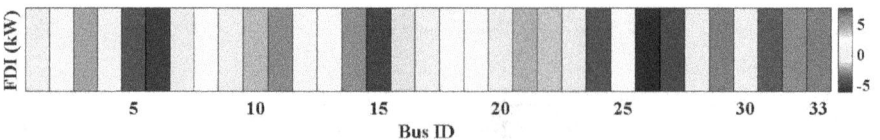

FIGURE 17.8 The FDI vector for load centers on modified IEEE 33-bus system (i.e., Figure 17.5).

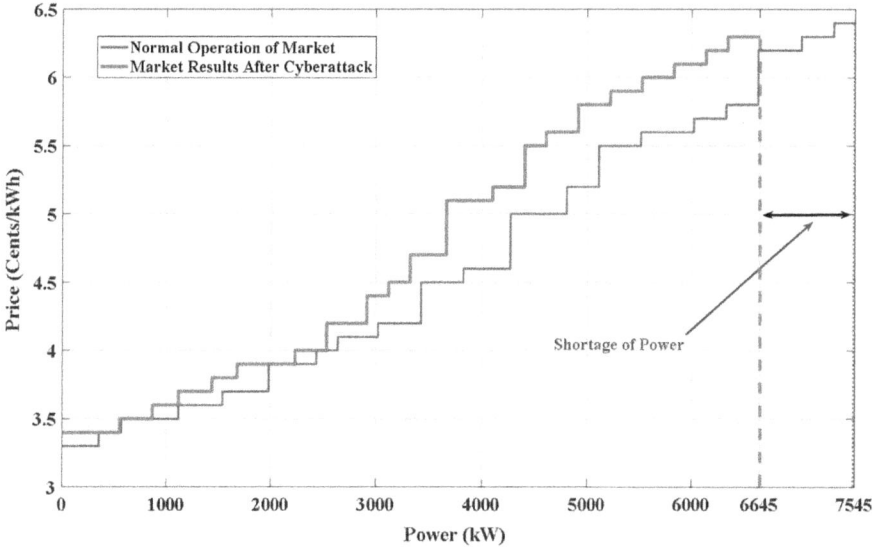

FIGURE 17.9   The schedule of market winners before and after the cyberattack targeting EV aggregators.

the cyberattack is not detected, the market will be cleared with a 900 kW shortage of power because the attacker maximizes the EV aggregators' objective function (i.e., (17.1)).

From Figure 17.9, it can be concluded that the patterns of market winners before and after the cyberattacks targeting EV aggregators are quite different, which negatively affects the normal operation of the distribution grid. In other words, the blue curve (i.e., market clearance without targeting the EV aggregators) indicates 21 steps for 21 winners out of 25 participants (see Figure 17.5); however, the red curve, which displays the malicious market winners, only has 20 steps, meaning that 5 agents have not been successful in the competition due to their expensive prices after compromising the submitted packages of EV aggregators by the attacker (refer to Figure 17.6).

To demonstrate the negative impacts of the launched cyberattacks (i.e., Figures 17.6–17.9) on the power flowing into the branches, Figure 17.10 depicts the magnitude of the current related to all the distribution branches. Based on this figure, one can infer that branches #5, #7, #12, #20, #22, #23, and #29 experience different rates of congestion; however, the relays of branches #5, #12, and #22 are set to trip the breakers and isolate the branches since they are, respectively, loaded up to almost 106%, 116%, and 108% of their rated currents after the cyberattack. In this regard, it is worth mentioning that although experiencing a current magnitude higher than 80% of the rated current in a typical branch of a distribution-level power grid is considered congestion, more than 105% of the rated current will result in tripping the corresponding breaker, consequently leading to a power outage for the downstream part of the system.

Referring to Figure 17.10, the consequence of the FDI cyberattack (i.e., (17.1)–(17.14)) results in power shortages for different parts of the case study due to the

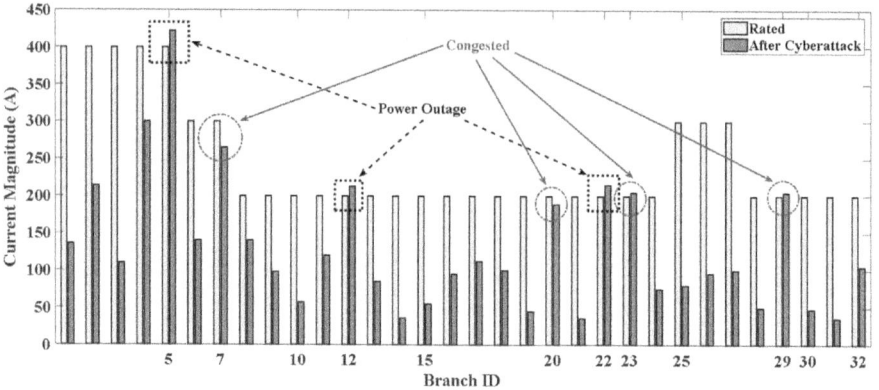

FIGURE 17.10   The current profile for all the branches after the cyberattack at 3:00 PM.

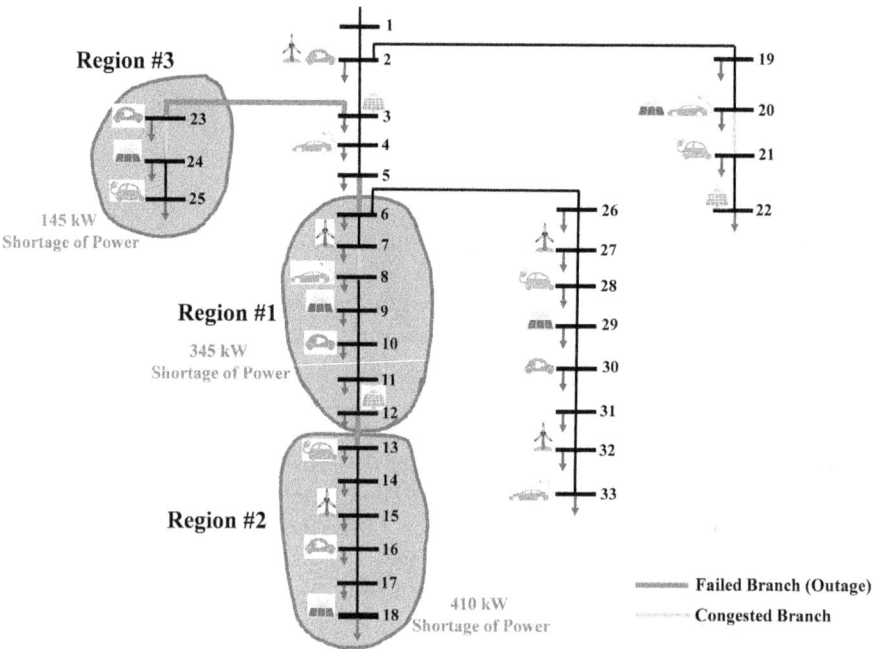

FIGURE 17.11   The status of the IEEE 33-bus distribution grid after the FDI cyberattack targeting EV aggregators.

intentional power outages. To obtain a better perspective, Figure 17.11 displays the IEEE 33-bus case study after the cyberattack.

From Figure 17.11, it can be concluded that since the relays of branches #5, #12, and #22 are set to trip the corresponding breakers if the cyberattack is neither prevented nor detected (see the red branches displayed in Figure 17.11), three regions (i.e., Regions #1, #2, and #3) in the system experience isolation from the main grid (see the pink regions in Figure 17.11). Hence, these three regions need to rely on

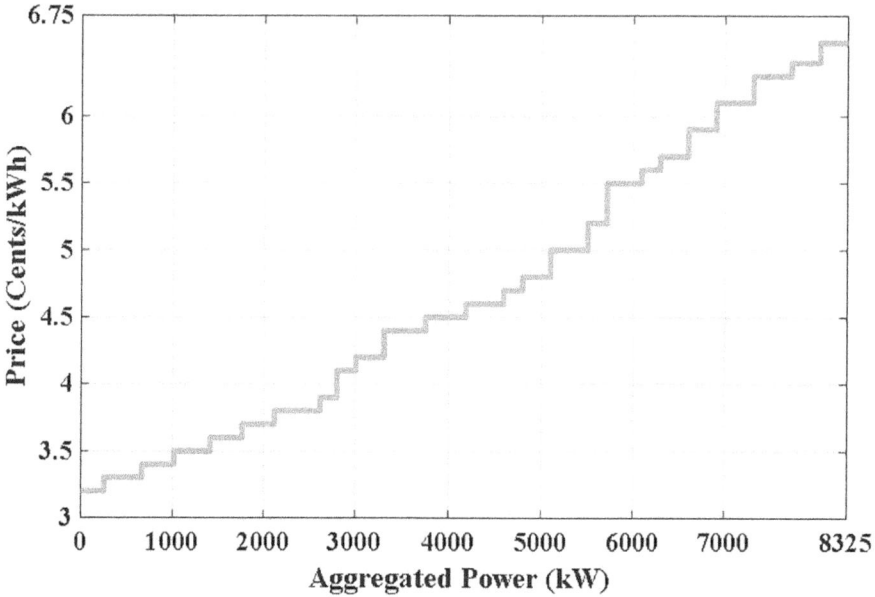

FIGURE 17.12 The schedule of the secondary market's winners to ameliorate the impacts of cyberattack.

their local generation; however, the level of power production is less than the corresponding demands for Regions #1–#3. As a result, Regions #1, #2, and #3 suffer from a 345 kW, 410 kW, and 145 kW shortage of power, respectively. This is where the importance of the proposed RAS comes under the spotlight by providing a suitable reaction mechanism to the FDI attack via taking advantage of the EVs as mobile energy storage devices throughout the distribution system. To this end, DMO runs the secondary, back-up distribution-level electricity market to mitigate the impacts of the successful cyberattack. Therefore, all of the market participants will be asked to submit their updated packages in a new repository designed to protect the new data from attackers' access. The new winning participants in the back-up market are shown in Figure 17.12, presenting a valid (i.e., non-malicious) schedule for the market participants.

According to Figure 17.12, it can be perceived that all of the market participants (i.e., 25 aggregators) contribute to serving the demand of the system to reduce the reliance of the distribution network on the up-stream grid. To have a better perspective about the status of EV aggregators, Table 17.2 presents the available power/capacity for all of the EV aggregators at 3:00 PM.

Based on Table 17.2, one can infer that 1) EV aggregators #20, #21, #28, and #33 can take advantage of the extra power of their clients to mitigate the cyberattack at 3:00 PM and 2) EV aggregator #4 can purchase 212 kW from PV aggregator #20 (see Figure 17.13). Therefore, as demonstrated in Figure 17.13, a 345 kW power shortage in Region #1 can be supplied by traveling EVs from bus #28 and #33 to the affected Region #1 (see the blue dashed-line travel path in Figure 17.13). Likewise, the 410 kW and 145 kW power shortages in Regions #2 and #3 can be, respectively,

TABLE 17.2
The Status of EV Aggregators after Running the Back-Up Market.

EV Aggregator ID	Extra Power (kW)	Extra Capacity (kW)
2	-	-
4	-	235
8	-	5
10	-	-
13	45	-
16	15	-
20	162	-
21	410	-
23	-	-
24	8	12
28	240	-
30	24	-
33	135	-

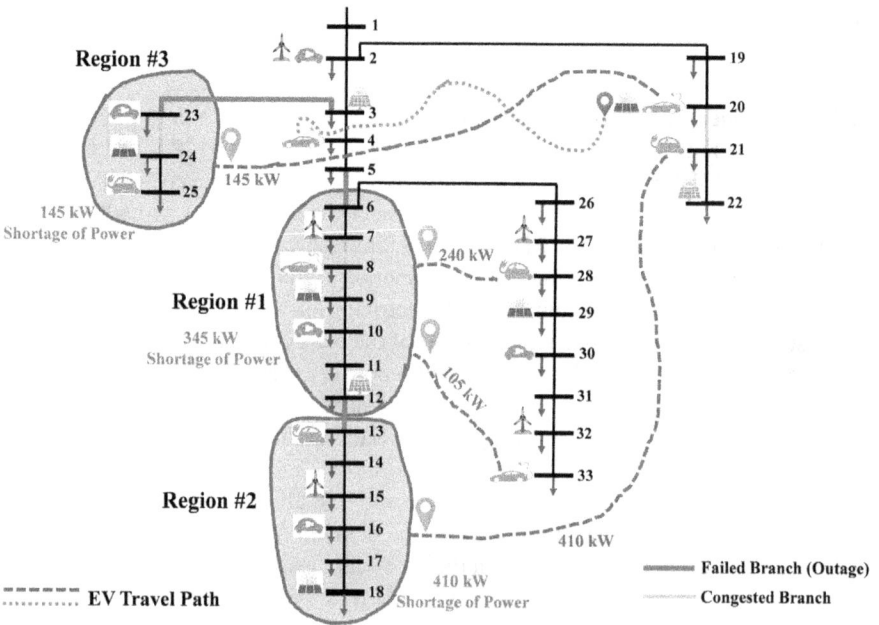

FIGURE 17.13 IEEE 33-bus distribution grid after applying the RAS.

supplied by traveling EVs from bus #21 and #20 to the affected regions (refer to the blue dashed-line travel path depicted in Figure 17.13). Furthermore, EV aggregator #4 sends several EVs to purchase the extra affordable power of PV aggregator #20 (see the pink dotted-line travel path in Figure 17.13).

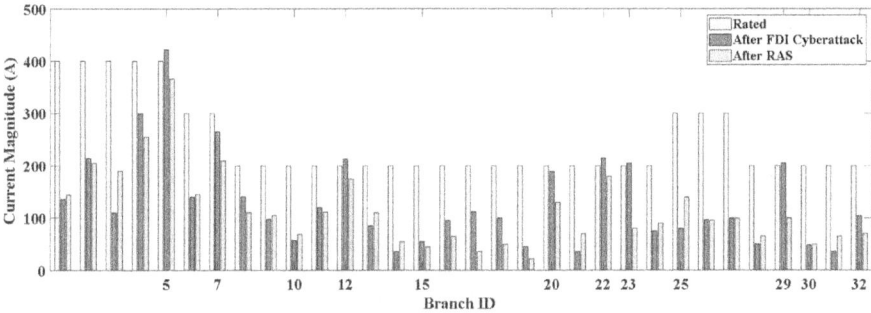

FIGURE 17.14 The current profile for all the branches after applying the proposed RAS to handle the cyberattack.

From Figure 17.13, it can be concluded that the negative impacts of the FDI cyber-attack from the power production point of view are alleviated without performing expensive procedures to handle the attack. The current profile of the targeted system after applying the proposed RAS is illustrated in Figure 17.14. From this figure, it can be inferred that although the magnitude of the current for several branches (e.g., branches #3, #9, #10, #13, #14, #21, #24, #25, #30, and #31) is increased after applying the RAS, none of the branches is overloaded, congested, or failed. In other words, all the branches are far away from congestion, and this is where the importance of such RASs comes under the spotlight to mitigate the impacts of cyberattacks targeting smart grids.

17.5 CONCLUSIONS

In the context of false data injection (FDI) cyberattacks, a multi-objective attack framework was examined in this chapter to target the electric vehicle (EV) aggrega-tors in a unified power and transportation system, resulting in shortage of power in some regions, severe congestions, and power outages. The effectiveness of the pro-posed remedial action framework was validated on the IEEE 33-bus distribution net-work modified to contain EV aggregators, photovoltaic (PV) aggregators, and wind turbine (WT) aggregators as the participants of the electricity market. The obtained simulation results confirmed that targeting the data control unit of only a group of EV aggregators and compromising their submitted packages can noticeably affect the normal operation of the electricity market as well as the entire distribution system. In addition, the results verified that solving the attacker's objective functions as a multi-objective optimization problem and determining a set of Pareto-optimal solu-tions provide the attacker a suitable degree of freedom to increase and decrease the severity of the cyberattack. Furthermore, the obtained results associated with the pro-posed remedial action scheme (RAS) demonstrated that the negative consequences of the examined cyberattack can be mitigated via 1) selling the extra energy stored in the EVs to the market to be utilized for regions with shortages of power and 2) purchasing the available power from the wind turbines by the qualified EVs (i.e., mobile energy storage systems) to increase the contribution of renewable-based generation units.

REFERENCES

[1] E. Ibanez, et al., "National energy and transportation systems: interdependencies within a long term planning model," *IEEE Energy 2030 Conference*, Atlanta, GA, USA, 2008, pp. 1–8, doi: 10.1109/ENERGY.2008.4781023.

[2] A. Moradi Amani and M. Jalili, "Power grids as complex networks: Resilience and reliability analysis," *IEEE Access*, vol. 9, pp. 119010–119031, Sep. 2021, doi: 10.1109/ACCESS.2021.3107492.

[3] G.K. Venayagamoorthy and S.F. Belcher, "Smart grid and electric transportation," *International IEEE Conference on Intelligent Transportation Systems*, St. Louis, MO, USA, 2009, pp. 1–2, doi: 10.1109/ITSC.2009.5309888.

[4] S. Acharya, Y. Dvorkin, H. Pandzic, and R. Karri, "Cybersecurity of smart electric vehicle charging: A power grid perspective," *IEEE Access*, vol. 8, pp. 214434–214453, Nov. 2020, doi: 10.1109/ACCESS.2020.3041074.

[5] M. Ahmed and A.K. Pathan, "False data injection attack (FDIA): An overview and new metrics for fair evaluation of its countermeasure," *Complex Adaptive Systems Modeling*, vol. 8, no. 4, pp. 1–14, Apr. 2020. doi: 10.1186/s40294-020-00070-w.

[6] C. Dong, H. Wang, D. Ni, Y. Liu, and Q. Chen, "Impact evaluation of cyber-attacks on traffic flow of connected and automated vehicles," *IEEE Access*, vol. 8, pp. 86824–86835, May 2020, doi: 10.1109/ACCESS.2020.2993254.

[7] H. Alnabulsi and R. Islam, "Protecting code injection attacks in intelligent transportation system," *IEEE International Conference On Trust, Security and Privacy in Computing and Communications/13th IEEE International Conference on Big Data Science and Engineering (TrustCom/BigDataSE)*, Rotorua, New Zealand, 2019, pp. 799–806. doi: 10.1109/TrustCom/BigDataSE.2019.00116.

[8] J. Petit and S.E. Shladover, "Potential cyberattacks on automated vehicles," *IEEE Transactions on Intelligent Transportation Systems*, vol. 16, no. 2, pp. 546–556, Apr. 2015, doi: 10.1109/TITS.2014.2342271.

[9] G. Comert, M. Rahman, M. Islam, and M. Chowdhury, "Change point models for real-time cyber attack detection in connected vehicle environment," *IEEE Transactions on Intelligent Transportation Systems*, Early Access, Sep. 2021, doi: 10.1109/TITS.2021.3113675.

[10] M.M. Rana, "IoT-based electric vehicle state estimation and control algorithms under cyber attacks," *IEEE Internet of Things Journal*, vol. 7, no. 2, pp. 874–881, Feb. 2020, doi: 10.1109/JIOT.2019.2946093.

[11] J. Harvey and S. Kumar, "A survey of intelligent transportation systems security: challenges and solutions," *International Conference on Big Data Security on Cloud (BigDataSecurity), International Conference on High Performance and Smart Computing, (HPSC), and International Conference on Intelligent Data and Security (IDS)*, Baltimore, MD, USA, 2020, pp. 263–268. doi: 10.1109/BigDataSecurity-HPSC-IDS49724.2020.00055.

[12] N.N. Tran, H.R. Pota, Q.N. Tran, and J. Hu, "Designing constraint-based false data-injection attacks against the unbalanced distribution smart grids," *IEEE Internet of Things Journal*, vol. 8, no. 11, pp. 9422–9435, Jun. 2021. doi: 10.1109/JIOT.2021.3056649.

[13] S. Lakshminarayana, A. Kammoun, M. Debbah, and H.V. Poor, "Data-driven false data injection attacks against power grids: A random matrix approach," *IEEE Transactions on Smart Grid*, vol. 12, no. 1, pp. 635–646, Jan. 2021. doi: 10.1109/TSG.2020.3011391.

[14] D.-T. Peng, J. Dong, and Q. Peng, "Overloaded branch chains induced by false data injection attack in smart grid," *IEEE Signal Processing Letters*, vol. 27, pp. 426–430, Feb. 2020. doi: 10.1109/LSP.2020.2974097.

[15] L. Che, X. Liu, Z. Li, and Y. Wen, "False data injection attacks induced sequential outages in power systems," *IEEE Transactions on Power Systems*, vol. 34, no. 2, pp. 1513–1523, Mar. 2019, doi: 10.1109/TPWRS.2018.2871345.

[16] E. Naderi, S. Pazouki, and A. Asrari, "A remedial action scheme against false data injection cyberattacks in smart transmission systems: Application of thyristor-controlled series capacitor (TCSC)," *IEEE Transactions on Industrial Informatics*, vol. 18, no. 4, pp. 2297–2309, Apr. 2022. doi: 10.1109/TII.2021.3092341.

[17] E. Naderi and A. Asrari, "Hardware-in-the-loop experimental validation for a lab-scale microgrid targeted by cyberattacks," *International Conference on Smart Grid (icSmartGrid)*, Setubal, Portugal, 2021, pp. 57–62. doi: 10.1109/icSmartGrid52357.2021.9551023.

[18] E. Naderi, M. Ansari, J.P. Trimble, and A. Asrari, "Experimental validation of a market-based remedial action coping with cyberattackers targeting renewable-based microgrids," *IEEE PES/IAS PowerAfrica*, Nairobi, Kenya, 2021, pp. 1–5. doi: 10.1109/PowerAfrica52236.2021.9543289.

[19] F.V. Cerna, M. Pourakbari-Kasmaei, L.S.S. Pinheiro, E. Naderi, M. Lehtonen, and J. Vontreras, "Intelligent energy management in a prosumer community considering the load factor enhancement," *Energies*, vol. 14, no. 12, p. 3624, Jun. 2021. doi:10.3390/en14123624.

[20] A. Azizivahed, E. Naderi, H. Narimani, M. Fathi, and M.R. Narimani, "A new bi-objective approach to energy management in distribution networks with energy storage systems," *IEEE Transactions on Sustainable Energy*, vol. 9, no. 1, pp. 56–64, Jan. 2018. doi: 10.1109/TSTE.2017.2714644.

[21] M. Cui and J. Wang, "Deeply hidden moving-target-defense for cybersecure unbalanced distribution systems considering voltage stability," *IEEE Transaction on Power Systems*, vol. 36, no. 3, 2021. doi: 10.1109/TPWRS.2020.3031256.

[22] J. Chen, G. Liang, Z. Cai, C. Hu, Y. Xu, F. Luo, and J. Zhao, "Impact analysis of false data injection attacks on power system static security assessment," *J. Mod. Power Syst. Clean Energy*, vol. 4, pp. 496–505, Jul. 2016. https://doi.org/10.1007/s40565-016-0223-6

[23] J. Khazaei, "Stealthy cyberattacks on loads and distributed generation aimed at multi-transmission line congestions in smart grids," *IEEE Transactions on Smart Grid*, vol. 12, no. 3, pp. 2518–2528, May 2021. doi: 10.1109/TSG.2020.3038045.

[24] A. Eldewahi and E. Basheir, "Authenticated key agreement protocol for virtual private network based on certificateless cryptography," *International Conference on Computing, Electrical, and Electronic Engineering (ICCEEE)*, Khartoum, Sudan, 2013, pp. 269–273. doi: 10.1109/ICCEEE.2013.6633946.

[25] I.A. Yari, B. Abdullahi, and S.A. Adeshina, "Towards a framework of configuring and evaluating ModSecurity WAF on Tomcat and Apache web servers," *International Conference on Electronics, Computer and Computation (ICECCO)*, Abuja, Nigeria, 2019, pp. 1–7. doi: 10.1109/ICECCO48375.2019.9043209.

[26] E. Naderi and A. Asrari, "Approaching optimal power flow from attacker's standpoint to launch false data injection cyberattack," *IEEE Green Energy and Smart Systems Conference (IGESSC)*, Long Beach, CA, USA, 2020, pp. 1–6, doi: 10.1109/IGESSC50231.2020.9285056.

[27] M.H. Abbasi, M. Taki, A. Rajabi, L. LI, and J. Zhang, "Risk-constrained offering strategies for a large-scale price-maker electric vehicle demand aggregator," *IET Smart Grid*, vol. 3, no. 6, pp. 860–869, Dec. 2020, doi: 10.1049/iet-stg.2019.0210.

[28] E. Naderi, M. Pourakbari-Kasmaei, F.V. Cerna, and M. Lehtonen, "A novel hybrid self-adaptive heuristic algorithm to handle single- and multi-objective optimal power flow problem," *Electrical Power and Energy Systems*, vol. 125, p. 106492, Feb. 2021, doi: 10.1016/j.ijepes.2020.106492.

[29] E. Naderi, A. Azizivahed, and A. Asrari, "A step toward cleaner energy production: A water saving-based optimization approach for economic dispatch in modern power systems," *Electric Power Systems Research*, vol. 204, p. 107689, Mar. 2022. doi: 10.1016/j.epsr.2021.107689.

[30] J. Khazaei and A. Asrari, "Second-order cone programming relaxation of stealthy cyberattacks resulting in overvoltages in cyber-physical power systems," *IEEE Systems Journal*, 2021, Early Access, doi: 10.1109/JSYST.2021.3108635.

[31] MATLAB and Simulink. [Online]. Available from: www.mathworks.com/products/matlab.html.

[32] The General Algebraic Modeling System. [Online]. Available from: www.gams.com/.

[33] CPLEX Solver Manual [Online]. Available from: www.gams.com/latest/docs/S_MAIN.html.

[34] GAMS-MATLAB Interface, "GDXMRW Manual" [Online]. Available from: www.gams.com/latest/docs/T_GDXMRW.html.

[35] S. Mirjalili, S.M. Mirjalili, and A. Lewis, "Grey wolf optimizer," *Advances in Engineering Software*, vol. 69, pp. 46–61, Mar. 2014. doi: 10.1016/j.advengsoft.2013.12.007.

[36] J. Kennedy and R. Eberhart, "Particle swarm optimization," *Proceedings of ICNN'95: International Conference on Neural Networks*, 1995, pp. 1942–1948. doi: 10.1109/ICNN.1995.488968.

[37] E. Naderi, M. Pourakbari-Kasmaei, and H. Abdi, "An efficient particle swarm optimization algorithm to solve optimal power flow problem integrated with FACTS devices," *Applied Soft Computing*, vol. 80, pp. 243–262, Jul. 2019. doi: 10.1016/j.asoc.2019.04.012.

[38] H. Narimani, A. Azizivahed, E. Naderi, M. Fathi, and M.R. Narimani, "A practical approach for reliability-oriented multi-objective unit commitment problem," *Applied Soft Computing*, vol. 85, p. 105786, Dec. 2019, doi: 10.1016/j.asoc.2019.105786.

[39] Kyocera Solar Website. Accessed: Sep. 12, 2020. [Online]. Available: www.kyocerasolar.eu/.

[40] Xantrex Power Inverter Website. Accessed: Sep. 12, 2020. [Online]. Available: www.xantrex.com/.

[41] Wind-Turbine-Models Website. Accessed: Apr. 10, 2019. [Online]. Available: https://en.wind-turbine-models.com/.

[42] EIA Website. Accessed: Apr. 10, 2019. [Online]. Available: www.eia.gov/.

[43] Weather Carbondale Website. Accessed: Apr. 10, 2019. [Online]. Available: www.meteoblue.com/.

[44] M.Q. Duong, T.D. Pham, T.T. Nguyen, A.T. Doan, and H.V. Tran, "Determination of optimal location and sizing of solar photovoltaic distribution generation units in radial distribution systems," *Energies*, vol. 12, no. 174, pp. 1–25. doi: 10.3390/en12010174.

[45] A. Asrari, M. Ansari, J. Khazaei, and P. Fajri, "A market framework for decentralized congestion management in smart distribution grids considering collaboration among electric vehicle aggregators," *IEEE Transactions on Smart Grid*, vol. 11, no. 2, pp. 1147–1158, Mar. 2020. doi: 10.1109/TSG.2019.2932695.

18 Hybrid Power Systems for Smart Marine Power Grids

Demonstration and Case Study

*Meysam Gheisarnejad, Mahdi Mosayebi,
Mehdi Rafiei, Jalil Boudjadar, and
Mohammad Hassan Khooban*

CONTENTS

DOI: 10.1201/9781003293989-18

18.1 CHALLENGES AND OPPORTUNITIES IN MARINE APPLICATIONS

Despite the great benefits of maritime hybrid electrification, there are some challenges preventing the widespread implementation of this technology. These obstacles include operational and technical issues, as well as economic factors.

18.1.1 Economical Perspectives

Nowadays, using fossil fuels to generate electricity has led to some major problems. Pollution is one of the major drawbacks of fossil fuels because they emit carbon dioxide and other harmful air pollutants when burned. So, the use of clean and renewable energies has been increased in shipboard power systems to make a hybrid architecture. Generally, in order to reduce the marine's pollution with changes in the power generation system, three different approaches are implemented. One of them is the hybridization of the electrical systems. Typically, hybrid electrical systems in marine applications consist of several forms of renewable energy to generate electricity. A very general structure of a hybrid electrical system may consist of a combination of renewable energy resources, diesel generators, and energy storage systems (ESSs) depending on the availability of these resources. However, a typical hybrid electrical system in marine applications may consist of photovoltaic systems (PVs)-ESSs-diesel, PVs-wind-fuel cells (FCs), PV-FCs-ESSs, and PV-wind-FCs-ESSs combinations [1, 2].

The utilization of energy storage systems with diesel generators make it possible to store the power in low peak times and use it in high peak times, which leads to improving the energy efficiency and reducing the emissions [3–5]. Using a hybrid energy storage system based on batteries and super-capacitors for the casting-off and docking maneuvers of a marine vessel provide a reduction of 1.39 kg CO_2 during the maneuvers and total emissions and consumption by 20%. Moreover, utilization of a hybrid propulsion plant with a hybrid power supply can save 6% additional fuel consumption. A hybrid energy system based on photovoltaics, batteries, diesel generators, and cold ironing is assessed in [6, 7].

From an economical point of view, the hybrid electrification of marine applications can affect the cost of energy, overall efficiency, lifecycle cost, cost-effective operation, and optimal energy management in these applications. It can be shown that through a hybrid power system and advanced control approaches fuel consumption and emission reduction can be up to 10–35% [1, 2, 8, 9].

18.1.2 Environmental Restrictions

Nowadays, fossil fuels are used in the marine industry and cause emissions such as CO_2 or NOx. These emissions have high environmental effects, like greenhouse gases (GHGs) and air pollution. It has been stated that the marine industry is responsible for up to 5% of worldwide CO_2 emissions. However, it has been predicted that the total emissions due to marine fuel consumption can cause the growth in CO_2 and NO_X emissions by 2030 if no action is taken to solve this problem. Therefore, without any preventive program in the marine industry, its emissions would increase significantly.

On the other hand, according to 2019 data from the International Energy Agency (IEA), the marine industry accounted for about 2% of global energy-related CO_2 emissions [10]. Moreover, to prevent pollution from the marine industry, the International Maritime Organization (IMO) has adopted a variety of restrictions and regulations under Annex VI (titled Regulations for the Prevention of Air Pollution from Ships) of the International Convention. In addition, in April 2018, the IMO aimed to reduce absolute GHG emissions by at least 50% by 2050 to bring the industry in line with the climate targets of the Paris Agreement.

Although marine industry operations have great importance in the European marine sector economy, the vessels in the European marine sector do not have the required standards and specifications to decrease environmental concerns. Many vessels and ferries that have been used in the European transport sector have not been updated to new standards and restrictions for the last two decades [11, 12]. Therefore, these ships are a critical source of carbon air-polluting emissions.

One solution for this problem is increasing the electrification of ships with a hybridization process. A hybrid energy system based on photovoltaics (PVs), energy storage systems (ESSs) like batteries, diesel generators, and cold ironing can decrease CO_2 and NO_x emissions. In this regard, different combinations of renewable energy sources (RESs) and conventional sources like diesel generators (DGs) can be used in hybrid electric systems in marine applications. These combinations include fuel cells (FCs), photovoltaic systems (PVs), wind turbines (WT), sea wave energy (SWE), energy storage systems (ESSs), and cold ironing (CI), which can be used for supplying the demand in these ships [13].

18.1.3 Technological Limitations

The hybrid AC/DC systems are used in marine applications to combines the benefits of AC and DC architectures, simultaneously. The main advantage of the hybrid AC/DC systems is that each AC and DC subsystem can facilitate the direct utilization of both AC- and DC-based renewable energy sources (RESs), ESSs and loads [14, 15]. This hybridization provides an efficient structure for the integration of RESs, FCs, and ESS units with minimum modifications, which can reduce the total cost of the system. However, the use of hybrid systems in marine applications raises some technical challenges.

These technical issues are related to the efficiency of power electronic converters, control approaches, protection strategies, energy management, stability, reliability, and communication technology. Furthermore, fast charging of marine batteries during the short port stays is also a technical challenge that can be solved with the development of a charging infrastructure.

Another important technological challenge of hybrid systems in marine applications is its power management and control strategy [16–19]. Due to continuously supplying the loads and consumers in hybrid systems, the reliability assessment is a key design in modern energy management. Hence, interruptions as a consequence of any failure in the power sources, protection system, and loads in the hybrid system's marine application can affect system reliability and reduce the overall performance of the energy management and efficiency of a hybrid energy system.

From the control point of view, the presence of a time-delay ship in hybrid power plants in marine applications generally deteriorates the controller performance and highly threatens the system stability of these systems [20, 21]. Therefore, considering the effects of delays in control, a design that can enhance the controller complexity from the theoretical point of view is necessary.

Protection of the hybrid systems in marine applications is another field that requires more attention from several aspects, such as suitable grounding architecture, fault current limiting technique, fast fault detection approach, and a proper circuit breaker. Moreover, a significant challenge associated with the protection of hybrid systems is that in islanded mode of operation the magnitude of short-circuit currents is too low.

Converter's efficiency is an important technological challenge in the hybridization process of marine applications. Power conversion in the hybrid systems in marine applications is always subject to losses. In the power converters, conversion losses are caused by resistive components, inductive components, and switches (switching losses). Therefore, minimizing these losses through optimal converter design or optimal converter control is subject to ongoing research.

18.1.4 Opportunities Coming from Hybridization

The utilization of the hybrid system in marine applications provide an optimal solution to combine the main advantages of both AC and DC power systems. The key advantages of the AC power systems are easy connection to the AC sources, such as diesel generators, fast and reliable detection and clearing of the faults with the existing protection devices, and lower bus voltage due to the universal standards of the voltage level. On the other hand, DC power systems have advantages in terms of system efficiency, cost, and system size. Due to the reduced number of power electronic converters, the overall efficiency is improved. Additionally, simple control, no reactive power control, no harmonic issues, no synchronization of power sources, and the easy interface between power sources and electronic loads are the most benefits of these power architectures.

The main feature of the hybrid systems is that the two-power network is connected in the distribution power grid, which facilitated the integration of both AC- and DC-based distributed generation, RESs, EESs, and loads. Moreover, with the hybrid systems, the integration of these sources and loads can be developed with minimum modifications of the current distribution grid, which reduces the total cost.

The most important advantages and opportunities of the utilization of hybrid systems in marine applications are:

- Reliability: reliability is one of the main concerns in the power system's marine application. Due to the use of renewable energy sources in hybrid power systems, the intermittency of these sources, as well as the failure of the component, can degrade the reliability of the whole system. The use of diesel generators and ESSs in a hybrid architecture can improve the reliability of the systems and suppress the intermittency of the RESs.
- Efficiency: due to the limitation of fuels and energy for long maritime missions, the efficiency of power systems is very important. Several components in the power system, such as power electronic converters, cables, power switches, and power sources have a role to improve efficiency. In the

hybrid systems in marine applications, the use of a DC power grid as part of the system can significantly increase efficiency. Moreover, in the hybrid power systems the number of power electronic converters are decreased, which can improve the efficiency of the power system.

- Cost-effective configuration: in the hybrid systems the power sources and loads are directly connected to the AC or DC bus based on their voltage type. This can remove the extra power of electronic converters. Therefore, the implementation cost of the hybrid power systems in marine applications is decreased due to the reduction in the number of power electronic converters.

- Efficient integration: devices based on the type of the working voltage (AC or DC) are directly connected to the power system with a minimum number of interface power converters, which can reduce the conversion stages and therefore the energy losses. This feature makes hybrid power systems suitable for the integration of DC-based power sources like PVs, FCs, and ESSs, while maintaining the AC-based power sources connected to the AC network.

- Simple control: in hybrid systems there is no need for the synchronization of power sources and energy storage units, as they are directly connected to both the AC and DC networks. Therefore, the control strategy for this device is simplified.

- Economic feasibility: by the addition of an advanced power electronic converter to the current power grid and the communication network for connection of devices the marine hybrid system can be developed. Therefore, due to the main power converter, the overall cost becomes higher than pure AC systems. However, with the increase in the number of attached devices, the investment of developing a marine hybrid system will be returned faster, as the number of total interface power converters is decreased.

- Ancillary services: Like the AC power systems, the hybrid AC-DC marine systems can provide ancillary services, such as reactive power compensation, power factor correction, voltage and frequency regulation, improvement in power quality by reducing effects of voltage sags, and uninterruptable power supply.

- Environment friendly: Marine hybrid systems can reduce the emissions of GHGs with the installation of renewable energy sources that are free from pollution.

18.2 MARINE HYBRID SYSTEM'S ENERGY TECHNOLOGIES

Recently, the application of sustainable and clean technologies for supplying the required power of the marine systems in the marine industry have drawn much attention. Various combinations of sustainable resources like FCs, solar panels, and batteries can be utilized to meet the demand of the ship systems.

18.2.1 Gray Energies in Marine Hybrid Systems

In many medium and high-voltage hybrid systems in marine applications, diesel generators are still a size- and cost-effective option. In marine applications the availability of power generation is a concern. Thus, the utilization of at least one diesel generator is preferred to improve overall system reliability, efficiency, and operating mode [22].

Due to its high efficiency, fast starting speed, and low maintenance, the diesel generator has been a proper backup option in hybrid power systems in marine applications. The voltage and power fluctuation raised by PVs, FCs, wind turbines (WTs), and/or varying or pulsed loads can be eliminated by using diesel generators [23]. The overall of diagram of a diesel generator power plant is shown in Figure 18.1. As depicted in Figure 18.1, the diesel power generator consists of a first-order governor and a first-order diesel generator, which is followed by two saturation functions.

The diesel system employs as the backup utility in hybrid systems. In other words, when the voltage and power of the hybrid system are balanced, no power is injected. Meanwhile, if a lower (higher) power is required, then a negative (positive) is supplied by the diesel system [24–26].

However, the diesel generators, which are used in the hybrid ship power systems emit significant NOx and CO_2 into the environment. Moreover, the optimum fuel utilization factor of these generators is just 40%, and the rest of the fuel's stored energy is wasted through the exhaust or heat. Therefore, utilizing green energies in marine hybrid systems is a solution to decrease the emission of GHGs and air pollution.

18.2.2 Green Energies in Marine Hybrid Systems

In marine vessel systems, the main consumption of the electric energies provided by ship generator units is dedicated to the propulsion electric engines. In addition, the integrated power ship systems consume a significant amount of energy because numerous pieces of equipment are required for each vessel. Some of these devices consist of cargo units (like pumps, cargo load) and other requirements, such as electric heaters, control devices, and compressors. For this purpose, various integration of RESs, such as solar panels, storage batteries, and cold ironing, are adopted for supplying the demand in such systems [27–29]. Fully hybrid electric propulsion units are regarded as an effective technology that leads to considerable increases in the stability and efficiency of ships. A high penetration of sustainable energies (e.g., PV and SWE, etc.) enhance

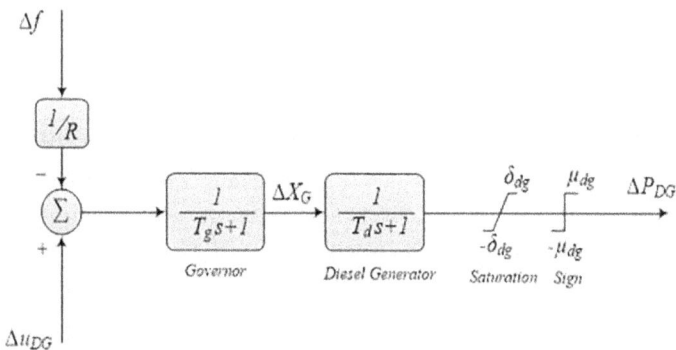

FIGURE 18.1   Diesel power system model.

the challenges of system regulation and unwanted increase in the cost of electricity due to the uncertainty of sun irradiation and wave energy. The ability to combine various units of electric generators, sustainable energies, and storage systems also makes it possible to comply with ship power efficiency directives, which are not accessible by any type. Hence, the deployment of innovative energy management schemes perfectly designed for ships with various components has become an urgent necessity [19, 30].

18.2.3 ENERGY STORAGE IN MARINE HYBRID SYSTEMS

The various forms of energy, such as electrochemical, electromagnetic, and thermal, are able to store energy at various capacities. The energy storage system can be classified based on the stored energy form, as depicted in Figure 18.2 [31].

To analyze the specifications of energy storage systems, various factors such as energy rating, efficiency, lifetime, and capital cost are often considered. The overview of the technical properties of the energy storage systems is furnished in Table 18.1 [31].

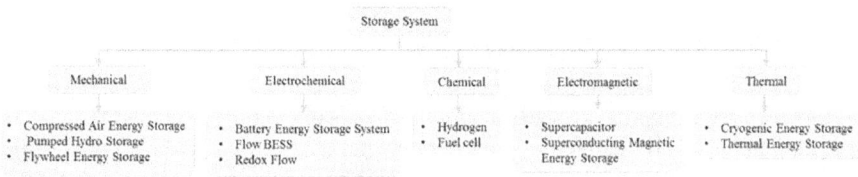

FIGURE 18.2 Classification of various storage components.

TABLE 18.1
Technical Specifications of Energy Storage Systems.

System	Energy rating		Efficiency (%)	Lifetime		Capital cost	
	Power rating (MW)	Discharge time typical		In years	In cycles	$ (kW h)	$ (kW h-per cycle)
PSH	100–5000	1–24 h+	70–80	>50	>15,000	5–100	0.1–1.4
CAES	5–300	1–24 h+	41–75	>25	>10,000	2–50	2–4
FES	0–0.25	s—h	80–90	15–20	104–107	1000–5000	3–25
LA	0–20	s—h	75–90	3–15	250–1500	200–400	20–100
NiCd	0–40	s—h	60–80	5–20	1500–3000	800–1500	20–100
Li-on	0–0.1	min—h	65–75	5–100	600–1200	600–2500	15–100
NaS	0.05–8	s—h	70–85	10–15	2500–4500	300–500	8–20
VRB	0.03–3	s—10 h	60–75	5–20	>10,000	150–1000	5–80
ZnBr	0.05–2	s—10 h	65–75	5–10	1000–3650	150–1000	5–80
FC	0–50	s—24	34–44	10–30	103–104	–	6000–20,000
SC	0–0.3	ms—1 h	85–98	4–12	104–105	300–2000	2–20
SMES	0.1–10	ms—8 s	75–80	–	–	1000–10,000	–

18.2.3.1 Superconducting Magnetic Energy Storage

The superconducting magnetic energy storage (SMES) is a promising storage technology that comprises a superconductive coil, a refrigerator, and a power conditioning system. This type of storage technology is comparable to super-capacitors, but with a feature of higher power density. SMES can be combined with wind farms [32, 33] to ameliorate the quality power and dynamic stability. However, the SMES has high sensitivity to temperature variations. So far, the implementation of SMES is limited, and only a few low-capacity SMESs are adopted for commercial applications [34, 35].

18.2.3.2 Super-Capacitors (SC)

Super-capacitors (SC) are widely adopted due to some of their advantages, such as ultrahigh power densities, high reliability, and a longer life cycle than batteries. SCs are made from two porous electrodes located in an electrolyte medium. SCs are adopted in applications that need many fast charge/discharge cycles, where they are utilized for energy smoothing and burst-mode power delivery [31, 36, 37].

18.2.3.3 Flywheel Energy Storage System

Flywheel has existed as a means to store energy for thousands of years as one of the oldest mechanical energy storage systems. These technologies cover a wide range of applications that were realized by rotating objects, like the water wheel, hand mills, and others. In spite of significant advances during its early versions, the implementation of flywheels has been negligible and has decreased with advancements of the power grid. However, with the recent advances in materials and power electronic equipment, the flywheel energy storage system (FESS) has emerged as a comprehensive option for energy-saving applications [38].

18.2.3.4 Pumped-Storage Hydroelectricity (PSH)

The conventional pumped-storage hydroelectricity (PSH) is one of the most well-known types of hydroelectric storage and is configured by two water reservoirs. At peak hours, the turbines pump the water stored in the upper reservoir to produce energy. When demand is low, excess production in the plant is utilized to pump the water from the lower reservoir to a higher elevation [39, 40].

Among the various forms of storage technologies, PSH has the highest energy rating, durability, and efficiency. The main application of this unit is for variable-speed pumps where the systems are able also to stabilize the grid frequency. Since these technologies have the ability to optimize operational efficiency by turning speed, they can offer a wide range of operating work in the generation state.

18.2.3.5 Compressed Air Energy Storage

Compressed air energy storage (CAES) is a technology to store energy based on the fundamentals of conventional gas turbine systems. In this unit, electrical compressors are adopted for the air-compressing process and stored in an underground structure, such as an abandon mine. At a utility scale, the power produced during the period of off-peak can be used to satisfy the higher demand during the peak load periods. From Table 18.2, it is evident that the high-power rating of the CAES makes it a proper option for wind farms, especially in the energy-management context. However, the application of CAES is also restricted by topographical situations [41].

18.2.3.6 Battery Energy Storage System

The battery energy storage system (BESS) stores electricity by a chemical reaction, which is applied to anode and cathode electrodes. The BESS is made from a series of low-voltage/low-power cells, which are joined in parallel to reach a desired electrical property. Up to now, different kinds of batteries have been constructed for industrial applications, such as lead acid (LA), nickel cadmium (NiCd), nickel metal hybrid (NiMH), lithium ion (Li-ion) and sodium sulphur (NaS).

The flow battery is another kind of battery technology adopted to store energy in electrically active species. Generally, extra electrolytes are adopted in tanks and pumped by a reactor cell. The energy rating and power rating are determined by the amount of electrolytes and the active region of the cell stack. The vanadium redox battery (VRB) and zinc bromine (ZnBr) are the most popular flow batteries in the industries [31].

18.3 MARINE HYBRID SYSTEMS

One of the most critical concerns over the last decades is that of coping with reducing the adverse effects of communities on the environment. Despite maritime transportation being regarded as an environmentally friendly form of transport, the increase in pollution, which has resulted from the increase in international trade, has led to an awareness of the effect of ships on environmental pollution. The most important pollutant factors of ships are sulfur oxides (SO_2), nitrogen oxides (NO_x), GHGs, particulate matter (PM) (especially particles with 10 micrometers and below (PM_{10}) and particles with 2.5 micrometers and below ($PM_{2.5}$)), which pollute and damage the environment [42, 43].

It is predicted that due to the fast growth of the e-commerce market, the world will face serious problems in relation to ship emissions in the next 10–40 years. The maritime industry, as one of the major sectors of transportation, is responsible for 14% of global emissions, and it has become the fourth major release of the world. Although the vessels are recognized as the biggest pollutants, transporting by maritime is one of the cleanest kinds of transportation and is more preferred than other forms of transportation, such as air and road transport [44].

The increasing energy demand, along with rising concerns of emissions from the maritime area, has necessitated the combination of sustainable energy technologies (e.g., PV, tidal, geothermal, WT, etc.) into the power systems [45, 46]. The integration of such modern technologies with diesel generators offers many benefits from a security and reliability perspective. In addition, by reducing emissions and energy waste, hybrid technologies decrease generator wear, reduce maintenance and breakdown costs, generate smoother electricity, and prevent temporary loads on engines. For instance, a solar power system is implemented by humans chiefly based on PV and photothermal methods that are frequently adopted in microgrids, aerospace, radio transmitters, lighting for roadways, and other applications.

Because the cargoes are often shipped, the ship fuel consumption and emissions are very high, particularly for ocean-going vessels. However, some renewable resources, like wave and wind energy, in the ocean are enormous, so installing such energy resources has a critical role in reducing the fuel consumption of large ocean-going vessels, addressing concerns related to the violations of pollution laws [47]. Up to now, the sustainable resources have been adopted for ships of all sizes comprising hybrid propulsion and on-board and shore-side applications. Sustainable resources can be combined by retrofitting fleets or using them to build and design new ships, with fewer new ships to generate 100% zero-emissions power for hybrid propulsion in the long term. Some programs offer the potential for immediate developments, even if the share of sustainable resources in the power integration of the transportation area is restricted in the short to medium term. Among the different types of ships adopted in the international ocean, a large flat deck at the top of the vertical space on the sides of the ship is relatively large and provides the space conditions for installation of large-scale WT and PV systems.

Due to the weather-dependent features of sustainable resources, unreliable power and unwanted rising of costs is created by the uncertainty of such energy resources. For a PV/WT/tidal/diesel-integrated system, an energy storage device is required. The installation of storage systems in the planning of a PV/WT/tidal/diesel-integrated ship system is the best option to ameliorate the quality of power and offer uninterrupted energy to the system. When the sustainable resources generate little power, the battery system meets the required power according to the load demand. Consequently, the battery hybrid power generation for ships can both reduce costs and reduce GHG emissions. Adopting PV/WT/tidal/diesel/battery storage provides reliable energy and decreases maintenance costs.

18.3.1 STATE-OF-THE-ART SYSTEMS IN MARINE APPLICATIONS

Electrical systems of large-scale ships that utilize a fully hybrid electric propulsion unit is an important option to improve the efficiency of the ship and reduce GHG emissions. The range of propulsion power is illustrated in Table 18.2, where it is between 150 kW and 100 MW.

Optimal management of the integrated ship power system has been frequently studied, as reported in [48]. For instance, in [30], a model predictive control (MPC) is employed for the energy-management problem in the electric ship to optimize the match between energy-saving battery and ship generators when the system is subjected to high-power ramp rate situations. A sub hourly energy-management technique has been employed for all electric ships with various components, including FCs, batteries, and PV cells. The authors of [49] developed a power-flow dispatching for a large green ship, including diesel, PV, storage unit, and cold ironing. By considering many constraints in the system modeling (such as PV output and storage capacity) applied with a penalty in the cost function, the energy management of the maritime system is optimally solved by defining an unconstrained and large-size optimization problem. In this regard, Balsamo *et al.* [50] proposed a novel programming strategy for optimal energy management based on a reduced optimization problem of an all-electric ship supplied with an integrated battery system. A quantitative analysis based on statistic coefficients of a storage unit has been made to analyze the effect of the storage losses during the system operation.

TABLE 18.2
Propulsion Power Requirements for Various Ship Applications.

Vessel class/application	Power requirement, approximate, MW
Ferries	
Small ferries	0.24–1.0
Large ferries, inland waters	2–12
Large ferries, open seas and high speed	20–44
Freight vessels	
Inland freight vessels, USA, towed and pushed barges	0.13–1.0
Inland/river vessels, Europe,	
Class 1—Vb (self-propelled) and push-tow barges	0.2–4.5
River cruise vessel, Class VIb	<2
Tugs, Europe, Class 1	10
Inland container vessel, Mississippi River	11.5
(proposed, 235 m, up to 2960 TEU)	
Cold ironing (ports)	
Container ships (5–13k TEU)	0.2–0.9
Cold ironing; tankers	Up to 3 MW
Cruise ships	
Cruise ship, hotel load	3–10
Cruise ship, maneuvering in port	20
Cruise ship, emergent power	2–4
Cruise ship, propulsion	25–97

Much research has centered on equipment of the electric ferry boats, such as cold ironing, pollution control, FCs, and utilization of sustainable resources. Cold ironing is a source of practical and useful energy to power when carrying ships on land or in the port in which energy is needed to ship via ferry to microgrids or the electricity grid onshore. In recent years, various works have been carried out to optimize the utilization of cold ironing. In this context, the considerable impact of cold ironing on bus voltages and power efficiency in the coastal area and power distribution network are studied in [51]. Ballini and Bozzo [52] addressed the potential of cold ironing technology at new cruise ships to deal with adverse impacts of air emissions from hotel yachts. An energy-management system (EMS) for an integrated power plant of a ferry boat, including various unites like a proton exchange membrane fuel cell (PEMFC), Li-ion battery bank, and cold ironing has been developed to enhance efficiency and decrease operation costs. To reduce hazardous air pollutants (e.g., carbon dioxide (CO_2), nitrogen oxides (NO_x), sulfur oxides (SO_x)), and system component costs, a source of cold ironing is integrated with liquefied natural gas (LNG) [53].

In addition to innovative technologies for conventional motors, fuel cells are regarded as a promising solution to ameliorate the utilization of alternative ship fuels. In comparison with the ship combustion engine, which produces the electricity with efficiencies of 33–35%, FCs have a higher level of efficiency (more than 60%), though their energy capacity is limited and cannot meet all maritime power requirements. Over the last few decades, focus on enhancing the efficiencies of such technologies has led to significant progress that has increased their applicability [54]. To deploy the potential of FCs for maritime systems, exergy analysis was performed for both a proton methanol membrane fuel cell and a direct methanol fuel cell [55]. In [56], a life cycle assessment (LCA) was performed for an onboard auxiliary maritime power system with a molten carbonate fuel cell. The generation of the manufacturing of molten carbonate fuel cell units have shown higher environmental effects than diesel engines. Therefore, recycling molten carbonate fuel cells is important in order to ameliorate the environmental impacts of such technologies [57]. A testbed of an integrated ship power system including a molten carbonate fuel cell (100 kW), a lead-acid battery system (30 kW), and a diesel unit (50 kW) was configured to reduce fuel consumption and CO_2 emissions by handling the charge of the base load in the marine power system.

Accordingly, numerous research works have focused on the utilization of advanced control methodology to ameliorate the performance of integrated power systems in the marine vessel. For instance, in [24], an adaptive non-integer fuzzy PD+I controller in a ship power system with WT, PV, SWE, and storage system was used. A fast-intelligent reconfiguration scheme [58] is proposed for an electric ship's power system in a multi-objective optimization manner. An analytical approach based on gain and phase margin values is proposed in [59] to improve the stability of shipboard microgrid systems. Since such integrated systems are prone to undesired frequency fluctuations due to the randomness of sustainable energies, a virtual gain and phase margin examiner was utilized to obtain the desired stabilization characterization. In [60], an adaptive control technique is developed based on neural networks (NN) for efficient regulation of actuated/ unactuated state coefficients, which doesn't need to any linearizing operations. In this study, the suggested control law automatically compensates for uncertainties of ship-mounted crane systems.

18.3.2 Case Study Assessment of Marine Hybrid Systems

18.3.2.1 Safety Improvement of Emission-Free Ships [61]

In this work, a hybrid power system of an all-electric ship (AES) is considered as a case study that includes a molten carbonate fuel cell, a PV system, and a lithium-ion battery. In addition, this integrated system is equipped with other units, like convertors, terminals, and wiring. The configuration of an integrated power system is illustrated in Figure 18.3.

In this test system, the studied boat is a bay tour ferry, which is adopted for transferring the passengers for a one-day tour. The characterizations of the integrated ship system are furnished in Table 18.3.

The systematic reliability centered maintenance (SRCM), which concentrates on the integrated power system, is adopted to achieve a proper maintenance plan of the system's components in the all-electric ship.

Based on the decision-making of reliability centered maintenance, the appropriate maintenance task should be found for all components after identifying all potential states of failure and their consequences. All information corresponding to the states

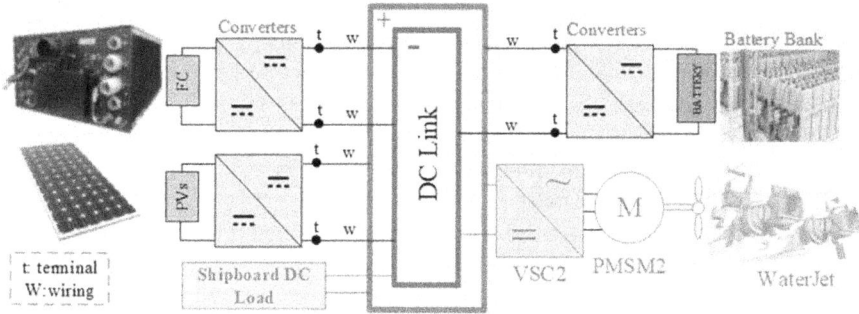

FIGURE 18.3 Illustration of integrated ship power system.

TABLE 18.3
Parameters of the Boat System.

Parameter	Value
Type	Passenger
Length (m)	20
Width (m)	8.2
Draft (m)	1.25
Displacement (t)	140
Fuel cells (kW)	3×100
Hydrogen tanks	6×125 kg at 500 bar
Photovoltaic (kWp)	2.7
Battery packs (kWh)	2×116
Electric propulsion motors (kW)	2×200

of failure and their consequences is borrowed from participants in the hydrogen fuel cell and battery system for the ship project [62]. An SRCM worksheet is adopted, which demonstrates the results of decisions made by experts as "Y" and "N," which means Yes and No to decision-making questions [63]. For an effective maintenance task, the parameters optimized for integrated systems can be described as follows:

- The plan of on-condition operation for the fuel cells;
- The plan of failure-finding operation for the fuel cells;
- The plan of on-condition operation for the batteries;
- The plan of restoration operation for the batteries;
- The plan of on-condition operation for the PVs;
- The plan of restoration operation for the PVs;
- The plan of on-condition operation for the convertors;
- The plan of on-condition operation for the wiring/terminal;
- The plan of restoration operation for the wiring/terminal.

The maintenance plans of the system's components are obtained by the SRCM scheme. Then, the Markov profile of the integrated system is adopted to determine the reliability of the system. Based on the maintenance tasks for each component and the Markov process, the optimization problem is formulated by the maximizing reliability index (MTTF) and minimizing costs. For application to the marine industry, the maintenance planning is characterized by two standards, including DNVGL-ST-0033 [64] and DNVGL-ST-0373 [65].

The efficiency of the suggested maintenance technique is evaluated using real-world data of the integrated marine power system. The outcomes demonstrated that with the application of proposed SRCM, the reliability, cost, and safety are ameliorated, which makes it an applicable scheme in practical environments.

18.3.2.2 Cost-Effective Power Scheduling in Zero-Emission Ferry Ships [13]

In this research paper, an energy management strategy for an integrated power system of a ferry boat is developed to enhance the efficiency and decrease the operation cost. Four distinct test systems based on an integrated energy management scheme have been studied for the ferry boat. The configuration of the combined power system for the studied ferry boat is depicted in Figure 18.4. In this work, a ferry boat equipped with two proton exchange membrane fuel cell units with 200 kW and a proton exchange membrane fuel cell unit with 100 kW capacity is considered. Moreover, 20 hydrogen storages with 18.8 kg content from Luxer-GMT with 5,000 psi have been accommodated on the marine system. In addition, the studied ship is equipped with two electromotors with 250 kW

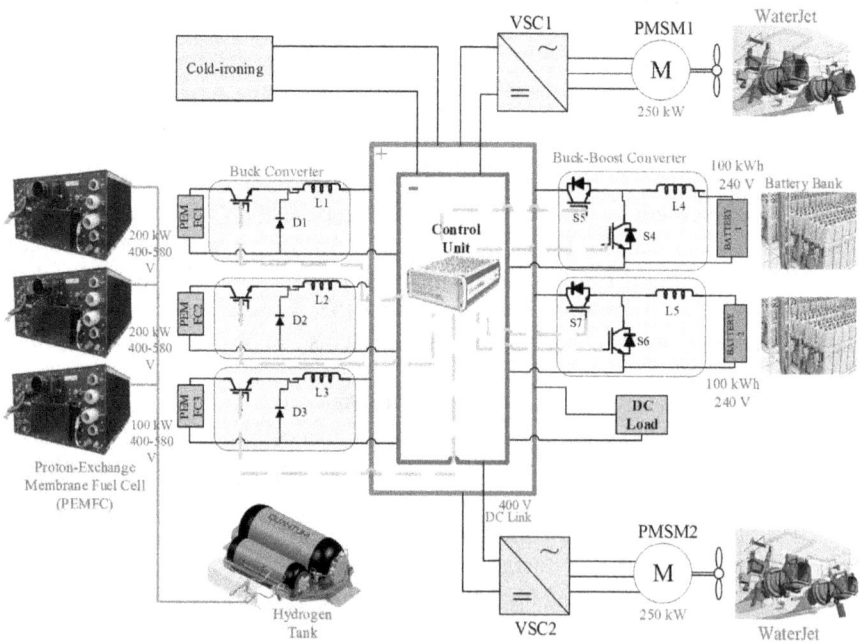

FIGURE 18.4   Structure of the integrated power system.

rated power. A room of batteries is also embedded to store and manage surplus energy generation of fuel cell output. Li-ion batteries with charge capacity of 200 kWh are adopted in the ferry boat to compensate the load power.

To model the various components in the energy management system, a single fuel cell system is considered, and its capacity is equal to the sum of the capacities of all proton exchange membrane fuel cell units. The power flow of the boat at the time of sailing is depicted in Figure 18.5. P_1 is the produced energy by the fuel cells for supplying the boat loads, P_2 is the produced energy of the fuel cells for charging the batteries, and P_3 is the discharged energy by the batteries.

Figure 18.6 illustrates the condition under which cold ironing can be operated while the ship is at the harbor. P_4 is the consumed energy the boat loads through cold ironing, and P_5 denotes the stored power in the batteries through cold ironing.

For optimal energy management, it is aimed at optimizing the operation cost and meeting the operation constraints of the component. For this purpose, the objective function is defined with considering the operation constraints of FCs, batteries, and cold ironing.

The profiles of produced energy of all resources in the studied boat is depicted in Figure 18.7. The energy produced by the FC meet all required loads of the ship without the utilization of sustainment resources or other conventional fossil fuels.

The power flow from the FC and cold ironing to supply the boat load is represented by P_1 and P_4, and the power flow from the FC and cold ironing to charge the battery are represented by P_2 and P_5. P_3 demonstrates the discharge power of the battery.

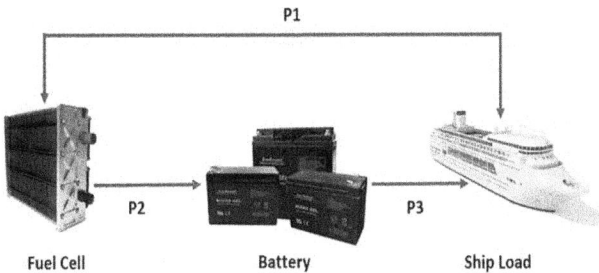

FIGURE 18.5 Fuel cell/battery/ship's load architecture.

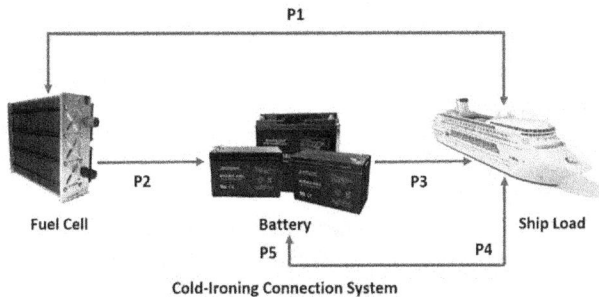

FIGURE 18.6 Configuration of the fuel cell/battery/cold ironing.

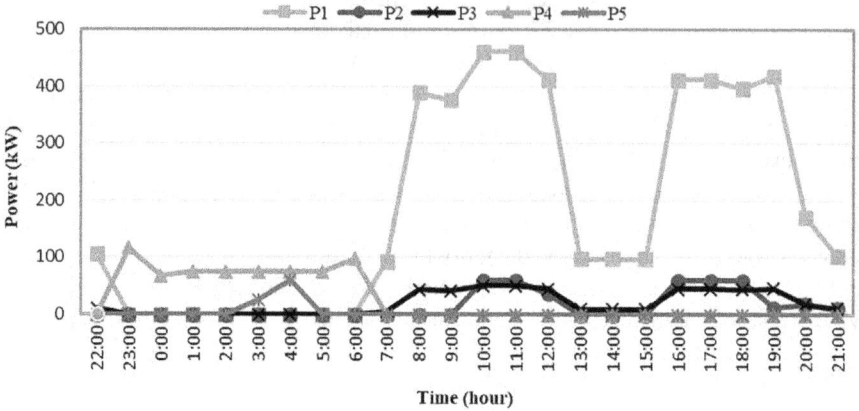

FIGURE 18.7   Energy management of integrated ship system.

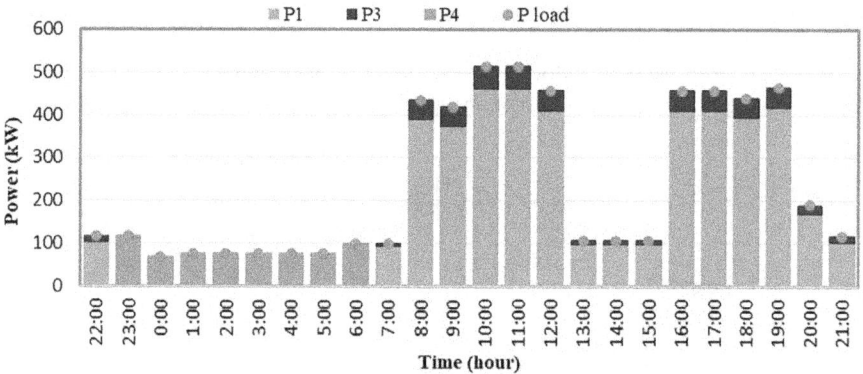

FIGURE 18.8   The needed power of the load supplied by various resources.

Injected power to the boat load by various units are depicted in Figure 18.8. According to Figure 18.8, at least 10% of the load energy is provided through the battery units discharging power (P_3). Since the cold ironing has a lower energy price than the FC at (1–6) and (23–24) hours, the ship is supplied by cold ironing (P_4) while the boat is in the harbor; also, the cold ironing can charge the battery bank (P_5) if required.

REFERENCES

[1] A. J. Sorensen *et al.*, "Toward safer, smarter, and greener ships: Using hybrid marine power plants," *IEEE Electrification Magazine*, vol. 5, no. 3, pp. 68–73, 2017.

[2] A. Roy, F. Auger, F. Dupriez-Robin, S. Bourguet, and Q. T. Tran, "Electrical power supply of remote maritime areas: A review of hybrid systems based on marine renewable energies," *Energies*, vol. 11, no. 7, p. 1904, 2018.

[3] Z. Wang, R. Carriveau, D. S. K. Ting, W. Xiong, and Z. Wang, "A review of marine renewable energy storage," *International Journal of Energy Research*, vol. 43, no. 12, pp. 6108–6150, 2019.

[4] F. Balsamo, C. Capasso, D. Lauria, and O. Veneri, "Optimal design and energy management of hybrid storage systems for marine propulsion applications," *Applied Energy*, vol. 278, p. 115629, 2020.

[5] A. N. Einrí, G. M. Jónsdóttir, and F. Milano, "Modeling and control of marine current turbines and energy storage systems," *IFAC-PapersOnLine*, vol. 52, no. 4, pp. 425–430, 2019.

[6] M. Kalikatzarakis, R. Geertsma, E. Boonen, K. Visser, and R. Negenborn, "Ship energy management for hybrid propulsion and power supply with shore charging," *Control Engineering Practice*, vol. 76, pp. 133–154, 2018.

[7] N. Bennabi, J. Charpentier, H. Menana, J.-Y. Billard, and P. Genet, "Hybrid propulsion systems for small ships: Context and challenges," in *2016 XXII International Conference on Electrical Machines (ICEM)*, 2016, pp. 2948–2954: IEEE.

[8] E. A. Sciberras, B. Zahawi, D. J. Atkinson, A. Breijs, and J. H. van Vugt, "Managing shipboard energy: A stochastic approach special issue on marine systems electrification," *IEEE Transactions on Transportation Electrification*, vol. 2, no. 4, pp. 538–546, 2016.

[9] A. Aktaş and Y. Kırçiçek, "A novel optimal energy management strategy for offshore wind/marine current/battery/ultracapacitor hybrid renewable energy system," *Energy*, vol. 199, p. 117425, 2020.

[10] S. O'Neill, "Global CO2 emissions level off in 2019, with a drop predicted in 2020," *Engineering (Beijing, China)*, 2020.

[11] A. Pfeifer, P. Prebeg, and N. Duić, "Challenges and opportunities of zero emission shipping in smart islands: A study of zero emission ferry lines," *eTransportation*, vol. 3, p. 100048, 2020.

[12] S. Anwar, M. Y. I. Zia, M. Rashid, G. Z. D. Rubens, and P. Enevoldsen, "Towards ferry electrification in the maritime sector," *Energies (19961073)*, vol. 13, no. 24, 2020.

[13] A. Letafat *et al.*, "An efficient and cost-effective power scheduling in zero-emission ferry ships," *Complexity*, vol. 2020, 2020.

[14] M. Mosayebi, S. M. Sadeghzadeh, M. Gheisarnejad, and M. H. Khooban, "Intelligent and fast model-free sliding mode control for shipboard DC microgrids," *IEEE Transactions on Transportation Electrification*, 2020.

[15] R. Heydari, M. Gheisarnejad, M. H. Khooban, T. Dragicevic, and F. Blaabjerg, "Robust and fast voltage-source-converter (VSC) control for naval shipboard microgrids," *IEEE Transactions on Power Electronics*, vol. 34, no. 9, pp. 8299–8303, 2019.

[16] C. S. Edrington, G. Ozkan, B. Papari, and D. Perkins, "Distributed adaptive power management for medium voltage ship power systems," *Journal of Marine Engineering & Technology*, pp. 1–16, 2021.

[17] K. Kwon, D. Park, and M. K. Zadeh, "Load frequency-based power management for shipboard DC hybrid power systems," in *2020 IEEE 29th International Symposium on Industrial Electronics (ISIE)*, 2020, pp. 142–147: IEEE.

[18] C. S. Edrington *et al.*, "Distributed energy management for ship power systems with distributed energy storage," *Journal of Marine Engineering & Technology*, vol. 19, no. supl, pp. 31–44, 2020.

[19] S. Hasanvand, M. Rafiei, M. Gheisarnejad, and M.-H. Khooban, "Reliable power scheduling of an emission-free ship: Multiobjective deep reinforcement learning," *IEEE Transactions on Transportation Electrification*, vol. 6, no. 2, pp. 832–843, 2020.

[20] S. Faddel, T. A. Youssef, and O. Mohammed, "Decentralized controller for energy storage management on MVDC ship power system with pulsed loads," in *2018 IEEE Transportation Electrification Conference and Expo (ITEC)*, 2018, pp. 254–259: IEEE.

[21] M. Shojaie, N. Elsayad, and O. Mohammed, "Design of an all-GaN bidirectional DC-DC converter for medium voltage DC ship power systems using series-stacked GaN modules," in *2018 IEEE Applied Power Electronics Conference and Exposition (APEC)*, 2018, pp. 2155–2161: IEEE.

[22] J. K. Kambrath *et al.*, "Modeling and control of marine diesel generator system with active protection," *IEEE Transactions on Transportation Electrification*, vol. 4, no. 1, pp. 249–271, 2017.

[23] C. Ghenai, M. Bettayeb, B. Brdjanin, and A. K. Hamid, "Hybrid solar PV/PEM fuel cell/diesel generator power system for cruise ship: A case study in Stockholm, Sweden," *Case Studies in Thermal Engineering*, vol. 14, p. 100497, 2019.

[24] M.-H. Khooban, T. Dragicevic, F. Blaabjerg, and M. Delimar, "Shipboard microgrids: A novel approach to load frequency control," *IEEE Transactions on Sustainable Energy*, vol. 9, no. 2, pp. 843–852, 2017.

[25] M.-H. Khooban, T. Niknam, M. Shasadeghi, T. Dragicevic, and F. Blaabjerg, "Load frequency control in microgrids based on a stochastic noninteger controller," *IEEE Transactions on Sustainable Energy*, vol. 9, no. 2, pp. 853–861, 2017.

[26] M.-H. Khooban, "Secondary load frequency control of time-delay stand-alone microgrids with electric vehicles," *IEEE Transactions on Industrial Electronics*, vol. 65, no. 9, pp. 7416–7422, 2017.

[27] R. Tang, Z. Wu, and X. Li, "Optimal operation of photovoltaic/battery/diesel/cold-ironing hybrid energy system for maritime application," *Energy*, vol. 162, pp. 697–714, 2018.

[28] J. Prousalidis, G. Antonopoulos, C. Patsios, A. Greig, and R. Bucknall, "Green shipping in emission controlled areas: Combining smart grids and cold ironing," in *2014 International Conference on Electrical Machines (ICEM)*, 2014, pp. 2299–2305: IEEE.

[29] N. Vahabzad, B. Mohammadi-Ivatloo, and A. Anvari-Moghaddam, "Optimal energy scheduling of a solar-based hybrid ship considering cold-ironing facilities," *IET Renewable Power Generation*, vol. 15, no. 3, pp. 532–547, 2021.

[30] T. Van Vu, D. Gonsoulin, F. Diaz, C. S. Edrington, and T. El-Mezyani, "Predictive control for energy management in ship power systems under high-power ramp rate loads," *IEEE Transactions on Energy Conversion*, vol. 32, no. 2, pp. 788–797, 2017.

[31] H. Zhao, Q. Wu, S. Hu, H. Xu, and C. N. Rasmussen, "Review of energy storage system for wind power integration support," *Applied Energy*, vol. 137, pp. 545–553, 2015.

[32] S. Abrazeh, A. Parvaresh, S.-R. Mohseni, M. J. Zeitouni, M. Gheisarnejad, and M. H. Khooban, "Nonsingular terminal sliding mode control with wltra-local model and single input interval type-2 fuzzy logic control for pitch control of wind turbines," *IEEE/CAA Journal of Automatica Sinica*, vol. 8, no. 3, pp. 690–700, 2021.

[33] A. Parvaresh, S. Abrazeh, S.-R. Mohseni, M. J. Zeitouni, M. Gheisarnejad, and M.-H. Khooban, "A novel deep learning backstepping controller-based digital twins technology for pitch angle control of variable speed wind turbine," *Designs*, vol. 4, no. 2, p. 15, 2020.

[34] P. Mukherjee and V. Rao, "Superconducting magnetic energy storage for stabilizing grid integrated with wind power generation systems," *Journal of Modern Power Systems and Clean Energy*, vol. 7, no. 2, pp. 400–411, 2019.

[35] X. Li and S. Wang, "A review on energy management, operation control and application methods for grid battery energy storage systems," *CSEE Journal of Power and Energy Systems*, 2019.

[36] S. Satpathy, S. Das, and B. K. Bhattacharyya, "How and where to use super-capacitors effectively, an integration of review of past and new characterization works on super-capacitors," *Journal of Energy Storage*, vol. 27, p. 101044, 2020.

[37] X. Zhang *et al.*, "Highly active N, S co-doped hierarchical porous carbon nanospheres from green and template-free method for super capacitors and oxygen reduction reaction," *Electrochimica Acta*, vol. 318, pp. 272–280, 2019.

[38] M. E. Amiryar and K. R. Pullen, "A review of flywheel energy storage system technologies and their applications," *Applied Sciences*, vol. 7, no. 3, p. 286, 2017.

[39] J. Jurasz, P. B. Dąbek, B. Kaźmierczak, A. Kies, and M. Wdowikowski, "Large scale complementary solar and wind energy sources coupled with pumped-storage hydro-electricity for Lower Silesia (Poland)," *Energy*, vol. 161, pp. 183–192, 2018.

[40] J. Menéndez, J. M. Fernández-Oro, M. Galdo, and J. Loredo, "Efficiency analysis of underground pumped storage hydropower plants," *Journal of Energy Storage*, vol. 28, p. 101234, 2020.

[41] M. Zeynalian, A. H. Hajialirezaei, A. R. Razmi, and M. Torabi, "Carbon dioxide capture from compressed air energy storage system," *Applied Thermal Engineering*, vol. 178, p. 115593, 2020.

[42] H. Lee, D. Park, S. Choo, and H. T. Pham, "Estimation of the non-greenhouse gas emissions inventory from ships in the port of incheon," *Sustainability*, vol. 12, no. 19, p. 8231, 2020.

[43] P. Serra and G. Fancello, "Towards the IMO's GHG goals: A critical overview of the perspectives and challenges of the main options for decarbonizing international shipping," *Sustainability*, vol. 12, no. 8, p. 3220, 2020.

[44] T. Caravella, C. Austell, C. Brady-Alvarez, and S. Elsaiah, "Hybrid maritime microgrids: A quest for future onboard integrated marine power systems," *Innovation in Energy Systems: New Technologies for Changing Paradigms*, p. 139, 2019.

[45] M. Gheisarnejad, P. Karimaghaee, J. Boudjadar, and M.-H. Khooban, "Real-time cellular wireless sensor testbed for frequency regulation in smart grids," *IEEE Sensors Journal*, vol. 19, no. 23, pp. 11656–11665, 2019.

[46] M. H. Khooban and M. Gheisarnejad, "Islanded microgrid frequency regulations concerning the integration of tidal power units: Real-time implementation," *IEEE Transactions on Circuits and Systems II: Express Briefs*, vol. 67, no. 6, pp. 1099–1103, 2019.

[47] M. Jahanshahi Zeitouni, A. Parvaresh, S. Abrazeh, S.-R. Mohseni, M. Gheisarnejad, and M.-H. Khooban, "Digital twins-assisted design of next-generation advanced controllers for power systems and electronics: Wind turbine as a case study," *Inventions*, vol. 5, no. 2, p. 19, 2020.

[48] F. Kanellos, "Optimal power management with GHG emissions limitation in all-electric ship power systems comprising energy storage systems," *IEEE Transactions on Power Systems*, vol. 29, no. 1, pp. 330–339, 2013.

[49] R. Tang, X. Li, and J. Lai, "A novel optimal energy-management strategy for a maritime hybrid energy system based on large-scale global optimization," *Applied Energy*, vol. 228, pp. 254–264, 2018.

[50] F. Balsamo, C. Capasso, G. Miccione, and O. Veneri, "Hybrid storage system control strategy for all-electric powered ships," *Energy Procedia*, vol. 126, pp. 1083–1090, 2017.

[51] E. A. Sciberras, B. Zahawi, and D. J. Atkinson, "Electrical characteristics of cold ironing energy supply for berthed ships," *Transportation Research Part D: Transport and Environment*, vol. 39, pp. 31–43, 2015.

[52] F. Ballini and R. Bozzo, "Air pollution from ships in ports: The socio-economic benefit of cold-ironing technology," *Research in Transportation Business & Management*, vol. 17, pp. 92–98, 2015.

[53] E. A. Sciberras, B. Zahawi, D. J. Atkinson, A. Juandó, and A. Sarasquete, "Cold ironing and onshore generation for airborne emission reductions in ports," *Proceedings of the Institution of Mechanical Engineers, Part M: Journal of Engineering for the Maritime Environment*, vol. 230, no. 1, pp. 67–82, 2016.

[54] H. Xing, C. Stuart, S. Spence, and H. Chen, "Fuel cell power systems for maritime applications: Progress and perspectives," *Sustainability*, vol. 13, no. 3, p. 1213, 2021.

[55] T. Leo, J. Durango, and E. Navarro, "Exergy analysis of PEM fuel cells for marine applications," *Energy*, vol. 35, no. 2, pp. 1164–1171, 2010.

[56] S. Alkaner and P. Zhou, "A comparative study on life cycle analysis of molten carbon fuel cells and diesel engines for marine application," *Journal of Power Sources*, vol. 158, no. 1, pp. 188–199, 2006.

[57] G. Roh, H. Kim, H. Jeon, and K. Yoon, "Fuel consumption and CO2 emission reductions of ships powered by a fuel-cell-based hybrid power source," *Journal of Marine Science and Engineering*, vol. 7, no. 7, p. 230, 2019.

[58] P. Mitra and G. K. Venayagamoorthy, "Implementation of an intelligent reconfiguration algorithm for an electric ship's power system," *IEEE Transactions on Industry Applications*, vol. 47, no. 5, pp. 2292–2300, 2011.

[59] B. Yildirim, M. Gheisarnejad, and M. H. Khooban, "Delay-dependent stability analysis of modern shipboard microgrids," *IEEE Transactions on Circuits and Systems I: Regular Papers*, vol. 68, no. 4, pp. 1693–1705, 2021.

[60] T. Yang, N. Sun, H. Chen, and Y. Fang, "Neural network-based adaptive antiswing control of an underactuated ship-mounted crane with roll motions and input dead zones," *IEEE Transactions on Neural Networks and Learning Systems*, vol. 31, no. 3, pp. 901–914, 2019.

[61] M. A. Igder, M. Rafiei, J. Boudjadar, and M.-H. Khooban, "Reliability and safety improvement of emission-free ships: Systemic reliability centered maintenance," *IEEE Transactions on Transportation Electrification*, 2020.

[62] Z. Karami, Q. Shafiee, Y. Khayat, M. Yaribeygi, T. Dragicevic, and H. Bevrani, "Decentralized model predictive control of DC microgrids with constant power load," *IEEE Journal of Emerging and Selected Topics in Power Electronics*, 2019.

[63] J. Moubray, *Reliability-centered maintenance*. Industrial Press Inc., 2001.

[64] A. Merabet, K. A. Tawfique, M. A. Islam, S. Enebeli, and R. Beguenane, "Wind turbine emulator using OPAL-RT real-time HIL/RCP laboratory," in *2014 26th International Conference on Microelectronics (ICM)*, 2014, pp. 192–195: IEEE.

[65] A. Delavari, P. Brunelle, and I. Kamwa, "Real-time closed-loop PQ control of NPC multi-level converter using OPAL-RT and speedgoat simulators," in *2018 IEEE Electrical Power and Energy Conference (EPEC)*, 2018, pp. 1–5: IEEE.

19 Distributed Optimization and Its Application in Electricity Grids, Including Electrical Vehicles

Amir Abolfazl Suratgar and
Mohammad Bagher Menhaj

CONTENTS

These days the use of electric vehicles (EVs) increases every day. The main reason is there is systematic optimization of cost, pollution and welfare. To charge, EVs use the electricity network. Connecting EVs to grids causes some critical problems. Usually for energy dispatch in grids, an optimal power flow problem (OPF) is solved, and the solution is then used for optimal dispatch [14]. In the near future, the usage of EVs will dramatically increase; therefore, charging and connecting them all will require the solution of the OPF in the grid, especially in micro grids. These problems can be considered from different points of view. First, it can be considered a distributed optimization problem with unknown nonlinear non-affine dynamics. Second, it may be considered a hybrid switching system. If there are few EVs, the problem of plug and play can be considered a robust optimal control one. The EVs' connection is considered an uncertainty. The recent point of view is not suitable because plenty of EVs will be used in the near future. In this chapter, we will present some distributed optimization methods for unknown systems with input constraints, as well as for an optimal consensus problem.

DOI: 10.1201/9781003293989-19

19.1 DISTRIBUTED OPTIMAL CONTROL FOR ELECTRICITY NETWORKS

Electricity networks can be considered continuous-time non-affine nonlinear systems with unknown dynamics. In this section, an extended adaptive dynamic programming (EADP) will be presented. A proper design of optimal control law is fully investigated through the framework of Hamilton–Jacobi–Bellman (HJB). The proposed EADP algorithm is then used to solve iteratively the HJB equation associated with each subsystem. The proposed distributed optimal control system mainly consisting of three weighted basis functions for each subsystem is fully developed to handle the issue associated with the availability of unknown dynamics.

Consider a framework, which consists of N nonlinear non-affine subsystems, with dynamics as

$$\dot{x} = f_i\left(x_i,u_i\right) + \sum_{\substack{i \neq j \\ j=1}}^{N} \Delta_{ij}\left(x_i,x_j\right), x_i\left(0\right) = 0, i = 1,\ldots,N, \tag{19.1}$$

where index i represents the subsystem number, N is the number of subsystems, $x_i \in \Omega_i \in R^{ni}$ represents the ith subsystem state vector and $u_i : R^+ \to R^{m_i}$ is a piecewise continuous control input of the ith subsystem. $f_i : R^{n_i} \times R^{m_i} \to R^{n_i}$ is a local Lipschitz function satisfying $f_i\left(0,0\right) = 0$.

The overall interconnected system can be written as

$$\dot{x} = f\left(x,u\right), x\left(0\right) = x_0, \tag{19.2}$$

where $f\left(0,0\right) = 0, f\left(x,u\right) = \left[f_1\left(x_1,u_1\right) + \sum_{j=2}^{N}\Delta_{1j},\ldots, f_N\left(x_N,u_N\right) + \sum_{j=1}^{N-1}\Delta_{Nj}\right]^T$.

It is assumed that the interconnected system in (19.2) is stabilizable in the sense that there exists a continuous control u that stabilizes the overall interconnected system asymptotically on a set that includes the origin. It is further assumed that the communication systems of the electricity network are perfect and there are no complications associated with communication links, such as delay or packet loss.

The global cost function is as follows:

$$\bar{J}\left(x;u\right) = \sum_{i=1}^{N}J_i\left(x_0;u\right), \tag{19.3}$$

where

$$J_i\left(x_0;u\right) = \int_0^\infty r_i\left(x,u_i\right)dt. \tag{19.4}$$

The common choice for r_i can be described as

$$r_i(x, u_i) = Q_i(x) + u_i^T R_i u_i, \tag{19.5}$$

where $u = \begin{bmatrix} u_1^T & \dots & \dots & u_N^T \end{bmatrix} \in D$ is defined to be admissible, i.e., if u_i represents a piecewise continuous function; u stabilizes asymptotically the origin of system (19.2); and $J_i = \int_0^\infty r_i(x, u_i) dt < \infty, \forall x_0 \in \Omega, x_0 < \infty.$

In order to find an optimal control policy, the HJB equation is obtained as:

$$H\left(x, \left(V_x^*\right)^T, u^*\right) = V_x^* f\left(x, u^*\right) + r\left(x, u^*\right) = 0$$

In HJB equations we obviously have the following inequality for an optimal policy:

$$H\left(x, \left(V_x^*\right)^T, u^*\right) < H\left(x, \left(V_x^*\right)^T, u\right), u^* \neq u \tag{19.6}$$

Afanasiev et al. [2] proposed the following theorem, which illustrates how by solving the HJB equation, an optimal control law is obtained.

Theorem 1. *Consider the system given by (19.2). Suppose that there is a unique continuously differentiable V* ∈ V satisfying*

$$\inf \left\{ V_x^* f(x, u) + r(x, u) \right\} = 0, \tag{19.7}$$

*and **u*** ∈ D such that*

$$\inf_{u \in D} \left\{ V_x^* f(x, u) + r(x, u) \right\} = V_x^* f\left(x, u^*\right) + r\left(x, u^*\right) \tag{19.8}$$

*then **u*** and V*(**x**) = J(**x**; **u***) are called the optimal control and the corresponding optimal value function, respectively.*

In the following, Theorem 1 is extended to design a distributed optimal controller for the system described in (19.1) with the cost function (19.3) based on the extended ADP method, which will be fully developed in the following.

A distributed control system consisting of several coupled subsystems is concerned with designing an individual control policy for each subsystem. The proposed algorithm is distributed in the sense that we can find the local control inputs separately while the control inputs utilize local and other subsystem information. The following Lemma 1 and Theorem 2 summarize the distributed version of optimization.

Lemma 1: *Consider system (19.2) in which **u**_i is a control input for subsystem i such that u ∈ D and V_i ∈ C^1 is the corresponding Bellman function as*

$$V_i\left(s, x(s)\right) = \inf_{\substack{u_i \\ s<t}} \left\{\int_s^\infty r_i\left(x, u_i\right) dt\right\}, u \in D \tag{19.9}$$

then the following equality, which is called the Bellman equation, holds

$$\inf_{u_i}\left\{\frac{\partial V_i}{\partial t} + \sum_{j=1}^N \frac{\partial V_i}{\partial x_j}\dot{x}_j + r_i\left(x, u_i\right)\right\} = 0, u \in D \tag{19.10}$$

For detail of proof, see [1].

Theorem 2. *Suppose there exists a unique continuously differentiable solution $Vi*, \forall i \in 1, 2, \ldots N$ such that*

$$\inf_{u_i}\left\{\frac{\partial V_i^*}{\partial t} + \sum_{j=1}^N \frac{\partial V_i^*}{\partial x_j}\dot{x}_j + r_i\left(x, u_i\right)\right\} = 0, u \in D \tag{19.11}$$

and a control input for each subsystem, $u*i$, such that $u* \in D$ and

$$\inf_{u_i}\left\{\frac{\partial V_i^*}{\partial t} + \sum_{j=1}^N \frac{\partial V_i^*}{\partial x_j}\dot{x}_j + r_i\left(x, u_i\right)\right\} = \sum_{\substack{j=1 \\ j\neq i}}^N \frac{\partial V_i^*}{\partial x_j}\dot{x}_j + \frac{\partial V_i^*}{\partial x_i}\left(f_i\left(x_i, u_i^*\right) + \bar{\Delta}_i(x)\right)$$

$$+ r_i\left(x, u_i^*\right), u \in D \tag{19.12}$$

where $\bar{\Delta}_j(x) = \sum_{\substack{l=1 \\ j\neq l}}^N \Delta_{jl}\left(x_j, x_i\right)$ then

$$J\left(x_0; u\right) \geq \sum_{i=1}^N J_i\left(x_0; u_i^*\right), \forall u \in D \tag{19.13}$$

For detail of proof, see [1].

Corollary 1. *The results of Lemma 1 and Theorem 2 confirm that the distributed optimal state feedback control law for any subsystem in (19.1) with the ith cost function (19.4) equals*

$$u_i^*(x) = \arg\min\left\{H_i\left(x, \left(V_{ix}^*(x)\right)^T, v_i\right)\right\}, v \in D, \tag{19.14}$$

where the HJB equation for the ith subsystem has the following form

$$H_i\left(x,\left(V_{ix}^*(x)\right)^T,u_i^*\right) = \sum_{j=1}^{N} \frac{\partial V_i^*}{\partial x_j} \dot{x}_j + r_i\left(x,u_i^*\right) = 0. \tag{19.15}$$

According to Corollary 1, we can define the total HJB as

$$H(x) = V_x^* \dot{x} + \bar{r}\left(x,u^*\right) = 0, \tag{19.16}$$

where

$$V_x^* = \begin{bmatrix} \dfrac{\partial V_1^*}{\partial x_1} & \cdots & \dfrac{\partial V_1^*}{\partial x_N} \\ \vdots & \ddots & \vdots \\ \dfrac{\partial V_N^*}{\partial x_1} & \cdots & \dfrac{\partial V_N^*}{\partial x_N} \end{bmatrix}, \bar{r} = \begin{bmatrix} r_1\left(x,u_1^*\right) \\ \vdots \\ r_N\left(x,u_N^*\right) \end{bmatrix},$$

are the Jacobian matrix and the cost function vector, respectively. One should note that the HJB equation in (19.15) is a nonlinear partial differential equation with respect to Vi*, which cannot be solved analytically. In the next section, we introduce an iterative numerical method to provide a solution to the distributed optimal state feedback control problem in (19.14) and (19.15).

19.2 AN ITERATIVE IMPLEMENTATION OF THE DISTRIBUTED OPTIMAL CONTROL STRATEGY

In what follows, we introduce the EADP algorithm 1 to obtain numerically the optimal solution given in (19.14) and (19.15). This is illustrated by the following theorem.

Theorem 3. If $\left\{u_i^j\right\}_{j=0}^{\infty}$ and $\left\{V_i^j\right\}_{j=0}^{\infty}$ with initial condition $u^0 = \begin{bmatrix} u_1^0 & \cdots & u_N^0 \end{bmatrix} \in D$ satisfy the following equations

$$H_i\left(x,\left(V_{ix}^j(x)\right)^T,u^j(x)\right) = V_{ix}^j \dot{x} + r_i\left(x,u_i^j\right) = 0, V_i^j(0) = 0 \tag{19.17}$$

$$u_i^{j+1}(x) = \operatorname{armin}\left\{H_i\left(x,\left(V_{ix}^j(x)\right)^T,v_i(x)\right)\right\}, \tag{19.18}$$

and $V_i^j \to V_i^*, u_i^j \to u_i^*$.

then $u_i^{j+1} \in D, V_i^j \in V, V_i^{j+1} \leq V_i^j, \forall x \in \Omega;$

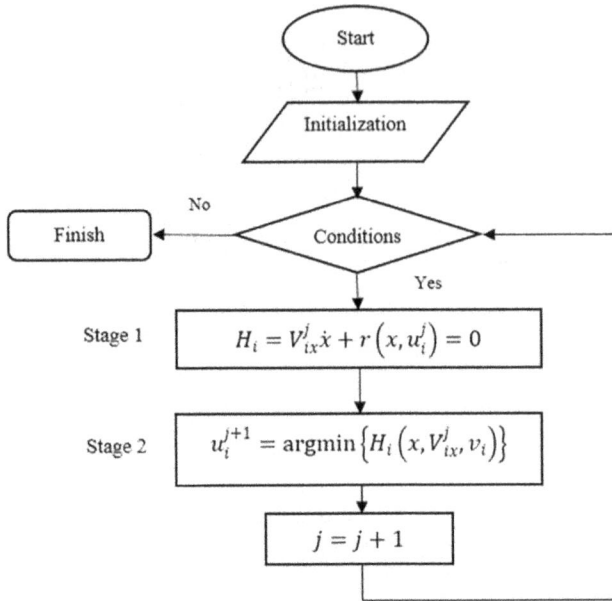

FIGURE 19.1 Flowchart of the proposed distributed optimization.

For detail of proof, see [1].

Based on Theorem 3, a block diagram representation of the proposed distributed optimization algorithm is given in Figure 19.1.

19.3 DISTRIBUTED OPTIMIZATION IN ELECTRICITY NETWORK SYSTEMS WITH UNKNOWN DYNAMICS

For systems with unknown dynamics, a computational intelligence-based technique is employed for the purpose of system identification [3,4,18]. A self-evolving non-linear consequent part recurrent type-2 fuzzy system is fully developed for system modeling and identification.

One of the most powerful tools in data analysis and inference is computational intelligence. Neural networks share lots of significant benefits, such as landmark computational ability, parallel processing and adaptation. The fuzzy systems are able to utilize the expert knowledge entitled "if-then rules" and possess actual parameter concepts [6,15,17]. As is well known, mathematical modeling is a substantial preliminary step in many control issues [16]. On the other hand, prediction, simulation and modeling of complicated systems established upon physical and chemical principles appear industrious in such a way that they will not yield consolidated mathematical forms.

Ten years after Zadeh presented type-1 fuzzy logic, he introduced type-2 fuzzy logic to resolve some problems of type 1. In type-1 fuzzy sets, the membership degree is a crisp number, but in type-2 fuzzy sets, the membership degree is a type-1 fuzzy number. In some systems, such as time-series prediction, the exact membership degree is determined in a very difficult manner due to their complexity and their noisy information [20]. So using type-2 fuzzy systems for describing behavior of these systems is very useful. Castro et al. [11] studied some shortcomings of type-1 fuzzy sets. Before 1998, research studies on type-2 fuzzy systems were very limited. Serious discussions of type-2 fuzzy logic and its applications were begun after publication of a book that included union and intersection of type-2 fuzzy sets [5]. Mendel [12] proposed comprehensive information of type-2 fuzzy systems computing such things as type reduction and defuzzification.

A general type-2 fuzzy set, $\tilde{A}$, can be characterized as follows:

$$\tilde{A} = \int_{x \in X} \mu_{\tilde{A}}(x)/x = \frac{\int_{x \in X}\left[\int_{\mu \in J_x} \frac{f_x(\mu)}{\mu}\right]}{x},$$

(19.19)

where $\mu_{\tilde{A}}(x)$ is a secondary membership function, J_x is the set of primary membership degrees of $x \in X$, with $\mu \in J_x$, and $f_x(\mu) \in [0,1]$ is a secondary membership degree. Figure 19.2 shows the Gaussian primary membership function and Gaussian secondary membership function. For example, if $m = 0$, $\sigma = 1$ and $x = 1$, then the degree of membership is 0.6; if this membership degree is also fuzzy, e.g., $\widetilde{0.6}$, then the primary membership is Gaussian type-1 fuzzy set with $m = 0$, $\sigma = 1$ and the secondary membership is Gaussian type-1 fuzzy set with $m = 0.6$, $\sigma = 0.1$.

Note that, when $f_x(\mu) = 1, \forall \mu \in J_x \subseteq [0,1]$ and the secondary MFs are interval sets, the fuzzy set can be called interval type-2 fuzzy sets. Here Gaussian membership functions are used to achieve type-2 fuzzy numbers.

In this case, a crisp number is fuzzified in two steps: first,

$$\mu_1 = \exp\left(-.5 * \frac{(x-M)^2}{\sigma_x^2}\right),$$

(19.20)

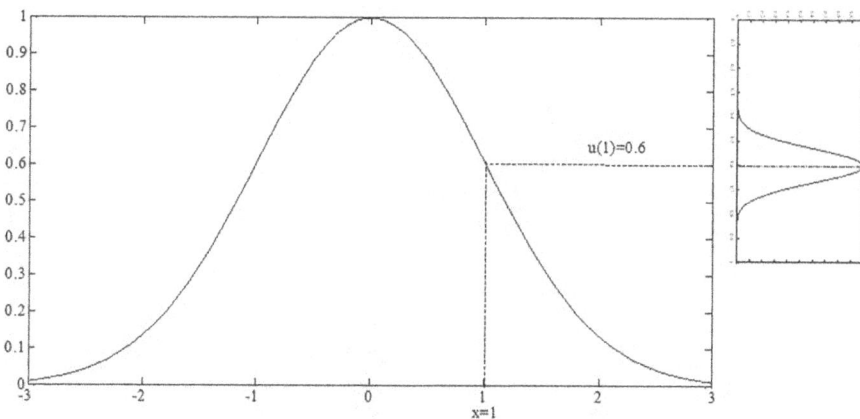

FIGURE 19.2 Gaussian primary and secondary membership functions.

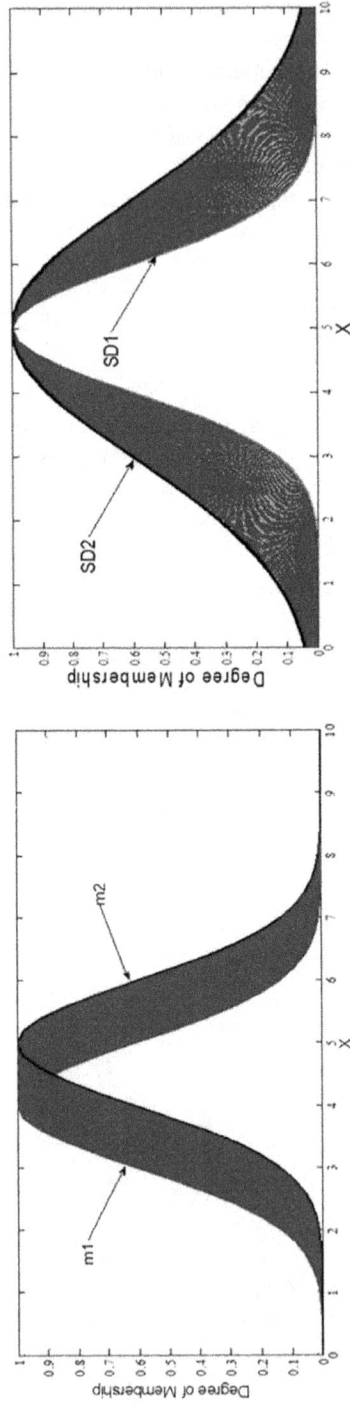

FIGURE 19.3   a) Uncertainty in standard deviation, b) uncertainty in mean.

where μ_1 is primary membership degree, and then

$$\mu_2\left(x,\mu_1\right)=\exp\left(-.5*\frac{\left(\mu_1\left(x\right)-a\right)^2}{\sigma_m^{\,2}}\right),\qquad(19.21)$$

where $\mu_2\left(x,\mu_1\right)$ is secondary degree and $a\in\left[0,1\right]$ is the domain of secondary membership function for each x, and σ_m is the secondary spread of the Gaussian membership function. A particular and simplified kind of general type-2 fuzzy set is an interval type-2 fuzzy set. Two cases of interval type-2 fuzzy sets are shown in Figure 19.3. In Figure 19.3a, a case of a fuzzy set characterized by a Gaussian membership function with mean m and a standard deviation in interval $[\sigma_1,\ \sigma_2]$ has been shown. Figure 19.3b shows a fuzzy set with a Gaussian membership function with a fixed standard deviation σ, but an uncertain mean, taking values in $[\,m_1\,,\ m_2]$.

In this study, Gaussian membership functions with fixed standard deviation σ and uncertain mean (Figure 19.3a) are used.

19.4 TYPE-2 FUZZY SYSTEMS

By defuzzification of the output of a type-1 fuzzy system we can get a crisp number, but output of a type-2 fuzzy system is a type-2 fuzzy set; hence, we must reduce the type of fuzzy sets from type 2 to 1; this process is called type reduction. Type reduction is an important topic in type-2 fuzzy systems [7]. The structure of a type-2 fuzzy system is shown in Figure 19.4.

As can be seen in Figure 19.4, a type-2 fuzzy system structure keeps the overall type-1 fuzzy system structure but only the "Type-reduction" block is added.

19.5 NCPR TYPE-2 FUZZY SYSTEM

This section formulates the nonlinear consequent part of recurrent type-2 fuzzy system. Knowing that 1) like type-1 TSK fuzzy sets, the output of type-2 TSK fuzzy systems is in general a polynomial of their inputs, and 2) in type-2 TSK fuzzy

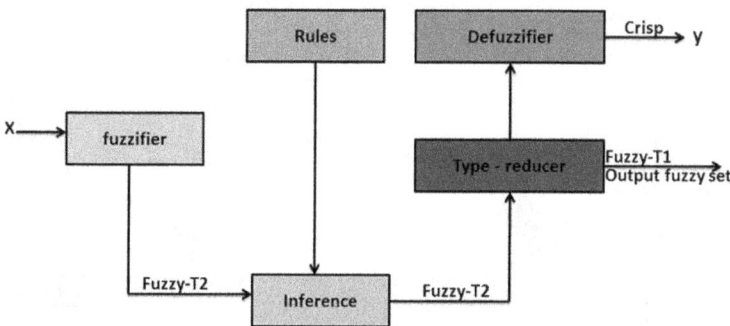

FIGURE 19.4 The structure of a type-2 fuzzy system.

systems the output and its coefficients are type-1 fuzzy sets [8], in this book a new
NCPRT2FS is introduced with a general structure depicted in Figure 19.4. As easily
observed, this system has seven layers. In general, in a first-order interval type-2
linear TSK model with M rules and n inputs, the kth rule can be described as:

$$R^k : if\ x_1\ is\ \tilde{A}_1^k\ and\ldots and\ x_n\ is\ \tilde{A}_n^k\ then\ \tilde{y}_k = C_{k,0} + C_{k,1}x_1 + \ldots + C_{k,n}x_n. \quad (19.22)$$

where $k = 1,\ldots,M$. $x_i\ (i = 1,\ldots,n)$ are inputs and $\tilde{y}_k$ is the output of the kth rule. $\tilde{y}_k$ is
an interval type-1 fuzzy set (since it is a linear combination of interval type-1 fuzzy
sets), $\tilde{A}_i^k$ are interval type-2 antecedent fuzzy sets, $C_{k,i} \in \left[c_{k,i} - s_{k,i}, c_{k,i} + s_{k,i}\right]$ are consequent interval type-1 fuzzy sets, $c_{k,i}$ denotes the center (mean) of $C_{k,i}$ and
$s_{k,i}$ denotes the spread of $c_{k,i}$.

In this book, the nonlinear consequent part is considered. The kth rule in
NCPRT2FS with two antecedent variables and three-time delay output in the consequent part is described as:

$$R^k : if\ x_1\ is\ \tilde{A}_1^k\ and\ x_2\ is\ \tilde{A}_2^k\ then$$

$$\tilde{y}_k = C_{k,0} + C_{k,1}x_1 + C_{k,2}x_2 + C_{k,3}y(t-1) + C_{k,4}x_1x_2 + C_{k,5}x_1y(t-1)$$
$$+ C_{k,6}x_2y(t-1) + C_{k,7}x_1y(t-2) + C_{k,8}x_2y(t-2) + C_{k,9}x_1y(t-3) \quad (19.23)$$
$$+ C_{k,10}x_2y(t-3)$$

It should be noted that the fuzzy rule (19.23) can be extended to n antecedent variables
and m time-delay output in consequent part that n can be designed with regard to degree
of nonlinearity and complexity of unknown system, which will be properly identified.

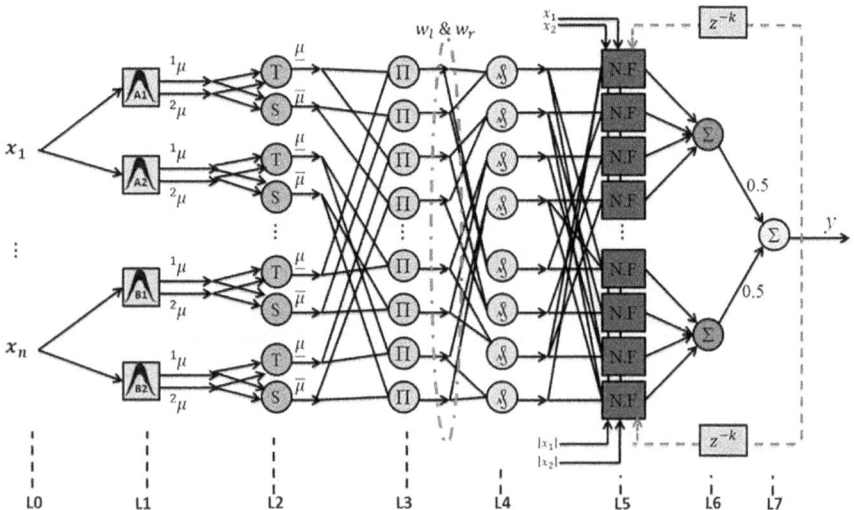

FIGURE 19.5   The structure of the proposed NCPRT2FS.

The layers in detail are as given next.

Layer 0: This layer is the input layer. The number of nodes in this layer is equal to the number of inputs.

Layer 1: This layer is a fuzzification layer. The output of this layer is determined as:

$$^1\mu_{k,i}\left(x_i,\left[\sigma_{k,i},{}^1m_{k,i}\right]\right)=e^{-.5\left(\frac{x_i-{}^1m_{k,i}}{\sigma_{k,i}}\right)^2} \tag{19.24}$$

$$^2\mu_{k,i}\left(x_i,\left[\sigma_{k,i},{}^2m_{k,i}\right]\right)=e^{-.5\left(\frac{x_i-{}^2m_{k,i}}{\sigma_{k,i}}\right)^2}, \tag{19.25}$$

where $m_{k,i}\in\left[{}^1m_{k,i},{}^2m_{k,i}\right]$ is an uncertain mean for the *kth* rule and *ith* input.

Layer 2: Every node in this layer is a circle labeled with T-norm and S-norm alternated:

$$\underline{\mu}_{k,i}\left(x_i\right)={}^1\mu_{k,i}\left(x_i\right)*{}^2\mu_{k,i}\left(x_i\right),\qquad k=1,2,\ldots,M,i=1,2,\ldots,n \tag{19.26}$$

$$\overline{\mu}_{k,i}\left(x_i\right)={}^1\mu_{k,i}\left(x_i\right)+{}^2\mu_{k,i}\left(x_i\right)-\underline{\mu}_{k,i}\left(x_i\right) \tag{19.27}$$

Layer 3: This is a rule layer. Each output node represents the lower ($\underline{f}^k$) and upper ($\overline{f}^k$) firing strength of a rule:

$$\underline{f}^k=\prod_{i=1}^n\underline{\mu}_{k,i};\qquad \overline{f}^k=\prod_{i=1}^n\overline{\mu}_{k,i} \tag{19.28}$$

Layer 4: Every node in this layer evaluates the left-most and right-most firing points denoted by

$$f_l^k=\frac{\overline{w}_l^k\overline{f}^k+\underline{w}_l^k\underline{f}^k}{\overline{w}_l^k+\underline{w}_l^k};\qquad f_r^k=\frac{\overline{w}_r^k\overline{f}^k+\underline{w}_r^k\underline{f}^k}{\overline{w}_r^k+\underline{w}_r^k}, \tag{19.29}$$

where *w is a vector of* adjustable weights.

Layer 5: This layer is used for consequent left–right firing points.

$$y_i^k=c_{k,0}+c_{k,1}x_1+c_{k,2}x_2+c_{k,3}y(t-1)+c_{k,4}x_1x_2+c_{k,5}x_1y(t-1)$$
$$+c_{k,6}x_2y(t-1)+c_{k,7}x_1y(t-2)+c_{k,8}x_2y(t-2)+c_{k,9}x_1y(t-3)$$

$$+c_{k,10}x_2y(t-3)-s_{k,0}-s_{k,1}|x_1|-s_{k,2}|x_2|-s_{k,3}|y(t-1)|$$
$$-s_{k,4}|x_1x_2|-s_{k,5}|x_1y(t-1)|-s_{k,6}|x_2y(t-1)|-s_{k,7}|x_1y(t-2)| \qquad (19.30)$$
$$-s_{k,8}|x_2y(t-2)|-s_{k,9}|x_1y(t-3)|-s_{k,10}|x_2y(t-3)|$$

$$y_r^k = c_{k,0}+c_{k,1}x_1+c_{k,2}x_2+c_{k,3}y(t-1)+c_{k,4}x_1x_2+c_{k,5}x_1y(t-1)$$
$$+c_{k,6}x_2y(t-1)+c_{k,7}x_1y(t-2)+c_{k,8}x_2y(t-2)+c_{k,9}x_1y(t-3)$$
$$+c_{k,10}x_2y(t-3)+s_{k,0}+s_{k,1}|x_1|$$
$$+s_{k,2}|x_2|+s_{k,3}|y(t-1)|+s_{k,4}|x_1x_2|+s_{k,5}|x_1y(t-1)|+s_{k,6}|x_2y(t-1)|$$
$$+s_{k,7}|x_1y(t-2)|+s_{k,8}|x_2y(t-2)|+s_{k,9}|x_1y(t-3)|+s_{k,10}|x_2y(t-3)| \qquad (19.31)$$

Layer 6: The two nodes in this layer evaluate the two end points, $\hat{y}_l$ and $\hat{y}_r$.

$$\hat{y}_l = \frac{\sum_{k=1}^M f_l^k y_l^k}{\sum_{k=1}^M f_l^k} \qquad (19.32)$$

$$\hat{y}_r = \frac{\sum_{k=1}^M f_r^k y_r^k}{\sum_{k=1}^M f_r^k} \qquad (19.33)$$

Layer 7: The single node in this layer computes the output.

$$\hat{y} = \frac{\hat{y}_l + \hat{y}_r}{2} \qquad (19.34)$$

In this study, we use type-2 fuzzy clustering for structure learning, in which an efficient rule and fuzzy set generation algorithm is proposed to generate fuzzy rules online and reduce the number of fuzzy sets in antecedent part [9]. Structure learning leads to simplicity of type-2 fuzzy system by reducing the total number of fuzzy rules. The task of structure learning is to generate new membership functions and rules as well as pruning additional membership functions and rules.

Geometrically, a rule corresponds to a cluster in the input space, and a rule firing strength can be regarded as the degree to which input data belong to the cluster. Since the firing strength in the NCPRT2FS is indeed represented by an interval, the center of the interval is then computed as

$$f_k = \frac{f^k + \bar{f}^k}{2}. \qquad (19.35)$$

And for generation a new membership function, find

$$\mu_{\tilde{A}_i^k} = \frac{\underline{\mu}_{\tilde{A}_i^k} + \overline{\mu}_{\tilde{A}_i^k}}{2}, \qquad i = 1, 2, \ldots, n. \tag{19.36}$$

For each incoming data $\vec{x} = \{x_1, \ldots, x_n\}$, calculate

$$I = \arg \max_{1 \le k \le M(t)} f_k . \tag{19.37}$$

For each newly generated rule, compute

$$I_i = \arg \max_{1 \le k \le k_i(t)} \mu_{\tilde{A}_i^k}, \qquad i = 1, 2, \ldots, n, \tag{19.38}$$

where $M(t)$ and $k_i(t)$ are the number of existing rules at time t and the number of fuzzy sets in input variable I, respectively. If $I \le \emptyset_{th}$, then a new rule is generated, where $\emptyset_{th} \in (01)$ is a pre-specified threshold [10]. If $I_i > \rho$, where $\rho \in [01]$, is a pre-specified threshold, then use the existing fuzzy set $\tilde{A}_i^{I_i}$ as the antecedent part of the new rule in input variable i. Otherwise, generate a new fuzzy set in input variable i and set $k_i(t+1) = k_i(t) +1$. Parameter ρ determines the number of fuzzy sets in each input variable. Fuzzy clustering is indeed a proper method for construction of a fuzzy model [11]. A novel type-2 fuzzy clustering method as given next is proposed in this chapter, which is an extension of Krishnapuram and Keller possibilistic C-mean (PCM) [12].

$$J_m(x, \tilde{\mu}, c) = min \left[\sum_{i=1}^{c} \sum_{j=1}^{N} \tilde{\mu}_{ij}^m D_{ij} + \sum_{i=1}^{c} \eta_i \sum_{j=1}^{N} \left(1 - \tilde{\mu}_{ij}\right)^m \right] \tag{19.39}$$

$$S.T : \begin{cases} 0 < \sum_{j=1}^{N} \tilde{\mu}_{ij} < N \\ \tilde{\mu}_{ij} \in [0,1] \quad \forall i, j, \\ \max \tilde{\mu}_{ij} > 0 \quad \forall j \end{cases} \tag{19.40}$$

where $\tilde{\mu}_{ij}$ is type-2 membership for the ith cluster in the jth data, D_{ij} is Euclidean distance of for the ith cluster's center to the jth data, η_i is positive numbers, c is the number of clusters and N is the number of input data. The distance to the cluster's center must be as low as possible (first term), and the membership values in each cluster should also be as large as possible (second term). The membership values for data in each cluster must lie in the interval [0,1], and their sum is limited to be smaller than the number of input data, as given in (19.21). It is desirable that η_i relate to the ith cluster and be of the order of D_{ij} [12].

$$\eta_i = \frac{\sum_{j=1}^{N} \tilde{\mu}_{ij}^m D_{ij}}{\sum_{j=1}^{N} \tilde{\mu}_{ij}^m} \qquad \forall i = 1, \ldots, c \tag{19.41}$$

Using (19.39) the optimal values of centers of the clusters are achieved. The initial uncertain mean $m_{k,i}$ and standard deviation $\sigma_{k,i}$ for the $k,(t+1)$ th interval type-2 fuzzy set in input variable i are

$$m_{k,i} \in \left[v_i - 0.1v_i, v_i + 0.1v_i \right] \tag{19.42}$$

$$\sigma_{k_i(t+1)i} = \beta \left| v_i - \frac{{}^1 m_{I_i,i} + {}^2 m_{I_i,i}}{2} \right|, \tag{19.43}$$

where v_i is the optimal value of the clusters center and $\beta > 0$ determines the overlap degree between two fuzzy sets. This chapter sets β at 0.5 so that the width of the new fuzzy set is half the distance between the average centers of the new fuzzy set and the fuzzy set I_i, and a suitable overlapping between these two fuzzy sets is generated.

The problem of determining an appropriate fuzzy set in each input variable is raised when a new rule is generated. One simple solution is to generate a new fuzzy set in each input variable for a new fuzzy rule. However, this method usually generates highly overlapping fuzzy sets in each input variable. An alternative method is proposed in [9], where identical widths are initially set to the n possible new fuzzy sets corresponding to the new rule.

In addition to assigning the initial antecedent parameters part, the initial consequent part of parameters should be also determined for a new generated rule. The initial consequent parameters are set to

$$\left[c_{k,0} - s_{k,0}, c_{k,0} + s_{k,0} \right] = \left[yd - 0.1, yd + 0.1 \right], \quad k = 1, 2, \dots, M, \tag{19.44}$$

where yd is the desired output for input $\vec{x} = \left\{ x_1, \dots, x_n \right\}$. All the other consequent parameters are zero.

Repeating the earlier process for every incoming piece of training data generates new rules, one after another, until the complete NCPRT2FS is finally constructed.

Gradient descent with adaptive learning rate backpropagation is used for the purpose of learning. In this method, for each input, the output of NCPRT2FS is calculated, then this output is compared with the target to calculate the error. Suppose the pair of input-output data is $\left\{ \left(x_p : t_p \right) \right\} \forall p = 1, \dots, q$, where p is the number of data and x and t are input and desired output, respectively. The error of NCPRT2FS output can be described as:

$$e_p = t_p - \hat{y}_p, \tag{19.45}$$

$$E_p = \frac{1}{2} e_p^2 = \frac{1}{2} \left(t_p - \hat{y}_p \right)^2 \tag{19.46}$$

$$E = \sum_{p=1}^{q} E_p \tag{19.47}$$

Equations (19.38)–(19.44) are used for updating the parameters of the consequent parts of rules.

$$^{new}c_{k,0} = {}^{old}c_{k,0} + \mu * 0.5 * e_p \left[\frac{f_l^k}{\sum_{k=1}^{M} f_l^k} + \frac{f_r^k}{\sum_{k=1}^{M} f_r^k} \right] \tag{19.48}$$

$$^{new}c_{k,i} = {}^{old}c_{k,i} + \eta * 0.5 * e_p \left[\frac{f_l^k}{\sum_{k=1}^{M} f_l^k} + \frac{f_r^k}{\sum_{k=1}^{M} f_r^k} \right] * x_i \quad i = 1,2 \tag{19.49}$$

$$^{new}c_{k,3} = {}^{old}c_{k,3} + \eta * 0.5 * e_p \left[\frac{f_l^k}{\sum_{k=1}^{M} f_l^k} + \frac{f_r^k}{\sum_{k=1}^{M} f_r^k} \right] * y(t-1) \tag{19.50}$$

$$^{new}c_{k,4} = {}^{old}c_{k,4} + \eta * 0.5 * e_p \left[\frac{f_l^k}{\sum_{k=1}^{M} f_l^k} + \frac{f_r^k}{\sum_{k=1}^{M} f_r^k} \right] * x_1 * x_2 \tag{19.51}$$

$$^{new}c_{k,5} = {}^{old}c_{k,5} + \eta * 0.5 * e_p \left[\frac{f_l^k}{\sum_{k=1}^{M} f_l^k} + \frac{f_r^k}{\sum_{k=1}^{M} f_r^k} \right] * x_1 * y(t-1) \tag{19.52}$$

$$^{new}c_{k,6} = {}^{old}c_{k,6} + \eta * 0.5 * e_p \left[\frac{f_l^k}{\sum_{k=1}^{M} f_l^k} + \frac{f_r^k}{\sum_{k=1}^{M} f_r^k} \right] * x_2 * y(t-1) \tag{19.53}$$

$$^{new}c_{k,7} = {}^{old}c_{k,7} + \eta * 0.5 * e_p \left[\frac{f_l^k}{\sum_{k=1}^{M} f_l^k} + \frac{f_r^k}{\sum_{k=1}^{M} f_r^k} \right] * x_1 * y(t-2) \tag{19.54}$$

$$^{new}c_{k,8} = {}^{old}c_{k,8} + \eta * 0.5 * e_p \left[\frac{f_l^k}{\sum_{k=1}^{M} f_l^k} + \frac{f_r^k}{\sum_{k=1}^{M} f_r^k} \right] * x_2 * y(t-2) \tag{19.55}$$

$$^{new}c_{k,9} = {}^{old}c_{k,9} + \eta * 0.5 * e_p \left[\frac{f_l^k}{\sum_{k=1}^{M} f_l^k} + \frac{f_r^k}{\sum_{k=1}^{M} f_r^k} \right] * x_1 * y(t-3) \tag{19.56}$$

$$^{new}c_{k,10} = {}^{old}c_{k,10} + \eta * 0.5 * e_p \left[\frac{f_l^k}{\sum_{k=1}^{M} f_l^k} + \frac{f_r^k}{\sum_{k=1}^{M} f_r^k} \right] * x_2 * y(t-3) \qquad (19.57)$$

$$^{new}s_{k,0} = {}^{old}s_{k,0} + \eta * 0.5 * e_p \left[\frac{f_l^k}{\sum_{k=1}^{M} f_l^k} - \frac{f_r^k}{\sum_{k=1}^{M} f_r^k} \right] \qquad (19.58)$$

$$^{new}s_{k,i} = {}^{old}s_{k,i} + \eta * 0.5 * e_p \left[\frac{f_l^k}{\sum_{k=1}^{M} f_l^k} - \frac{f_r^k}{\sum_{k=1}^{M} f_r^k} \right] * |x_i| \qquad i = 1,2 \qquad (19.59)$$

$$^{new}s_{k,3} = {}^{old}s_{k,3} + \eta * 0.5 * e_p \left[\frac{f_l^k}{\sum_{k=1}^{M} f_l^k} - \frac{f_r^k}{\sum_{k=1}^{M} f_r^k} \right] * |y(t-1)| \qquad (19.60)$$

$$^{new}s_{k,4} = {}^{old}s_{k,4} + \eta * 0.5 * e_p \left[\frac{f_l^k}{\sum_{k=1}^{M} f_l^k} - \frac{f_r^k}{\sum_{k=1}^{M} f_r^k} \right] * |x_1 x_2| \qquad (19.61)$$

$$^{new}s_{k,5} = {}^{old}s_{k,5} + \eta * 0.5 * e_p \left[\frac{f_l^k}{\sum_{k=1}^{M} f_l^k} - \frac{f_r^k}{\sum_{k=1}^{M} f_r^k} \right] * |x_1 * y(t-1)| \qquad (19.62)$$

$$^{new}s_{k,6} = {}^{old}s_{k,6} + \eta * 0.5 * e_p \left[\frac{f_l^k}{\sum_{k=1}^{M} f_l^k} - \frac{f_r^k}{\sum_{k=1}^{M} f_r^k} \right] * |x_2 * y(t-1)| \qquad (19.63)$$

$$^{new}s_{k,7} = {}^{old}s_{k,7} + \eta * 0.5 * e_p \left[\frac{f_l^k}{\sum_{k=1}^{M} f_l^k} - \frac{f_r^k}{\sum_{k=1}^{M} f_r^k} \right] * |x_1 * y(t-2)| \qquad (19.64)$$

$$^{new}s_{k,8} = {}^{old}s_{k,8} + \eta * 0.5 * e_p \left[\frac{f_l^k}{\sum_{k=1}^{M} f_l^k} - \frac{f_r^k}{\sum_{k=1}^{M} f_r^k} \right] * |x_2 * y(t-2)| \qquad (19.65)$$

$$^{new}s_{k,9} = {}^{old}s_{k,9} + \eta * 0.5 * e_p \left[\frac{f_l^k}{\sum_{k=1}^{M} f_l^k} - \frac{f_r^k}{\sum_{k=1}^{M} f_r^k} \right] * |x_1 * y(t-3)| \qquad (19.66)$$

$$^{new}s_{k,10} = {}^{old}s_{k,10} + \eta * 0.5 * e_p \left[\frac{f_l^k}{\sum_{k=1}^{M} f_l^k} - \frac{f_r^k}{\sum_{k=1}^{M} f_r^k} \right] * |x_2 * y(t-3)| \qquad (19.67)$$

where η is the learning rate.

Equations (19.68)–(19.71) are used for updating the left–right-most firing set points.

$$^{new}\underline{w}_l^k = {}^{old}\underline{w}_l^k + \eta * .5 * e_p \cdot \frac{y_l^k - \breve{y}_l}{\sum_{j=1}^M f_l^j} * \frac{\underline{f}^k - f_l^k}{\overline{w}_l^k + \underline{w}_l^k} \qquad (19.68)$$

$$^{new}\overline{w}_l^k = {}^{old}\overline{w}_l^k + \eta * .5 * e_p * \frac{y_l^k - \breve{y}_l}{\sum_{j=1}^M f_l^j} * \frac{\overline{f}^k - f_l^k}{\overline{w}_l^k + \underline{w}_l^k} \qquad (19.69)$$

$$^{new}\underline{w}_r^k = {}^{old}\underline{w}_r^k + \eta * .5 * e_p * \frac{y_r^k - \breve{y}_r}{\sum_{j=1}^M f_r^j} * \frac{\underline{f}^k - f_r^k}{\overline{w}_r^k + \underline{w}_r^k} \qquad (19.70)$$

$$^{new}\overline{w}_r^k = {}^{old}\overline{w}_r^k + \eta * .5 * e_p \cdot \frac{y_r^k - \breve{y}_r}{\sum_{j=1}^M f_r^j} * \frac{\overline{f}^k - f_r^k}{\overline{w}_r^k + \underline{w}_r^k} \qquad (19.71)$$

And, finally, the equations for updating the antecedent parameters can be described as:

$$^1m_{k,i}^{new} = {}^1m_{k,i}^{old} + \eta * .5 * e_p \left[\frac{y_l^k - \breve{y}_l}{\sum_{j=1}^M f_l^j} * \frac{\partial f_l^k}{\partial {}^1 m_{k,i}} + \frac{y_r^k - \breve{y}_r}{\sum_{j=1}^M f_r^j} * \frac{\partial f_r^k}{\partial {}^1 m_{k,i}} \right] \qquad (19.72)$$

$$^2m_{k,i}^{new} = {}^2m_{k,i}^{old} + \eta * .5 * e_p \left[\frac{y_l^k - \breve{y}_l}{\sum_{j=1}^M f_l^j} * \frac{\partial f_l^k}{\partial {}^2 m_{k,i}} + \frac{y_r^k - \breve{y}_r}{\sum_{j=1}^M f_r^j} * \frac{\partial f_r^k}{\partial {}^2 m_{k,i}} \right] \qquad (19.73)$$

$$\sigma_{k,i}^{new} = \sigma_{k,i}^{old} + \eta * .5 * e_p \left[\frac{y_l^k - \breve{y}_l}{\sum_{j=1}^M f_l^j} * \frac{\partial f_l^k}{\partial \sigma_{k,i}} + \frac{y_r^k - \breve{y}_r}{\sum_{j=1}^M f_r^j} * \frac{\partial f_r^k}{\partial \sigma_{k,i}} \right] \qquad (19.74)$$

where,

$$\frac{\partial f_l^k}{\partial {}^1 m_{k,i}} = \frac{\overline{w}_l^k \cdot \left[\overline{f}^k - {}^2\mu_{k,i} \cdot \prod_{l=1, l \neq i}^n \left(\overline{\mu}_{k,l} \right) \right] + \underline{w}_l^k \cdot \underline{f}^k}{\overline{w}_l^k + \underline{w}_l^k} \cdot \frac{x_i - {}^1 m_{k,i}}{\left(\sigma_{k,i} \right)^2} \qquad (19.75)$$

$$\frac{\partial f_l^k}{\partial {}^2 m_{k,i}} = \frac{\overline{w}_l^k \cdot \left[\overline{f}^k - {}^1\mu_{k,i} \cdot \prod_{l=1, l \neq i}^n \left(\overline{\mu}_{k,l} \right) \right] + \underline{w}_l^k \cdot \underline{f}^k}{\overline{w}_l^k + \underline{w}_l^k} \cdot \frac{x_i - {}^2 m_{k,i}}{\left(\sigma_{k,i} \right)^2} \qquad (19.76)$$

$$\frac{\partial f_l^k}{\partial \sigma_{k,i}} = \frac{\overline{w}_l^k \cdot \left[\left(\overline{f}^k - {}^2\mu_{k,i} \cdot \prod_{l=1,l\neq i}^{n} \left(\overline{\mu}_{k,l} \right) \right) \cdot \frac{\left(x_i - {}^1 m_{k,i} \right)^2}{\left(\sigma_{k,i} \right)^3} \right]}{\overline{w}_l^k + \underline{w}_l^k}$$

$$+ \frac{\overline{w}_l^k \cdot \left[\left(\overline{f}^k - {}^1 m_{k,i} \cdot \prod_{l=1,l\neq i}^{n} \left(\overline{m}_{k,l} \right) \right) \cdot \frac{\left(x_i - {}^2 m_{k,i} \right)^2}{\left(s_{k,i} \right)^3} \right]}{\overline{w}_l^k + \underline{w}_l^k} + \frac{\underline{w}_l^k \cdot \underline{f}^k \cdot \left[\frac{\left(x_i - {}^1 m_{k,i} \right)^2 + \left(x_i - {}^2 m_{k,i} \right)^2}{\left(s_{k,i} \right)^3} \right]}{\overline{w}_l^k + \underline{w}_l^k} \quad (19.77)$$

$$\frac{\partial f_r^k}{\partial {}^1 m_{k,i}} = \frac{\overline{w}_r^k \cdot \left[\overline{f}^k - {}^2\mu_{k,i} \cdot \prod_{l=1,l\neq i}^{n} \left(\overline{\mu}_{k,l} \right) \right] + \underline{w}_r^k \cdot \underline{f}^k}{\overline{w}_r^k + \underline{w}_r^k} \cdot \frac{x_i - {}^1 m_{k,i}}{\left(\sigma_{k,i} \right)^2} \quad (19.78)$$

$$\frac{\partial f_r^k}{\partial {}^2 m_{k,i}} = \frac{\overline{w}_r^k \cdot \left[\overline{f}^k - {}^1\mu_{k,i} \cdot \prod_{l=1,l\neq i}^{n} \left(\overline{\mu}_{k,l} \right) \right] + \underline{w}_r^k \cdot \underline{f}^k}{\overline{w}_r^k + \underline{w}_r^k} \cdot \frac{x_i - {}^2 m_{k,i}}{\left(\sigma_{k,i} \right)^2} \quad (19.79)$$

$$\frac{\partial f_r^k}{\partial \sigma_{k,i}} = \frac{\overline{w}_r^k \cdot \left[\left(\overline{f}^k - {}^2\mu_{k,i} \cdot \prod_{l=1,l\neq i}^{n} \left(\overline{\mu}_{k,l} \right) \right) \cdot \frac{\left(x_i - {}^1 m_{k,i} \right)^2}{\left(\sigma_{k,i} \right)^3} \right]}{\overline{w}_r^k + \underline{w}_r^k}$$

$$+ \frac{\overline{w}_r^k \cdot \left[\left(\overline{f}^k - {}^1\mu_{k,i} \cdot \prod_{l=1,l\neq i}^{n} \left(\overline{\mu}_{k,l} \right) \right) \cdot \frac{\left(x_i - {}^2 m_{k,i} \right)^2}{\left(\sigma_{k,i} \right)^3} \right]}{\overline{w}_r^k + \underline{w}_r^k} \quad (19.80)$$

$$+ \frac{\underline{w}_r^k \cdot \underline{f}^k \cdot \left[\frac{\left(x_i - {}^1 m_{k,i} \right)^2 + \left(x_i - {}^2 m_{k,i} \right)^2}{\left(\sigma_{k,i} \right)^3} \right]}{\overline{w}_r^k + \underline{w}_r^k}$$

19.6 LEARNING CONVERGENCE ANALYSIS

A Lyapunov function is then used to guarantee learning algorithm convergence [13]. This function is given in the following equation.

$$V_p(k) = E_p(k) = \frac{1}{2} e_p^2(k) = \frac{1}{2} \left(t_p(k) - \check{y}_p(k) \right)^2 \quad (19.81)$$

Equation (19.82) shows the change of the Layapunov function.

$$\Delta V_p(k) = V_p(k+1) - V_p(k) = \frac{1}{2}\left(e_p^2(k+1) - e_p^2(k)\right) \qquad (19.82)$$

Next, the moment error is calculated by

$$e_p(k+1) = e_p(k) + \Delta e_p(k) \cong e_p(k) + \left[\frac{\partial e_p(k)}{\partial W}\right]^T \Delta W \qquad (19.83)$$

In Equation (19.83), ΔW is the parameter correction term, where $W = \left[\sigma_{k,i}, \,^1m_{k,i}, \,^2m_{k,i}, c_{k,i}, s_{k,i}\right]$

A backpropagation (BP) algorithm is used to update the unknown parameters in IT2-NTSK-FNN, as described in Equation (19.84).

$$W(k+1) = W(k) + \Delta W(k) = W(k) + \eta * \left(-\frac{\partial E_p(k)}{\partial W}\right), \qquad (19.84)$$

where

$$\frac{\partial E_p(k)}{\partial W} = -e_p(k) * \frac{\partial \breve{y}}{\partial W}. \qquad (19.85)$$

Equation (19.82) can be rewritten as:

$$
\begin{aligned}
\Delta V_p(k) &= \frac{1}{2}\left(e_p^2(k+1) - e_p^2(k)\right) \\
&= \frac{1}{2}\left[\left(e_p(k+1) - e_p(k)\right)\right] * \left[\left(e_p(k+1) + e_p(k)\right)\right] \\
&= \frac{1}{2}\Delta e_p(k)\left[2\left(e_p(k)\right) + \Delta e_p(k)\right] \\
&= \Delta e_p(k)\left[e_p(k) + \frac{1}{2}\Delta e_p(k)\right] \qquad (19.86) \\
&= \left[\frac{\partial e_p(k)}{\partial W}\right]^T * \eta * e_p(k) * \frac{\partial \breve{y}(k)}{\partial W} * \left\{e_p(k) + \frac{1}{2}\left[\frac{\partial e_p(k)}{\partial W}\right]^T * \eta * e_p(k) * \frac{\partial \breve{y}(k)}{\partial W}\right\} \\
&= -\left[\frac{\partial \breve{y}(k)}{\partial W}\right]^T * \eta * e_p(k) * \frac{\partial \breve{y}(k)}{\partial W} * \left\{e_p(k) - \frac{1}{2}\left[\frac{\partial \breve{y}(k)}{\partial W}\right]^T * \eta * e_p(k) * \frac{\partial \breve{y}(k)}{\partial W}\right\} \\
&= -\eta * \left(e_p(k)\right)^2 \left|\frac{\partial \breve{y}(k)}{\partial W}\right|^2 * \left[1 - \frac{1}{2}\eta * \left|\frac{\partial \breve{y}(k)}{\partial W}\right|^2\right]
\end{aligned}
$$

In order that $\Delta V_p(k) < 0$, equation (19.87) must be satisfied

$$0 < \eta < \frac{2}{max \left| \dfrac{\partial \tilde{y}(k)}{\partial W} \right|^2} \tag{19.87}$$

If for every parameter $W = \left[\sigma_{k,i}, {}^1m_{k,i}, {}^2m_{k,i}, c_{k,i}, s_{k,i} \right]$, η holds in equation (19.87) then convergence is guaranteed. In this chapter η is initially chosen as

$$\eta = \frac{1}{max \left| \dfrac{\partial \tilde{y}(k)}{\partial W} \right|^2}$$

After all k pair data were applied, to design a stable and fast-learning algorithm, the learning rate is then chosen as

$$\begin{cases} if & \dfrac{RMSE(l)}{RMSE(l-1)} < 1 & \rightarrow & \eta(l) = \eta(l-1) \\[4mm] if & \dfrac{RMSE(l)}{RMSE(l-1)} \geq 1 & \rightarrow & \eta(l) = 0.9 \times \eta(l-1) \end{cases},$$

where $RMSE$ stands for root mean square error and l is the number of iterations.

For each system, the structure of the system and the NCPRT2FS-based identifier are shown in Figure 19.6.

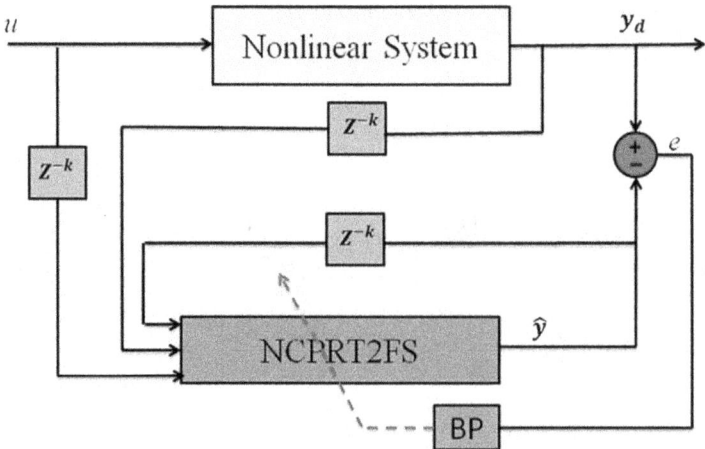

FIGURE 19.6　The structure of the system and the NCPRT2FS-based identifier.

The inputs to the NCPRT2FS-based identifier are the input and one delay output of the system. The parameters of the NCPRT2FS structure must be tuned by minimizing a proper cost function of plant output y_d and the identifier output $\check{y}$ for all input values of x.

Example 1: [13] Real data of a 660 kw wind turbine have been taken from Iran Renewable Energy Organization (SUNA).[1] The model of the wind turbine is S47–660 kw, made by VESTAS (Denmark):

Cut-in wind speed: 4 m/s
Rated wind speed: 15 m/s
Cut-out wind speed: 25 m/s
Survival wind speed: 60 m/s
Rotor:
Diameter: 47 m
Swept area: 1.735 m²
Number of blades: 3
Rotor speed, max: 28.5 U/min

FIGURE 19.7   Manjil and Rudbar Wind Farm.

Tipspeed:	70.1 m/s
Type:	22.9
Material:	GFK
Generator:	
Generator:	
Type:	Asynchronous
Number:	1.0
Speed, max:	1.650 U/min
Voltage:	400 V
Grid connection:	Thyristor
Grid frequency:	50 Hz

In this example, $u(k)$, $k = 1,\ldots,365$ is wind speed that is fed to the wind turbine system and obtains the 365 samples of $y(k)$ that is the output power of the wind turbine. Figure 19.8 shows the identification results of the NCPRT2FS. Here

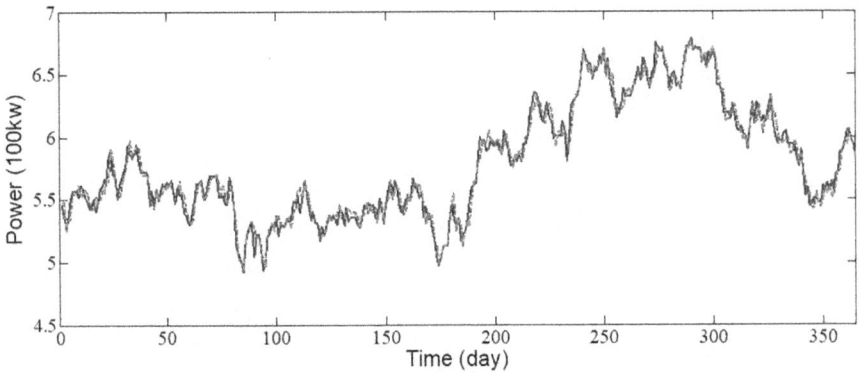

FIGURE 19.8   Identification results of the NCPRT2FS for wind turbines.

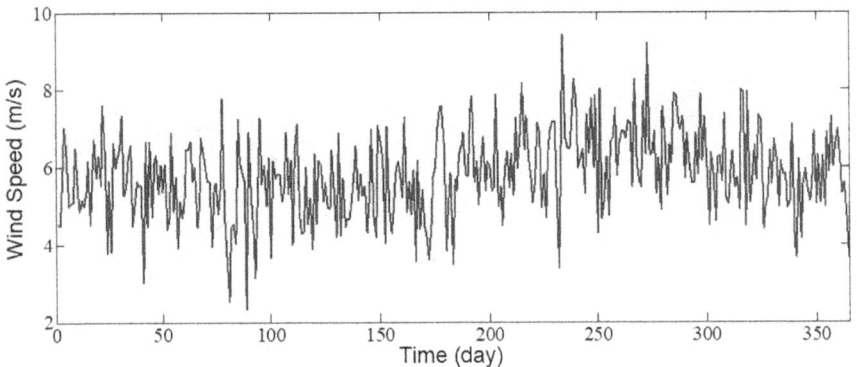

FIGURE 19.9   Wind speed of a place in Ilam for a year.

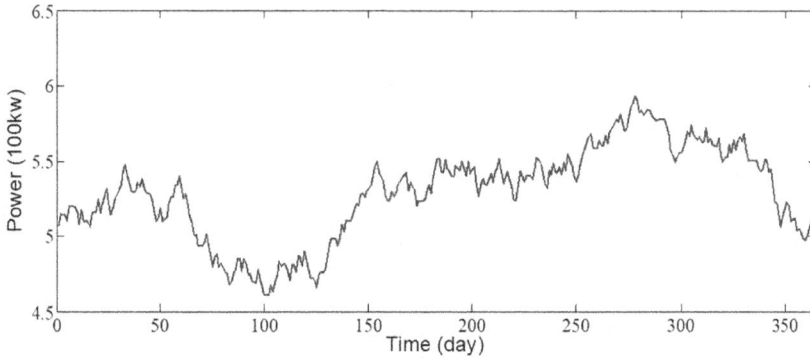

FIGURE 19.10   Predicted wind power of a place in Ilam for a year.

TABLE 19.1
The Final Values of NCPRT2FS Parameters

		$^1m_{ij}$	$^2m_{ij}$	σ_{ij}
u(k)		$^1m_{11} = 3.62$	$^2m_{11} = 4.32$	$\sigma_{11} = 0.38$
		$^1m_{21} = 6.13$	$^2m_{21} = 7.02$	$\sigma_{21} = 1.1$
Antecedent		$^1m_{31} = 8.19$	$^2m_{31} = 9.51$	$\sigma_{31} = 0.89$
parameters **y(k-1)**		$^1m_{12} = 4.93$	$^2m_{12} = 5.12$	$\sigma_{12} = 0.21$
		$^1m_{22} = 5.34$	$^2m_{22} = 5.66$	$\sigma_{22} = 0.09$
		$^1m_{32} = 5.81$	$^2m_{32} = 5.98$	$\sigma_{32} = 0.36$
		$^1m_{42} = 6.11$	$^2m_{42} = 6.48$	$\sigma_{42} = 0.18$

		$^1m_{ij}$	$^2m_{ij}$	σ_{ij}
Fourth layer adaptive weights	$\overline{w}_r^1 = 1.92$	$\underline{w}_r^1 = 1.5$	$\overline{w}_l^1 = 1.0$	$\underline{w}_l^1 = 0.63$
	$\overline{w}_r^2 = 1.66$	$\underline{w}_r^2 = 0.92$	$\overline{w}_l^2 = 0.71$	$\underline{w}_l^2 = 0.06$
	$\overline{w}_r^3 = 0.8$	$\underline{w}_r^3 = 0.7$	$\overline{w}_l^3 = 0.56$	$\underline{w}_l^3 = 0.43$
	$\overline{w}_r^4 = 1.87$	$\underline{w}_r^4 = 0.94$	$\overline{w}_l^4 = 0.85$	$\underline{w}_l^4 = 0.77$

	Rule 1	Rule 2	Rule 3	Rule 4	Rule 1	Rule 2	Rule 3	Rule 4
	$s_{1,0}=0.4$	$s_{2,0}=0.33$	$s_{3,0}=0.27$	$s_{4,0}=0.52$	$c_{1,0}=1$	$c_{2,0}=1.4$	$c_{3,0}=1$	$c_{4,0}=1.4$
	$s_{1,1}=0.55$	$s_{2,1}=0.39$	$s_{3,1}=0.48$	$s_{4,1}=0.43$	$c_{1,1}=1.1$	$c_{2,1}=1$	$c_{3,1}=1$	$c_{4,1}=1$
	$s_{1,2}=1$	$s_{2,2}=1$	$s_{3,2}=1$	$s_{4,2}=1$	$c_{1,2}=1$	$c_{2,2}=1.32$	$c_{3,2}=0.81$	$c_{4,2}=0.93$
	$s_{1,3}=0.43$	$s_{2,3}=0.39$	$s_{3,3}=0.65$	$s_{4,3}=.9$	$c_{1,3}=1$	$c_{2,3}=1$	$c_{3,3}=1.65$	$c_{4,3}=1.82$
Consequent	$s_{1,4}=0.62$	$s_{2,4}=1$	$s_{3,4}=1$	$s_{4,4}=1$	$c_{1,4}=1$	$c_{2,4}=1.09$	$c_{3,4}=1$	$c_{4,4}=1$
parameters	$s_{1,5}=0.87$	$s_{2,5}=0.1$	$s_{3,5}=1$	$s_{4,5}=1$	$c_{1,5}=1.1$	$c_{2,5}=1$	$c_{3,5}=1.55$	$c_{4,5}=1.9$
	$s_{1,6}=1$	$s_{2,6}=1$	$s_{3,6}=1$	$s_{4,6}=1$	$c_{1,6}=1$	$c_{2,6}=1$	$c_{3,6}=1$	$c_{4,6}=1$
	$s_{1,7}=0.69$		$s_{3,7}=0.31$	$s_{4,7}=0.06$	$c_{1,7}=0.8$	$c_{2,7}=0.72$	$c_{3,7}=0.67$	$c_{4,7}=0.81$
	$s_{1,8}=0.96$	$s_{2,8}=0.11$	$s_{3,8}=0.54$	$s_{4,8}=0.21$	$c_{1,8}=1.1$	$c_{2,8}=1$	$c_{3,8}=0.92$	$c_{4,8}=0.59$
	$s_{1,9}=0.3$	$s_{2,9}=0.32$	$s_{3,9}=0.36$	$s_{4,9}=0.98$	$c_{1,9}=0.95$	$c_{2,9}=0.77$	$c_{3,9}=1$	$c_{4,9}=1$
	$s_{1,10}=0.35$	$s_{2,10}=0.31$	$s_{3,10}=0.54$	$s_{4,10}=0.5$	$c_{1,10}=1$	$c_{2,10}=0.44$	$c_{3,10}=0.64$	$c_{4,10}=0.89$

the plant output (solid line) and the NCPRT2FS identifier output (dashed line) are shown.

Trained NCPRT2FS is used to calculate the wind power in a place of Ilam.[2] Figure 19.9 shows the wind speed of Ilam for a year. Figure 19.10 shows the predicted wind power in Ilam.

The final values of NCPRT2FS parameters are shown in Table 19.1.

Example 2: A real solar cell system is shown in Figure 19.11.

In this example, $u(k)$, $k = 1,\ldots,600$ is solar radiation that is fed to the real solar cell system and the 600 samples of $y(k)$ obtained. The other details are the same as the proposed NCPRT2FS in Example 1. Figure 19.12 shows the identification results of the NCPRT2FS for three solar radiations. Here the plant output (solid line) and the NCPRT2FS identifier output (dashed line) is shown.

After structure learning, for NCPRT2FS three rules are generated and the RMSE value for the NCPRT2FS and IT2-TSK-FNN for training and test are shown in Table 19.3. The final values of NCPRT2FS parameters are shown in Table 19.2.

The trained NCPRT2FS is then used to calculate the solar power of Ilam. Figure 19.13 shows the solar radiation of Ilam for a year. Figure 19.14 shows the predicted solar power in Ilam.

Finally, Table 19.3 presents comparison results of the proposed method with those of the method developed in [19].

Simulation results easily approve that the proposed NCPRT2FS has high performances in function approximation and system identification. This study also shows that the number of rules of the proposed NCPRT2FS is less than that of the other method, the accuracy of identification is also better, but the training time that was achieved by the average of running the program 10 times (computer processor: Dual

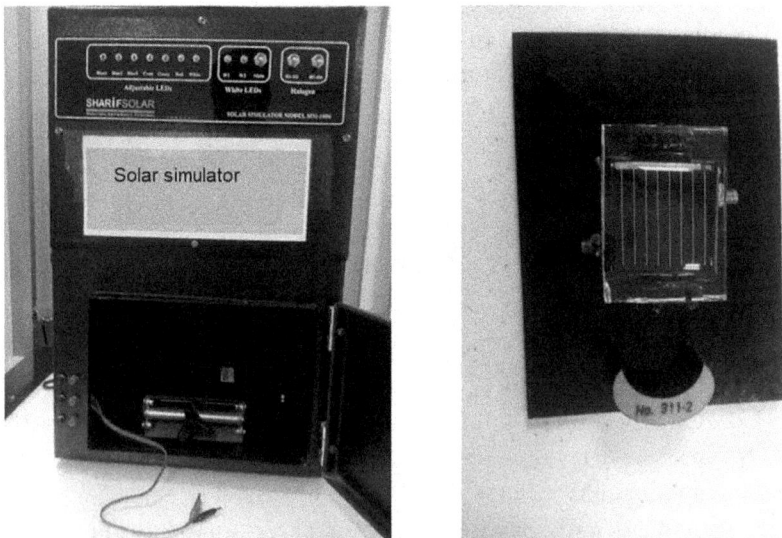

FIGURE 19.11 (a) Experimental solar cell testing system, (b) a solar cell.

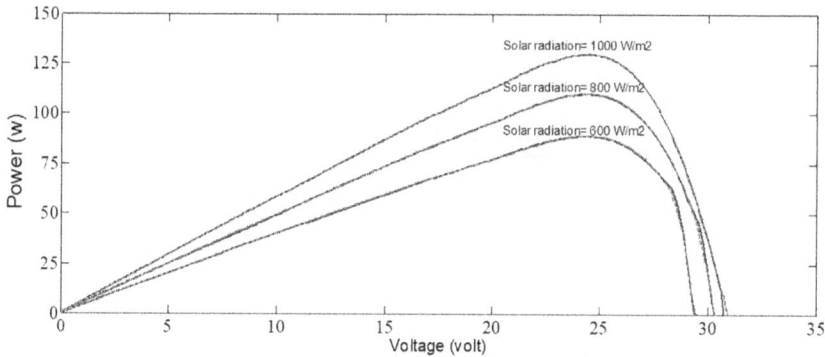

FIGURE 19.12 Identification results of the NCPRT2FS for solar cell.

TABLE 19.2
The Final Values of NCPRT2FS Parameters

		$^1m_{ij}$	$^2m_{ij}$	σ_{ij}
	u(k)	$^1m_{11} = 251$	$^2m_{11} = 332$	$\sigma_{11} = 43$
Antecedent		$^1m_{21} = 598$	$^2m_{21} = 615$	$\sigma_{21} = 12$
parameters		$^1m_{31} = 798$	$^2m_{31} = 949$	$\sigma_{31} = 211$
	y(k-1)	$^1m_{12} = 69$	$^2m_{12} = 75$	$\sigma_{12} = 11$
		$^1m_{22} = 82$	$^2m_{22} = 89$	$^2m_{32} = 97$
		$^1m_{32} = 93$	$^2m_{32} = 97$	$\sigma_{32} = 3$
Fourth layer	$\overline{w}_r^1 = 0.2$	$\underline{w}_r^1 = 0.06$	$\overline{w}_l^1 = 0.12$	$\underline{w}_l^1 = 0.09$
adaptive	$\overline{w}_r^2 = 1.8$	$\underline{w}_r^2 = 1$	$\overline{w}_l^2 = 1.42$	$\underline{w}_l^2 = 0.98$
weights	$\overline{w}_r^3 = 0.57$	$\underline{w}_r^3 = 0.21$	$\overline{w}_l^3 = 1.93$	$\underline{w}_l^3 = 1.1$

	Rule 1	Rule 2	Rule 3	Rule 1	Rule 2	Rule 3
	$s_{1,0} = 0.1$	$s_{2,0} = 0.84$	$s_{3,0} = 1$	$c_{1,0} = 0.56$	$c_{2,0} = 1$	$c_{3,0} = 1.22$
	$s_{1,1} = 0.32$	$s_{2,1} = 0.39$	$s_{3,1} = 0.37$	$c_{1,1} = .94$	$c_{2,1} = 1.6$	$c_{3,1} = 1$
	$s_{1,2} = 1$	$s_{2,2} = 1$	$s_{3,2} = 0.61$	$c_{1,2} = 1$	$c_{2,2} = 1$	$c_{3,2} = 1$
	$s_{1,3} = 0.22$	$s_{2,3} = 1.2$	$s_{3,3} = 0.5$	$c_{1,3} = 1$	$c_{2,3} = 1.77$	$c_{3,3} = 1.2$
Consequent	$s_{1,4} = 0.1$	$s_{2,4} = 0.42$	$s_{3,4} = 1$	$c_{1,4} = 1.61$	$c_{2,4} = 0.6$	$c_{3,4} = 1.63$
parameters	$s_{1,5} = 0.47$	$s_{2,5} = 1$	$s_{3,5} = 1$	$c_{1,5} = 1.3$	$c_{2,5} = 1$	$c_{3,5} = 2$
	$s_{1,6} = 0.1$	$s_{2,6} = 1$	$s_{3,6} = 1$	$c_{1,6} = 1$	$c_{2,6} = 1.11$	$c_{3,6} = 1$
	$s_{1,7} = 1.2$	$s_{2,7} = 1$	$s_{3,7} = .19$	$c_{1,7} = 1.1$	$c_{2,7} = 1.5$	$c_{3,7} = 0.88$
	$s_{1,8} = 1$	$s_{2,8} = 0.36$	$s_{3,8} = 0.69$	$c_{1,8} = 1.6$	$c_{2,8} = .89$	$c_{3,8} = 0.91$
	$s_{1,9} = 1$	$s_{2,9} = 0.28$	$s_{3,9} = 0.11$	$c_{1,9} = 1.53$	$c_{2,9} = 0.95$	$c_{3,9} = 0.48$
	$s_{1,10} = 0.55$	$s_{2,10} = 0.35$	$s_{3,10} = 0.5$	$c_{1,10} = 0.88$	$c_{2,10} = 1$	$c_{3,10} = 1$

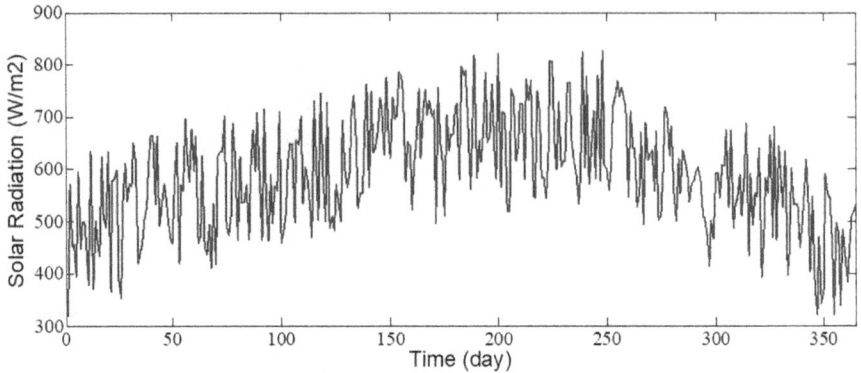

FIGURE 19.13   Solar radiation of Ilam.

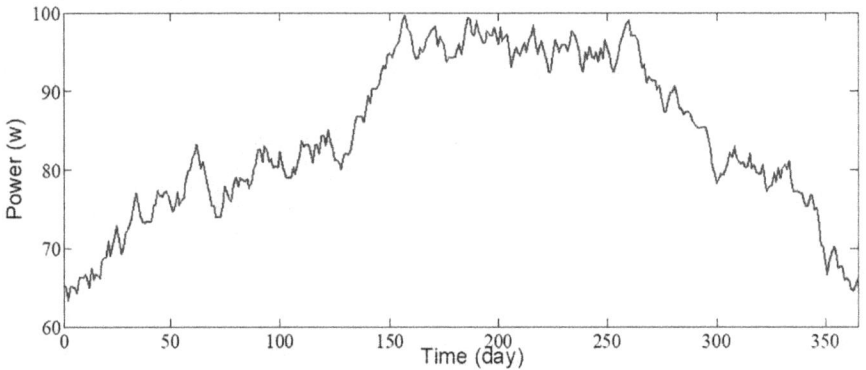

FIGURE 19.14   Predicted solar power in ILAM for a year.

TABLE 19.3
Comparison between Results of the NCPRT2FS Method and the Method of [45]

Example	Method of [19]			NCPRT2FS		
-	Rules	Run time (s)	RMSE	Rules	Run time (s)	RMSE
1	4	4	0.0159	4	6	0.0057
2	5	4	0.00759	3	7	0.0013

CPU T3200 @ 2.00 GHz 2.00 GHz, RAM: 2.00 GB and MATLAB 2011a) is a bit longer.

NOTES

1. www.suna.org.ir/en/home
2. A city in the west of Islamic Republic of Iran.

REFERENCES

[1] Farzanegan, B.; Suratgar, A.A.; Menhaj, M.B.; Zamani, M. Distributed optimal control for continuous-time nonaffine nonlinear interconnected systems. International Journal of Control 2021. https://doi.org/10.1080/00207179.2021.1976420.

[2] Afanasiev, V.N.; Kolmanovskii, V.; Nosov, V. Mathematical Theory of Control Systems Design. Springer Science & Business Media, 2013. Originally published by Kluwer Academic Publishers,1996.

[3] Tavoosi, J.; Suratgar, A.A.; Menhaj, M.B. Stability analysis of a class of MIMO recurrent type-2 fuzzy systems. Int. J. Fuzzy Syst. 2017, 19, 895–908.

[4] Jahangiri, F.; Doustmohammadi, A.; Menhaj, M.B. An adaptive wavelet differential neural networks based identifier and its stability analysis. Neurocomputing 2012, 77, 12–19.

[5] Suratgar, A.A.; Nikravesh, S.K. A new method for linguistic modeling with stability analysis and applications. Intell. Autom. Soft Comput. 2009, 15, 329–342.

[6] Li, C.; Wang, L.; Zhang, G.; Wang, H.; Shang, F. Functional-type single-input-rule-modules connected neural fuzzy system for wind speed prediction. J. Autom. Sin. 2017, 4, 751–762.

[7] Karakuş, O.; Kuruoğlu, E.E.; Altınkaya, M.A. One-day ahead wind speed/power prediction based on polynomial autoregressive model. IET Renew. Power Gener. 2017, 11, 1430–1439.

[8] Tian, Y.; Wang, B.; Zhu, D.; Wu, F. Takagi: Sugeno fuzzy generalised predictive control of a time-delay non-linear hydro-turbine governing system. IET Renew. Power Gener. 2019, 13, 2338–2345.

[9] Morshedizadeh, M.; Kordestani, M.; Carriveau, R.; Ting, D.S.; Saif, M. Power production prediction of wind turbines using a fusion of MLP and ANFIS networks. IET Renew. Power Gener. 2018, 12, 1025–1033.

[10] Tavoosi, J. A new type-2 fuzzy systems for flexible-joint robot arm control. Aut J. Model. Simul. 2019, 51. doi:10.22060/miscj.2019.14478.5108.

[11] Castro, J.R.; Castillo, O.; Martínez, L.G. Interval type-2 fuzzy logic toolbox. Eng. Lett. 2007, 15, 1.

[12] Mendel, J.M. Uncertain Rule-Based Fuzzy Logic Systems: Introduction and New Directions. Prentice-Hall: Upper New Jersey River, NJ, USA, 2001.

[13] Tavoosi, J.; Suratgar, A.A.; Menhaj, M.B.; Mosavi, A.; Mohammadzadeh, A.; Ranjbar, E. Modeling renewable energy systems by a self-evolving nonlinear consequent part recurrent type-2 fuzzy system for power prediction. Sustainability 2021, 13, 3301.

[14] Saadati Moghadam, A.; Suratgar, A.A.; Hesamzadeh, M.R.; Nikravesh, S.K.Y. Multi-objective ACOPF using distributed gradient dynamics. International Journal of Electrical Power & Energy System 2022 October, 141.

[15] Padar, N.; Suratgar, A.A.; Menhaj, M.B. Modeling and Fuzzy Predictive Voltage Control of VSC-Based Microgrids-11th Smart Grid Conference (SGC), 2021.

[16] Asgari, S.; Suratgar, A.A.; Kazemi, M.G. Feedforward fractional order PID load frequency control of microgrid using harmony search algorithm. Iranian Journal of Science and Technology, 2021.

[17] Asgari, S., Menhaj, M.B., Suratgar, A.A., Kazemi, M.G. A disturbance observer based fuzzy feedforward proportional integral load frequency control of microgrids. International Journal of Engineering, 2021.

[18] Farzanegan, B.; Zamani, M.; Suratgar, A.A.; Menhaj, M.B. A neuro-observer-based optimal control for nonaffine nonlinear systems with control input saturations. Control Theory and Technology, 2021, 19(2), 283–294.

[19] Tavoosi, J.; Badamchizadeh, M.A. A class of type-2 fuzzy neural networks for non-linear dynamical system identification. Neural Computing & Application 2013, 23(3), 707–717.

[20] Ruano, A.E. Intelligent Control Systems Using Computational Intelligence Techniques. Institution of Engineering and Technology, 2005.

Index

Note: Page locators in **bold** indicate a table. Page locators in *italics* indicate a figure.

For Product Safety Concerns and Information please contact our EU
representative GPSR@taylorandfrancis.com
Taylor & Francis Verlag GmbH, Kaufingerstraße 24, 80331 München, Germany